MW01077078

Cecilia Yu

LINEAR ALGEBRA
FOR
EVERYONE

GILBERT STRANG

Massachusetts Institute of Technology

WELLESLEY - CAMBRIDGE PRESS

Box 812060 Wellesley MA 02482

Linear Algebra for Everyone
Copyright ©2020 by Gilbert Strang
ISBN 978-1-7331466-3-0

All rights reserved. No part of this book may be reproduced or stored or transmitted by any means, including photocopying, without written permission from Wellesley - Cambridge Press. Translation in any language is strictly prohibited — authorized translations are arranged by the publisher.

LaTeX typesetting by Ashley C. Fernandes (info@problemsolvingpathway.com)
Printed in the United States of America

9 8 7 6 5 4 3 2 1

QA184.2 .S773 2020 | DDC 512/.5–dc23

Other texts from Wellesley - Cambridge Press

Linear Algebra and Learning from Data, 2019, Gilbert Strang ISBN 978-0-6921963-8-0

Introduction to Linear Algebra, 5th Ed., 2016, Gilbert Strang ISBN 978-0-9802327-7-6

Computational Science and Engineering, Gilbert Strang ISBN 978-0-9614088-1-7

Wavelets and Filter Banks, Gilbert Strang and Truong Nguyen ISBN 978-0-9614088-7-9

Introduction to Applied Mathematics, Gilbert Strang ISBN 978-0-9614088-0-0

Calculus Third Edition, Gilbert Strang ISBN 978-0-9802327-5-2

Algorithms for Global Positioning, Kai Borre & Gilbert Strang ISBN 978-0-9802327-3-8

Essays in Linear Algebra, Gilbert Strang ISBN 978-0-9802327-6-9

Differential Equations and Linear Algebra, Gilbert Strang ISBN 978-0-9802327-9-0

An Analysis of the Finite Element Method, 2008 edition, Gilbert Strang and George Fix ISBN 978-0-9802327-0-7

Wellesley - Cambridge Press
Box 812060, Wellesley MA 02482 USA
www.wellesleycambridge.com
LAFEeveryone@gmail.com

Gilbert Strang's page : **math.mit.edu/~gs**
For orders : **math.mit.edu/weborder.php**
Outside US/Canada : **www.cambridge.org**
Select books, India : **www.wellesleypublishers.com**

The website for this book (with Solution Manual) is **math.mit.edu/everyone**
2019 book : **Linear Algebra and Learning from Data (math.mit.edu/learningfromdata)**
2016 book : **Introduction to Linear Algebra, 5th Edition (math.mit.edu/linearalgebra)**
2014 book : **Differential Equations and Linear Algebra (math.mit.edu/dela)**

Linear Algebra is included in MIT's OpenCourseWare **ocw.mit.edu/courses/mathematics**
Those videos (including 18.06SC and 18.065) are also on **www.youtube.com/mitocw**
18.06 Linear Algebra 18.06SC with problem solutions 18.065 Learning from Data

MATLAB® is a registered trademark of The MathWorks, Inc.

The cover design was created by Gail Corbett and Lois Sellers : **lsellersdesign.com**

Table of Contents

Preface

This is a linear algebra textbook with a new start. Chapter 1 begins as usual with vectors. We see their linear combinations and dot products. Then the new ideas come with matrices. Let me illustrate those ideas right away by an example.

Suppose we are given a 3 by 3 matrix A with columns $\boldsymbol{a}_1, \boldsymbol{a}_2, \boldsymbol{a}_3$:

$$\boldsymbol{A} = \begin{bmatrix} & & \\ \boldsymbol{a}_1 & \boldsymbol{a}_2 & \boldsymbol{a}_3 \\ & & \end{bmatrix} = \begin{bmatrix} 1 & 2 & 3 \\ 3 & 4 & 7 \\ 4 & 2 & 6 \end{bmatrix}.$$

Those columns are three-dimensional vectors. The first vectors $\boldsymbol{a}_1$ and $\boldsymbol{a}_2$ connect the center point $(0,0,0)$ to the points $(1,3,4)$ and $(2,4,2)$. The picture shows those points in 3-dimensional space (xyz space). The key to this matrix is the third vector $\boldsymbol{a}_3$ going to the point $(3,7,6)$.

When I look at those vectors, I see something exceptional. Adding columns 1 and 2 produces column 3. In other words $\boldsymbol{a}_1 + \boldsymbol{a}_2 = \boldsymbol{a}_3$. In a 3-dimensional picture, $\boldsymbol{a}_1$ and $\boldsymbol{a}_2$ go from the center point $(0,0,0)$ to the points $(1,3,4)$ and $(2,4,2)$. The picture shows how to add those vectors. It is normal that all combinations of two vectors will fill up a plane. (The plane is actually infinite, we just drew the part between the vectors.) What is really exceptional is that **the third point $\boldsymbol{a}_3 = (\mathbf{3},\mathbf{7},\mathbf{6})$ lies on this plane of $\boldsymbol{a}_1$ and $\boldsymbol{a}_2$.**

Most points don't lie on that plane. Most vectors $\boldsymbol{a}_3$ are *not* combinations of $\boldsymbol{a}_1$ and $\boldsymbol{a}_2$. Most 3 by 3 matrices have *independent* columns. Then the matrix will be invertible. But these three columns are *dependent* because they lie on the same plane: $\boldsymbol{a}_1 + \boldsymbol{a}_2 = \boldsymbol{a}_3$.

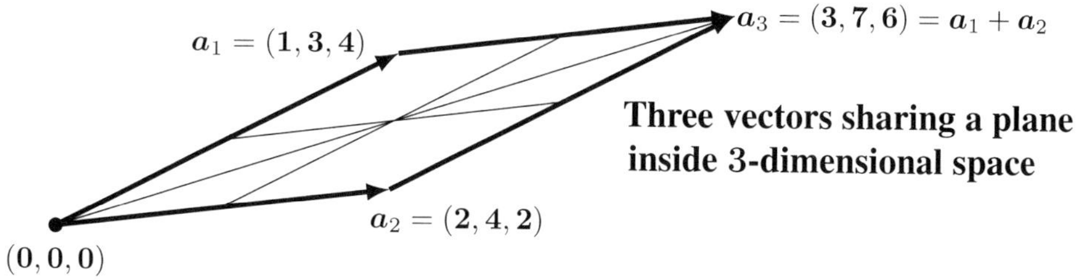

Three vectors sharing a plane inside 3-dimensional space

That picture reveals the most important fact about those three vectors $\boldsymbol{a}_1, \boldsymbol{a}_2, \boldsymbol{a}_3$. But we need the right language to describe it. Our goal is to see the idea, and to get better and better at expressing it. Here are three steps in a good direction.

1 **Idea in words** Column 3 is the sum of column 1 and column 2

2 **Idea in symbols** $\boldsymbol{a}_3 = \boldsymbol{a}_1 + \boldsymbol{a}_2$

3 **Matrix times vector** $\begin{bmatrix} \\ a_3 \\ \\ \end{bmatrix} = \begin{bmatrix} & & \\ a_1 & & a_2 \\ & & \end{bmatrix} \begin{bmatrix} \mathbf{1} \\ \mathbf{1} \end{bmatrix}$

Step 3 shows how a matrix C multiplies a vector $\boldsymbol{x}$. The columns $\boldsymbol{a}_1$ and $\boldsymbol{a}_2$ in C multiply the numbers $x_1 = 1$ and $x_2 = 1$ in $\boldsymbol{x}$. *The output $C\boldsymbol{x}$ is a combination of the columns.* Here that column combination is $\boldsymbol{a}_1 + \boldsymbol{a}_2$.

One more crucial step allows several combinations at once. This is the way forward. We cannot take this step with pictures or words. **A matrix multiplies a matrix.**

4 **Matrix times matrix** $\boldsymbol{A} = \boldsymbol{CR}$ is $\begin{bmatrix} & & \\ a_1 & a_2 & a_3 \\ & & \end{bmatrix} = \begin{bmatrix} & \\ a_1 & a_2 \\ & \end{bmatrix} \begin{bmatrix} 1 & 0 & 1 \\ 0 & 1 & 1 \end{bmatrix}$

Those three columns of $\boldsymbol{R}$ give three combinations in $\boldsymbol{A}$ of the two columns in $\boldsymbol{C}$

Column 1: $\mathbf{1}\boldsymbol{a}_1 + \mathbf{0}\boldsymbol{a}_2 = \boldsymbol{a}_1$ **Column 2**: $\mathbf{0}\boldsymbol{a}_1 + \mathbf{1}\boldsymbol{a}_2 = \boldsymbol{a}_2$

Column 3: $\mathbf{1}\boldsymbol{a}_1 + \mathbf{1}\boldsymbol{a}_2 = (\mathbf{1}, \mathbf{3}, \mathbf{4}) + (\mathbf{2}, \mathbf{4}, \mathbf{2}) = (\mathbf{3}, \mathbf{7}, \mathbf{6}) = \boldsymbol{a}_3$

That matrix multiplication $\boldsymbol{A} = \boldsymbol{CR}$ displays major information:

A has dependent columns : the combination of columns $1 + 2 - 3$ gives $(0, 0, 0)$

A also has dependent rows: some combination of its rows will give $(0, 0, 0)$!

The "*column space*" of this A is only a plane and not the whole 3D space

The "*row space*" of this A is also a plane and not the whole 3D space

The square matrix A has no inverse. Its determinant is zero. It is unusual.

Subspaces of Vector Spaces

In Chapters 1 to 4, the organizing ideas are **vector spaces**. The columns of A are in m-dimensional space $\mathbf{R}^m$. The rows are in n-dimensional space $\mathbf{R}^n$. But the action is in **four subspaces inside $\mathbf{R}^m$ and $\mathbf{R}^n$**. The matrix A multiplies $\boldsymbol{x}_{\text{row}}$ in its *row space* and $\boldsymbol{x}_{\text{null}}$ in its *nullspace*. The output is $A\boldsymbol{x}_{\text{row}} = \boldsymbol{b}$ in its *column space* because $A\boldsymbol{x}_{\text{null}} = \textit{zero}$. Then the complete solution to $A\boldsymbol{x} = \boldsymbol{b}$ has a row space part and a nullspace part:

$$\boxed{\boldsymbol{x} = \boldsymbol{x}_{\textbf{row}} + \text{ any } \boldsymbol{x}_{\textbf{null}} \qquad A\boldsymbol{x} = A\boldsymbol{x}_{\textbf{row}} + A\boldsymbol{x}_{\textbf{null}} = \boldsymbol{b} + \mathbf{0} = \boldsymbol{b}}$$

Every linear equation $A\boldsymbol{x} = \boldsymbol{b}$ is solved this way: particular $\boldsymbol{x}_{\text{row}}$ + any homogeneous $\boldsymbol{x}_{\text{null}}$.

The Plan for the Book

That example is part of a new start. I believe it is a better start (for the reader and the course). By working with specific matrices to introduce the algebra, the subject unfolds. Chapter 1 develops the matrix equation $A = C$ times R. C takes independent vectors like $\boldsymbol{a}_1$ and $\boldsymbol{a}_2$ from the column space of A. R takes independent vectors from the row space of A. Those two "vector spaces" are at the center of linear algebra. We meet them properly in Chapter 3. But you will know about independence from examples in Chapter 1.

The big step is to factor A into C times R. Matrix multiplication is a crucial operation, and Chapter 1 ends with four different ways to do it—seen on the back cover of the book. This sets the path to all the great factorizations of linear algebra:

$\boldsymbol{A = LU}$	Chapter 2 solves n equations $A\boldsymbol{x} = \boldsymbol{b}$ in n unknowns: A is square
$\boldsymbol{A = CR}$	Chapter 3 reduces to r independent columns and r independent rows
$\boldsymbol{A = QR}$	Chapter 4 changes the columns of A into perpendicular columns of Q
$\boldsymbol{S = Q\Lambda Q^{\mathrm{T}}}$	Chapter 6 has **eigenvectors** in $\boldsymbol{Q}$. The **eigenvalues** of S are in $\boldsymbol{\Lambda}$
$\boldsymbol{A = U\Sigma V^{\mathrm{T}}}$	Chapter 7 has **singular vectors** in $\boldsymbol{U}$ and $\boldsymbol{V}$ and **singular values** in $\boldsymbol{\Sigma}$

The columns of A are in m-dimensional space $\mathbf{R}^m$, the rows are in $\mathbf{R}^n$. The m by n matrix multiplies vectors $\boldsymbol{x}$ in $\mathbf{R}^n$ to produce $A\boldsymbol{x}$ in $\mathbf{R}^m$. But the real action of A is seen in the four fundamental subspaces.

Chapter 2 only allows one solution $\boldsymbol{x}_{\text{row}}$. The matrix A is square and invertible. Chapter 3 finds every solution to $A\boldsymbol{x} = \boldsymbol{b}$ by adding every $\boldsymbol{x}_{\text{null}}$. Chapter 4 deals with equations that don't have a solution (because $\boldsymbol{b}$ has a piece from the mysterious fourth subspace). I hope you will like the "big picture of linear algebra" on page 124: **all four subspaces.**

Those five factorizations are a perfect way to organize and remember linear algebra. The eigenvalues in the matrix $\boldsymbol{\Lambda}$ and the singular values in $\boldsymbol{\Sigma}$ come from S and A in a beautiful way—but not a simple way. Those numbers in $S\boldsymbol{x} = \lambda\boldsymbol{x}$ and $A\boldsymbol{v} = \sigma\boldsymbol{u}$ see deeply into the symmetric matrix S and the m by n matrix A. Often S appears in engineering and physics. Often A is a matrix of data. And data is now coming from everywhere.

Please don't miss Tim Baumann's page 272 on compressing photographs by the SVD.

Chapter 5 explains the amazing formulas produced by determinants. Amazing but unfortunately difficult to compute! We solve equations $A\boldsymbol{x} = \boldsymbol{b}$ before (not after) we find the determinant of A. Those equations ask us to produce $\boldsymbol{b}$ from the columns of A.

$$\boldsymbol{Ax} = \boldsymbol{x}_1\,(\textbf{column 1}) + \boldsymbol{x}_2\,(\textbf{column 2}) + \cdots + \boldsymbol{x}_n\,(\textbf{column } \boldsymbol{n}) = \textbf{ right side } b.$$

In principle, determinants lead to eigenvalues (Chapter 6) and singular values (Chapter 7). In practice, we look for other ways to find and use those important numbers. And yet determinants tell us about geometry too—like the volume of a tilted box in n dimensions.

A short course can go directly from dimensions in 3.5 to eigenvalues for 2×2 matrices.

Final Chapter : Learning from Data

We added Chapter 8 on **data science**. The data often comes in a rectangular matrix A. Each row measures n properties of a sample. That number n is large, and the number of samples is often very large. Out of this big matrix, applied linear algebra has to find out what is important. The goal is to produce understanding that leads to a decision.

In machine learning the output is a nonlinear function of the input. Deep learning aims to find that function from the training data. It produces (often with giant computations) a learning function $\boldsymbol{F}(\boldsymbol{x}, \boldsymbol{v})$. The input vectors $\boldsymbol{v}$ contain the features of each training sample (many $\boldsymbol{v}$'s from many samples). The vector $\boldsymbol{x}$ contains the weights assigned to those features. And hidden inside F is a nonlinear function that mixes the inputs and the weights. A favorite is the simple function ReLU (y) = maximum of 0 and y : a *ramp function*.

Optimizing the weights to learn from the training data $\boldsymbol{v}$ is the big computation. When that is well done, we can input new samples that the system has never seen. The success of deep learning is that $F(\boldsymbol{x}, \boldsymbol{v})$ is often close to the correct output. The system has identified an image, or translated a sentence, or chosen a winning move.

This is *piecewise linear algebra*, a mixture of weight matrices and ReLU. It is included with no expectation of testing students. This is a chapter to use later, in any way you want. You could experiment with the website **playground.tensorflow.org**.

Machine learning has become important and powerful—based on linear algebra and calculus (optimizing the weights) and on statistics (controlling the mean and variance). *But it is not required or even expected to be part of the course. What I hope is that the faster start allows you to reach eigenvalues and singular values*—those are true highlights of this subject. Chapters 2 to 7 confirm the jump of intuition near the end of Chapter 1 :

> If all columns of A lie on an r-dimensional plane, then all rows of A also lie on a (usually different) r-dimensional plane. That fact has far-reaching consequences.

This is a textbook for a normal linear algebra class—to explain the key ideas of this beautiful subject to *everyone*. As clearly as I can. Thank you.

Gilbert Strang

After writing this introduction, I looked back at the opening example. That 3 by 3 matrix A has dependent columns (column 1 + column 2 = column 3). By linear algebra, A must also have dependent rows. The numbers to show this are $-\mathbf{5}$ and $\mathbf{3}$:

$$-\mathbf{5}\,(\text{row } 1) + \mathbf{3}\,(\text{row } 2) = \text{row } 3 \text{ of } A \qquad -\mathbf{5}\,(\mathbf{1},\mathbf{2},\mathbf{3}) + \mathbf{3}\,(\mathbf{3},\mathbf{4},\mathbf{7}) = (\mathbf{4},\mathbf{2},\mathbf{6}).$$

The number of independent rows equals the number of independent columns. *Wonderful* !

Websites for Linear Algebra

The dedicated website for this textbook is **math.mit.edu/everyone**. Several key sections of the book can be downloaded. That site also has brief solutions to the problem sets. (For homework , the instructor may ask for more detailed solutions.) Every class will find a balance between learning the essential ideas of this subject and practicing the steps on small matrices.

The most beautiful website for linear algebra is **3Blue1Brown.com**, created by Grant Sanderson. He chose that name based on an unusual genetic feature of his eyes.

For blackboard lectures you could go to the OpenCourseWare site **ocw.mit.edu** created by MIT. The videos for Math 18.06 and Math 18.06SC and Math 18.065 have had millions of viewers, very often accessed through YouTube. 18.06 is MIT's large linear algebra course. The 18.06SC videos include problem solutions by the instructors. 18.065 is the newer course that leads to Chapter 8 of this book and to **math.mit.edu/learningfromdata**.

Important to add : The "*new start*" in this book was first tested in 18.065—which begins with a substantial review of linear algebra. Two more online materials were added in 2020 to the 18.06 site :

> A 70-minute video outlines the new start in teaching and learning linear algebra. That "2020 vision" has guided Sections 1.3 and 1.4 of this new book. I am convinced that working with the columns of actual matrices is a direct route to understanding linear independence and column spaces and matrix multiplication. The column-row factorization $\boldsymbol{A} = \boldsymbol{CR}$ is at the heart of solving linear equations $\boldsymbol{Ax} = \boldsymbol{b}$.
>
> Professor Raj Rao has developed a very successful course on computational linear algebra and machine learning at the University of Michigan. The key idea is to implement (in class !) what you learn. The website **mynerva.io** describes the course and the online textbook and future plans. That textbook is complementary to this one.

Properly used, the Web has become truly valuable to students and all readers. It provides a different way to explain linear algebra, and it is alive ! Please use the video lectures to see the flow of the course. And please use the book to capture that flow and hold it and practice it and understand it.

This book begins with independent vectors from the column space and row space : $\boldsymbol{A} = \boldsymbol{CR}$. The book ends with *orthogonal* vectors from those spaces : the $\boldsymbol{u}$'s and $\boldsymbol{v}$'s in the SVD with $A\boldsymbol{v} = \sigma\boldsymbol{u}$. In between this we have the central ideas of linear algebra.

Computational Linear Algebra

We need to solve large problems quickly and accurately. Linear algebra is often the key to computations in engineering and science and management and finance. This page aims to provide an update on hard power and soft power—computers and math.

The 500 fastest computers in the world are listed on **top500.org/news**. The leaders are mostly paid for and controlled by governments, and we can give their locations: Kobe, Oak Ridge, Livermore, Wuxi, Guangzhou. The next group includes Italy, Switzerland, Germany, France, Korea, Australia, Taiwan, UAE, UK, Russia, Spain, Saudi Arabia, Finland, Norway, and Brazil. The scientific interest is in their speed and special processors. A good source of information is **top500.org/resources/frequently-asked-questions**.

The benchmark tests are linear algebra problems—factoring a big matrix into L times U (lower triangular times upper triangular) and then solving a system of linear equations $A\boldsymbol{x} = \boldsymbol{b}$. The actual test is in the High Performance LINPACK benchmark. Those problems are at the heart of Chapters 2 and 3, now speeded up by parallel processing.

The top machine achieves 415 petaflops $=$ 415 times 10^{15} double precision floating point operations per second. This is with extremely careful coding for special hardware. The important point is that ordinary computers have also seen a tremendous increase in speed. And there are comparable savings from better algorithms.

Numerical Linear Algebra

This is the subject of major research: *fast algorithms for matrix computations*. The first was thousands of years ago with "elimination". For that idea and any new one, part of the test is to count the steps: in this case $n^3/3$ for n linear equations with n unknowns. To compute eigenvalues and now singular values, big progress brought that also to cn^3. Favorite textbooks among many good ones are

- L. N. Trefethen and David Bau, *Numerical Linear Algebra*, SIAM (1997)
- Gene Golub and Charles van Loan, *Matrix Computations*, Johns Hopkins (2013).

For the mathematics of linear algebra (not focused on computation) we mention

- Roger Horn and Charles Johnson, *Matrix Analysis*, Cambridge (2013)
- R. Bhatia, *Matrix Analysis*, Springer and *Positive Definite Matrices*, Princeton

Randomized Numerical Linear Algebra

The usual algorithms meet a barrier when the problem becomes very large. We can hardly store and read all the data. Even matrix multiplication AB becomes impossible. Eigenvalues hit their own barrier at matrix size $n = 10^4$ or 10^5.

A solution has been found if the matrix has enough structure. Random samples tell us about the matrix! This is a deep and beautiful subject—see a masterpiece by P. G. Martinsson and J. A. Tropp in *Acta Numerica* (2020) and arXiv: 2002.01387.

Gratitude for Help

This book was written during the months of lockdown for the coronavirus. Life was limited but time for writing was nearly unlimited. Difficult for our society but perfect for an author. The idea for a new and more active start had just gone into a new video for Math 18.06 on OpenCourseWare. (The matrix multiplication $A = CR$ in Section 1.4 is part of the idea, with independent columns of A going into C.) Developing that idea into this textbook has been exciting every morning.

The time was right but help was needed. It came in the best possible way. My good friend Ashley C. Fernandes in Mumbai received handwritten pages every day. Then he returned LaTeX pages overnight. Those pages went back and forth many times. Working with Ashley has made *Linear Algebra for Everyone* possible; this is our eighth book. I am truly grateful for these happy months.

Another good fortune has been help from Daniel Drucker. He is the most careful reader I know. Let me leave you to decide on this one : "To be really picky, in the Preface you say that the three columns lie *on* the same plane, but in the figure you say that the three vectors are *in* a plane." I won't do that again. It made my day when Dan liked the small matrices on the front cover. The goals of the text are clarity and simplicity :

> The basic ideas of matrix multiplication evolve step by step in Chapter 1.
>
> Columns of CR are combinations of columns of C
>
> There are four different ways to multiply AB (see the back cover)
>
> The key property is AB times $C = A$ times BC

This is the tenth cover created by two artists : Gail Corbett and Lois Sellers. You might have seen the rectangles for the four subspaces on *Introduction to Linear Algebra*. Before that came *Essays on Linear Algebra* with Alberto Giacometti's "Walking Man" and *Calculus* with a famous curve painted by Jasper Johns. Perhaps the most beautiful was the photograph on the finite element book. These are very happy memories for an author.

The whole idea of helping students is beautiful.

My greatest gratitude is to my wife Jill and our sons David and John and Robert.

This book is dedicated to them.

Dictionary of Matrices

A good way to tell you what this book contains is to name the matrices you will meet. This wide variety of matrices is a special feature of linear algebra.

Chapter 1

Identity matrix I

Column basis C

Row basis R

Rank 1 matrix $\boldsymbol{uv}^{\mathrm{T}}$

Chapter 2

Elimination matrix E

Lower triangular L

Upper triangular U

Inverse matrix A^{-1}

Transpose matrix A^{T}

Permutation P

Fourier matrix F

Chapter 3

Echelon matrix R

Free matrix F

Special solutions S

Mixing matrix $M = W^{-1}$

Incidence matrix A

Laplacian matrix L

Pseudoinverse A^{+}

Chapter 4

Orthogonal matrix Q

Projection matrix P

Least squares $A^{\mathrm{T}}A$

Upper triangular R

Reflection matrix H

Chapter 5

Cofactor matrix C

Change of basis B

Tilted box E

House matrix H

Chapter 6

Symmetric matrix S

Eigenvectors X

Eigenvalues Λ

Fibonacci matrix F

Jordan matrix J

Similar matrix BAB^{-1}

Exponential e^{At}

Chapter 7

Singular values Σ

Left singular vectors U

Right singular vectors V

Compressed matrix A_k

Sample covariance AA^{T}

Hilbert matrix H

Chapter 8

Weight matrix A

Convolution C

Jacobian matrix J

Hessian matrix H

Covariance matrix V

Shift matrix S

1 Vectors and Matrices

1.1 Linear Combinations of Vectors

1.2 Lengths and Angles from Dot Products

1.3 Matrices and Column Spaces

1.4 Matrix Multiplication and $A = CR$

The heart of linear algebra is in two operations—both with vectors. We add vectors to get $\boldsymbol{v} + \boldsymbol{w}$. We multiply them by numbers c and d to get $c\boldsymbol{v}$ and $d\boldsymbol{w}$. Combining those two operations (adding $c\boldsymbol{v}$ to $d\boldsymbol{w}$) gives the ***linear combination*** $c\boldsymbol{v} + d\boldsymbol{w}$.

Linear combination $$c\boldsymbol{v} + d\boldsymbol{w} = c\begin{bmatrix} 1 \\ 1 \end{bmatrix} + d\begin{bmatrix} 2 \\ 3 \end{bmatrix} = \begin{bmatrix} c + 2d \\ c + 3d \end{bmatrix}$$

Linear combinations are all-important in this subject! Sometimes we want one particular combination, the specific choice $c = 2$ and $d = 1$ that produces $c\boldsymbol{v} + d\boldsymbol{w} = (4, 5)$. Other times we want *all the combinations* of $\boldsymbol{v}$ and $\boldsymbol{w}$. Combinations that produce the zero vector have special importance. Of course $0\boldsymbol{v} + 0\boldsymbol{w}$ is always the zero vector.

The vectors $c\boldsymbol{v}$ lie along a line. When $\boldsymbol{w}$ is not on that line, **the combinations** $c\boldsymbol{v} + d\boldsymbol{w}$ **fill a complete two-dimensional plane**. Starting from three vectors $\boldsymbol{u}, \boldsymbol{v}, \boldsymbol{w}$ in three-dimensional space, their combinations $c\boldsymbol{u} + d\boldsymbol{v} + e\boldsymbol{w}$ are likely to fill the whole space—but not always. The vectors and their combinations could lie in a plane or on a line. This is a key problem: **describe all combinations of n given vectors.**

Next step: Put two vectors into the columns of a matrix A or B. Then multiplying a matrix by a vector $\boldsymbol{x}$ exactly produces a linear combination of the columns:

$$A\boldsymbol{x} = \begin{bmatrix} 1 & 2 \\ 1 & 3 \end{bmatrix}\begin{bmatrix} c \\ d \end{bmatrix} = \begin{bmatrix} c + 2d \\ c + 3d \end{bmatrix} \qquad B\boldsymbol{x} = \begin{bmatrix} 1 & 3 \\ 1 & 3 \end{bmatrix}\begin{bmatrix} c \\ d \end{bmatrix} = \begin{bmatrix} c + 3d \\ c + 3d \end{bmatrix}$$

Again those combinations $A\boldsymbol{x}$ fill a plane. The outputs from $B\boldsymbol{x}$ only fill a line.

The first example had "independent columns". The second example has "dependent columns". Chapter 1 explains these central ideas, on which everything builds. Linear algebra moves from 2 columns in 2 dimensions to n columns in m dimensions. Your mental picture stays correct—and we end by *multiplying matrices*.

1.1 ***Vector addition*** $\boldsymbol{v} + \boldsymbol{w}$ ***and linear combinations*** $c\boldsymbol{v} + d\boldsymbol{w}$.

1.2 ***The dot product*** $\boldsymbol{v} \cdot \boldsymbol{w}$ ***of two vectors and the length*** $\|\boldsymbol{v}\| = \sqrt{\boldsymbol{v} \cdot \boldsymbol{v}}$.

1.3 ***Matrix A times vector*** x ***is a combination of the columns of A.***

1.4 ***Matrix A times matrix B is*** $\begin{bmatrix} A\boldsymbol{b}_1 & \cdots & A\boldsymbol{b}_n \end{bmatrix}$. ***Multiply A times each column of B.***

1.1 Linear Combinations of Vectors

1 $3\boldsymbol{v}+5\boldsymbol{w}$ is a typical **linear combination** $c\boldsymbol{v}+d\boldsymbol{w}$ of the vectors $\boldsymbol{v}$ and $\boldsymbol{w}$.

2 For $\boldsymbol{v}=\begin{bmatrix}1\\1\end{bmatrix}$ and $\boldsymbol{w}=\begin{bmatrix}2\\3\end{bmatrix}$ that combination is $3\begin{bmatrix}1\\1\end{bmatrix}+5\begin{bmatrix}2\\3\end{bmatrix}=\begin{bmatrix}3+10\\3+15\end{bmatrix}=\begin{bmatrix}13\\18\end{bmatrix}$.

3 The combinations $c\begin{bmatrix}1\\1\end{bmatrix}+d\begin{bmatrix}2\\3\end{bmatrix}$ fill the **whole $\boldsymbol{xy}$ plane**. They produce every $\begin{bmatrix}x\\y\end{bmatrix}$.

4 The combinations $c\begin{bmatrix}1\\1\\1\end{bmatrix}+d\begin{bmatrix}2\\3\\4\end{bmatrix}$ fill a **plane in $\boldsymbol{xyz}$ space**. Same plane for $\begin{bmatrix}1\\1\\1\end{bmatrix},\begin{bmatrix}3\\4\\5\end{bmatrix}$.

Arithmetic starts with numbers. We operate on those numbers in two essential ways:

Addition $2+3=5$ **Multiplication** $(4)\,(5)=20$

Subtracting 3 is just the inverse of adding 3. Subtract $5-3$ to recover 2.
Dividing by 5 is just the inverse of multiplying by 5. Divide 20 by 5 to recover 4.
Combining addition and multiplication leads to $(2)\,(3+4)=(2)\,(3)+(2)\,(4)$.

Linear algebra moves addition and multiplication into higher dimensions. Instead of working with single numbers, we work with **vectors**. The vector $\boldsymbol{v}=(3,1,7)$ is a string of three numbers. It is a "3-dimensional vector". The good way is to write $\boldsymbol{v}$ as a column vector. Then we can add two column vectors $\boldsymbol{v}$ and $\boldsymbol{w}$:

$$\textbf{Vector addition}\quad \boldsymbol{v}+\boldsymbol{w}=\begin{bmatrix}3\\1\\7\end{bmatrix}+\begin{bmatrix}4\\5\\2\end{bmatrix}=\begin{bmatrix}7\\6\\9\end{bmatrix}\quad \text{(add each pair of components)} \tag{1}$$

Subtracting $\boldsymbol{w}$ is just the inverse of adding $\boldsymbol{w}$, so that $\boldsymbol{v}+\boldsymbol{w}-\boldsymbol{w}$ recovers $\boldsymbol{v}$.

$$\textbf{Vector subtraction}\quad (\boldsymbol{v}+\boldsymbol{w})-\boldsymbol{w}=\begin{bmatrix}7\\6\\9\end{bmatrix}-\begin{bmatrix}4\\5\\2\end{bmatrix}=\begin{bmatrix}3\\1\\7\end{bmatrix}=\boldsymbol{v} \tag{2}$$

What about multiplication of vectors? We could multiply each pair of components. If we add the results, we arrive at the "**dot product**" $\boldsymbol{v}\cdot\boldsymbol{w}=31$:

$$\boldsymbol{v}\cdot\boldsymbol{w}=\begin{bmatrix}3\\1\\7\end{bmatrix}\cdot\begin{bmatrix}4\\5\\2\end{bmatrix}=\begin{matrix}\mathbf{(3)\,(4)+(1)\,(5)+(7)\,(2)}=12+5+14=\mathbf{31}\\ \textbf{Dot product}=\mathbf{31}\end{matrix}. \tag{3}$$

The dot product $\boldsymbol{v}\cdot\boldsymbol{w}$ is a useful way to multiply vectors. In the next section $\boldsymbol{v}\cdot\boldsymbol{v}$ reveals the length of $\boldsymbol{v}$, and $\boldsymbol{v}\cdot\boldsymbol{w}$ reveals the angle between $\boldsymbol{v}$ and $\boldsymbol{w}$. But a more important multiplication in linear algebra is $A\boldsymbol{v}$ = *matrix times vector*.

The output from Av is a vector not a number: Matrix A times vector $\boldsymbol{v}$ equals vector $A\boldsymbol{v}$. The matrix A is a rectangle of numbers: *m rows and n columns.* A 2 by 3 matrix multiplies a vector $\boldsymbol{v}$ with $n = 3$ components.

Matrix times vector
$$A\boldsymbol{v} = \begin{bmatrix} 4 & 5 & 2 \\ 1 & 2 & 1 \end{bmatrix} \begin{bmatrix} 3 \\ 1 \\ 7 \end{bmatrix} = \begin{bmatrix} 4 \cdot 3 + 5 \cdot 1 + 2 \cdot 7 \\ 1 \cdot 3 + 2 \cdot 1 + 1 \cdot 7 \end{bmatrix} = \begin{bmatrix} \mathbf{31} \\ \mathbf{12} \end{bmatrix}. \quad (4)$$

Please notice: A has 2 rows. **A times $\boldsymbol{v}$ involved 2 dot products.** The first component in $A\boldsymbol{v}$ used row 1 of A. The result was 31. The second component of $A\boldsymbol{v}$ used row 2 of A:

$$(\text{row 2 of } A) \cdot (\text{column vector } \boldsymbol{v}) = 1 \cdot 3 + 2 \cdot 1 + 1 \cdot 7 = \mathbf{12}$$

This is the usual way to multiply A times $\boldsymbol{v}$: dot products of the rows of A with $\boldsymbol{v}$. But Section 1.3 will explain a better way to understand $A\boldsymbol{v}$. Computing row · column is fine, but understanding $A\boldsymbol{v}$ becomes clearer with *linear combinations of column vectors.*

Let me show one "linear combination" because this is the fundamental operation on vectors. *Multiply vectors by numbers like 2 and 4 and add the results:*

Linear combination
$c\boldsymbol{v} + d\boldsymbol{w} = 2\boldsymbol{v} + 4\boldsymbol{w}$
$$2\begin{bmatrix} 3 \\ 1 \\ 7 \end{bmatrix} + 4\begin{bmatrix} 1 \\ 2 \\ 1 \end{bmatrix} = \begin{bmatrix} 6 \\ 2 \\ 14 \end{bmatrix} + \begin{bmatrix} 4 \\ 8 \\ 4 \end{bmatrix} = \begin{bmatrix} 10 \\ 10 \\ 18 \end{bmatrix}. \quad (5)$$

Those combinations go into the big step: **Multiply a matrix by a matrix.** I would like to save that step for Section 1.4. We have explained three ways to multiply, involving numbers and vectors and matrices:

1. Number times vector ($c\boldsymbol{v}$) **2.** Vector · vector ($\boldsymbol{v} \cdot \boldsymbol{w}$) **3.** Matrix times vector ($A\boldsymbol{v}$)

Those are in Sections 1.1 and 1.2 and 1.3. Then AB is matrix multiplication in Section 1.4.

Let me also say: A times $\boldsymbol{v}$ can use the rows of a matrix A or the columns of A. There are m row vectors in A and there are n column vectors. Both ways use the same mn numbers. A major key to linear algebra comes from the connections between two ideas.

Dot products with rows of A

$$\begin{bmatrix} 3 & 4 \\ 5 & 6 \end{bmatrix} \begin{bmatrix} x \\ y \end{bmatrix} = \begin{bmatrix} 3x + 4y \\ 5x + 6y \end{bmatrix}$$

Combinations of columns of A

$$\begin{bmatrix} 3 & 4 \\ 5 & 6 \end{bmatrix} \begin{bmatrix} x \\ y \end{bmatrix} = x\begin{bmatrix} 3 \\ 5 \end{bmatrix} + y\begin{bmatrix} 4 \\ 6 \end{bmatrix}$$

Linear Combinations

Combining addition with scalar multiplication produces a "**linear combination**" of $\boldsymbol{v}$ and $\boldsymbol{w}$. Multiply $\boldsymbol{v}$ by c and multiply $\boldsymbol{w}$ by d. Then add $c\boldsymbol{v} + d\boldsymbol{w}$.

The sum of $c\boldsymbol{v}$ ***and*** $d\boldsymbol{w}$ ***is a*** *linear combination* $c\boldsymbol{v} + d\boldsymbol{w}$.

Four special linear combinations are: sum, difference, zero, and a scalar multiple $c\boldsymbol{v}$:

$$\begin{aligned} 1\boldsymbol{v} + 1\boldsymbol{w} &= \text{sum of vectors } (4,2) + (-1,2) = (3,4) \\ 1\boldsymbol{v} - 1\boldsymbol{w} &= \text{difference of vectors } (4,2) - (-1,2) = (5,0) \\ 0\boldsymbol{v} + 0\boldsymbol{w} &= \textbf{\textit{zero vector }} \mathbf{(0,0)} \\ c\boldsymbol{v} + 0\boldsymbol{w} &= \text{vector } c\boldsymbol{v} \text{ in the direction of } \boldsymbol{v} \end{aligned}$$

The zero vector is always a possible combination from $c = d = 0$. Every time we see a "space" of vectors, that zero vector will be included. This big view, taking *all the combinations* of $\boldsymbol{v}$ and $\boldsymbol{w}$, is linear algebra at work.

The figures show how you can visualize vectors. For algebra, we just need the components (like 4 and 2). That vector $\boldsymbol{v}$ is represented by an arrow. The arrow goes $v_1 = 4$ units to the right and $v_2 = 2$ units up. It ends at the point whose x, y coordinates are $4, 2$. This point is another representation of the vector—so we have three ways to describe $\boldsymbol{v}$:

Represent vector $\boldsymbol{v}$ Two numbers Arrow from $(0,0)$ Point in the plane

We add using the numbers. We visualize $\boldsymbol{v} + \boldsymbol{w}$ using arrows for $\boldsymbol{v}$ and $\boldsymbol{w}$ and $\boldsymbol{v} + \boldsymbol{w}$.

Vector addition (head to tail) ***At the end of*** $\boldsymbol{v}$***, place the start of*** $\boldsymbol{w}$.

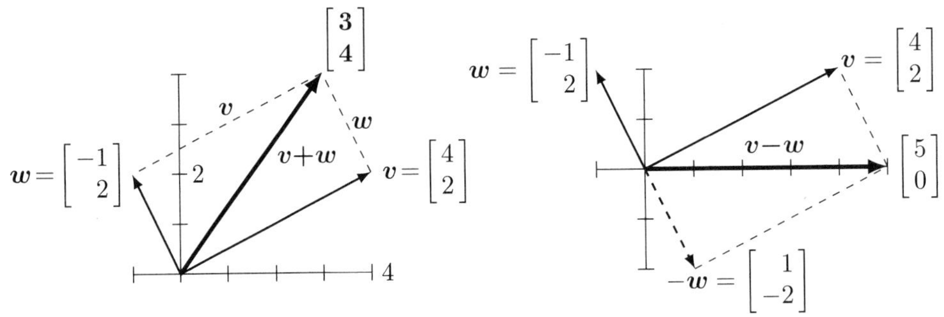

$$\boldsymbol{v} + \boldsymbol{w} = \begin{bmatrix}4\\2\end{bmatrix} + \begin{bmatrix}-1\\2\end{bmatrix} = \begin{bmatrix}3\\4\end{bmatrix} \qquad \boldsymbol{v} - \boldsymbol{w} = \begin{bmatrix}4\\2\end{bmatrix} - \begin{bmatrix}-1\\2\end{bmatrix} = \begin{bmatrix}5\\0\end{bmatrix}$$

Figure 1.1: Vector addition $\boldsymbol{v} + \boldsymbol{w} = (3,\ 4)$ produces the diagonal of a parallelogram. On the right: The reverse of $\boldsymbol{w}$ is $-\boldsymbol{w}$. The linear combination is $\boldsymbol{v} - \boldsymbol{w} = (5,\ 0)$.

We travel along $\boldsymbol{v}$ and then along $\boldsymbol{w}$. Or we take the diagonal shortcut along $\boldsymbol{v} + \boldsymbol{w}$. We could also go along $\boldsymbol{w}$ and then $\boldsymbol{v}$. In other words, $\boldsymbol{w} + \boldsymbol{v}$ **gives the same answer as** $\boldsymbol{v} + \boldsymbol{w}$. These are different ways along the parallelogram (in this example it is a rectangle).

Vectors in Three Dimensions

A vector with two components corresponds to a point in the xy plane. The components of $\boldsymbol{v}$ are the coordinates of the point: $x = v_1$ and $y = v_2$. The arrow ends at this point (v_1, v_2), when it starts from $(0, 0)$. Now we allow vectors to have three components (v_1, v_2, v_3).

The xy plane is replaced by three-dimensional xyz space. Here are typical vectors (still column vectors but with three components):

$$\boldsymbol{v} = \begin{bmatrix} 1 \\ 1 \\ -1 \end{bmatrix} \quad \text{and} \quad \boldsymbol{w} = \begin{bmatrix} 2 \\ 3 \\ 4 \end{bmatrix} \quad \text{and} \quad \boldsymbol{v} + \boldsymbol{w} = \begin{bmatrix} 3 \\ 4 \\ 3 \end{bmatrix}.$$

The vector $\boldsymbol{v}$ corresponds to an arrow in 3-space. Usually the arrow starts at the "origin", where the xyz axes meet and the coordinates are $(0, 0, 0)$. The arrow ends at the point with coordinates v_1, v_2, v_3. There is a perfect match between the ***column vector*** and the ***arrow from the origin*** and the ***point where the arrow ends***.

The vector (x, y) in the plane (with 2 numbers) is different from $(x, y, 0)$ in 3-space!

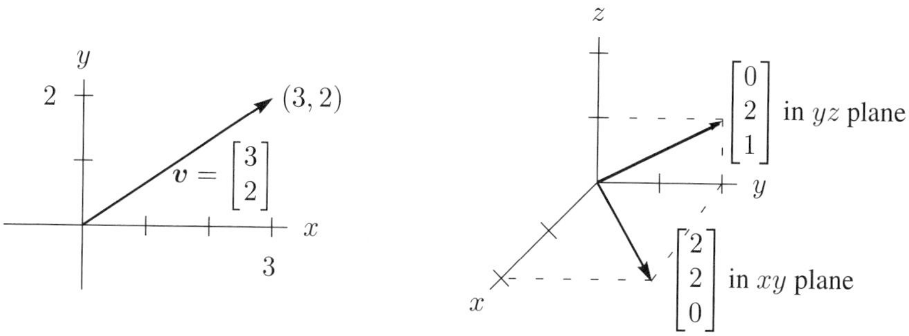

Figure 1.2: Vectors $\begin{bmatrix} x \\ y \end{bmatrix}$ and $\begin{bmatrix} x \\ y \\ z \end{bmatrix}$ correspond to points (x, y) and (x, y, z) in 2D and 3D.

$$\textbf{\textit{Space saver: From now on}} \quad \boldsymbol{v} = \begin{bmatrix} 1 \\ 1 \\ -1 \end{bmatrix} \quad \textbf{\textit{is also written as}} \quad \boldsymbol{v} = (1, 1, -1).$$

The reason for the row form (in parentheses) is to save space. But $\boldsymbol{v} = (1, 1, -1)$ is not a row vector! It is in actuality a column vector, just temporarily lying down. The row vector $[\,1 \;\; 1 \;\; -1\,]$ is absolutely different, even though it has the same three components. That 1 by 3 row vector $\boldsymbol{v}^{\mathrm{T}} = [\,1 \;\; 1 \;\; -1\,]$ is the "transpose" of the 3 by 1 column vector $\boldsymbol{v}$.

In three dimensions, $\boldsymbol{v} + \boldsymbol{w}$ is still found a component at a time. The sum has components $v_1 + w_1$ and $v_2 + w_2$ and $v_3 + w_3$. You see how to add vectors in 4 or 5 or n dimensions. When $\boldsymbol{w}$ starts at the end of $\boldsymbol{v}$, the third side is $\boldsymbol{v} + \boldsymbol{w}$. The other way around the parallelogram is $\boldsymbol{w} + \boldsymbol{v}$. *Do the four sides all lie in the same plane? Yes.* And the sum $\boldsymbol{v} + \boldsymbol{w} - \boldsymbol{v} - \boldsymbol{w}$ goes completely around to produce the ______ vector.

A typical linear combination of three vectors in three dimensions is $\boldsymbol{u} + 4\boldsymbol{v} - 2\boldsymbol{w}$:

Linear combination
Multiply by $1, 4, -2$
Then add $\boldsymbol{u} + 4\boldsymbol{v} - 2\boldsymbol{w}$

$$\begin{bmatrix}1\\0\\3\end{bmatrix} + 4\begin{bmatrix}1\\2\\1\end{bmatrix} - 2\begin{bmatrix}2\\3\\-1\end{bmatrix} = \begin{bmatrix}\mathbf{1}\\\mathbf{2}\\\mathbf{9}\end{bmatrix}.$$

The Important Question: All Combinations

For one vector $\boldsymbol{u}$, the only linear combinations are the multiples $c\boldsymbol{u}$. For two vectors, the combinations are $c\boldsymbol{u} + d\boldsymbol{v}$. For three vectors, the combinations are $c\boldsymbol{u} + d\boldsymbol{v} + e\boldsymbol{w}$. Will you take the big step from *one* combination to **all combinations**? **Every c and d and e are allowed**. Suppose the vectors $\boldsymbol{u}, \boldsymbol{v}, \boldsymbol{w}$ are in three-dimensional space:

1. What is the picture of *all* combinations $c\boldsymbol{u}$?
2. What is the picture of *all* combinations $c\boldsymbol{u} + d\boldsymbol{v}$?
3. What is the picture of *all* combinations $c\boldsymbol{u} + d\boldsymbol{v} + e\boldsymbol{w}$?

The answers depend on the particular vectors $\boldsymbol{u}, \boldsymbol{v}$, and $\boldsymbol{w}$. If they are typical nonzero vectors (chosen at random), here are the three answers. This is a key to our subject:

1. The combinations $c\boldsymbol{u}$ fill a ***line through*** $(\mathbf{0}, \mathbf{0}, \mathbf{0})$.
2. The combinations $c\boldsymbol{u} + d\boldsymbol{v}$ fill a ***plane through*** $(\mathbf{0}, \mathbf{0}, \mathbf{0})$.
3. The combinations $c\boldsymbol{u} + d\boldsymbol{v} + e\boldsymbol{w}$ fill ***three-dimensional space***.

The zero vector $(0, 0, 0)$ is on the line because c can be zero. It is on the plane because c and d could both be zero. The line of vectors $c\boldsymbol{u}$ is infinitely long (forward and backward). It is the **plane of all $c\boldsymbol{u} + d\boldsymbol{v}$** (combining two vectors $\boldsymbol{u}, \boldsymbol{v}$ in three-dimensional space) that I especially ask you to think about.

Adding all $c\boldsymbol{u}$ on one line to all $d\boldsymbol{v}$ on the other line fills in the plane in Figure 1.3.

When we include a third vector $\boldsymbol{w}$, the multiples $e\boldsymbol{w}$ give a third line. **Suppose that third line along $\boldsymbol{w}$ is not in the plane of $\boldsymbol{u}$ and $\boldsymbol{v}$**. Then combining all $e\boldsymbol{w}$ with all $c\boldsymbol{u}+d\boldsymbol{v}$ fills up the whole three-dimensional space: $c\boldsymbol{u} + d\boldsymbol{v} + e\boldsymbol{w}$ matches every point in 3D.

This is the typical situation! **Line**, then **plane**, then **space**. But other possibilities exist. When $\boldsymbol{w}$ happens to be $c\boldsymbol{u} + d\boldsymbol{v}$, that third vector $\boldsymbol{w}$ is in the plane of the first two. The combinations of $\boldsymbol{u}, \boldsymbol{v}, \boldsymbol{w}$ will not go outside that $\boldsymbol{uv}$ plane. We do not get the full three-dimensional space. Please think about the special cases in Problem 1 of the PSet.

Line containing all cu

Figure 1.3

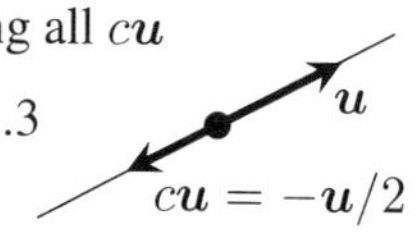

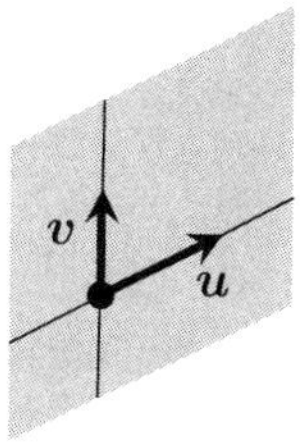

Plane from all $cu + dv$

■ WORKED EXAMPLES ■

1.1 A The linear combinations of $v = (1,1,0)$ and $w = (0,1,1)$ fill a plane in $\mathbf{R}^3$. *Describe that plane.* Find a vector that is *not* a combination of v and w—not on the plane.

Solution The plane of v and w contains all combinations $cv + dw$. The vectors in that plane allow any c and d. The plane of Figure 1.3 fills in between the two lines.

$$\text{Combinations} \quad cv + dw = c\begin{bmatrix}1\\1\\0\end{bmatrix} + d\begin{bmatrix}0\\1\\1\end{bmatrix} = \begin{bmatrix}c\\c+d\\d\end{bmatrix} \text{ fill a plane.}$$

Four vectors in that plane are $(0,0,0)$ and $(2,3,1)$ and $(5,7,2)$ and $(\pi, 2\pi, \pi)$. The second component $c + d$ is always the sum of the first and third components. Like most vectors, $(1,2,3)$ *is* ***not*** *in the plane, because* 2 *does not equal* $1 + 3$.

Another description of this plane through $(0,0,0)$ is to know that $n = (1,-1,1)$ is **perpendicular** to the plane. Section 1.2 will confirm that 90° angle by testing dot products.

1.1 B Find two equations for c and d so that **the linear combination** $cv + dw$ **equals** b:

$$v = \begin{bmatrix}2\\-1\end{bmatrix} \qquad w = \begin{bmatrix}-1\\2\end{bmatrix} \qquad b = \begin{bmatrix}1\\0\end{bmatrix}.$$

Solution In applying mathematics, many problems have two parts. Here we are asked for the modeling part (the equations). Chapter 2 is devoted to the solution part (finding c and d). Our example fits into a fundamental model for linear algebra:

Vector equation Find 2 numbers c and d so that $cv + dw = b$.

For $n = 2$ we will soon find a formula for c and d. The "elimination method" in Chapter 2 succeeds far beyond $n = 1000$ column vectors. At that point we must use matrices!

Vector equation $cv + dw = b$	$c\begin{bmatrix}2\\5\end{bmatrix} + d\begin{bmatrix}-1\\-3\end{bmatrix} = \begin{bmatrix}1\\1\end{bmatrix}$	gives	$2c - d = 1$ $5c - 3d = 1$	$c = 2$ $d = 3$

Vector addition produces two equations. The graph of each equation produces a line. Two lines cross at the solution. Why not see this also as a **matrix equation** $Ax = b$, since that is where we are going:

$$\begin{matrix}\textbf{2 by 2 matrix}\\ Ax = b\end{matrix} \qquad \begin{bmatrix}\mathbf{2} & \mathbf{-1}\\ \mathbf{5} & \mathbf{-3}\end{bmatrix}\begin{bmatrix}c\\d\end{bmatrix} = \begin{bmatrix}1\\1\end{bmatrix}.$$

Problem Set 1.1

1 Under what conditions on a, b, c, d is $\begin{bmatrix} c \\ d \end{bmatrix}$ a multiple m of $\begin{bmatrix} a \\ b \end{bmatrix}$? Start with the two equations $c = ma$ and $d = mb$. By eliminating m, find one equation connecting a, b, c, d. You can assume no zeros in these numbers.

2 Going around a triangle from $(0,0)$ to $(5,0)$ to $(0,12)$ to $(0,0)$, what are those three vectors $\boldsymbol{u}, \boldsymbol{v}, \boldsymbol{w}$? What is $\boldsymbol{u} + \boldsymbol{v} + \boldsymbol{w}$? What are their lengths $||\boldsymbol{u}||$ and $||\boldsymbol{v}||$ and $||\boldsymbol{w}||$? The length squared of a vector $\boldsymbol{u} = (u_1, u_2)$ is $||\boldsymbol{u}||^2 = u_1^2 + u_2^2$.

Problems 3–9 are about addition of vectors and linear combinations.

3 Describe geometrically (line, plane, or all of $\mathbf{R}^3$) all linear combinations of

$$\text{(a)} \begin{bmatrix} 1 \\ 2 \\ 3 \end{bmatrix} \text{ and } \begin{bmatrix} 3 \\ 6 \\ 9 \end{bmatrix} \qquad \text{(b)} \begin{bmatrix} 1 \\ 0 \\ 0 \end{bmatrix} \text{ and } \begin{bmatrix} 0 \\ 2 \\ 3 \end{bmatrix} \qquad \text{(c)} \begin{bmatrix} 2 \\ 0 \\ 0 \end{bmatrix} \text{ and } \begin{bmatrix} 0 \\ 2 \\ 2 \end{bmatrix} \text{ and } \begin{bmatrix} 2 \\ 2 \\ 3 \end{bmatrix}$$

4 Draw $\boldsymbol{v} = \begin{bmatrix} 4 \\ 1 \end{bmatrix}$ and $\boldsymbol{w} = \begin{bmatrix} -2 \\ 2 \end{bmatrix}$ and $\boldsymbol{v}+\boldsymbol{w}$ and $\boldsymbol{v}-\boldsymbol{w}$ in a single xy plane.

5 If $\boldsymbol{v} + \boldsymbol{w} = \begin{bmatrix} 5 \\ 1 \end{bmatrix}$ and $\boldsymbol{v} - \boldsymbol{w} = \begin{bmatrix} 1 \\ 5 \end{bmatrix}$, compute and draw the vectors $\boldsymbol{v}$ and $\boldsymbol{w}$.

6 From $\boldsymbol{v} = \begin{bmatrix} 2 \\ 1 \end{bmatrix}$ and $\boldsymbol{w} = \begin{bmatrix} 1 \\ 2 \end{bmatrix}$, find the components of $3\boldsymbol{v} + \boldsymbol{w}$ and $c\boldsymbol{v} + d\boldsymbol{w}$.

7 Compute $\boldsymbol{u} + \boldsymbol{v} + \boldsymbol{w}$ and $2\boldsymbol{u} + 2\boldsymbol{v} + \boldsymbol{w}$. How do you know $\boldsymbol{u}, \boldsymbol{v}, \boldsymbol{w}$ lie in a plane?

These lie in a plane because $\boldsymbol{w} = c\boldsymbol{u} + d\boldsymbol{v}$. Find c and d

$$\boldsymbol{u} = \begin{bmatrix} 1 \\ 2 \\ 3 \end{bmatrix}, \quad \boldsymbol{v} = \begin{bmatrix} -3 \\ 1 \\ -2 \end{bmatrix}, \quad \boldsymbol{w} = \begin{bmatrix} 2 \\ -3 \\ -1 \end{bmatrix}.$$

8 Every combination of $\boldsymbol{v} = (1, -2, 1)$ and $\boldsymbol{w} = (0, 1, -1)$ has components that add to ____. Find c and d so that $c\boldsymbol{v} + d\boldsymbol{w} = (3, 3, -6)$. Why is $(3, 3, 6)$ impossible?

9 In the xy plane mark all nine of these linear combinations:

$$c \begin{bmatrix} 2 \\ 1 \end{bmatrix} + d \begin{bmatrix} 0 \\ 1 \end{bmatrix} \quad \text{with} \quad c = 0, 1, 2 \quad \text{and} \quad d = 0, 1, 2.$$

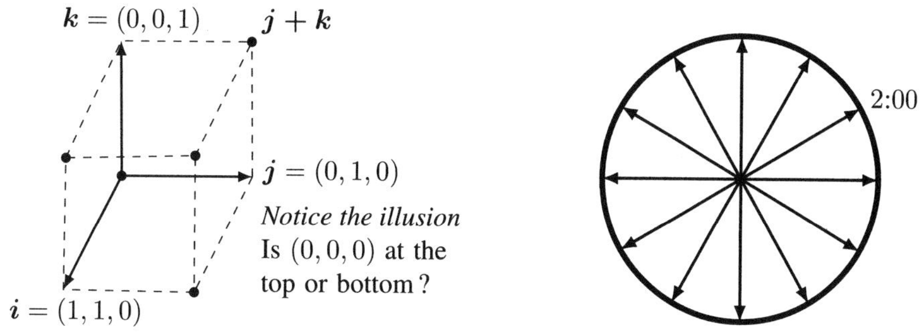

Figure 1.4: Unit cube from $\boldsymbol{i}, \boldsymbol{j}, \boldsymbol{k}$ and twelve clock vectors : all lengths $= 1$.

Problems 10–14 are about special vectors on cubes and clocks in Figure 1.4.

10 If three corners of a parallelogram are $(1, 1)$, $(4, 2)$, and $(1, 3)$, what are all three of the possible fourth corners? Draw two of them.

11 Four corners of this unit cube are $(0, 0, 0)$, $(1, 0, 0)$, $(0, 1, 0)$, $(0, 0, 1)$. What are the other four corners? Find the coordinates of the center point of the cube. The center points of the six faces are ______. The cube has how many edges?

12 *Review Question.* In xyz space, where is the plane of all linear combinations of $\boldsymbol{i} = (1, 0, 0)$ and $\boldsymbol{i} + \boldsymbol{j} = (1, 1, 0)$?

13 (a) What is the sum $\boldsymbol{V}$ of the twelve vectors that go from the center of a clock to the hours 1:00, 2:00, ..., 12:00?

(b) If the 2:00 vector is removed, why do the 11 remaining vectors add to 8:00?

(c) What are the x, y components of that 2:00 vector $\boldsymbol{v} = (\cos\theta, \sin\theta)$?

14 Suppose the twelve vectors start from 6:00 at the bottom instead of $(0, 0)$ at the center. The vector to 12:00 is doubled to $(0, 2)$. The new twelve vectors add to ____.

15 Draw vectors $\boldsymbol{u}$, $\boldsymbol{v}$, $\boldsymbol{w}$ so that their combinations $c\boldsymbol{u} + d\boldsymbol{v} + e\boldsymbol{w}$ fill only a line. Find vectors $\boldsymbol{u}$, $\boldsymbol{v}$, $\boldsymbol{w}$ so that their combinations $c\boldsymbol{u} + d\boldsymbol{v} + e\boldsymbol{w}$ fill only a plane.

16 What combination $c\begin{bmatrix} 1 \\ 2 \end{bmatrix} + d\begin{bmatrix} 3 \\ 1 \end{bmatrix}$ produces $\begin{bmatrix} 14 \\ 8 \end{bmatrix}$? Express this question as two equations for the coefficients c and d in the linear combination.

Problems 17–18 go further with linear combinations of $\boldsymbol{v}$ and $\boldsymbol{w}$ (Figure 1.5a).

17 Figure 1.5a shows $\frac{1}{2}\boldsymbol{v} + \frac{1}{2}\boldsymbol{w}$. Mark the points $\frac{3}{4}\boldsymbol{v} + \frac{1}{4}\boldsymbol{w}$ and $\frac{1}{4}\boldsymbol{v} + \frac{1}{4}\boldsymbol{w}$ and $\boldsymbol{v} + \boldsymbol{w}$. Draw the line of all combinations $c\boldsymbol{v} + d\boldsymbol{w}$ that have $c + d = 1$.

18 Restricted by $0 \le c \le 1$ and $0 \le d \le 1$, shade in all the combinations $c\boldsymbol{v} + d\boldsymbol{w}$. Restricted only by $c \ge 0$ and $d \ge 0$ draw the "cone" of all combinations $c\boldsymbol{v} + d\boldsymbol{w}$.

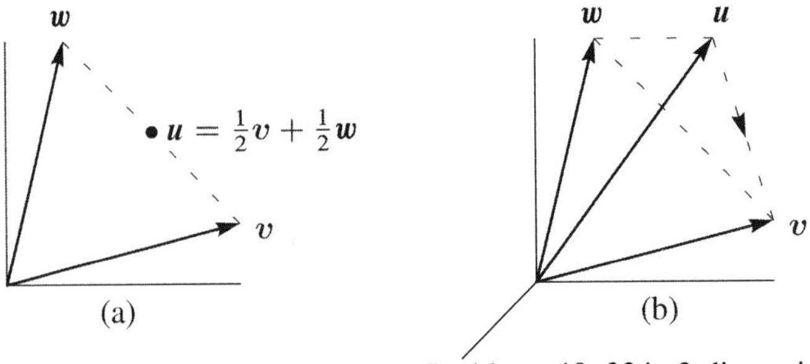

Figure 1.5: Problems **17–18** in a plane Problems **19–22** in 3-dimensional space

Problems 19–22 deal with u, v, w in three-dimensional space (see Figure 1.5b).

19 Locate $\frac{1}{3}\boldsymbol{u} + \frac{1}{3}\boldsymbol{v} + \frac{1}{3}\boldsymbol{w}$ and $\frac{1}{2}\boldsymbol{u} + \frac{1}{2}\boldsymbol{w}$ in Figure 1.5b. Challenge problem: Under what restrictions on c, d, e, will the combinations $c\boldsymbol{u} + d\boldsymbol{v} + e\boldsymbol{w}$ fill in the dashed triangle? To stay in the triangle, one requirement is $c \geq 0, d \geq 0, e \geq 0$.

20 The three dashed lines in the triangle are $\boldsymbol{v} - \boldsymbol{u}$ and $\boldsymbol{w} - \boldsymbol{v}$ and $\boldsymbol{u} - \boldsymbol{w}$. Their sum is ______. Draw the head-to-tail addition around a plane triangle of $(3, 1)$ plus $(-1, 1)$ plus $(-2, -2)$.

21 Shade in the pyramid of combinations $c\boldsymbol{u} + d\boldsymbol{v} + e\boldsymbol{w}$ with $c \geq 0, d \geq 0, e \geq 0$ and $c + d + e \leq 1$. Mark the vector $\frac{1}{2}(\boldsymbol{u} + \boldsymbol{v} + \boldsymbol{w})$ as inside or outside this pyramid.

22 If you look at *all* combinations of those $\boldsymbol{u}, \boldsymbol{v}$, and $\boldsymbol{w}$, is there any vector that can't be produced from $c\boldsymbol{u} + d\boldsymbol{v} + e\boldsymbol{w}$? Different answer if $\boldsymbol{u}, \boldsymbol{v}, \boldsymbol{w}$ are all in ______.

Challenge Problems

23 How many corners does a cube have in 4 dimensions? How many 3D faces? How many edges? A typical corner is $(0, 0, 1, 0)$. A typical edge goes to $(0, 1, 0, 0)$.

24 Find *two different combinations* of the three vectors $\boldsymbol{u} = (1, 3)$ and $\boldsymbol{v} = (2, 7)$ and $\boldsymbol{w} = (1, 5)$ that produce $\boldsymbol{b} = (0, 1)$. Slightly delicate question: If I take any three vectors $\boldsymbol{u}, \boldsymbol{v}, \boldsymbol{w}$ in the plane, will there always be two different combinations that produce $\boldsymbol{b} = (0, 1)$?

25 The linear combinations of $\boldsymbol{v} = (a, b)$ and $\boldsymbol{w} = (c, d)$ fill the plane unless ______. Find four vectors $\boldsymbol{u}, \boldsymbol{v}, \boldsymbol{w}, \boldsymbol{z}$ with four components each so that their combinations $c\boldsymbol{u} + d\boldsymbol{v} + e\boldsymbol{w} + f\boldsymbol{z}$ produce all vectors (b_1, b_2, b_3, b_4) in four-dimensional space.

26 Write down three equations for c, d, e so that $c\boldsymbol{u} + d\boldsymbol{v} + e\boldsymbol{w} = \boldsymbol{b}$. Can you somehow find c, d, e for this $\boldsymbol{b}$?

$$\boldsymbol{u} = \begin{bmatrix} 2 \\ -1 \\ 0 \end{bmatrix} \quad \boldsymbol{v} = \begin{bmatrix} -1 \\ 2 \\ -1 \end{bmatrix} \quad \boldsymbol{w} = \begin{bmatrix} 0 \\ -1 \\ 2 \end{bmatrix} \quad \boldsymbol{b} = \begin{bmatrix} 1 \\ 0 \\ 0 \end{bmatrix}.$$

1.2 Lengths and Angles from Dot Products

1 The "dot product" of $\boldsymbol{v} = \begin{bmatrix} 1 \\ 2 \end{bmatrix}$ and $\boldsymbol{w} = \begin{bmatrix} 4 \\ 6 \end{bmatrix}$ is $\boldsymbol{v} \cdot \boldsymbol{w} = (1)(4) + (2)(6) = 4 + 12 = \mathbf{16}$.

2 The length squared of $\boldsymbol{v} = (1,3,2)$ is $\boldsymbol{v} \cdot \boldsymbol{v} = 1 + 9 + 4 = 14$. **The length is** $||\boldsymbol{v}|| = \sqrt{\mathbf{14}}$.

3 The unit vector $\boldsymbol{u} = \dfrac{\boldsymbol{v}}{||\boldsymbol{v}||} = \dfrac{\boldsymbol{v}}{\sqrt{14}} = \dfrac{1}{\sqrt{14}}(1,3,2)$ has $||\boldsymbol{u}||^2 = \dfrac{1}{14} + \dfrac{9}{14} + \dfrac{4}{14} = \mathbf{1}$.

4 $\boldsymbol{v} = (1,3,2)$ is perpendicular to $\boldsymbol{w} = (4,-4,4)$ because $\boldsymbol{v} \cdot \boldsymbol{w} = \mathbf{0}$.

5 The angle $\theta = 45°$ between $\boldsymbol{v} = \begin{bmatrix} 1 \\ 0 \end{bmatrix}$ and $\boldsymbol{w} = \begin{bmatrix} 1 \\ 1 \end{bmatrix}$ has $\cos\theta = \dfrac{\boldsymbol{v} \cdot \boldsymbol{w}}{||\boldsymbol{v}|| \; ||\boldsymbol{w}||} = \dfrac{1}{(1)(\sqrt{2})}$.

6 All angles have $|\cos\theta| \leq 1$. All vectors have **Schwarz inequality** $|\boldsymbol{v} \cdot \boldsymbol{w}| \leq ||\boldsymbol{v}|| \; ||\boldsymbol{w}||$ | **Triangle inequality** $||\boldsymbol{v}+\boldsymbol{w}|| \leq ||\boldsymbol{v}|| + ||\boldsymbol{w}||$.

The dot product $\boldsymbol{v} \cdot \boldsymbol{v}$ tells us the squared length $||\boldsymbol{v}||^2$ of a vector $\boldsymbol{v}$.
The dot product $\boldsymbol{v} \cdot \boldsymbol{w}$ tells us the angle between two vectors $\boldsymbol{v}$ and $\boldsymbol{w}$.

The length $||\boldsymbol{v}||$ **is given by** $||\boldsymbol{v}||^2 = \boldsymbol{v} \cdot \boldsymbol{v} = v_1^2 + v_2^2 + \cdots + v_n^2$. (1)

In two dimensions, this is the Pythagoras formula $a^2 + b^2 = c^2$ for a right triangle. The sides have $a^2 = v_1^2$ and $b^2 = v_2^2$. The hypotenuse has $||v||^2 = v_1^2 + v_2^2 = a^2 + b^2$. That formula for length squared matches ordinary plane geometry.

To reach n dimensions, we can add one dimension at a time. Figure 1.6 shows $\boldsymbol{w} = (1,2,3)$ in three dimensions. Now the right triangle has sides $(1,2,0)$ and $(0,0,3)$. Those vectors add to $\boldsymbol{w}$. The first side is in the xy plane, the second side goes up the perpendicular z axis . For this triangle in 3D with hypotenuse $\boldsymbol{w} = (1,2,3)$, the law $a^2 + b^2 = c^2$ becomes $(1^2 + 2^2) + (3^2) = 14 = ||\boldsymbol{w}||^2$.

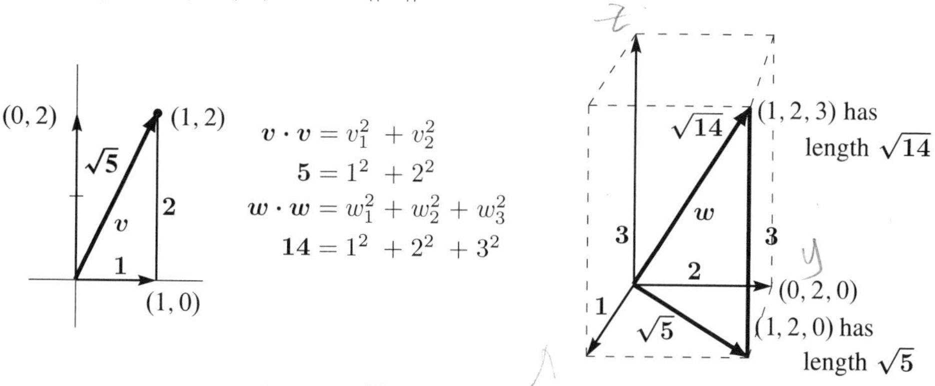

Figure 1.6: The length $\sqrt{\boldsymbol{v} \cdot \boldsymbol{v}} = \sqrt{5}$ in a plane and $\sqrt{\boldsymbol{w} \cdot \boldsymbol{w}} = \sqrt{14}$ in three dimensions.

The length of a four-dimensional vector would be $\sqrt{v_1^2 + v_2^2 + v_3^2 + v_4^2}$. Thus the vector $(1, 1, 1, 1)$ has length $\sqrt{1^2 + 1^2 + 1^2 + 1^2} = 2$. This is the diagonal through a unit cube in four-dimensional space. That diagonal in n dimensions has length $\sqrt{n}$.

We use the words **unit vector** when the length of the vector is 1. Divide $\boldsymbol{v}$ by $||\boldsymbol{v}||$.

> **A unit vector $\boldsymbol{u}$ has length $||\boldsymbol{u}|| = 1$. If $\boldsymbol{v} \neq \mathbf{0}$, then $\boldsymbol{u} = \dfrac{\boldsymbol{v}}{||\boldsymbol{v}||}$ is a unit vector.**

Example 1 The standard unit vectors along the x and y axes are written $\boldsymbol{i}$ and $\boldsymbol{j}$. In the xy plane, the unit vector that makes an angle "theta" with the x axis is $(\cos\theta, \sin\theta)$:

$$\textbf{Unit vectors} \quad \boldsymbol{i} = \begin{bmatrix} 1 \\ 0 \end{bmatrix} \quad \text{and} \quad \boldsymbol{j} = \begin{bmatrix} 0 \\ 1 \end{bmatrix} \quad \text{and} \quad \boldsymbol{u} = \begin{bmatrix} \cos\theta \\ \sin\theta \end{bmatrix}.$$

When $\theta = 0$, the horizontal vector $\boldsymbol{u}$ is $\boldsymbol{i}$. When $\theta = 90°$, the vertical vector is $\boldsymbol{j}$. For any angle, $\boldsymbol{u} =$ **$(\cos\theta, \sin\theta)$ is a unit vector because $\boldsymbol{u} \cdot \boldsymbol{u} = \cos^2\theta + \sin^2\theta = 1$.**

In three dimensions the standard unit vectors are $\boldsymbol{i} = (1, 0, 0)$ and $\boldsymbol{j} = (0, 1, 0)$ and $\boldsymbol{k} = (0, 0, 1)$. In four dimensions, one example of a unit vector is $\boldsymbol{u} = (\frac{1}{2}, \frac{1}{2}, \frac{1}{2}, \frac{1}{2})$. Or you could start with the vector $\boldsymbol{v} = (1, 5, 5, 7)$. Then $||\boldsymbol{v}||^2 = 1 + 25 + 25 + 49 = 100$. So $\boldsymbol{v}$ has length 10 and $\boldsymbol{u} = \boldsymbol{v}/10$ is a unit vector.

The word "**unit**" is always indicating that some measurement equals "one". The unit price is the price for one item. A unit cube has sides of length one. A unit circle is a circle with radius one. Now we see the meaning of a "unit vector": length = 1.

Perpendicular Vectors

Suppose the angle between $\boldsymbol{v}$ and $\boldsymbol{w}$ is $90°$. Its cosine is zero. That produces a valuable test $\boldsymbol{v} \cdot \boldsymbol{w} = \mathbf{0}$ for perpendicular vectors.

> **Perpendicular vectors have $\boldsymbol{v} \cdot \boldsymbol{w} = 0$. Then $||\boldsymbol{v} + \boldsymbol{w}||^2 = ||\boldsymbol{v}||^2 + ||\boldsymbol{w}||^2$.** (2)

This is the most important special case. It has brought us back to $90°$ angles and lengths $a^2 + b^2 = c^2$. The algebra for perpendicular vectors ($\boldsymbol{v} \cdot \boldsymbol{w} = 0 = \boldsymbol{w} \cdot \boldsymbol{v}$) is easy:

$$||\boldsymbol{v} + \boldsymbol{w}||^2 = (\boldsymbol{v} + \boldsymbol{w}) \cdot (\boldsymbol{v} + \boldsymbol{w}) = \boldsymbol{v} \cdot \boldsymbol{v} + \boldsymbol{v} \cdot \boldsymbol{w} + \boldsymbol{w} \cdot \boldsymbol{v} + \boldsymbol{w} \cdot \boldsymbol{w} = ||\boldsymbol{v}||^2 + ||\boldsymbol{w}||^2. \quad (3)$$

Two terms were zero. Please notice that $||\boldsymbol{v} - \boldsymbol{w}||^2$ is also equal to $||\boldsymbol{v}||^2 + ||\boldsymbol{w}||^2$.

Example 2 The vector $\boldsymbol{v} = (1, 1)$ is at a $45°$ angle with the x axis
The vector $\boldsymbol{w} = (1, -1)$ is at a $-45°$ angle with the x axis

So the angle between $(1, 1)$ and $(1, -1)$ is $90°$. Their dot product is $\boldsymbol{v} \cdot \boldsymbol{w} = 1 - 1 = 0$. This right triangle has $||\boldsymbol{v}||^2 = 2$ and $||\boldsymbol{w}||^2 = 2$ and $||\boldsymbol{v} - \boldsymbol{w}||^2 = ||\boldsymbol{v} + \boldsymbol{w}||^2 = 4$.

Example 3 The vectors $\boldsymbol{v} = (4, 2)$ and $\boldsymbol{w} = (-1, 2)$ have a *zero* dot product:

Dot product is zero
Vectors are perpendicular
$$\begin{bmatrix} 4 \\ 2 \end{bmatrix} \cdot \begin{bmatrix} -1 \\ 2 \end{bmatrix} = -4 + 4 = 0.$$

Put a weight of 4 at the point $x = -1$ (left of zero) and a weight of 2 at the point $x = 2$ (right of zero). The x axis will balance on the center point like a see-saw. The weights balance because the dot product is $(4)(-1) + (2)(2) = 0$.

This example is typical of engineering and science. The vector of weights is $(w_1, w_2) = (4, 2)$. The vector of distances from the center is $(v_1, v_2) = (-1, 2)$. The weights times the distances, $w_1 v_1$ and $w_2 v_2$, give the "moments". The equation for the see-saw to balance is $\boldsymbol{w} \cdot \boldsymbol{v} = w_1 v_1 + w_2 v_2 = 0$.

Example 4 The unit vectors $\boldsymbol{v} = (1, 0)$ and $\boldsymbol{u} = (\cos\theta, \sin\theta)$ have $\boldsymbol{v} \cdot \boldsymbol{u} = \cos\theta$. Now we are connecting the dot product to the angle between vectors.

Cosine of the angle $\boldsymbol{\theta}$ The cosine formula is easy to remember for unit vectors:

If $\|\|\boldsymbol{v}\|\| = 1$ and $\|\|\boldsymbol{u}\|\| = 1$, the angle θ between $\boldsymbol{v}$ and $\boldsymbol{u}$ has $\boldsymbol{\cos\theta = v \cdot u}$.

In mathematics, zero is always a special number. For dot products, it means that *these two vectors are perpendicular*. The angle between them is $90°$. The clearest example of perpendicular vectors is $\boldsymbol{i} = (1, 0)$ along the x axis and $\boldsymbol{j} = (0, 1)$ up the y axis. Again the dot product is $\boldsymbol{i} \cdot \boldsymbol{j} = 0 + 0 = 0$. The cosine of $90°$ is zero.

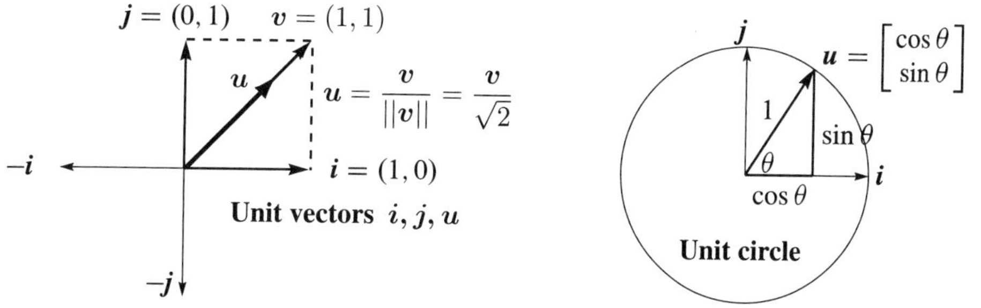

Figure 1.7: The coordinate vectors $\boldsymbol{i}$ and $\boldsymbol{j}$. The unit vector $\boldsymbol{u}$ at angle $45°$ divides $\boldsymbol{v} = (1, 1)$ by its length $||\boldsymbol{v}|| = \sqrt{2}$. The unit vector $\boldsymbol{u} = (\cos\theta, \sin\theta)$ is at angle θ.

Example 5 Dot products enter in economics and business. We have three goods to buy. Their prices are (p_1, p_2, p_3) for each unit—this is the price vector $\boldsymbol{p}$. The quantities we buy are (q_1, q_2, q_3). *Buying q_1 units at the price p_1 brings in $q_1 p_1$*. The total cost becomes quantities q times prices p. ***This is the dot product $\boldsymbol{q \cdot p}$ in three dimensions***:

$$\textit{Cost} = (q_1, q_2, q_3) \cdot (p_1, p_2, p_3) = q_1 p_1 + q_2 p_2 + q_3 p_3 = \textbf{\textit{dot product}}.$$

A zero dot product means that "the books balance". Total sales equal total purchases if $\boldsymbol{q} \cdot \boldsymbol{p} = 0$. Then $\boldsymbol{p}$ is perpendicular to $\boldsymbol{q}$ (in three-dimensional space). A supermarket with thousands of goods goes quickly into high dimensions.

Spreadsheets have become essential in management. They compute linear combinations and dot products. What you see on the screen is a matrix.

The Angle Between Two Vectors

We know that perpendicular vectors have $\boldsymbol{v} \cdot \boldsymbol{w} = 0$. The dot product is zero when the angle is 90°. Our next step is to connect all dot products to angles. The dot product $\boldsymbol{v} \cdot \boldsymbol{w}$ finds the angle between any two nonzero vectors $\boldsymbol{v}$ and $\boldsymbol{w}$.

Example 6 The unit vectors $\boldsymbol{v} = (\cos\alpha, \sin\alpha)$ and $\boldsymbol{w} = (\cos\beta, \sin\beta)$ have $\boldsymbol{v} \cdot \boldsymbol{w} = \cos\alpha\cos\beta + \sin\alpha\sin\beta$. In trigonometry this is the formula for $\cos(\beta - \alpha)$. Figure 1.8 shows that the angle between the unit vectors $\boldsymbol{v}$ and $\boldsymbol{w}$ is $\beta - \alpha$.

The dot product $\boldsymbol{w} \cdot \boldsymbol{v}$ ***equals*** $\boldsymbol{v} \cdot \boldsymbol{w}$. The order of $\boldsymbol{v}$ and $\boldsymbol{w}$ makes no difference.

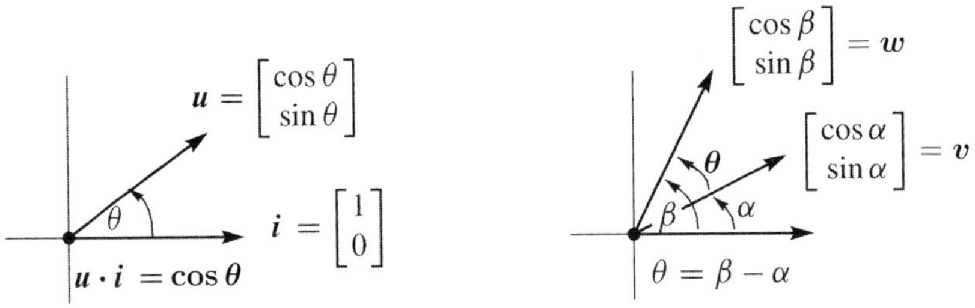

Figure 1.8: Unit vectors: $\boldsymbol{u} \cdot \boldsymbol{U} = \cos\theta$. The angle between the vectors is θ.

Suppose $\boldsymbol{v} \cdot \boldsymbol{w}$ is **not zero**. It may be positive, it may be negative. The sign of $\boldsymbol{v} \cdot \boldsymbol{w}$ immediately tells whether we are below or above a right angle. The angle is less than 90° when $\boldsymbol{v} \cdot \boldsymbol{w}$ is positive. The angle is above 90° when $\boldsymbol{v} \cdot \boldsymbol{w}$ is negative. The right side of Figure 1.9 shows a typical vector $\boldsymbol{v} = (3, 1)$. The angle with $\boldsymbol{w} = (1, 3)$ is less than 90° because $\boldsymbol{v} \cdot \boldsymbol{w} = 6$ is *positive*.

The borderline is where vectors are perpendicular to $\boldsymbol{v}$. On that dividing line between plus and minus, $\boldsymbol{w}_2 = (1, -3)$ is perpendicular to $\boldsymbol{v} = (3, 1)$. Their dot product is zero. Then $\boldsymbol{w}_3$ goes beyond a 90° angle with $\boldsymbol{v}$. The test becomes $\boldsymbol{v} \cdot \boldsymbol{w}_3 < 0$: *negative*.

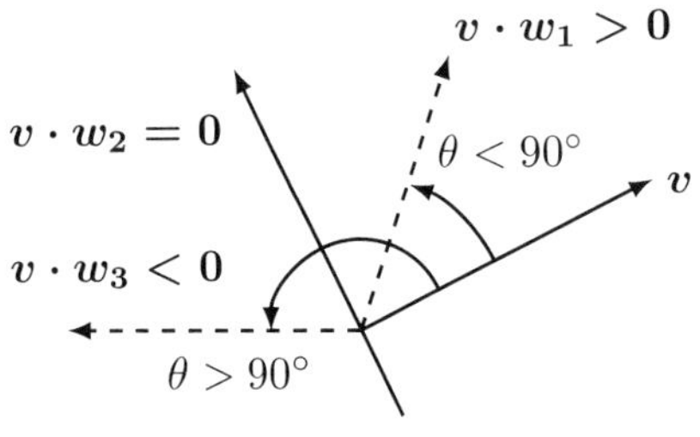

Figure 1.9: Small angle $\boldsymbol{v} \cdot \boldsymbol{w_1} > \mathbf{0}$. Right angle $\boldsymbol{v} \cdot \boldsymbol{w_2} = \mathbf{0}$. Large angle $\boldsymbol{v} \cdot \boldsymbol{w_3} < \mathbf{0}$.

The dot product reveals the exact angle $\boldsymbol{\theta}$. To repeat: For unit vectors $\boldsymbol{u}$ and $\boldsymbol{U}$, ***the dot product*** $\boldsymbol{u} \cdot \boldsymbol{U}$ ***is the cosine of*** $\boldsymbol{\theta}$. This remains true in n dimensions.

Unit vectors $\boldsymbol{u}$ **and** $\boldsymbol{U}$ **at angle** θ **have** $\boldsymbol{u} \cdot \boldsymbol{U} = \cos\theta$. **Certainly** $|\boldsymbol{u} \cdot \boldsymbol{U}| \leq \mathbf{1}$.

Remember that $\cos\theta$ is never greater than 1. It is never less than -1. *The dot product of unit vectors is between* -1 *and* 1. **The cosine of** θ **is revealed by** $\boldsymbol{u} \cdot \boldsymbol{U}$.

What if $\boldsymbol{v}$ and $\boldsymbol{w}$ are not unit vectors? Divide by their lengths to get $\boldsymbol{u} = \boldsymbol{v}/\|\boldsymbol{v}\|$ and $\boldsymbol{U} = \boldsymbol{w}/\|\boldsymbol{w}\|$. Then the dot product of those unit vectors $\boldsymbol{u}$ and $\boldsymbol{U}$ gives $\cos\theta$.

COSINE FORMULA If $\boldsymbol{v}$ and $\boldsymbol{w}$ are nonzero vectors then $\dfrac{\boldsymbol{v}\cdot\boldsymbol{w}}{\|\boldsymbol{v}\|\,\|\boldsymbol{w}\|} = \cos\theta.$ (4)

Whatever the angle, this dot product of $\boldsymbol{v}/\|\boldsymbol{v}\|$ with $\boldsymbol{w}/\|\boldsymbol{w}\|$ never exceeds one. That is the ***"Schwarz inequality"*** $|\boldsymbol{v}\cdot\boldsymbol{w}| \le \|\boldsymbol{v}\|\,\|\boldsymbol{w}\|$ for dot products—or more correctly the Cauchy-Schwarz-Buniakowsky inequality. It was found in France and Germany and Russia (and maybe elsewhere—it is the most important inequality in mathematics).

Since $|\cos\theta|$ never exceeds 1, the cosine formula in (4) gives two great inequalities.

SCHWARZ INEQUALITY	$\lvert\boldsymbol{v}\cdot\boldsymbol{w}\rvert \le \|\boldsymbol{v}\|\,\|\boldsymbol{w}\|$
TRIANGLE INEQUALITY	$\|\boldsymbol{v}+\boldsymbol{w}\| \le \|\boldsymbol{v}\| + \|\boldsymbol{w}\|$

Example 7 Find $\cos\theta$ for $\boldsymbol{v} = \begin{bmatrix} 2 \\ 1 \end{bmatrix}$ and $\boldsymbol{w} = \begin{bmatrix} 1 \\ 2 \end{bmatrix}$ and check both inequalities.

The dot product is $\boldsymbol{v}\cdot\boldsymbol{w} = \mathbf{4}$. Both $\boldsymbol{v}$ and $\boldsymbol{w}$ have length $\sqrt{\mathbf{5}}$. So $\|\boldsymbol{v}\|\,\|\boldsymbol{w}\| = \mathbf{5}$.

$$\cos\theta = \frac{\boldsymbol{v}\cdot\boldsymbol{w}}{\|\boldsymbol{v}\|\,\|\boldsymbol{w}\|} = \frac{4}{\sqrt{5}\sqrt{5}} = \frac{4}{5}.$$

The Schwarz inequality is $4 < 5$. By the triangle inequality, side $3 = \|\boldsymbol{v}+\boldsymbol{w}\|$ is less than side $1 +$ side 2. With $\boldsymbol{v}+\boldsymbol{w} = (3,3)$ the three sides are $\sqrt{18} < \sqrt{5}+\sqrt{5}$. Square this inequality to get $18 < 20$. This confirms the triangle inequality.

Example 8 The dot product of $\boldsymbol{v} = (a,b)$ and $\boldsymbol{w} = (b,a)$ is $2ab$. Their lengths are $\|\boldsymbol{v}\| = \|\boldsymbol{w}\| = \sqrt{a^2+b^2}$. The Schwarz inequality $\boldsymbol{v}\cdot\boldsymbol{w} \le \|\boldsymbol{v}\|\,\|\boldsymbol{w}\|$ is $\boldsymbol{2ab \le a^2+b^2}$.

For any numbers a^2 and b^2, *geometric mean* $\boldsymbol{ab} \le$ *arithmetic mean* $\frac{1}{2}\left(\boldsymbol{a}^2+\boldsymbol{b}^2\right)$.

The triangle inequality comes directly from the Schwarz inequality ! Finally, here is a proof of the Schwarz inequality that doesn't use angles. Every vector $\boldsymbol{u}$ has $0 \le \boldsymbol{u}\cdot\boldsymbol{u}$. We apply this to the vectors $\boldsymbol{u} = \|\boldsymbol{v}\|\boldsymbol{w} \pm \|\boldsymbol{w}\|\boldsymbol{v}$:

$$0 \le \boldsymbol{u}\cdot\boldsymbol{u} = \|\boldsymbol{v}\|^2\,\boldsymbol{w}\cdot\boldsymbol{w} \pm 2\,\|\boldsymbol{v}\|\,\|\boldsymbol{w}\|\,\boldsymbol{w}\cdot\boldsymbol{v} + \|\boldsymbol{w}\|^2\,\boldsymbol{v}\cdot\boldsymbol{v} \ \text{ means that}$$

$$2\,\|\boldsymbol{v}\|^2\|\boldsymbol{w}\|^2 \ge 2\,\|\boldsymbol{v}\|\,\|\boldsymbol{w}\|\,|\boldsymbol{w}\cdot\boldsymbol{v}|. \tag{5}$$

Divide by $2\,\|\boldsymbol{v}\|\,\|\boldsymbol{w}\|$. Then $|\boldsymbol{v}\cdot\boldsymbol{w}| \le \|\boldsymbol{v}\|\,\|\boldsymbol{w}\|$ is the Schwarz inequality. It leads to

$$\|\boldsymbol{v}+\boldsymbol{w}\|^2 = \boldsymbol{v}\cdot\boldsymbol{v} + \boldsymbol{v}\cdot\boldsymbol{w} + \boldsymbol{w}\cdot\boldsymbol{v} + \boldsymbol{w}\cdot\boldsymbol{w} \le \|\boldsymbol{v}\|^2 + 2\|\boldsymbol{v}\|\,\|\boldsymbol{w}\| + \|\boldsymbol{w}\|^2. \tag{6}$$

The square root is $\|\boldsymbol{v}+\boldsymbol{w}\| \le \|\boldsymbol{v}\| + \|\boldsymbol{w}\|$. **Side 3 cannot exceed Side 1 + Side 2.**

■ WORKED EXAMPLES ■

1.2 A For the vectors $\boldsymbol{v} = (3,4)$ and $\boldsymbol{w} = (4,3)$ test the Schwarz inequality on $\boldsymbol{v} \cdot \boldsymbol{w}$ and the triangle inequality on $\|\boldsymbol{v} + \boldsymbol{w}\|$. Find $\cos\theta$ for the angle between $\boldsymbol{v}$ and $\boldsymbol{w}$.

Solution The dot product is $\boldsymbol{v} \cdot \boldsymbol{w} = (3)(4) + (4)(3) = 24$. The length of $\boldsymbol{v}$ is $\|\boldsymbol{v}\| = \sqrt{9+16} = 5$ and also $\|\boldsymbol{w}\| = 5$. The sum $\boldsymbol{v} + \boldsymbol{w} = (7,7)$ has length $7\sqrt{2} < 10$.

Schwarz inequality $\quad |\boldsymbol{v} \cdot \boldsymbol{w}| \le \|\boldsymbol{v}\|\,\|\boldsymbol{w}\|$ is $24 < 25$.

Triangle inequality $\quad \|\boldsymbol{v} + \boldsymbol{w}\| \le \|\boldsymbol{v}\| + \|\boldsymbol{w}\|$ is $7\sqrt{2} < 5 + 5$.

Cosine of angle $\quad \cos\theta = \dfrac{24}{25}$ Thin angle from $\boldsymbol{v} = (3,4)$ to $\boldsymbol{w} = (4,3)$

1.2 B Which $\boldsymbol{v}$ and $\boldsymbol{w}$ give *equality* $|\boldsymbol{v} \cdot \boldsymbol{w}| = \|\boldsymbol{v}\|\,\|\boldsymbol{w}\|$ and $\|\boldsymbol{v} + \boldsymbol{w}\| = \|\boldsymbol{v}\| + \|\boldsymbol{w}\|$?

Equality: One vector is a multiple of the other as in $\boldsymbol{w} = c\boldsymbol{v}$. Then the angle is $0°$ or $180°$. In this case $|\cos\theta| = 1$ and $|\boldsymbol{v} \cdot \boldsymbol{w}|$ *equals* $\|\boldsymbol{v}\|\,\|\boldsymbol{w}\|$. If the angle is $0°$, as in $\boldsymbol{w} = 2\boldsymbol{v}$, then $\|\boldsymbol{v} + \boldsymbol{w}\| = \|\boldsymbol{v}\| + \|\boldsymbol{w}\|$ (both sides give $3\|\boldsymbol{v}\|$). This $\boldsymbol{v}, 2\boldsymbol{v}, 3\boldsymbol{v}$ triangle is extremely thin.

1.2 C Find a unit vector $\boldsymbol{u}$ in the direction of $\boldsymbol{v} = (3,4)$. Find a unit vector $\boldsymbol{U}$ that is perpendicular to $\boldsymbol{u}$. There are two possibilities for $\boldsymbol{U}$.

Solution For a unit vector $\boldsymbol{u}$, divide $\boldsymbol{v}$ by its length $\|\boldsymbol{v}\| = 5$. For a perpendicular vector $\boldsymbol{V}$ we can choose $(-4,3)$ or $(4,-3)$. For a *unit* vector $\boldsymbol{U}$, divide $\boldsymbol{V}$ by its length $\|\boldsymbol{V}\| = 5$.

Problem Set 1.2 1-16.

1 Calculate the dot products $\boldsymbol{u} \cdot \boldsymbol{v}$ and $\boldsymbol{u} \cdot \boldsymbol{w}$ and $\boldsymbol{u} \cdot (\boldsymbol{v} + \boldsymbol{w})$ and $\boldsymbol{w} \cdot \boldsymbol{v}$:

$$\boldsymbol{u} = \begin{bmatrix} -.6 \\ .8 \end{bmatrix} \qquad \boldsymbol{v} = \begin{bmatrix} 4 \\ 3 \end{bmatrix} \qquad \boldsymbol{w} = \begin{bmatrix} 1 \\ 2 \end{bmatrix}.$$

2 Compute the lengths $\|\boldsymbol{u}\|$ and $\|\boldsymbol{v}\|$ and $\|\boldsymbol{w}\|$ of those vectors. Check the Schwarz inequalities $|\boldsymbol{u} \cdot \boldsymbol{v}| \le \|\boldsymbol{u}\|\,\|\boldsymbol{v}\|$ and $|\boldsymbol{v} \cdot \boldsymbol{w}| \le \|\boldsymbol{v}\|\,\|\boldsymbol{w}\|$.

3 Find unit vectors in the directions of $\boldsymbol{v}$ and $\boldsymbol{w}$ in Problem 1, and find the cosine of the angle θ. Choose vectors $\boldsymbol{a}, \boldsymbol{b}, \boldsymbol{c}$ that make $0°$, $90°$, and $180°$ angles with $\boldsymbol{w}$.

4 For any *unit* vectors $\boldsymbol{v}$ and $\boldsymbol{w}$, find the dot products (actual numbers) of

(a) $\boldsymbol{v}$ and $-\boldsymbol{v}$ (b) $\boldsymbol{v} + \boldsymbol{w}$ and $\boldsymbol{v} - \boldsymbol{w}$ (c) $\boldsymbol{v} - 2\boldsymbol{w}$ and $\boldsymbol{v} + 2\boldsymbol{w}$

5 Find unit vectors $\boldsymbol{u}_1$ and $\boldsymbol{u}_2$ in the directions of $\boldsymbol{v} = (1,3)$ and $\boldsymbol{w} = (2,1,2)$. Find unit vectors $\boldsymbol{U}_1$ and $\boldsymbol{U}_2$ that are perpendicular to $\boldsymbol{u}_1$ and $\boldsymbol{u}_2$.

6 (a) Describe every vector $\boldsymbol{w} = (w_1, w_2)$ that is perpendicular to $\boldsymbol{v} = (2, -1)$.

(b) All vectors perpendicular to $\boldsymbol{V} = (1, 1, 1)$ lie on a ______ in 3 dimensions.

(c) The vectors perpendicular to both $(1, 1, 1)$ and $(1, 2, 3)$ lie on a ______.

7 Find the angle θ (from its cosine) between these pairs of vectors:

(a) $\boldsymbol{v} = \begin{bmatrix} 1 \\ \sqrt{3} \end{bmatrix}$ and $\boldsymbol{w} = \begin{bmatrix} 1 \\ 0 \end{bmatrix}$ (b) $\boldsymbol{v} = \begin{bmatrix} 2 \\ 2 \\ -1 \end{bmatrix}$ and $\boldsymbol{w} = \begin{bmatrix} 2 \\ -1 \\ 2 \end{bmatrix}$

(c) $\boldsymbol{v} = \begin{bmatrix} 1 \\ \sqrt{3} \end{bmatrix}$ and $\boldsymbol{w} = \begin{bmatrix} -1 \\ \sqrt{3} \end{bmatrix}$ (d) $\boldsymbol{v} = \begin{bmatrix} 3 \\ 1 \end{bmatrix}$ and $\boldsymbol{w} = \begin{bmatrix} -1 \\ -2 \end{bmatrix}$.

8 True or false (give a reason if true or find a counterexample if false):

(a) If $\boldsymbol{u} = (1, 1, 1)$ is perpendicular to $\boldsymbol{v}$ and $\boldsymbol{w}$, then $\boldsymbol{v}$ is parallel to $\boldsymbol{w}$.

(b) If $\boldsymbol{u}$ is perpendicular to $\boldsymbol{v}$ and $\boldsymbol{w}$, then $\boldsymbol{u}$ is perpendicular to $\boldsymbol{v} + 2\boldsymbol{w}$.

(c) If $\boldsymbol{u}$ and $\boldsymbol{v}$ are perpendicular unit vectors then $\|\boldsymbol{u} - \boldsymbol{v}\| = \sqrt{2}$. *Yes!*

9 The slopes of the arrows from $(0, 0)$ to (v_1, v_2) and (w_1, w_2) are v_2/v_1 and w_2/w_1. **Suppose the product** $v_2 w_2 / v_1 w_1$ **of those slopes is** -1. Show that $\boldsymbol{v} \cdot \boldsymbol{w} = 0$ and the vectors are perpendicular. (The line $y = 4x$ is perpendicular to $y = -\frac{1}{4}x$.)

10 Draw arrows from $(0, 0)$ to the points $\boldsymbol{v} = (1, 2)$ and $\boldsymbol{w} = (-2, 1)$. Multiply their slopes. That answer is a signal that $\boldsymbol{v} \cdot \boldsymbol{w} = 0$ and the arrows are ______.

11 If $\boldsymbol{v} \cdot \boldsymbol{w}$ is negative, what does this say about the angle between $\boldsymbol{v}$ and $\boldsymbol{w}$? Draw a 3-dimensional vector $\boldsymbol{v}$ (an arrow), and show where to find all $\boldsymbol{w}$'s with $\boldsymbol{v} \cdot \boldsymbol{w} < 0$.

12 With $\boldsymbol{v} = (1, 1)$ and $\boldsymbol{w} = (1, 5)$ choose a number c so that $\boldsymbol{w} - c\boldsymbol{v}$ is perpendicular to $\boldsymbol{v}$. Then find the formula for c starting from *any* nonzero $\boldsymbol{v}$ and $\boldsymbol{w}$.

13 Find nonzero vectors $\boldsymbol{u}, \boldsymbol{v}, \boldsymbol{w}$ that are perpendicular to $(1, 1, 1, 1)$ and to each other.

14 The geometric mean of $x = 2$ and $y = 8$ is $\sqrt{xy} = 4$. The arithmetic mean is larger: $\frac{1}{2}(x+y) =$ ______. This would come from the Schwarz inequality for $\boldsymbol{v} = (\sqrt{2}, \sqrt{8})$ and $\boldsymbol{w} = (\sqrt{8}, \sqrt{2})$. Find $\cos\theta$ for this $\boldsymbol{v}$ and $\boldsymbol{w}$.

15 **How long is the vector** $\boldsymbol{v} = (1, 1, \ldots, 1)$ **in** 9 **dimensions**? Find a unit vector $\boldsymbol{u}$ in the same direction as $\boldsymbol{v}$ and a unit vector $\boldsymbol{w}$ that is perpendicular to $\boldsymbol{v}$.

16 What are the cosines of the angles α, β, θ between the vector $(1, 0, -1)$ and the unit vectors $\boldsymbol{i}$, $\boldsymbol{j}$, $\boldsymbol{k}$ along the axes? Check the formula $\cos^2\alpha + \cos^2\beta + \cos^2\theta = 1$.

Problems 17–25 lead to the main facts about lengths and angles in triangles.

17 The parallelogram with sides $\boldsymbol{v} = (4, 2)$ and $\boldsymbol{w} = (-1, 2)$ is a rectangle. Check the Pythagoras formula $a^2 + b^2 = c^2$ which is for ***right triangles only***:

$$(\text{length of } \boldsymbol{v})^2 + (\text{length of } \boldsymbol{w})^2 = (\text{length of } \boldsymbol{v} + \boldsymbol{w})^2.$$

18 (Rules for dot products) These equations are simple but useful:

(1) $\boldsymbol{v} \cdot \boldsymbol{w} = \boldsymbol{w} \cdot \boldsymbol{v}$ **(2)** $\boldsymbol{u} \cdot (\boldsymbol{v} + \boldsymbol{w}) = \boldsymbol{u} \cdot \boldsymbol{v} + \boldsymbol{u} \cdot \boldsymbol{w}$ **(3)** $(c\boldsymbol{v}) \cdot \boldsymbol{w} = c(\boldsymbol{v} \cdot \boldsymbol{w})$

Use **(2)** with $\boldsymbol{u} = \boldsymbol{v} + \boldsymbol{w}$ to prove $\|\boldsymbol{v} + \boldsymbol{w}\|^2 = \boldsymbol{v} \cdot \boldsymbol{v} + 2\boldsymbol{v} \cdot \boldsymbol{w} + \boldsymbol{w} \cdot \boldsymbol{w}$.

19 The "Law of Cosines" comes from $(\boldsymbol{v} - \boldsymbol{w}) \cdot (\boldsymbol{v} - \boldsymbol{w}) = \boldsymbol{v} \cdot \boldsymbol{v} - 2\boldsymbol{v} \cdot \boldsymbol{w} + \boldsymbol{w} \cdot \boldsymbol{w}$:

$$\textbf{Cosine Law} \qquad \|\boldsymbol{v} - \boldsymbol{w}\|^2 = \|\boldsymbol{v}\|^2 - 2\|\boldsymbol{v}\|\,\|\boldsymbol{w}\| \cos\theta + \|\boldsymbol{w}\|^2.$$

Draw a triangle with sides $\boldsymbol{v}$ and $\boldsymbol{w}$ and $\boldsymbol{v} - \boldsymbol{w}$. Which of the angles is θ?

20 The ***triangle inequality*** says: (length of $\boldsymbol{v} + \boldsymbol{w}$) $\leq$ (length of $\boldsymbol{v}$) + (length of $\boldsymbol{w}$). Problem 18 found $\|\boldsymbol{v} + \boldsymbol{w}\|^2 = \|\boldsymbol{v}\|^2 + 2\boldsymbol{v} \cdot \boldsymbol{w} + \|\boldsymbol{w}\|^2$. Increase that $\boldsymbol{v} \cdot \boldsymbol{w}$ to $\|\boldsymbol{v}\|\,\|\boldsymbol{w}\|$ to show that $\|$**side 3**$\|$ cannot exceed $\|$**side 1**$\|$ + $\|$**side 2**$\|$:

$$\textbf{Triangle inequality} \qquad \|\boldsymbol{v} + \boldsymbol{w}\|^2 \leq (\|\boldsymbol{v}\| + \|\boldsymbol{w}\|)^2 \quad \textbf{or} \quad \boxed{\|\boldsymbol{v} + \boldsymbol{w}\| \leq \|\boldsymbol{v}\| + \|\boldsymbol{w}\|}$$

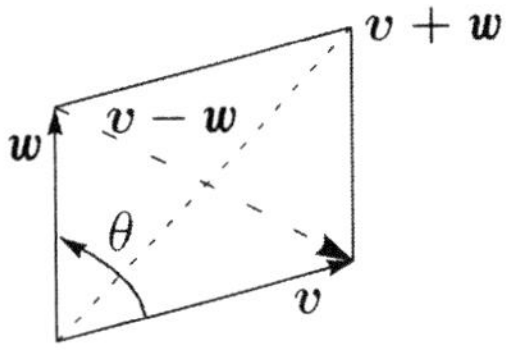

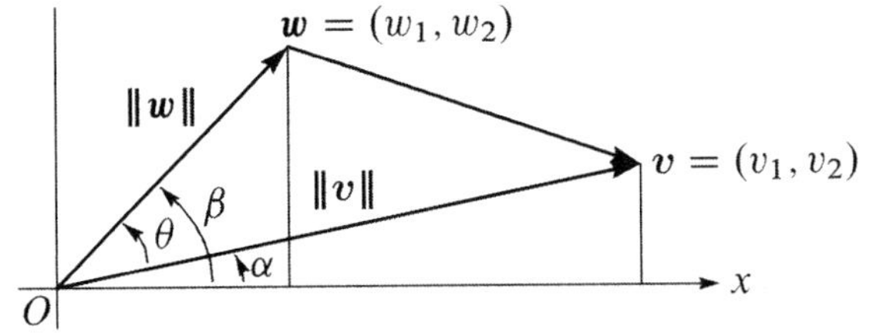

21 The Schwarz inequality $|\boldsymbol{v} \cdot \boldsymbol{w}| \leq \|\boldsymbol{v}\|\,\|\boldsymbol{w}\|$ by algebra instead of trigonometry:

(a) Multiply out both sides of $(v_1 w_1 + v_2 w_2)^2 \leq (v_1^2 + v_2^2)(w_1^2 + w_2^2)$.

(b) Show that the difference between those two sides equals $(v_1 w_2 - v_2 w_1)^2$. This cannot be negative since it is a square—so the inequality is true.

22 One-line proof of the inequality $|\boldsymbol{u} \cdot \boldsymbol{U}| \leq 1$ for unit vectors (u_1, u_2) and (U_1, U_2):

$$|\boldsymbol{u} \cdot \boldsymbol{U}| \leq |u_1|\,|U_1| + |u_2|\,|U_2| \leq \frac{u_1^2 + U_1^2}{2} + \frac{u_2^2 + U_2^2}{2} = 1.$$

Put $(u_1, u_2) = (.6, .8)$ and $(U_1, U_2) = (.8, .6)$ in that whole line and find $\cos\theta$.

23 Why is $|\cos\theta|$ never greater than 1? Find $\cos\theta$ in an equilateral triangle.

24 Show that the squared diagonal lengths $||\boldsymbol{v}+\boldsymbol{w}||^2 + ||\boldsymbol{v}-\boldsymbol{w}||^2$ add to the sum of four squared side lengths $2||\boldsymbol{v}||^2 + 2||\boldsymbol{w}||^2$.

25 (*Recommended*) If $||\boldsymbol{v}|| = 5$ and $||\boldsymbol{w}|| = 3$, what are the smallest and largest possible values of $||\boldsymbol{v} - \boldsymbol{w}||$? What are the smallest and largest possible values of $\boldsymbol{v} \cdot \boldsymbol{w}$?

Challenge Problems

26 Can three vectors in the xy plane have $\boldsymbol{u} \cdot \boldsymbol{v} < 0$ and $\boldsymbol{v} \cdot \boldsymbol{w} < 0$ and $\boldsymbol{u} \cdot \boldsymbol{w} < 0$? I don't know how many vectors in xyz space can have all negative dot products. (Four of those vectors in the plane would certainly be impossible ...).

27 Find 4 perpendicular unit vectors of the form $\left(\pm\frac{1}{2}, \pm\frac{1}{2}, \pm\frac{1}{2}, \pm\frac{1}{2}\right)$: Choose $+$ or $-$.

28 Using $\boldsymbol{v} = \mathsf{randn}(3,1)$ in MATLAB, create a random unit vector $\boldsymbol{u} = \boldsymbol{v}/||\boldsymbol{v}||$. Using $V = \mathsf{randn}(3,30)$ create 30 more random unit vectors $\boldsymbol{U}_j$. What is the average size of the dot products $|\boldsymbol{u} \cdot \boldsymbol{U}_j|$? In calculus, the average is $\int_0^\pi |\cos\theta| d\theta/\pi = 2/\pi$.

29 In the xy plane, when could four vectors $\boldsymbol{v}_1, \boldsymbol{v}_2, \boldsymbol{v}_3, \boldsymbol{v}_4$ not be the four sides of a quadrilateral ?

1.3 Matrices and Column Spaces

1 $A = \begin{bmatrix} 1 & 2 \\ 3 & 4 \\ 5 & 6 \end{bmatrix}$ is a **3 by 2 matrix**: $m = 3$ rows and $n = 2$ columns. Rank 2.

2 The 3 components of Ax are dot products of the 3 rows of A with the vector $\boldsymbol{x}$:

Row at a time
A times x

$$\begin{bmatrix} 1 & 2 \\ 3 & 4 \\ 5 & 6 \end{bmatrix} \begin{bmatrix} 7 \\ 8 \end{bmatrix} = \begin{bmatrix} 1 \cdot 7 + 2 \cdot 8 \\ 3 \cdot 7 + 4 \cdot 8 \\ 5 \cdot 7 + 6 \cdot 8 \end{bmatrix} = \begin{bmatrix} 23 \\ 53 \\ 83 \end{bmatrix}.$$

3 $\begin{bmatrix} 1 & 2 \\ 3 & 4 \\ 5 & 6 \end{bmatrix} \begin{bmatrix} 7 \\ 8 \end{bmatrix}$ is also a **combination of the columns** $\quad Ax = \mathbf{7} \begin{bmatrix} 1 \\ 3 \\ 5 \end{bmatrix} + \mathbf{8} \begin{bmatrix} 2 \\ 4 \\ 6 \end{bmatrix}$.

4 The **column space of A** contains **all combinations** $A\boldsymbol{x} = x_1\boldsymbol{a}_1 + x_2\boldsymbol{a}_2$ of the columns.

5 **Rank one matrices**: All columns of A (and all combinations $A\boldsymbol{x}$) lie on one line.

Sections 1.1 and 1.2 explained the mechanics of vectors—linear combinations, dot products, lengths, and angles. We have vectors in $\mathbf{R}^2$ and $\mathbf{R}^3$ and every $\mathbf{R}^n$.

Section 1.3 begins the algebra of m by n matrices: our true goal. A typical matrix A is a rectangle of m times n numbers—$\boldsymbol{m}$ **rows and** $\boldsymbol{n}$ **columns**. If m equals n then A is a "square matrix". The examples below are 3 by 3 matrices.

$$\begin{bmatrix} \mathbf{1} & 0 & 0 \\ 0 & \mathbf{1} & 0 \\ 0 & 0 & \mathbf{1} \end{bmatrix} \qquad \begin{bmatrix} \mathbf{2} & 0 & 0 \\ 0 & \mathbf{4} & 0 \\ 0 & 0 & \mathbf{5} \end{bmatrix} \qquad \begin{bmatrix} 2 & \mathbf{1} & \mathbf{-3} \\ \mathbf{0} & 4 & 7 \\ \mathbf{0} & \mathbf{0} & 5 \end{bmatrix} \qquad \begin{bmatrix} 2 & \mathbf{1} & \mathbf{-3} \\ \mathbf{1} & 4 & \mathbf{7} \\ \mathbf{-3} & \mathbf{7} & 5 \end{bmatrix}$$

Identity matrix — **Diagonal** matrix — **Triangular** matrix — **Symmetric** matrix

We often think of the columns of A as vectors $\boldsymbol{a}_1, \boldsymbol{a}_2, \ldots, \boldsymbol{a}_n$. Each of those n vectors is in m-dimensional space. In this example the $\boldsymbol{a}$'s have $m = 3$ components each:

$m = 3$ rows
$n = 4$ columns
3 by 4 matrix

$$A = \begin{bmatrix} \boldsymbol{a}_1 & \boldsymbol{a}_2 & \boldsymbol{a}_3 & \boldsymbol{a}_4 \end{bmatrix} = \begin{bmatrix} \mathbf{-1} & \mathbf{1} & 0 & 0 \\ 0 & \mathbf{-1} & \mathbf{1} & 0 \\ 0 & 0 & \mathbf{-1} & \mathbf{1} \end{bmatrix}$$

This example is a "difference matrix" because multiplying A times $\boldsymbol{x}$ produces a vector $A\boldsymbol{x}$ of differences. *How does an m by n matrix A multiply an n by 1 vector $\boldsymbol{x}$?* There are two ways to the same answer—we work with the rows of A or we work with the columns.

The **row picture** of $A\boldsymbol{x}$ will come from **dot products** of $\boldsymbol{x}$ with the rows of A.
The **column picture** will come from **linear combinations** of the columns of A.

Row picture of $A\boldsymbol{x}$ Each row of A multiplies the column vector $\boldsymbol{x}$. Those multiplications *row times column* are dot products ! The first dot product comes from row 1 of A :

$$(\textbf{row 1})\cdot\boldsymbol{x}=(-1,1,0,0)\cdot(x_1,x_2,x_3,x_4)=\boldsymbol{x_2-x_1}.$$

It takes m times n small multiplications to find the $m=3$ dot products that go into $A\boldsymbol{x}$.

Three dot products

$$A\boldsymbol{x}=\begin{bmatrix}-1&1&0&0\\0&-1&1&0\\0&0&-1&1\end{bmatrix}\begin{bmatrix}x_1\\x_2\\x_3\\x_4\end{bmatrix}=\begin{bmatrix}\text{row }1\cdot\boldsymbol{x}\\\text{row }2\cdot\boldsymbol{x}\\\text{row }3\cdot\boldsymbol{x}\end{bmatrix}=\begin{bmatrix}\boldsymbol{x_2-x_1}\\\boldsymbol{x_3-x_2}\\\boldsymbol{x_4-x_3}\end{bmatrix}\qquad(1)$$

Notice well that each row of A has the same number of components as the vector $\boldsymbol{x}$. Four columns multiply x_1 to x_4. Otherwise multiplying $A\boldsymbol{x}$ would be impossible.

Column picture of $A\boldsymbol{x}$ The matrix A times the vector $\boldsymbol{x}$ is a **combination of the columns of** $\boldsymbol{A}$. The n columns are multiplied by the n numbers in $\boldsymbol{x}$. Then add those column vectors $x_1\boldsymbol{a}_1,\ldots,x_n\boldsymbol{a}_n$ to find the vector $A\boldsymbol{x}$:

$$\boxed{A\boldsymbol{x}=x_1(\text{column }\boldsymbol{a}_1)+x_2(\text{column }\boldsymbol{a}_2)+x_3(\text{column }\boldsymbol{a}_3)+x_4(\text{column }\boldsymbol{a}_4)}\qquad(2)$$

This combination of n columns involves exactly the same multiplications as dot products of $\boldsymbol{x}$ with the m rows. But it is higher level ! We have a vector equation instead of three dot products. You see the same $A\boldsymbol{x}$ in equations (1) and (3).

Combination of columns

$$A\boldsymbol{x}=x_1\begin{bmatrix}-1\\0\\0\end{bmatrix}+x_2\begin{bmatrix}1\\-1\\0\end{bmatrix}+x_3\begin{bmatrix}0\\1\\-1\end{bmatrix}+x_4\begin{bmatrix}0\\0\\1\end{bmatrix}=\begin{bmatrix}\boldsymbol{x_2-x_1}\\\boldsymbol{x_3-x_2}\\\boldsymbol{x_4-x_3}\end{bmatrix}\qquad(3)$$

Let me admit something right away. If I have numbers in A and $\boldsymbol{x}$, and I want to compute $A\boldsymbol{x}$, then I tend to use dot products: the row picture. But if I want to *understand* $A\boldsymbol{x}$, the column picture is better. "The column vector $A\boldsymbol{x}$ is a combination of the columns of A."

We are aiming for a picture of not just one combination $A\boldsymbol{x}$ of the columns (from a particular $\boldsymbol{x}$). What we really want is a picture of **all combinations of the columns** (from multiplying A by all vectors $\boldsymbol{x}$). This figure shows one combination $2\boldsymbol{a}_1+\boldsymbol{a}_2$ and then it tries to show the plane of all combinations $x_1\boldsymbol{a}_1+x_2\boldsymbol{a}_2$ (for every x_1 and x_2).

Figure 1.10: A linear combination of $\boldsymbol{a}_1$ and $\boldsymbol{a}_2$. All linear combinations fill a plane.

The next important words are **independence**, **dependence**, and **column space**.

Here is a key point! Columns of A might not contribute anything new. They might be combinations of earlier columns (which we already included). Examples 1 and 2 show columns that give a new direction, and columns that are combinations of previous columns.

Example 1 Independent columns $A_1 = \begin{bmatrix} 1 & 0 & 0 \\ 2 & 4 & 0 \\ 3 & 5 & 6 \end{bmatrix}$ *Each column gives a new direction.* Their combinations fill 3D space $\mathbf{R}^3$.

If we look at all combinations of the columns, we see all vectors (b_1, b_2, b_3): *3D space.* The first column $x_1(1,2,3)$ allows us to match any number b_1. Then $x_2(0,4,5)$ leaves b_1 alone and we can match any number b_2. Finally $x_3(0,0,6)$ doesn't touch b_1 and b_2 and allows us to match any b_3. We have found x_1, x_2, x_3 so that $A_1\boldsymbol{x} = \boldsymbol{b}$.

Independence means: The only combination of columns that produces $A\boldsymbol{x} = (0,0,0)$ is $\boldsymbol{x} = (0,0,0)$. The columns are **independent** when each new column is a vector that we don't already have as a combination of previous columns. That word "independent" will be important.

Example 2 Dependent columns $A_2 = \begin{bmatrix} 1 & 2 & 3 \\ 1 & 4 & 5 \\ 6 & 0 & 6 \end{bmatrix}$ **Column 1 + column 2 = column 3** **Their combinations don't fill 3D space** $\begin{matrix} 1+2=3 \\ 1+4=5 \\ 6+0=6 \end{matrix}$

The opposite of independent is "*dependent*". These three columns of A_2 are **dependent.** Column 3 is in the plane of columns 1 and 2. Nothing new from column 3.

I usually test independence going from left to right. The column $(1,1,6)$ is no problem. Column 2 is *not* a multiple of column 1 and $(2,4,0)$ gives a new direction. But column 3 is the sum of columns 1 and 2. The third column vector $(3,5,6)$ is not independent of $(1,1,6)$ and $(2,4,0)$. We only have two independent columns.

If I went from right to left, I would start with independent columns 3 and 2. Then column 1 is a combination (column 3 minus column 2). Either way we find that the three columns are in the **same plane**: two independent columns produce a plane in 3D.

That plane is the column space of this matrix: Plane = all combinations of the columns.

Dependent columns in Example 2 | column 1 + column 2 − column 3 is $(\mathbf{0},\mathbf{0},\mathbf{0})$. |

Example 3 $A_3 = \begin{bmatrix} 1 & 3 & 4 \\ 2 & 6 & 8 \\ 5 & 15 & 20 \end{bmatrix}$ Now $\boldsymbol{a}_2$ is 3 times $\boldsymbol{a}_1$. And $\boldsymbol{a}_3$ is 4 times $\boldsymbol{a}_1$. **Every pair of columns is dependent.**

This example is important. You could call it an extreme case. All three columns of A_3 lie on the **same line** in 3-dimensional space. That line consists of all column vectors $(c, 2c, 5c)$—all the multiples of $(1,2,5)$. Notice that $c = 0$ gives the point $(0,0,0)$.

That line in 3D is the column space for this matrix A_3. The line contains all vectors $A_3\boldsymbol{x}$. By allowing every vector $\boldsymbol{x}$, we fill in the column space of A_3—and here we only filled one line. That is almost the smallest possible column space.

| The **column space of A** is the set of all vectors $A\boldsymbol{x}$: **All combinations of the columns.** |

Thinking About the Column Space of A

"Vector spaces" are a central topic. Examples are coming unusually early. They give you a chance to see what linear algebra is about. The combinations of all columns produce the column space, but you only need r independent columns. So we start with column 1, and go **from left to right** in identifying independent columns. Here are two examples A_4 and A_5.

$$A_4 = \begin{bmatrix} 1 & 1 & 1 & 1 \\ 0 & 1 & 1 & 1 \\ 0 & 0 & 1 & 1 \\ 0 & 0 & 0 & 1 \end{bmatrix} \qquad A_5 = \begin{bmatrix} \mathbf{1} & \mathbf{1} & 0 & 0 \\ 0 & \mathbf{1} & \mathbf{1} & 0 \\ 0 & 0 & \mathbf{1} & \mathbf{1} \\ \mathbf{1} & 0 & 0 & \mathbf{1} \end{bmatrix}$$

A_4 has four independent columns. For example, column 4 is not a combination of columns $1, 2, 3$. There are no dependent columns in A_4. Triangular matrices like A_4 are easy *provided the main diagonal has no zeros*. Here the diagonal is $1, 1, 1, 1$.

A_5 is not so easy. Columns 1 and 2 and 3 are independent. The big question is whether column 4 is independent—or is it a combination of columns $1, 2, 3$? To match the final 1 in column 4, that combination will have to start with column 1.

To cancel the 1 in the top left corner of A_5, we need *minus* the second column. Then we need *plus* column 3 so that -1 and $+1$ in row 2 will also cancel. Now we see what is true about this matrix A_5 :

$$\textbf{Column 4 of } \boldsymbol{A_5} = \textbf{Column 1} - \textbf{Column 2} + \textbf{Column 3.} \tag{4}$$

So column 4 of A_5 is *a combination of* columns $1, 2, 3$. A_5 has only 3 independent columns.

The next step is to "visualize" the column space—all combinations of the four columns. That word is in quotes because the task may be impossible. I don't think that drawing a 4-dimensional figure would help (possibly this is wrong). The first matrix A_4 is a good place to start, because its column space is the full 4-dimensional space $\mathbf{R}^4$.

Do you see why $\mathbf{C}(A_4) = \mathbf{R}^4$? If we look to algebra, we see that every vector $\boldsymbol{v}$ in $\mathbf{R}^4$ is a combination of the columns. By writing $\boldsymbol{v}$ as (v_1, v_2, v_3, v_4), we can literally show the exact combination that produces every vector $\boldsymbol{v}$ from A_4 :

$$\boldsymbol{v} = \begin{bmatrix} v_1 \\ v_2 \\ v_3 \\ v_4 \end{bmatrix} = (v_1 - v_2)\begin{bmatrix} 1 \\ 0 \\ 0 \\ 0 \end{bmatrix} + (v_2 - v_3)\begin{bmatrix} 1 \\ 1 \\ 0 \\ 0 \end{bmatrix} + (v_3 - v_4)\begin{bmatrix} 1 \\ 1 \\ 1 \\ 0 \end{bmatrix} + (v_4)\begin{bmatrix} 1 \\ 1 \\ 1 \\ 1 \end{bmatrix} \tag{5}$$

This says that $\boldsymbol{v}$ is a combination of the columns. More than that, equation (5) shows what the combination is. **We have solved the four equations $\boldsymbol{A_4 x = v}$!** The four unknowns in $\boldsymbol{x} = (x_1, x_2, x_3, x_4)$ are now known in the four parentheses of equation (5).

Geometrically, every vector $\boldsymbol{v}$ is a combination of the 4 columns of A_4. Here is one way to look at A_4. The first column $(1, 0, 0, 0)$ is responsible for a line in 4-dimensional space. That line contains every vector $(c_1, 0, 0, 0)$. The second column is responsible for another line, containing every vector $(c_2, c_2, 0, 0)$. *If you add every vector* $(c_1, 0, 0, 0)$ *to every vector* $(c_2, c_2, 0, 0)$, you get a 2-dimensional plane inside 4-dimensional space.

That was the first two columns. The main rule of linear algebra is *keep going*. The last two columns give two more directions in $\mathbf{R}^4$, and they are *independent of the first two*. At the end, equation (5) shows how **every 4-dimensional vector is a combination of the four columns of A_4**. The column space of A_4 is all of $\mathbf{R}^4$.

If we attempt the same plan for the matrix A_5, the first 3 columns cooperate. But column 4 of A_5 is a combination of columns $1, 2, 3$. Those three columns combine to give a *three-dimensional subspace* inside $\mathbf{R}^4$. Column 4 happens to be *in that subspace*.

That three-dimensional subspace is the whole column space $\mathbf{C}(A_5)$. We can only solve $A_2\boldsymbol{x} = \boldsymbol{v}$ when $\boldsymbol{v}$ is in $\mathbf{C}(A_5)$. The matrix A_5 only has three independent columns.

I always write $\mathbf{C}(A)$ for the column space of A. When A has m rows, the columns are vectors in m-dimensional space $\mathbf{R}^m$. The column space might fill all of $\mathbf{R}^m$ or it might not. For $m = 3$, here are all four possibilities for column spaces in 3-dimensional space:

1.	The whole space $\mathbf{R}^3$	3 independent columns
2.	A plane in $\mathbf{R}^3$ going through $(0, 0, 0)$	2 independent columns
3.	A line in $\mathbf{R}^3$ going through $(0, 0, 0)$	1 independent column
4.	The single point $(0, 0, 0)$ in $\mathbf{R}^3$ (when A is a matrix of zeros !)	

Here are simple matrices to show those four possibilities for the column space $\mathbf{C}(A)$:

$$\begin{bmatrix} \mathbf{1} & 0 & 0 \\ 0 & \mathbf{1} & 0 \\ 0 & 0 & \mathbf{1} \end{bmatrix} \qquad \begin{bmatrix} \mathbf{1} & 0 & 0 \\ 0 & \mathbf{1} & 0 \\ 0 & 0 & 0 \end{bmatrix} \qquad \begin{bmatrix} \mathbf{1} & 0 & 0 \\ 0 & 0 & 0 \\ 0 & 0 & 0 \end{bmatrix} \qquad \begin{bmatrix} 0 & 0 & 0 \\ 0 & 0 & 0 \\ 0 & 0 & 0 \end{bmatrix}$$

$\mathbf{C}(A) = \mathbf{R}^3 = xyz$ space $\quad \mathbf{C}(A) = xy$ plane $\quad \mathbf{C}(A) = x$ axis $\quad \mathbf{C}(A) =$ one point $(0, 0, 0)$

Author's note The words "column space" have not appeared in Chapter 1 of my previous books. I thought the idea of a *space* was too important to come so soon. Now I think that the best way to understand such an important idea is to see it early and often. It is examples more than definitions that make ideas clear—in mathematics as in life.

Here is a succession of questions. With practice in the next section 1.4, you will find the keys to the answers. They give a real understanding of any matrix A.

1. How many columns of A are independent? That number r is the **"rank"** of A.
2. Which are the first r independent columns? They are a **"basis"** for the column space.
3. What combinations of those r basic columns produce the remaining $n - r$ columns?
4. Write A as an m by r column matrix C times an r by n matrix R : $\boldsymbol{A = CR}$.
5. (Amazing) The r rows of R are a basis for the **row space** of A: combinations of rows.

Section I.4 will explain how to multiply those matrices C and R. The result is $A = CR$. C contains columns from A. Please notice that the rows of R *do not* come directly from A.

Matrices of Rank One

Now we come to the building blocks for all matrices. *Every matrix of rank r is the sum of r matrices of rank one.* For a rank one matrix, all column vectors lie along the same line. That line through $(0,0,0)$ is the whole column space of the rank one matrix.

Example $A_6 = \begin{bmatrix} 1 & 3 & -2 \\ 4 & 12 & -8 \\ 2 & 6 & -4 \end{bmatrix}$ has rank $\boldsymbol{r} = \mathbf{1}$. All columns: same direction!

Columns 2 and 3 are multiples of the first column $\boldsymbol{a}_1 = (1,4,2)$. Column 2 is $3\boldsymbol{a}_1$ and column 3 is $-2\boldsymbol{a}_1$. The column space $\mathbf{C}(A_6)$ is only the line of all vectors $c\boldsymbol{a}_1 = (c, 4c, 2c)$.

Here is a wonderful fact about any rank one matrix. You may have noticed the rows of A_6. **All the rows are multiples of one row**. When the column space is a single line in m-dimensional space, the row space is a single line in n-dimensional space. All rows of this matrix A_6 are multiples of ____.

An example like A_6 raises a basic question. **If all columns are in the same direction, why does it happen that all rows are in the same direction?** To find an answer, look first at this 2 by 2 matrix. Column 2 is m times column 1 so the column rank is 1.

$$A = \begin{bmatrix} a & ma \\ b & mb \end{bmatrix} \qquad \text{Is row 2 a multiple of row 1?}$$

Yes! The second row (b, mb) is $\frac{b}{a}$ times the first row (a, ma). If the column rank is 1, then the row rank is 1. To cover every possibility we have to check the case when $a = 0$. Then the first row $\begin{bmatrix} 0 & 0 \end{bmatrix}$ is 0 times row 2. So the row space is the line through row 2.

Our 2 by 2 proof is complete. Let me look next at this 3 by 3 matrix of rank 1:

$$A = \begin{bmatrix} a & ma & pa \\ b & mb & pb \\ c & mc & pc \end{bmatrix} \qquad \begin{array}{l} \text{Column 2 is } m \text{ times column 1} \\ \text{Column 3 is } p \text{ times column 1} \\ \text{Rows 2 and 3 are } b/a \text{ and } c/a \text{ times row 1} \end{array}$$

This matrix does not have two independent columns. Is it the same for the rows of A? Is row 2 in the same direction as row 1? *Yes.* Is row 3 in the same direction as row 1? *Yes. The rule still holds.* The row rank of this A is also 1 (equal to the column rank).

Let me jump from rank one matrices to all matrices. At this point we could make a guess: It looks possible that **row rank equals column rank for every matrix**. If A has r independent columns, then A has r independent rows. A wonderful fact!

I believe that this is the first great theorem in linear algebra. So far we have only seen the case of rank one matrices. The next section 1.4 will explain matrix multiplication AB and lead us toward an understanding of "row rank = column rank" for all matrices.

Problem Set 1.3

This chapter introduces column spaces. But we don't yet have a computational system to decide independence or dependence of column vectors. So these problems stay with whole numbers and small matrices.

1 Describe the column space of these matrices : a point, a line, a plane, all of 3D.

$$A_1 = \begin{bmatrix} 2 & 2 \\ 1 & 1 \\ 5 & 6 \end{bmatrix} \quad A_2 = \begin{bmatrix} 1 & 0 & 0 \\ 1 & 1 & 0 \\ 1 & 1 & 1 \end{bmatrix} \quad A_3 = \begin{bmatrix} 1 & 5 \\ 2 & 10 \\ 1 & 5 \end{bmatrix} \quad A_4 = \begin{bmatrix} 0 & 0 \\ 0 & 0 \\ 0 & 0 \end{bmatrix}$$

2 Find a combination of the columns that produces $(0,0,0)$: column space = *plane*.

Dependent columns
$$A_1 = \begin{bmatrix} 1 & 2 & 3 \\ 4 & 5 & 6 \\ 7 & 8 & 9 \end{bmatrix} \qquad A_2 = \begin{bmatrix} 1 & 4 & 7 \\ 2 & 5 & 8 \\ 3 & 6 & 9 \end{bmatrix}$$

3 Describe the column spaces in $\mathbf{R}^3$ of B and C :

$$B = \begin{bmatrix} 1 & 2 \\ 2 & 1 \\ 3 & 3 \end{bmatrix} \qquad C = \begin{bmatrix} B & -B \end{bmatrix} \quad \text{(3 by 4 block matrix)}$$

4 Multiply Ax and By and Iz using dot products as in (rows of A) $\cdot\, x$:

$$Ax = \begin{bmatrix} 2 & 1 & 2 \\ 4 & 2 & 4 \\ 0 & 1 & 0 \end{bmatrix} \begin{bmatrix} 1 \\ 2 \\ 5 \end{bmatrix} \quad By = \begin{bmatrix} 1 & 0 & 0 \\ 1 & 1 & 0 \\ 1 & 1 & 1 \end{bmatrix} \begin{bmatrix} 4 \\ 4 \\ 10 \end{bmatrix} \quad Iz = \begin{bmatrix} 1 & 0 & 0 \\ 0 & 1 & 0 \\ 0 & 0 & 1 \end{bmatrix} \begin{bmatrix} z_1 \\ z_2 \\ z_3 \end{bmatrix}$$

5 Multiply the same A times x and B times y and I times z using combinations of the columns of A and B and I, as in $Ax = 1(\text{column } 1) + 2(\text{column } 2) + 5(\text{column } 3)$.

6 In Problem 4, how many independent columns does A have ? How many independent columns in B ? How many independent columns in $A + B$?

7 Can you find A and B (both with two independent columns) so that $A + B$ has (a) 1 independent column (b) No independent columns (c) 4 independent columns

8 The "column space" of a matrix contains all combinations of the columns. Describe the column spaces in $\mathbf{R}^3$ of A and B and C :

$$A = \begin{bmatrix} 1 & 0 & 0 \\ 0 & 1 & 0 \\ 0 & 0 & 1 \end{bmatrix} \quad B = \begin{bmatrix} 2 & 4 \\ 1 & 2 \\ 2 & 4 \end{bmatrix} \quad C = \begin{bmatrix} 1 & 0 & 1 & 2 \\ 0 & 2 & 2 & 4 \\ 0 & 2 & 2 & 4 \end{bmatrix}$$

9 Find a 3 by 3 matrix A with 3 independent columns and all nine entries = 1 or 2. (What is the maximum possible number of 1's ?)

10 Complete A and B so that they are rank one matrices. What are the column spaces of A and B ? What are the row spaces of A and B ?

$$A = \begin{bmatrix} 3 & \\ 5 & 15 \end{bmatrix} \qquad B = \begin{bmatrix} 1 & 2 & -5 \\ 4 & & \end{bmatrix}$$

11 Suppose A is a 5 by 2 matrix with columns $\boldsymbol{a}_1$ and $\boldsymbol{a}_2$. We include one more column to produce B (5 by 3). Do A and B have the same column space if

(a) the new column is the zero vector ? (b) the new column is $(1, 1, 1)$?

(c) the new column is the difference $\boldsymbol{a}_2 - \boldsymbol{a}_1$?

12 Explain this important sentence. It connects column spaces to linear equations.

> $A\boldsymbol{x} = \boldsymbol{b}$ has a solution vector $\boldsymbol{x}$ if the vector $\boldsymbol{b}$ is in the column space of A.

The equation $A\boldsymbol{x} = \boldsymbol{b}$ looks for a combination of columns of A that produces $\boldsymbol{b}$. What vector will solve $A\boldsymbol{x} = \boldsymbol{b}$ for these right hand sides $\boldsymbol{b}$?

$$\begin{bmatrix} 1 & 3 \\ 2 & 4 \end{bmatrix} \begin{bmatrix} x_1 \\ x_2 \end{bmatrix} = \begin{bmatrix} 4 \\ 6 \end{bmatrix} \text{ or } \begin{bmatrix} -2 \\ -2 \end{bmatrix} \text{ or } \begin{bmatrix} 1 \\ 1 \end{bmatrix}$$

13 Find two 3 by 3 matrices A and B with the same column space = the plane of all vectors perpendicular to $(1, 1, 1)$. What is the column space of $A + B$?

14 Which numbers q would leave A with two independent columns ?

$$A = \begin{bmatrix} 1 & 0 & 2 \\ 3 & 1 & 9 \\ 5 & 0 & q \end{bmatrix} \quad A = \begin{bmatrix} 1 & 4 & 7 \\ 2 & 5 & 8 \\ 3 & 6 & q \end{bmatrix} \quad A = \begin{bmatrix} 1 & 1 & 2 \\ 2 & 2 & 4 \\ 0 & 0 & q \end{bmatrix}$$

15 Suppose A times $\boldsymbol{x}$ equals $\boldsymbol{b}$. If you add $\boldsymbol{b}$ as an extra column of A, explain why the rank r (number of independent columns) stays the same.

16 *True or false*

(a) If the 5 by 2 matrices A and B have independent columns, so does $A + B$.

(b) If the m by n matrix A has independent columns, then $m \geq n$.

(c) A random 3 by 3 matrix almost surely has independent columns.

17 If A and B have rank 1, what are the possible ranks of $A + B$? Give an example of each possibility.

18 Find the linear combination $3\boldsymbol{s}_1 + 4\boldsymbol{s}_2 + 5\boldsymbol{s}_3 = \boldsymbol{b}$. Then write $\boldsymbol{b}$ as a matrix-vector multiplication $S\boldsymbol{x}$, with 3, 4, 5 in $\boldsymbol{x}$. Compute the three dot products (row of S) $\cdot\, \boldsymbol{x}$:

$$\boldsymbol{s}_1 = \begin{bmatrix} 1 \\ 1 \\ 1 \end{bmatrix} \quad \boldsymbol{s}_2 = \begin{bmatrix} 0 \\ 1 \\ 1 \end{bmatrix} \quad \boldsymbol{s}_3 = \begin{bmatrix} 0 \\ 0 \\ 1 \end{bmatrix} \text{ go into the columns of } S.$$

19 Solve these equations $S\boldsymbol{y} = \boldsymbol{b}$ with $\boldsymbol{s}_1, \boldsymbol{s}_2, \boldsymbol{s}_3$ in the columns of the sum matrix S:

$$\begin{bmatrix} 1 & 0 & 0 \\ 1 & 1 & 0 \\ 1 & 1 & 1 \end{bmatrix} \begin{bmatrix} y_1 \\ y_2 \\ y_3 \end{bmatrix} = \begin{bmatrix} 1 \\ 1 \\ 1 \end{bmatrix} \text{ and } \begin{bmatrix} 1 & 0 & 0 \\ 1 & 1 & 0 \\ 1 & 1 & 1 \end{bmatrix} \begin{bmatrix} y_1 \\ y_2 \\ y_3 \end{bmatrix} = \begin{bmatrix} 1 \\ 4 \\ 9 \end{bmatrix}.$$

The sum of the first 3 odd numbers is ____. The sum of the first 10 is ____.

20 Solve these three equations for y_1, y_2, y_3 in terms of c_1, c_2, c_3:

$$S\boldsymbol{y} = \boldsymbol{c} \qquad \begin{bmatrix} 1 & 0 & 0 \\ 1 & 1 & 0 \\ 1 & 1 & 1 \end{bmatrix} \begin{bmatrix} y_1 \\ y_2 \\ y_3 \end{bmatrix} = \begin{bmatrix} c_1 \\ c_2 \\ c_3 \end{bmatrix}.$$

Write the solution $\boldsymbol{y}$ as a matrix A times the vector $\boldsymbol{c}$. A is the "*inverse matrix*" S^{-1}. Are the columns of S independent or dependent?

21 The three rows of this square matrix A are dependent. Then linear algebra says that the three columns must also be dependent. Find $\boldsymbol{x}$ in $A\boldsymbol{x} = \boldsymbol{0}$:

$$A = \begin{bmatrix} 1 & 2 & 3 \\ 3 & 5 & 6 \\ 4 & 7 & 9 \end{bmatrix} \qquad \begin{array}{l} \text{Row } 1 + \text{ row } 2 = \text{ row } 3 \\ \text{Two independent rows} \\ \text{Then only two independent columns} \end{array}$$

22 Which numbers c give dependent columns? Then a combination of columns is zero.

$$\begin{bmatrix} 1 & 1 & 0 \\ 3 & 2 & 1 \\ 7 & 4 & c \end{bmatrix} \qquad \begin{bmatrix} 1 & 0 & c \\ 1 & 1 & 0 \\ 0 & 1 & 1 \end{bmatrix} \qquad \begin{bmatrix} c & c & c \\ 2 & 1 & 5 \\ 3 & 3 & 6 \end{bmatrix} \qquad \begin{bmatrix} c & 1 \\ 4 & c \end{bmatrix}$$

23 If the columns combine into $A\boldsymbol{x} = \boldsymbol{0}$ then each row of A has **row** $\cdot\, \boldsymbol{x} = 0$:

$$\text{If} \begin{bmatrix} \boldsymbol{a}_1 & \boldsymbol{a}_2 & \boldsymbol{a}_3 \end{bmatrix} \begin{bmatrix} x_1 \\ x_2 \\ x_3 \end{bmatrix} = \begin{bmatrix} 0 \\ 0 \\ 0 \end{bmatrix} \quad \textbf{then by rows} \begin{bmatrix} \boldsymbol{r}_1 \cdot \boldsymbol{x} \\ \boldsymbol{r}_2 \cdot \boldsymbol{x} \\ \boldsymbol{r}_3 \cdot \boldsymbol{x} \end{bmatrix} = \begin{bmatrix} 0 \\ 0 \\ 0 \end{bmatrix}.$$

The three rows also lie in a plane. Why is that plane perpendicular to $\boldsymbol{x}$?

1.4 Matrix Multiplication and $A = CR$

1 To multiply AB we need *row length for* $A =$ *column length for* B.

2 The number in row i, column j of AB is (**row** $\boldsymbol{i}$ **of** $\boldsymbol{A}$) $\cdot$ (**column** $\boldsymbol{j}$ **of** $\boldsymbol{B}$).

3 By columns: **$\boldsymbol{A}$ times column $\boldsymbol{j}$ of $\boldsymbol{B}$ produces column $\boldsymbol{j}$ of $\boldsymbol{AB}$.**

4 Usually AB is different from BA. But always $(\boldsymbol{AB})\boldsymbol{C} = \boldsymbol{A}(\boldsymbol{BC})$.

5 If A has r independent columns, then $\boldsymbol{A} = \boldsymbol{CR} = (m \text{ by } r)\,(r \text{ by } n)$.

At this point we can multiply a matrix A times a vector $\boldsymbol{x}$ to produce $A\boldsymbol{x}$. Remember the row way and the column way. The output is a vector.

Row way Dot products of $\boldsymbol{x}$ with each row of A

Column way $A\boldsymbol{x} = x_1\boldsymbol{a}_1 + \cdots + x_n\boldsymbol{a}_n =$ combination of the columns of A

Now we come to the higher level operation in linear algebra: **Multiply two matrices.** We can multiply AB if their shapes are right. When A has n columns, B must have n rows.

If A is m by n and B is n by p, then AB is m by p: m columns and p rows.

The rules for AB will be an extension of the rules for $A\boldsymbol{x}$. We can think of the vector $\boldsymbol{x}$ as a matrix B with only one column. What we did for $A\boldsymbol{x}$ we will now do for AB.

The columns of B are vectors $\begin{bmatrix} \boldsymbol{x} & \boldsymbol{y} & \boldsymbol{z} \end{bmatrix}$. **The columns of $\boldsymbol{AB}$ are vectors** $\begin{bmatrix} A\boldsymbol{x} & A\boldsymbol{y} & A\boldsymbol{z} \end{bmatrix}$

In other words, **multiply $\boldsymbol{A}$ times each column of $\boldsymbol{B}$.** There are two ways to multiply A times a column vector, and those give *two ways to multiply AB*:

Dot products	(**row $\boldsymbol{i}$ of $\boldsymbol{A}$**) $\cdot$ (**column $\boldsymbol{j}$ of $\boldsymbol{B}$**) goes into row i, column j of AB
Combinations of columns of $\boldsymbol{A}$	Use the numbers from each column of B

We have dot products (numbers) or linear combinations of columns of A (vectors). For computing by hand, I would use the row way to find each number in AB. I "think" the column way to see the big picture: **Columns of $\boldsymbol{AB}$ are combinations of columns of $\boldsymbol{A}$.**

Example 1 Multiply $AB = \begin{bmatrix} \mathbf{1} & \mathbf{2} \\ \mathbf{3} & \mathbf{4} \end{bmatrix} \begin{bmatrix} \mathbf{5} & \mathbf{6} \\ \mathbf{7} & \mathbf{8} \end{bmatrix}$ both ways. How many steps?

The dot product (row 1 of A) $\cdot$ (column 1 of B) is $(\mathbf{1}, \mathbf{2}) \cdot (\mathbf{5}, \mathbf{7}) = 5 + 14 = \mathbf{19}\downarrow$

(Rows of $\boldsymbol{A}$) $\cdot$ (columns of $\boldsymbol{B}$) $$AB = \begin{bmatrix} \text{row } 1 \cdot \text{col } 1 & \text{row } 1 \cdot \text{col } 2 \\ \text{row } 2 \cdot \text{col } 1 & \text{row } 2 \cdot \text{col } 2 \end{bmatrix} = \begin{bmatrix} \mathbf{19} & \mathbf{22} \\ \mathbf{43} & \mathbf{50} \end{bmatrix}$$

$\boldsymbol{Ab_1}$ and $\boldsymbol{Ab_2}$ are combinations of the columns of $\boldsymbol{A}$ $$AB = \left[\mathbf{5}\begin{bmatrix} 1 \\ 3 \end{bmatrix} + \mathbf{7}\begin{bmatrix} 2 \\ 4 \end{bmatrix} \quad \mathbf{6}\begin{bmatrix} 1 \\ 3 \end{bmatrix} + \mathbf{8}\begin{bmatrix} 2 \\ 4 \end{bmatrix} \right] = \begin{bmatrix} \mathbf{19} & \mathbf{22} \\ \mathbf{43} & \mathbf{50} \end{bmatrix}$$

Either way, AB requires 8 multiplications for 2 by 2 matrices. Multiplying $A\boldsymbol{x}$ needed mn multiplications for an m by n matrix. When B has p columns, we have to multiply A times each column. Then *multiplying AB uses mnp multiplications* for (m by n) times (n by p). When $m = n = p = 2$, we have $2 \cdot 2 \cdot 2 = 8$ multiplications.

A difficult question: Could n by n matrices A and B be multiplied with fewer than n^3 small multiplications? *This is known to be possible* (allowing extra additions). Right now we don't know the smallest exponent E in the multiplication count n^E. We know that $E < 3$ and in fact $E < 2.373$. But $E = 2.0001$ may be impossible.

Here is a challenge question about matrix multiplication. *Explain why every vector in* $\mathbf{C}(AB)$—every combination of the columns of AB—*is also in the column space of A.*

Example 2 The **identity matrix** I has $AI = A$ and $IB = B$ if matrix sizes are right.

$$\boldsymbol{AI} = \begin{bmatrix} a & b \\ c & d \end{bmatrix} \begin{bmatrix} \mathbf{1} & 0 \\ 0 & \mathbf{1} \end{bmatrix} = \begin{bmatrix} a & b \\ c & d \end{bmatrix} = \boldsymbol{A} \qquad \text{for every } A$$

$$\boldsymbol{IB} = \begin{bmatrix} \mathbf{1} & 0 \\ 0 & \mathbf{1} \end{bmatrix} \begin{bmatrix} P & Q \\ R & S \end{bmatrix} = \begin{bmatrix} P & Q \\ R & S \end{bmatrix} = \boldsymbol{B} \qquad \text{for every } B$$

Example 3 The matrix $\boldsymbol{E} = \begin{bmatrix} 0 & \mathbf{1} \\ \mathbf{1} & 0 \end{bmatrix}$ will *exchange columns* or *exchange rows.*

$$\boldsymbol{AE} = \begin{bmatrix} a & b \\ c & d \end{bmatrix} \begin{bmatrix} 0 & \mathbf{1} \\ \mathbf{1} & 0 \end{bmatrix} = \begin{bmatrix} b & a \\ d & c \end{bmatrix} \qquad \text{Exchange columns of } A \ (E \text{ is on the right})$$

$$\boldsymbol{EB} = \begin{bmatrix} 0 & \mathbf{1} \\ \mathbf{1} & 0 \end{bmatrix} \begin{bmatrix} P & Q \\ R & S \end{bmatrix} = \begin{bmatrix} R & S \\ P & Q \end{bmatrix} \qquad \text{Exchange rows of } B \ (E \text{ is on the left})$$

Example 4 $AE \neq EA$ for most matrices: Exchange columns or exchange rows.

$$A = \begin{bmatrix} 1 & 2 \\ 3 & 4 \end{bmatrix} \qquad AE = \begin{bmatrix} 2 & 1 \\ 4 & 3 \end{bmatrix} \text{ is not the same as } EA = \begin{bmatrix} 3 & 4 \\ 1 & 2 \end{bmatrix}$$

This example shows an important fact: $\boldsymbol{AB}$ **can easily be different from** $\boldsymbol{BA}$. Matrix multiplication is not "*commutative*". We must keep matrices AB or ABC in order. There are certainly examples where $AB = BA$, but those special cases are not typical.

Example 5 Squaring the exchange matrix gives $E^2 = \begin{bmatrix} 0 & 1 \\ 1 & 0 \end{bmatrix} \begin{bmatrix} 0 & 1 \\ 1 & 0 \end{bmatrix} = I$. Why is this true?

Example 6 $\boldsymbol{AB}$ **times** $\boldsymbol{C}$ **equals** $\boldsymbol{A}$ **times** $\boldsymbol{BC}$. Matrix multiplication is "*associative*".

I include this here without proof, because it is so important: $(\boldsymbol{AB})\boldsymbol{C} = \boldsymbol{A}(\boldsymbol{BC})$. The order ABC of those matrices must stay the same! But we can multiply AB first or multiply BC first. Many many proofs in linear algebra depend on this simple fact.

Certainly the matrix sizes for A, B, C must match: (m by n) $\times$ (n by p) $\times$ (p by q). A special 2 by 2 case of this associative law would be $(EA)E = E(AE)$. Exchange the rows of A first, or exchange the columns of A first. The triple product EAE does both.

Rank One Matrices and $A = CR$

All columns of a rank one matrix lie on the same line. That line is the column space $\mathbf{C}(A)$. Examples in Section 1.3 pointed to a remarkable fact: *The rows also lie on a line.* When all the columns are in the same column direction, then all the rows are in the same row direction. Here is an example:

$$A = \begin{bmatrix} 1 & 2 & 10 & 100 \\ 3 & 6 & 30 & 300 \\ 2 & 4 & 20 & 200 \end{bmatrix} = \begin{array}{l}\text{rank one matrix} \\ \text{one independent column} \\ \text{one independent row!}\end{array}$$

All columns are multiples of $(1,3,2)$. All rows are multiples of $\begin{bmatrix} 1 & 2 & 10 & 100 \end{bmatrix}$. Only one independent row when there is only one independent column. *Why is this true?*

Our approach is through matrix multiplication. We factor A into C times R. For this special matrix, C has one column and R has one row. $\boldsymbol{CR}$ is $(3 \times 1)(1 \times 4)$.

$$\boldsymbol{A = CR} \qquad \begin{bmatrix} 1 & 2 & 10 & 100 \\ 3 & 6 & 30 & 300 \\ 2 & 4 & 20 & 200 \end{bmatrix} = \begin{bmatrix} \mathbf{1} \\ \mathbf{3} \\ \mathbf{2} \end{bmatrix} \begin{bmatrix} \mathbf{1} & \mathbf{2} & \mathbf{10} & \mathbf{100} \end{bmatrix} \tag{1}$$

The dot products (row of C) $\cdot$ (column of R) are just multiplications like 3 times 10. This is multiplication of thin matrices CR. Only 12 small multiplications.

The rows of A are numbers $1, 3, 2$ times the (only) row $\begin{bmatrix} 1 & 2 & 10 & 100 \end{bmatrix}$ of R. By factoring this special A into **one column times one row**, the conclusion jumps out:

If the column space of A is a line, the row space of A is also a line.

One column in C, one row in R. That is beautiful, but we are certainly not finished. Our big goal is to allow r columns in C and to find r rows in R. And to see $A = CR$.

C Contains Independent Columns

Suppose we go from left to right, looking for independent columns in any matrix A:

> If column 1 of A is not all zero, put it into the matrix C
>
> If column 2 of A is not a multiple of column 1, put it into C
>
> If column 3 of A is not a combination of columns 1 and 2, put it into C. *Continue.*

At the end C will have r columns taken from A. That number r is the **rank of A**. The n columns of A might be dependent. The r columns of C will surely be **independent**.

Independent	*No column of C is a combination of previous columns*
columns	*No combination of columns gives $C\boldsymbol{x} = \mathbf{0}$ except $\boldsymbol{x} =$ all zeros*

When those independent columns combine to give all columns, we have a **basis**.

$Cx = \mathbf{0}$ means that x_1(column 1 of C) + x_2(column 2 of C) + $\cdots$ = *zero vector*. With independent columns, this only happens if *all x's are zero*. Otherwise we can divide by the last nonzero coefficient x and that column would be a combination of the earlier columns—which our construction forbids. Therefore C has independent columns.

Example 7 $$A = \begin{bmatrix} \mathbf{2} & 6 & \mathbf{4} \\ \mathbf{4} & 12 & \mathbf{8} \\ \mathbf{1} & 3 & \mathbf{5} \end{bmatrix} \quad \text{leads to} \quad C = \begin{bmatrix} \mathbf{2} & \mathbf{4} \\ \mathbf{4} & \mathbf{8} \\ \mathbf{1} & \mathbf{5} \end{bmatrix}$$

Column 1 goes into C. Column 2 does not (3 times column 1). Column 3 goes into C.

Matrix Multiplication C times R

Now comes the new step to $A = CR$. R tells how to produce the columns of A from the columns of C. The first column of A is actually in C, so the first column of R just has 1 and 0. The third column of A is also in C, so the third column of R just has 0 and 1.

Rank 2
Notice I
inside R

$$A = \begin{bmatrix} 2 & 6 & 4 \\ 4 & 12 & 8 \\ 1 & 3 & 5 \end{bmatrix} = \begin{bmatrix} \mathbf{2} & \mathbf{4} \\ \mathbf{4} & \mathbf{8} \\ \mathbf{1} & \mathbf{5} \end{bmatrix} \begin{bmatrix} \mathbf{1} & ? & \mathbf{0} \\ \mathbf{0} & ? & \mathbf{1} \end{bmatrix} = \boldsymbol{CR}. \qquad (2)$$

Two columns of A went straight into C, so *part of R is the identity matrix*. The question marks are in column 2 because column 2 of A is *not in C*. It was not an independent column. Column 2 of A is 3 times column 1. *That number* 3 *goes into R*. Then R shows how to combine the two columns of C to get all three columns of the original A.

A is $m \times n$
C is $m \times r$
R is $r \times n$

$$\boxed{\boldsymbol{A = CR} \text{ is } \begin{bmatrix} 2 & 6 & 4 \\ 4 & 12 & 8 \\ 1 & 3 & 5 \end{bmatrix} = \begin{bmatrix} \mathbf{2} & \mathbf{4} \\ \mathbf{4} & \mathbf{8} \\ \mathbf{1} & \mathbf{5} \end{bmatrix} \begin{bmatrix} \mathbf{1} & \mathbf{3} & \mathbf{0} \\ \mathbf{0} & \mathbf{0} & \mathbf{1} \end{bmatrix}} \qquad (3)$$

This completes $A = CR$. The magic is now seen in the rows. **All the rows of A come from the rows of R.** This fact follows immediately from matrix multiplication CR:

Multiply CR
using rows of R

Row 1 of A is $\mathbf{2}$ (row 1 of R) + $\mathbf{4}$ (row 2 of R)
Row 2 of A is $\mathbf{4}$ (row 1 of R) + $\mathbf{8}$ (row 2 of R)
Row 3 of A is $\mathbf{1}$ (row 1 of R) + $\mathbf{5}$ (row 2 of R)

***R* has the same row space as *A*.** Combinations of rows of R produce every row of A. This A has only 2 independent rows, not 3. Two rows in R combine to give all rows of A.

Second example
of $A = CR$
from the front cover

$$\begin{bmatrix} 1 & 2 & 3 \\ 4 & 5 & 6 \\ 7 & 8 & 9 \end{bmatrix} = \begin{bmatrix} 1 & 2 \\ 4 & 5 \\ 7 & 8 \end{bmatrix} \begin{bmatrix} \mathbf{1} & \mathbf{0} & \mathbf{-1} \\ \mathbf{0} & \mathbf{1} & \mathbf{2} \end{bmatrix} \qquad (4)$$

When a column of A goes into C, a column of I goes into R. The "free" column $-1, 2$ of R tells us how to produce the *dependent* column of A from the *independent* columns in C. Column 3 of A is $\mathbf{-1}\,(1,4,7) + \mathbf{2}\,(2,5,8)$. And also: rows of A from rows of R.

Column j of A = C times **column j of R.** **Row i of A = row i of C** times **R.**

Question If all n columns of A are independent, then $C = A$. What matrix is R?

Answer This case of n independent columns has $\boldsymbol{R} = \boldsymbol{I}$ (identity matrix). The rank is n.

How to find R. Start with r independent columns of A going into C.

If column 3 of A = 2nd independent column in C, then column 3 of R is $\begin{matrix}\mathbf{0}\\ \mathbf{1}\end{matrix}$

$$\boldsymbol{A} = \begin{bmatrix} 1 & 2 & \mathbf{3} & 4 \\ 1 & 2 & \mathbf{4} & 5 \end{bmatrix} = \begin{bmatrix} 1 & \mathbf{3} \\ 1 & \mathbf{4} \end{bmatrix} \begin{bmatrix} 1 & 2 & \mathbf{0} & 1 \\ 0 & 0 & \mathbf{1} & 1 \end{bmatrix} = \boldsymbol{CR} \qquad \text{All three ranks} = 2$$

Dependent: If column 4 of A = columns $1 + 2$ of C, then column 4 of R is $\begin{matrix}\mathbf{1}\\ \mathbf{1}\end{matrix}$

R tells how to recover all columns of A from the independent columns in C.

Here is an informal proof that ***row rank of A equals column rank of A***

1. The r columns of C are independent (by their construction)
2. Every column of A is a combination of those r columns of C (because $A = CR$)
3. The r rows of R are independent (they contain the r by r matrix I)
4. Every row of A is a combination of those r rows of R (because $A = CR$)

Key facts

The r columns of C are a **basis** for the column space of A: **dimension r**

The r rows of R are a **basis** for the row space of A: **dimension r**

Those words "basis" and "dimension" are properly defined in Section 3.4! Section 3.2 will show how the same row matrix R can be constructed directly from the "reduced row echelon form" of A, by deleting any zero rows. Chapter 1 starts with independent columns of A, placed in C. *Chapter* 3 *starts with rows of A, and combines them into R.*

We are emphasizing CR because both matrices are important. C contains r independent columns of A. R tells how to combine those columns to give all columns of A. (R contains I, when columns of A are already in C.) Chapter 3 will produce R directly from A by *elimination*, the most used algorithm in computational mathematics. This will be the key to a fundamental problem: solving linear equations $A\boldsymbol{x} = \boldsymbol{b}$.

Why is Matrix Multiplication AB Defined This Way?

The definition of AB was chosen to produce this crucial equation: **$(\boldsymbol{AB})$ times $\boldsymbol{x}$ is equal to $\boldsymbol{A}$ times $\boldsymbol{Bx}$.** This leads to the all-important law $(\boldsymbol{AB})\boldsymbol{C} = \boldsymbol{A}(\boldsymbol{BC})$. We had no other reasonable choice for AB! Linear algebra will use these laws over and over. Let me show in three steps why that crucial equation $(AB)\boldsymbol{x} = A(B\boldsymbol{x})$ is correct:

$B\boldsymbol{x}$ is a combination $x_1\boldsymbol{b}_1 + x_2\boldsymbol{b}_2 + \cdots + x_n\boldsymbol{b}_n$ of the columns of B.

Matrix-vector multiplication is linear: $A(B\boldsymbol{x}) = x_1 A\boldsymbol{b}_1 + x_2 A\boldsymbol{b}_2 + \cdots + x_n(A\boldsymbol{b}_n)$.

We want this to agree with $(AB)\boldsymbol{x} = x_1$(column 1 of AB) $+ \cdots + x_n$(column n of AB).

Compare lines 2 and 3. *Column* 1 *of AB absolutely must equal A times column* 1 *of B.* This is our rule: When $B = \begin{bmatrix} \boldsymbol{x} & \boldsymbol{y} & \boldsymbol{z} \end{bmatrix}$ the columns of AB are $\begin{bmatrix} A\boldsymbol{x} & A\boldsymbol{y} & A\boldsymbol{z} \end{bmatrix}$.

Example 8 $A = \begin{bmatrix} 1 & 2 \\ 3 & 4 \end{bmatrix}$ $B = \begin{bmatrix} 5 & 6 \\ 7 & 8 \end{bmatrix}$ $\boldsymbol{x} = \begin{bmatrix} 1 \\ 0 \end{bmatrix}$ $(AB)\boldsymbol{x} = A(B\boldsymbol{x})$

When we show that $(AB)\boldsymbol{x} = A(B\boldsymbol{x})$, that fact will allow us to erase the parentheses. *We can just write* $AB\boldsymbol{x}$. For three matrices we can just write ABC.

To compare $(AB)\boldsymbol{x}$ with $A(B\boldsymbol{x})$, remember that Example 1 computed AB:

$$(AB)\boldsymbol{x} = \begin{bmatrix} 19 & 22 \\ 43 & 50 \end{bmatrix}\begin{bmatrix} 1 \\ 0 \end{bmatrix} = \begin{bmatrix} \mathbf{19} \\ \mathbf{43} \end{bmatrix}$$

$$A(B\boldsymbol{x}) = \begin{bmatrix} 1 & 2 \\ 3 & 4 \end{bmatrix}\begin{bmatrix} 5 & 6 \\ 7 & 8 \end{bmatrix}\begin{bmatrix} 1 \\ 0 \end{bmatrix} = \begin{bmatrix} 1 & 2 \\ 3 & 4 \end{bmatrix}\begin{bmatrix} 5 \\ 7 \end{bmatrix} = \begin{bmatrix} \mathbf{19} \\ \mathbf{43} \end{bmatrix}$$

The parentheses don't matter but the order ABC certainly does matter. The multiplications BAC and ACB almost always give different answers. In fact BAC may be impossible.

Columns of *A* times Rows of *B*

Before this chapter ends, I want to add this message. There is another way to multiply matrices (producing the same matrix AB as always). This way is not so well known, but it is powerful. **It multiplies columns of *A* times rows of *B*.** We will see it again and use it.

$$AB = \begin{bmatrix} \boldsymbol{a}_1 & \cdots & \boldsymbol{a}_n \end{bmatrix}\begin{bmatrix} \boldsymbol{b}_1^* \\ \vdots \\ \boldsymbol{b}_n^* \end{bmatrix} = \boldsymbol{a}_1\boldsymbol{b}_1^* + \boldsymbol{a}_2\boldsymbol{b}_2^* + \cdots + \boldsymbol{a}_n\boldsymbol{b}_n^*. \quad (5)$$

columns $\boldsymbol{a}_k$ rows $\boldsymbol{b}_k^*$ columns $\boldsymbol{a}_k$ times rows $\boldsymbol{b}_k^*$

Those matrices $\boldsymbol{a}_k\boldsymbol{b}_k^*$ are called *outer products*. We recognize that they have rank one: column times row. They are entirely different from dot products (rows times columns, also known as *inner products*). If A is an m by n matrix and B is an n by p matrix, adding columns times rows gives the same answer AB as rows times columns.

Actually they involve the same mnp small multiplications but in a different order!

(Row) · (Column)	mp dot products, n multiplications each	**total** $\boldsymbol{mnp}$
(Column) (Row)	n rank one matrices, mp multiplications each	**total** $\boldsymbol{mnp}$

Columns × Rows for *A* times *B*

$$\begin{bmatrix} 1 & 4 \\ 2 & 5 \\ 3 & 6 \end{bmatrix}\begin{bmatrix} 7 & 8 & 9 \\ 10 & 11 & 12 \end{bmatrix} = \begin{bmatrix} 1 \\ 2 \\ 3 \end{bmatrix}\begin{bmatrix} 7 & 8 & 9 \end{bmatrix} + \begin{bmatrix} 4 \\ 5 \\ 6 \end{bmatrix}\begin{bmatrix} 10 & 11 & 12 \end{bmatrix}$$

Rank 1 +Rank 1

$$= \begin{bmatrix} 7 & 8 & 9 \\ 14 & 16 & 18 \\ 21 & 24 & 27 \end{bmatrix} + \begin{bmatrix} 40 & 44 & 48 \\ 50 & 55 & 60 \\ 60 & 66 & 72 \end{bmatrix} = \begin{bmatrix} \mathbf{47} & \mathbf{52} & \mathbf{57} \\ \mathbf{64} & \mathbf{71} & \mathbf{78} \\ \mathbf{81} & \mathbf{90} & \mathbf{99} \end{bmatrix} \quad (6)$$

This example has $mnp = (3)(2)(3) = 18$. At the start of the second line you see the 18 multiplications (in two 3 by 3 matrices). Then 9 additions give the correct answer AB.

As we learned in this section, the rank of AB is 2. *Two independent columns, not three. Two independent rows, not three.* The next chapter will use different words. *AB has no inverse matrix: it is not invertible.* And later in the book: *The determinant of AB is zero.*

Note about the "echelon matrices" R and R_0

We were amazed to learn that the row matrix R in $A = CR$ is already a famous matrix in linear algebra! It is essentially the **"reduced row echelon form"** of the original A. MATLAB calls it **rref** (A) and includes $m - r$ zero rows. With the zero rows, we call it $\boldsymbol{R_0}$.

The factorization $\boldsymbol{A} = \boldsymbol{CR}$ is a big step in linear algebra. The Problem Set will look closely at the matrix R, its form is remarkable. R has the identity matrix in r columns. Then C multiplies each column of R to produce a column of A. $\boldsymbol{R_0}$ **comes in Chapter 3.**

Example 9 $$A = \begin{bmatrix} \boldsymbol{a}_1 & \boldsymbol{a}_2 & 3\boldsymbol{a}_1 + 4\boldsymbol{a}_2 \end{bmatrix} = \begin{bmatrix} \boldsymbol{a}_1 & \boldsymbol{a}_2 \end{bmatrix} \begin{bmatrix} \mathbf{1} & 0 & \mathbf{3} \\ 0 & \mathbf{1} & \mathbf{4} \end{bmatrix} = \boldsymbol{CR}.$$

Here $\boldsymbol{a}_1$ and $\boldsymbol{a}_2$ are the independent columns of A. The third column is dependent—a combination of $\boldsymbol{a}_1$ and $\boldsymbol{a}_2$. Therefore it is in the plane produced by columns 1 and 2. All three matrices A, C, R have rank $r = 2$.

We can try that new way (**columns** $\times$ **rows**) to quickly multiply CR in Example 9:

Columns of C times rows of R $$CR = \boldsymbol{a}_1 \begin{bmatrix} 1 & 0 & 3 \end{bmatrix} + \boldsymbol{a}_2 \begin{bmatrix} 0 & 1 & 4 \end{bmatrix} = \begin{bmatrix} \boldsymbol{a}_1 & \boldsymbol{a}_2 & 3\boldsymbol{a}_1 + 4\boldsymbol{a}_2 \end{bmatrix} = A$$

Four Ways to Multiply $AB = C$

$\begin{bmatrix} X & X \\ x & x \\ x & x \end{bmatrix}\begin{bmatrix} X & x & x & x \\ X & x & x & x \end{bmatrix}$	**(Row i of A)** $\cdot$ **(Column k of B)** $i = 1$ to 3 $\quad k = 1$ to 4	$=$ **Number C_{ik}** **12 numbers**
$\begin{bmatrix} X & X \\ X & X \\ X & X \end{bmatrix}\begin{bmatrix} X & x & x & x \\ X & x & x & x \end{bmatrix}$	$\boldsymbol{A}$ times **(Column k of B)** $k = 1$ to 4	$=$ **Column k of C** **4 columns**
$\begin{bmatrix} X & X \\ x & x \\ x & x \end{bmatrix}\begin{bmatrix} X & X & X & X \\ X & X & X & X \end{bmatrix}$	**(Row i of A)** times $\boldsymbol{B}$ $i = 1$ to 3	$=$ **Row i of C** **3 rows**
$\begin{bmatrix} X & x \\ X & x \\ X & x \end{bmatrix}\begin{bmatrix} X & X & X & X \\ x & x & x & x \end{bmatrix}$	**(Column j of A)** **(Row j of B)** $j = 1$ to 2	$=$ **Rank 1 Matrix** **2 matrices**

Problem Set 1.4

AB = m·p.

1 Construct this four-way table when A is m by n and B is n by p. How many dot products and columns and rows and rank one matrices go into AB? In all four cases the total count of small multiplications is mnp.

2 If all columns of $A = \begin{bmatrix} \boldsymbol{a} & \boldsymbol{a} & \boldsymbol{a} \end{bmatrix}$ contain the same $\boldsymbol{a} \neq \mathbf{0}$, what are C and R?

C·R = [a] · [1 1 1]

3 Multiply A times B (3 examples) using *dot products*: each row times each column.

$$\begin{bmatrix} 1 & 0 & 0 \\ 1 & 1 & 0 \\ 1 & 1 & 1 \end{bmatrix}\begin{bmatrix} 1 & 0 & 0 \\ -1 & 1 & 0 \\ 1 & -1 & 1 \end{bmatrix} \qquad \begin{bmatrix} 1 & 2 & 3 \end{bmatrix}\begin{bmatrix} 4 \\ 5 \\ 6 \end{bmatrix} \qquad \begin{bmatrix} 4 \\ 5 \\ 6 \end{bmatrix}\begin{bmatrix} 1 & 2 & 3 \end{bmatrix}$$

4 Test the truth of the associative law $(AB)C = A(BC)$.

(a) $\begin{bmatrix} 1 & 1 \end{bmatrix}\begin{bmatrix} 1 \\ 1 \end{bmatrix}\begin{bmatrix} 1 & 1 & 1 \end{bmatrix}$ (b) $\begin{bmatrix} 1 & 2 \\ 0 & 1 \end{bmatrix}\begin{bmatrix} 1 & 3 \\ 0 & 1 \end{bmatrix}\begin{bmatrix} 1 & 4 \\ 0 & 1 \end{bmatrix}$

5 Why is it impossible for a matrix A with 7 columns and 4 rows to have 5 independent columns? This is not a trivial or useless question.

6 Going from left to right, put each column of A into the matrix C if that column is not a combination of earlier columns:

$$A = \begin{bmatrix} 2 & -2 & 1 & 6 & 0 \\ 1 & -1 & 0 & 2 & 0 \\ 3 & -3 & 0 & 6 & 1 \end{bmatrix} \qquad C = \begin{bmatrix} 2 \\ 1 \\ 3 \end{bmatrix}$$

7 Find R in Problem 6 so that $A = CR$. If your C has r columns, then R has r rows. The 5 columns of R tell how to produce the 5 columns of A from the columns in C.

8 This matrix A has 3 independent columns. So C has the same 3 columns as A. What is the 3 by 3 matrix R so that $A = CR$? What is different about B?

$$\text{Upper triangular} \quad A = \begin{bmatrix} 2 & 2 & 2 \\ 0 & 4 & 4 \\ 0 & 0 & 6 \end{bmatrix} \qquad B = \begin{bmatrix} 2 & 2 & 2 \\ 0 & 0 & 4 \\ 0 & 0 & 6 \end{bmatrix}$$

9 Suppose A is a random 4 by 4 matrix. The probability is 1 that the columns of A are "independent". In that case, what are the matrices C and R in $A = CR$?

Note Random matrix theory has become an important part of applied linear algebra—especially for very large matrices when even multiplication AB is too expensive. An example of "*probability* 1" is choosing two whole numbers at random. The probability is 1 that they are different. But they could be the same! Problem 10 is another example of this type.

10 Suppose A is a random 4 by 5 matrix. With probability 1, what can you say about C and R in $A = CR$? In particular, which columns of A (going into C) are probably independent of previous columns, going from left to right?

11 Create your own example of a 4 by 4 matrix A of rank $r = 2$. Then factor A into $CR = (4 \text{ by } 2)\,(2 \text{ by } 4)$.

12 Factor these matrices into $A = CR = (m \text{ by } r)\,(r \text{ by } n)$: all ranks equal to r.

$$A_1 = \begin{bmatrix} 1 & 2 & 3 \\ 1 & 3 & 4 \end{bmatrix} \quad A_2 = \begin{bmatrix} 0 & 1 & 2 & 3 \\ 0 & 1 & 3 & 5 \end{bmatrix} \quad A_3 = \begin{bmatrix} 2 & 1 & 3 \\ 6 & 3 & 9 \end{bmatrix} \quad A_4 = \begin{bmatrix} 1 & 0 & 0 & 4 \\ 0 & 2 & 2 & 0 \end{bmatrix}$$

13 Starting from $C = \begin{bmatrix} 1 \\ 3 \end{bmatrix}$ and $R = \begin{bmatrix} 2 & 4 \end{bmatrix}$ compute CR and RC and CRC and RCR.

14 Complete these 2 by 2 matrices to meet the requirements printed underneath:

$$\begin{bmatrix} 3 & 6 \\ 5 & \end{bmatrix} \qquad \begin{bmatrix} 6 & \\ 7 & \end{bmatrix} \qquad \begin{bmatrix} 2 & \\ 3 & 6 \end{bmatrix} \quad \begin{bmatrix} 3 & 4 \\ & -3 \end{bmatrix}$$

rank one orthogonal columns rank 2 $A^2 = I$

15 Suppose $A = CR$ with independent columns in C and independent rows in R. Explain how each of these logical steps follows from $A = CR = (m \text{ by } r)\,(r \text{ by } n)$.

1. Every column of A is a combination of columns of C.
2. Every row of A is a combination of rows of R. What combination is row 1 ?
3. The number of columns of C = the number of rows of R (needed for CR ?).
4. *Column rank equals row rank.* The number of independent columns of A equals the number of independent rows in A.

16 (a) The vectors $AB\boldsymbol{x}$ produce the column space of AB. Show why this vector $AB\boldsymbol{x}$ is also in the column space of A. (Is $AB\boldsymbol{x} = A\boldsymbol{y}$ for some vector $\boldsymbol{y}$?) Conclusion: The column space of A *contains* the column space of AB.

(b) Choose nonzero matrices A and B so the column space of AB contains only the zero vector. This is the smallest possible column space.

17 True or false, with a reason (not easy):

(a) If 3 by 3 matrices A and B have rank 1, then AB will always have rank 1.

(b) If 3 by 3 matrices A and B have rank 3, then AB will always have rank 3.

(c) Suppose $AB = BA$ for every 2 by 2 matrix B. Then $A = \begin{bmatrix} c & 0 \\ 0 & c \end{bmatrix} = cI$ for some number c. Only those matrices A commute with every B.

18 Example 6 in this section mentioned a special case of the law $\boldsymbol{(AB)C = A(BC)}$.

$$A = C = \text{exchange matrix} \begin{bmatrix} 0 & 1 \\ 1 & 0 \end{bmatrix} \qquad B = \begin{bmatrix} 1 & 2 \\ 3 & 4 \end{bmatrix}.$$

(a) First compute AB (row exchange) and also BC (column exchange).

(b) Now compute the double exchanges: $(AB)C$ with rows first and $A(BC)$ with columns first. Verify that those double exchanges produce the same ABC.

19 Test the column-row multiplication in equation (5) to find AB and BA:

$$AB = \begin{bmatrix} 1 & 0 & 0 \\ 1 & 1 & 0 \\ 1 & 1 & 1 \end{bmatrix} \begin{bmatrix} 1 & 1 & 1 \\ 0 & 1 & 1 \\ 0 & 0 & 1 \end{bmatrix} \qquad BA = \begin{bmatrix} 1 & 1 & 1 \\ 0 & 1 & 1 \\ 0 & 0 & 1 \end{bmatrix} \begin{bmatrix} 1 & 0 & 0 \\ 1 & 1 & 0 \\ 1 & 1 & 1 \end{bmatrix}$$

20 How many small multiplications for $(AB)C$ and $A(BC)$ if those matrices have sizes $ABC = (4 \times 3)\,(3 \times 2)\,(2 \times 1)$? That choice affects the operation count.

Thoughts on Chapter 1

Most textbooks don't have a place for the author's thoughts. But a lot of decisions go into starting a new textbook. This chapter has intentionally jumped right into the subject, with discussion of independence and rank. There are so many good ideas ahead, and they take time to absorb, so why not get started ? Here are two questions that influenced the writing.

What makes this subject easy ? All the equations are linear.

What makes this subject hard ? So many equations and unknowns and ideas.

Book examples are small size. But if we want the temperature at many points of an engine, there is an equation at every point: easily $n = 1000$ unknowns.

I believe the key is to work right away with matrices. $A\boldsymbol{x} = \boldsymbol{b}$ is a perfect format to accept problems of all sizes. The linearity is built into the symbols $A\boldsymbol{x}$ and the rule is $A(\boldsymbol{x} + \boldsymbol{y}) = A\boldsymbol{x} + A\boldsymbol{y}$. Each of the m equations in $A\boldsymbol{x} = \boldsymbol{b}$ represents a flat surface:

$2x + 5y - 4z = 6$ is a plane in three-dimensional space

$2x + 5y - 4z + 7w = 9$ is a 3D plane (hyperplane ?) in four-dimensional space

Linearity is on our side, but there is a serious problem in visualizing 10 planes meeting in 11-dimensional space. Hopefully they meet along a line: dimension $11 - 10 = 1$. An 11th plane should cut through that line at one point (which solves all 11 equations). What the textbook and the notation must do is to keep the counting simple

Here is what we expect for a random m by n matrix A:

$\boldsymbol{m < n}$ Many solutions or no solutions to the m equations $A\boldsymbol{x} = \boldsymbol{b}$

$\boldsymbol{m = n}$ Probably one solution to the n equations $A\boldsymbol{x} = \boldsymbol{b}$

$\boldsymbol{m > n}$ Probably no solution: too many equations with only n unknowns in $\boldsymbol{x}$

But this count is not necessarily what we get! Columns of A can be combinations of previous columns: nothing new. An equation can be a combination of previous equations. **The rank r tells us the real size of our problem**, from independent columns and rows. The beautiful formula is $\boldsymbol{A} = \boldsymbol{CR} = (m \times r)\,(r \times n)$: three matrices of rank r.

Notice: The columns of A that go into C must produce the matrix I inside R.

We end with the great associative law $(\boldsymbol{AB})\,\boldsymbol{C} = \boldsymbol{A}\,(\boldsymbol{BC})$. Suppose C has 1 column:

AB has columns $A\boldsymbol{b}_1, \ldots, A\boldsymbol{b}_n$ and then $(AB)\boldsymbol{c}$ equals $c_1 A\boldsymbol{b}_1 + \cdots + c_n A\boldsymbol{b}_n$.

$B\boldsymbol{c}$ has one column $c_1\boldsymbol{b}_1 + \cdots + c_n\boldsymbol{b}_n$ and $A(B\boldsymbol{c}) = A\,(c_1\boldsymbol{b}_1 + \cdots + c_n\boldsymbol{b}_n)$.

Linearity gives equality of those two sums. This proves $(AB)\,\boldsymbol{c} = A\,(B\boldsymbol{c})$.

The same is true for every column of C. Therefore $(\boldsymbol{AB})\,\boldsymbol{C} = \boldsymbol{A}\,(\boldsymbol{BC})$.

2 Solving Linear Equations $Ax = b$

The matrices in this chapter are square: n by n. Then $A\boldsymbol{x} = \boldsymbol{b}$ gives n equations (one from each row of A). Those equations have n unknowns in the vector $\boldsymbol{x}$. Often but not always there is one solution $\boldsymbol{x}$ for each $\boldsymbol{b}$. In this case A has an **inverse A^{-1}** with $A^{-1}A = I$ and $AA^{-1} = I$. Multiplying $A\boldsymbol{x} = \boldsymbol{b}$ by A^{-1} produces the symbolic solution $\boldsymbol{x} = A^{-1}\boldsymbol{b}$.

This chapter aims to find that solution $\boldsymbol{x}$, but not by computing A^{-1}. (That would solve $A\boldsymbol{x} = \boldsymbol{b}$ for every possible $\boldsymbol{b}$.) We go forward column by column, assuming that A has independent columns. We only stop if this proves wrong. At the end we have triangular matrices L, U and $\boldsymbol{x}$ is easy to find.

$A\boldsymbol{x} = \boldsymbol{b}$ is a universal problem in science and engineering and every quantitative subject. There might be $n = 10$ equations—this is already beyond hand calculations. Many problems have $n = 1000$ or more—and *we certainly don't want to find* A^{-1}. What we do need is an efficient way to compute the solution vector $\boldsymbol{x}$.

Here is an idea that goes back thousands of years (to China). Each step of "elimination" will produce a zero in the matrix. The original A gradually changes into an *upper triangular* U. Half of this matrix will be zero. A simple elimination matrix E_{ij} produces one zero where row i meets column j. This is not exciting, it is just the natural way to simplify A.

To describe all these steps we need matrices. This is the point of linear algebra! There are **elimination matrices** like E to reach U. And we multiply U by an **inverse matrix** $L = E^{-1}$ to come back to A. Here are key matrices in this chapter of the book:

Coefficient matrix A	Upper triangular U	Lower triangular L
Elimination matrix E_{ij}	Overall elimination E	Inverse matrix A^{-1}
Permutation matrix P	Transpose matrix A^{T}	Symmetric matrix $S = S^{\mathrm{T}}$

Our goal is to explain the elimination steps from $\boldsymbol{A}$ **to** $\boldsymbol{EA} = \boldsymbol{U}$ **to** $\boldsymbol{A} = \boldsymbol{E}^{-1}\boldsymbol{U} = \boldsymbol{LU}$. (If the steps fail, this signals that $A\boldsymbol{x} = \boldsymbol{b}$ has no solution.) Every computer system has a code to find U and then $\boldsymbol{x}$. Those codes are used so often that elimination adds up to the greatest cost in all of scientific computing.

But the codes are highly engineered and we don't know a better way to find $\boldsymbol{x}$.

2.1 The Idea of Elimination

1 Elimination subtracts ℓ_{ij} times row j from row i, to turn A_{ij} into zero.

2 $A\boldsymbol{x} = \boldsymbol{b}$ becomes $U\boldsymbol{x} = \boldsymbol{c}$ (or else $A\boldsymbol{x} = \boldsymbol{b}$ is proved to have no solution).

3 $U\boldsymbol{x} = \boldsymbol{c}$ is solved by back substitution and possible row exchanges.

This chapter explains a systematic way to solve $A\boldsymbol{x} = \boldsymbol{b}$: *$n$ equations for n unknowns*. The n by n matrix A is given and the n by 1 column vector $\boldsymbol{b}$ is given. There may be *no vector* $\boldsymbol{x} = (x_1, x_2, \ldots, x_n)$ that solves $A\boldsymbol{x} = \boldsymbol{b}$, or there may be exactly *one solution*, or there may be *infinitely many* solution vectors $\boldsymbol{x}$. Our job is to decide among these three possibilities and to find all solutions. Here are examples with $n = 2$.

1 **Exactly one solution to $A\boldsymbol{x} = \boldsymbol{b}$.** In this case A has independent columns. The rank of A is 2. The only solution to $A\boldsymbol{x} = \boldsymbol{0}$ is $\boldsymbol{x} = 0$. A has an *inverse matrix* A^{-1}.

The best case has a square matrix A $(m = n)$ with independent columns. Then there is one solution $\boldsymbol{x}$ (one combination of the columns of A) for *every* vector $\boldsymbol{b}$.

Example with one solution $(\boldsymbol{x}, \boldsymbol{y}) = (\mathbf{1}, \mathbf{1})$ $\quad 2x + 3y = 5$
Independent columns $(\mathbf{2}, \mathbf{4})$ and $(\mathbf{3}, \mathbf{2})$ $\quad 4x + 2y = 6$

2 **No solution to $A\boldsymbol{x} = \boldsymbol{b}$.** In this case $\boldsymbol{b}$ is not a combination of the columns of A. In other words $\boldsymbol{b}$ is not in the column space of A. The rank of A is 1.

Example with no solution $\quad 2x + 3y = 6$
Dependent columns $(\mathbf{2}, \mathbf{4})$ and $(\mathbf{3}, \mathbf{6})$ $\quad 4x + 6y = 15$

Subtract 2 times the first equation from the second to get $0 = 3$. **No solution**.

3 There will be **infinitely many solutions** to $A\boldsymbol{X} = \boldsymbol{0}$ when the columns of A are not independent. This is the meaning of **dependent columns**—many ways to produce the zero vector $\boldsymbol{b} = \boldsymbol{0}$. Every $c\boldsymbol{X}$ gives $A(c\boldsymbol{X}) = \boldsymbol{0}$.

If there is one solution to $A\boldsymbol{x} = \boldsymbol{b}$ then we can add any solution to $A\boldsymbol{X} = \boldsymbol{0}$. All the vectors $\boldsymbol{x} + c\boldsymbol{X}$ solve the same equations, so we have many solutions.

$$A(\boldsymbol{x} + c\boldsymbol{X}) = A\boldsymbol{x} + cA\boldsymbol{X} = \boldsymbol{b} + \boldsymbol{0} = \boldsymbol{b}. \tag{1}$$

Example with infinitely many solutions $\quad 2x + 3y = 6$
***A* has dependent columns $\boldsymbol{b}$ is in $\mathbf{C}(\boldsymbol{A})$** $\quad 4x + 6y = 12$

Those equations $A\boldsymbol{x} = \boldsymbol{b}$ are solved by $x = 0, y = 2$. But there are more solutions because $X = (3, -2)$ solves $AX = 0$. Then $2X = (6, -4)$ also solves $A(2X) = 0$. All vectors cX can be added to the particular solution $\boldsymbol{x} = (0, 2)$ to produce more solutions:

$\boldsymbol{x} + c\boldsymbol{X} = (\mathbf{0} + \mathbf{3}\boldsymbol{c}, \mathbf{2} - \mathbf{2}\boldsymbol{c})$ *is a line of solutions to our two equations* $A\boldsymbol{x} = \boldsymbol{b}$.

This chapter will give a systematic way to decide between those possibilities $\mathbf{1}, \mathbf{2}, \mathbf{3}$: One solution, no solution, infinitely many solutions. This system is called **elimination**. It simplifies the matrix A without changing any solution $\boldsymbol{x}$ to the equation $A\boldsymbol{x} = \boldsymbol{b}$. We do the same operations to both sides of the equation, and those operations are reversible. Elimination keeps all solutions $\boldsymbol{x}$ and creates no new ones.

Let me show you the ideal case, when elimination produces an upper triangular matrix. That matrix is called U. Then $A\boldsymbol{x} = \boldsymbol{b}$ **leads to** $U\boldsymbol{x} = \boldsymbol{c}$, which we easily solve:

Elimination reaches U
Back substitution finds x

$$U\boldsymbol{x} = \boldsymbol{c} \text{ is } \begin{bmatrix} \mathbf{2} & 3 & 4 \\ 0 & \mathbf{5} & 6 \\ 0 & 0 & \mathbf{7} \end{bmatrix} \begin{bmatrix} x_1 \\ x_2 \\ x_3 \end{bmatrix} = \begin{bmatrix} 19 \\ 17 \\ 14 \end{bmatrix} = \boldsymbol{c}$$

That letter $\boldsymbol{U}$ stands for **upper triangular**. The matrix has all zeros below its diagonal. Highly important: **The "pivots" $\mathbf{2, 5, 7}$ on that main diagonal are not zero**. Then we can solve the equations by going from bottom to top: *find x_3 then x_2 then x_1*.

Back substitution	The last equation $7x_3 = 14$ gives $\boldsymbol{x_3} = \mathbf{2}$
Work upwards	The next equation $5x_2 + 6(2) = 17$ gives $\boldsymbol{x_2} = \mathbf{1}$
Upwards again	The first equation $2x_1 + 3(1) + 4(2) = 19$ gives $\boldsymbol{x_1} = \mathbf{4}$
Conclusion	The only solution to this example $U\boldsymbol{x} = \boldsymbol{c}$ is $\boldsymbol{x} = (4, 1, 2)$.
Special note	In solving for x_1, x_2, x_3 we needed to divide by the pivots $2, 5, 7$.

These pivots were probably not on the diagonal of the original matrix A (which we haven't seen). The pivots $2, 5, 7$ were discovered when "elimination" produced the lower triangular zeros in U. This crucial step from A to U is still to be explained!

Note We would not allow the number zero to be a pivot. That would destroy our plan because an equation like $0x_1 = 2$ or $0x_2 = 5$ or $0x_3 = 8$ has no solution. Back substitution will break down with a zero in any pivot position (on the diagonal of U). The test for independent columns (when $C = A$ in Chapter 1, and $R = I$, and $A = CR$ becomes $A = AI$) is $\boldsymbol{n}$ **nonzero pivots**.

Every square matrix A with independent columns (full rank) can be reduced to a triangular matrix U with nonzero pivots. *This is our job.* It is possible that we may need to put the equations $A\boldsymbol{x} = \boldsymbol{b}$ in a different order. We start with the usual case when elimination goes from A to U and back substitution finds the one and only solution vector $\boldsymbol{x}$ to $A\boldsymbol{x} = \boldsymbol{b}$.

Elimination in Each Column

First comes a matrix A (independent columns) that will require no row exchanges.

$$\text{The starting matrix is } A \quad \text{The first pivot is } 2 \qquad A = \begin{bmatrix} \mathbf{2} & 3 & 4 \\ 4 & 11 & 14 \\ 2 & 8 & 17 \end{bmatrix}. \tag{2}$$

The first pivot row is $\begin{bmatrix} 2 & 3 & 4 \end{bmatrix}$. Multiply that row by 2 and **subtract** from row 2:

$$\text{First step : } E_{21}A \text{ is} \quad \text{The multiplier was } 4/2 = 2 \qquad \begin{bmatrix} \mathbf{2} & 3 & 4 \\ \mathbf{0} & 5 & 6 \\ 2 & 8 & 17 \end{bmatrix} \tag{3}$$

This produced the desired zero in column 1. To produce another zero, we subtract row 1 from row 3. This completes elimination in column 1:

$$\text{Second step : } E_{31}E_{21}A \text{ is} \quad \text{The multiplier was } 2/2 = 1 \qquad \begin{bmatrix} \mathbf{2} & 3 & 4 \\ \mathbf{0} & \mathbf{5} & 6 \\ \mathbf{0} & 5 & 13 \end{bmatrix} \tag{4}$$

Move now to column 2 and row 2 (the second pivot row). The pivot is 5, on the diagonal. To eliminate the 5 below it, multiply row 2 by the number 1 and subtract from row 3.

$$\text{Final : } E_{32}E_{31}E_{21}A \text{ is triangular} \quad \text{The multiplier was } 5/5 = 1 \qquad U = \begin{bmatrix} \mathbf{2} & 3 & 4 \\ \mathbf{0} & \mathbf{5} & 6 \\ \mathbf{0} & \mathbf{0} & \mathbf{7} \end{bmatrix} \tag{5}$$

We reached the upper triangular U. Forward elimination is complete. Since U has $2, 5, 7$ on its diagonal we know that back substitution will succeed. Those steps came on the previous page. The columns of U are independent (and therefore the columns of the original A were independent, as we will see). The matrices A and U have full rank 3.

We can summarize the elimination steps when no row exchanges are involved.

> Use the first equation to produce zeros in column 1 below the first pivot.
>
> Use the new second equation to clear out column 2 below pivot 2 in row 2.
>
> *Continue to column* 3. The expected result is an upper triangular matrix U.

Elimination on A produces U. The same steps must be applied to the right side $\boldsymbol{b}$, and they produce a new right side $\boldsymbol{c}$. Then the new equations $U\boldsymbol{x} = \boldsymbol{c}$ (equivalent to the old equations $A\boldsymbol{x} = \boldsymbol{b}$) are solved by back substitution. This gives the solution $\boldsymbol{x}$.

Possible Breakdown of Elimination

Elimination can fail. *Zero can appear in a pivot position.* Subtracting that zero from lower rows will not clear out the column below the unwanted zero. Here is an example:

Zero in pivot 2 from elimination in column 1

$$A = \begin{bmatrix} 2 & 3 & 4 \\ 4 & 6 & 14 \\ 2 & 8 & 17 \end{bmatrix} \rightarrow \begin{bmatrix} 2 & 3 & 4 \\ 0 & \mathbf{0} & 6 \\ 0 & \mathbf{5} & 13 \end{bmatrix} = B$$

The cure is simple if it works. **Exchange row 2 with the zero for row 3 with the 5.** Then the second pivot is 5 and we can clear out the second column below that pivot. So elimination continues as normal after the row exchange by the matrix P.

Row exchange

$$PB = \begin{bmatrix} 1 & 0 & 0 \\ 0 & 0 & 1 \\ 0 & 1 & 0 \end{bmatrix} \begin{bmatrix} 2 & 3 & 4 \\ 0 & \mathbf{0} & 6 \\ 0 & 5 & 13 \end{bmatrix} = \begin{bmatrix} 2 & 3 & 4 \\ 0 & \mathbf{5} & 13 \\ 0 & 0 & 6 \end{bmatrix} = U.$$

For this small example, the row exchange is all we need. It produced the upper triangular U with nonzero pivots $2, 5, 6$. Normally there are more columns and rows to work on, before we reach U.

Caution! That row exchange was a success. This is what we hope for, to reach U with no zeros on its main diagonal. (The pivots $2, 5, 6$ are on the diagonal.) But a slightly different matrix A^* would lead to a bad situation: **no pivot is available in column 2.**

$$\boldsymbol{A^*} = \begin{bmatrix} 2 & 3 & 4 \\ 4 & 6 & 14 \\ 2 & 3 & 17 \end{bmatrix} \rightarrow \begin{bmatrix} 2 & 3 & 4 \\ 0 & \mathbf{0} & 6 \\ 0 & \mathbf{0} & 13 \end{bmatrix} = \boldsymbol{U^*} \qquad (6)$$

At this point elimination is helpless in column 2. *No pivot is available.* This misfortune tells us that **the matrix $\boldsymbol{A^*}$ did not have full rank**. Column 2 of U^* is in the same direction as column 1 of U^*. So column 2 of A^* is in the same direction as column 1 of A^*.

You see how dependent columns are systematically identified by elimination. There are nonzero solutions $\boldsymbol{X}$ to $A^*\boldsymbol{X} = \mathbf{0}$. The columns are not independent.

This example has column 2 $= \frac{3}{2}$ column 1. The solution vector $\boldsymbol{X}$ is $(3, -2, 0)$. *The equation $A^*\boldsymbol{x} = \boldsymbol{b}$ may or may not be solvable, depending on $\boldsymbol{b}$*: probably not.

.

Dependent or Independent Columns

This A^* looks like a failure of elimination: No second pivot. But it was a success because the problem was identified: dependent columns. The beauty of aiming for a triangular matrix U or U^* is that the diagonal entries tell us everything.

A triangular matrix $\boldsymbol{U}$ has full rank exactly when its main diagonal has no zeros.

In that case (square matrix with nonzero pivots) the columns of U are independent. Also the rows are independent. We can see this directly because elimination has simplified the original A to the triangular U.

How do we know that a zero on the diagonal of U^* leads to dependent columns?

$$U^* = \begin{bmatrix} * & * & * & * \\ 0 & * & * & * \\ 0 & 0 & \mathbf{0} & * \\ 0 & 0 & 0 & * \end{bmatrix} = \text{upper triangular with an extra zero on its diagonal}$$

1. The first three columns are dependent. **2.** The last two rows are dependent.

The Row Picture and the Column Picture

The next pictures will show those three possibilities for $A\boldsymbol{x} = \boldsymbol{b}$: *No* solution or *one* solution or *infinitely many* solutions. There are two ways to see this. We start with the *rows of* A and we graph the two equations: the row picture.

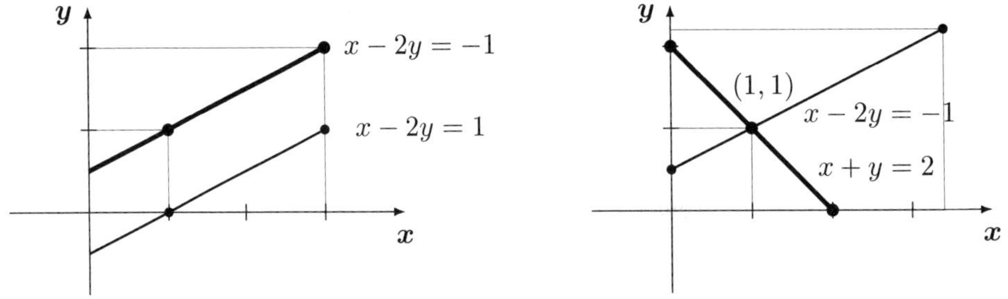

Figure 2.1: Parallel lines mean no solution. Top line twice means many solutions.

Intersecting lines give one solution. The solution is where the lines meet.

If we had three equations for x, y, and z, those two lines would change to three planes. Each plane like $2x + 4y + 3z = 9$ would be in 3-dimensional space. This row picture becomes hard to draw. The column picture is much easier in three or more dimensions.

The **column picture** just shows column vectors: columns of A and also the vector $\boldsymbol{b}$. We are not looking for points where these vectors meet. The goal of $A\boldsymbol{x} = \boldsymbol{b}$ is to **combine the columns** of A so as to produce the vector $\boldsymbol{b}$.

This is always possible when the columns of A (n vectors in n-dimensional space) are *independent*. Then the column space of A contains all vectors $\boldsymbol{b}$ in $\mathbf{R}^n$. There is exactly one combination $A\boldsymbol{x}$ of the columns that equals $\boldsymbol{b}$. Elimination finds that $\boldsymbol{x}$.

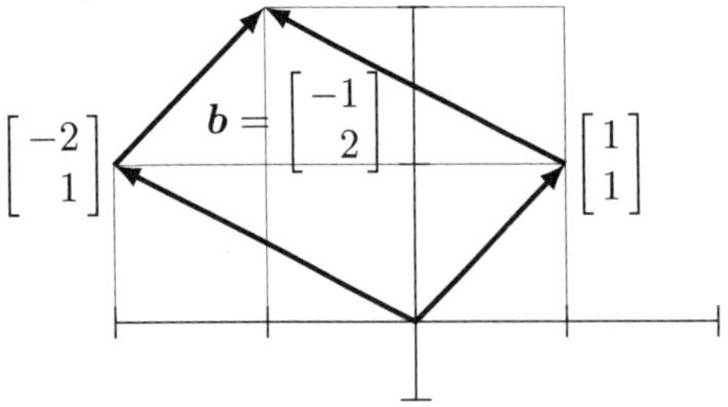

The columns of A are independent
Column 1 + Column 2 = $\boldsymbol{b}$
Then the solution is $x = 1, y = 1$
Construct $\boldsymbol{b}$ from the columns!

Figure 2.2: Column picture. The vector $\boldsymbol{b}$ is a combination $A\boldsymbol{x}$ of the columns of A.

Examples of Elimination and Permutation

This chapter will go on to express the whole process using matrices. An elimination matrix E will act on A. In case zero appears in a pivot position, a permutation matrix P is needed. The result is an upper triangular U and a new right hand side $\boldsymbol{c}$. Then $U\boldsymbol{x} = \boldsymbol{c}$ is solved by back substitution.

In reality a computer takes those steps ($\boldsymbol{x} = A\backslash\boldsymbol{b}$ in MATLAB). But it is good to solve a few examples—*not too many*—by hand. Then you see the steps to $U\boldsymbol{x} = \boldsymbol{c}$ and not only the solution $\boldsymbol{x}$. This page contains a variety of examples, hopefully to show the way.

$$A = \begin{bmatrix} 2 & 4 & -2 \\ 4 & 9 & -3 \\ -2 & -3 & 7 \end{bmatrix} \xrightarrow{E_{21}} \begin{bmatrix} 2 & 4 & -2 \\ \mathbf{0} & 1 & 1 \\ -2 & -3 & 7 \end{bmatrix} \xrightarrow{E_{31}} \begin{bmatrix} 2 & 4 & -2 \\ \mathbf{0} & 1 & 1 \\ \mathbf{0} & 1 & 5 \end{bmatrix} \xrightarrow{E_{32}} \begin{bmatrix} 2 & 4 & -2 \\ \mathbf{0} & 1 & 1 \\ \mathbf{0} & \mathbf{0} & 4 \end{bmatrix} = U$$

Those elimination steps E_{21} and E_{31} and E_{32} produced zeros in positions $(2,1)$ and $(3,1)$ and $(3,2)$. The matrices E have -2 and $+1$ and -1 in those positions. The same steps must be applied to the right hand side $\boldsymbol{b}$, to keep the equations correct.

$$\boldsymbol{b} = \begin{bmatrix} 2 \\ 8 \\ 10 \end{bmatrix} \to E_{21}\boldsymbol{b} = \begin{bmatrix} 2 \\ 4 \\ 10 \end{bmatrix} \to E_{31}E_{21}\boldsymbol{b} = \begin{bmatrix} 2 \\ 4 \\ 12 \end{bmatrix} \to E_{32}E_{31}E_{21}\boldsymbol{b} = E\boldsymbol{b} = \boldsymbol{c} = \begin{bmatrix} 2 \\ 4 \\ 8 \end{bmatrix}.$$

There is a better way to make sure that every operation on the matrix A (left side of equations) is also executed on $\boldsymbol{b}$ (right side of equations). The good way is to **include $\boldsymbol{b}$ as an extra column with $\boldsymbol{A}$**. The combination $\begin{bmatrix} A & \boldsymbol{b} \end{bmatrix}$ is called an **augmented matrix**.

$$\begin{bmatrix} A & \boldsymbol{b} \end{bmatrix} = \left[\begin{array}{cc|c} 3 & 2 & 7 \\ 6 & 5 & 16 \end{array}\right] \xrightarrow{E} \left[\begin{array}{cc|c} 3 & 2 & 7 \\ 0 & 1 & 2 \end{array}\right] = \begin{bmatrix} U & \boldsymbol{c} \end{bmatrix}. \tag{7}$$

Now we include an example that requires a permutation matrix P. It will exchange equations and avoid zero in the pivot. This example needs P in column 2.

Exchange rows 2 and 3

$$A = \begin{bmatrix} 1 & 1 & 1 \\ 2 & 2 & 3 \\ 0 & 4 & 5 \end{bmatrix} \xrightarrow{E} \begin{bmatrix} 1 & 1 & 1 \\ 0 & \mathbf{0} & 1 \\ 0 & \mathbf{4} & 5 \end{bmatrix} \xrightarrow{P} \begin{bmatrix} 1 & 1 & 1 \\ 0 & \mathbf{4} & 5 \\ 0 & \mathbf{0} & 1 \end{bmatrix} = \boldsymbol{U}$$

In the final description $PA = LU$ of elimination on A, all the E's will be moved to the right side. Each matrix in $E_{32}E_{31}E_{21}$ is **inverted**. Those inverses come in **reverse order** in $\boldsymbol{L} = E_{21}^{-1}E_{31}^{-1}E_{32}^{-1}$. The overall equation is $\boldsymbol{PA} = \boldsymbol{LU}$. Often no permutations are needed and elimination produces $\boldsymbol{A} = \boldsymbol{LU}$: the best equation of all.

That permutation P_{23} exchanged rows 2 and 3 when it was needed to avoid a zero pivot. But we could have exchanged rows 2 and 3 at the start. (Then E_{21} and E_{31} have to change places.) Section 2.4 will return to understand all the possible permutations of n rows. *There are $n\,!$ possible matrices P, including $P = I$ for no row exchanges.*

Problem Set 2.1

Problems 1–10 are about elimination on 2 by 2 systems.

1 What multiple ℓ_{21} of equation 1 should be subtracted from equation 2 ?

$$\begin{aligned} 2x + 3y &= 1 \\ 10x + 9y &= 11. \end{aligned}$$

After elimination, write down the upper triangular system and circle the two pivots. Use back substitution to find x and y (and check that solution).

2 If equation 1 is added to equation 2, which of these are changed: the planes in the row picture, the vectors in the column picture, the coefficient matrix, the solution ?

3 What multiple of equation 1 should be *subtracted* from equation 2 ?

$$\begin{aligned} 2x - 4y &= 6 \\ -x + 5y &= 0. \end{aligned}$$

After this elimination step, solve the triangular system. If the right side changes to $(-6, 0)$, what is the new solution?

4 What multiple ℓ of equation 1 should be subtracted from equation 2 to remove c ?

$$\begin{aligned} ax + by &= f \\ cx + dy &= g. \end{aligned}$$

The first pivot is a (assumed nonzero). Elimination produces what formula for the second pivot ? What is y ? The second pivot is missing when $ad = bc$: singular.

5 Choose a right side which gives no solution and another right side which gives infinitely many solutions. What are two of those solutions ?

Singular system $\quad 3x + 2y = 10 \quad$ and $\quad 6x + 4y =$ ____

6 Choose a coefficient b that makes this system singular. Then choose a right side g that makes it solvable. Find two solutions in that singular case.

$$\begin{aligned} 2x + by &= 16 \\ 4x + 8y &= g. \end{aligned}$$

7 For which numbers a does elimination break down (1) permanently (2) temporarily ? Solve for x and y after fixing the temporary breakdown by a row exchange.

$$\begin{aligned} ax + 3y &= -3 \\ 4x + 6y &= 6. \end{aligned}$$

8 For which three numbers k does elimination break down ? Which is fixed by a row exchange ? Is the number of solutions 0 or 1 or ∞ ? **Draw the 3 row pictures.**

$$\begin{aligned} kx + 3y &= 6 \\ 3x + ky &= -6. \end{aligned}$$

9 What test on b_1 and b_2 decides whether these two equations allow a solution? How many solutions will they have? Draw the column pictures for $\boldsymbol{b} = (1, 2)$ and $(1, 0)$.

$$\begin{aligned} 3x - 2y &= b_1 \\ 6x - 4y &= b_2. \end{aligned}$$

10 Draw the lines $x + y = 5$ and $x + 2y = 6$ and the equation $y =$ _____ that comes from elimination. Which line $5x - 4y = c$ goes through the solution of these equations?

Problems 11–20 study elimination on 3 by 3 systems (and possible failure).

11 (Recommended) A system of linear equations can't have exactly two solutions. *Why*?

(a) If (x, y, z) and (X, Y, Z) are two solutions, what is another solution?

(b) If 25 planes meet at two points, where else do they meet?

12 Reduce to upper triangular form by row operations. Then find x, y, z.

$$\begin{aligned} 2x + 3y + z &= 8 \\ 4x + 7y + 5z &= 20 \\ -2y + 2z &= 0 \end{aligned} \qquad\qquad \begin{aligned} 2x - 3y \phantom{{}+ z} &= 3 \\ 4x - 5y + z &= 7 \\ 2x - y - 3z &= 5 \end{aligned}$$

13 Which number d forces a row exchange, and what is the triangular system (not singular) for that d? Which d makes this system singular (no third pivot)?

$$\begin{aligned} 2x + 5y + z &= 0 \\ 4x + dy + z &= 2 \\ y - z &= 3. \end{aligned}$$

14 Which number b leads later to a row exchange? Which b leads to a missing pivot? In that singular case find a nonzero solution x, y, z.

$$\begin{aligned} x + by \phantom{{} - z} &= 0 \\ x - 2y - z &= 0 \\ y + z &= 0. \end{aligned}$$

15 (a) Construct a 3 by 3 system that needs 2 row exchanges to become triangular.
(b) Construct a system that needs a row exchange and breaks down later.

16 If rows 1 and 2 are the same, how far can you get with elimination (allowing row exchange)? If columns 1 and 2 are the same, which pivot is missing?

$$\textbf{Equal rows}\quad \begin{aligned} 2x - y + z &= 0 \\ 2x - y + z &= 0 \\ 4x + y + z &= 2 \end{aligned} \qquad\qquad \begin{aligned} 2x + 2y + z &= 0 \\ 4x + 4y + z &= 0 \\ 6x + 6y + z &= 2. \end{aligned}\quad \textbf{Equal columns}$$

17 Construct a 3 by 3 example that has 9 different coefficients on the left side, but rows 2 and 3 become zero in elimination. How many solutions to your system with $\boldsymbol{b} = (1, 10, 100)$ and how many with $\boldsymbol{b} = (0, 0, 0)$?

18 Which number q makes this system singular and which right side t gives it infinitely many solutions? Find the solution that has $z = 1$.

$$\begin{aligned} x + 4y - 2z &= 1 \\ x + 7y - 6z &= 6 \\ 3y + qz &= t. \end{aligned}$$

19 For which two numbers a will elimination fail on $A = \begin{bmatrix} a & 2 \\ a & a \end{bmatrix}$?

20 For which three numbers a will elimination fail to give three pivots?

$$A = \begin{bmatrix} a & 2 & 3 \\ a & a & 4 \\ a & a & a \end{bmatrix} \text{ is singular for three values of } a.$$

21 Look for a matrix that has row sums 4 and 8, and column sums 2 and s: The four equations are solvable only if $s =$ _____. Find two matrices with the correct row and column sums. Write down the 4 by 4 system $A\boldsymbol{x} = \boldsymbol{b}$ with $\boldsymbol{x} = (a, b, c, d)$ and make A triangular by elimination.

$$\text{Matrix} \begin{bmatrix} a & b \\ c & d \end{bmatrix} \qquad \begin{matrix} a + b = 4 & a + c = 2 \\ c + d = 8 & b + d = s \end{matrix} \qquad \begin{matrix} \text{4 equations} \\ \text{4 unknowns} \end{matrix}$$

22 Create a MATLAB command `A(2, : ) = ...` for the new row 2, to subtract 3 times row 1 from the existing row 2 if the 3 by 3 matrix A is already known.

23 Find experimentally the average 1st and 2nd and 3rd pivot sizes from MATLAB's $[L, U] =$ **lu** (**rand** (3)) with random entries between 0 and 1. The average of $U(1, 1)$ is above $\frac{1}{2}$ because **lu** picks the largest pivot in column 1.

24 If the last corner entry is $A(5, 5) = 11$ and the last pivot of A is $U(5, 5) = 4$, what different entry $A(5, 5)$ would have made A singular?

25 Suppose elimination takes A to U without row exchanges. Then row j of U is a combination of which rows of A? If $A\boldsymbol{x} = \boldsymbol{0}$, is $U\boldsymbol{x} = \boldsymbol{0}$? If $A\boldsymbol{x} = \boldsymbol{b}$, is $U\boldsymbol{x} = \boldsymbol{b}$? If A starts out lower triangular, what is the upper triangular U?

26 Start with 100 equations $A\boldsymbol{x} = \boldsymbol{0}$ for 100 unknowns $\boldsymbol{x} = (x_1, \ldots, x_{100})$. Suppose elimination reduces the 100th equation to $0 = 0$, so the system is "singular".

(a) Singular systems $A\boldsymbol{x} = \boldsymbol{0}$ have infinitely many solutions. This means that some linear combination of the 100 ***columns*** of A is _____.

(b) Invent a 100 by 100 singular matrix with no zero entries.

(c) Describe in words the row picture and column picture of your $A\boldsymbol{x} = \boldsymbol{0}$.

2.2 Elimination Matrices and Inverse Matrices

1 Elimination multiplies A by $E_{21}, \ldots, E_{n1}$ then $E_{32}, \ldots, E_{n2}, \ldots$ as A becomes $EA = U$.

2 In reverse order the inverses of the E's multiply U to recover $A = E^{-1}U$. This is LU.

3 $A^{-1}A = I$ and $(LU)^{-1} = U^{-1}L^{-1}$. Then $A\boldsymbol{x} = \boldsymbol{b}$ becomes $\boldsymbol{x} = A^{-1}\boldsymbol{b} = U^{-1}L^{-1}\boldsymbol{b}$.

All the steps of elimination can be done with matrices. Those steps can also be *undone* (inverted) with matrices. For a 3 by 3 matrix we can write out each step in detail—almost word for word. But for real applications, matrices are a much better way.

The basic elimination step is to subtract a multiple ℓ_{ij} of equation j from equation i. We always speak about *subtractions* as elimination proceeds. If the first pivot is $a_{11} = 3$ and below it is $a_{21} = -3$, we could just add equation 1 to equation 2. That produces zero. But we stay with subtraction: *subtract* $\ell_{21} = -1$ *times equation* 1 *from equation* 2. Same result. The inverse step is addition. Compare equation (10) to (11) to see it all.

Here is the matrix that subtracts 2 times row 1 from row 3: Rows 1 and 2 stay the same.

Elimination matrix E_{ij}
Row 3, column 1, multiplier 2

$$E_{31} = \begin{bmatrix} \mathbf{1} & 0 & 0 \\ 0 & \mathbf{1} & 0 \\ \mathbf{-2} & 0 & \mathbf{1} \end{bmatrix}$$

If no row exchanges are needed, then three elimination matrices E_{21} and E_{31} and E_{32} will produce three zeros below the diagonal. This changes A to the triangular U:

A is 3 by 3	**U is upper triangular**	**$E_{32}E_{31}E_{21}A = U$**	(1)

The number ℓ_{32} is affected by the ℓ_{21} and ℓ_{31} that came first. We subtract ℓ_{32} times *row* 2 *of* U (the final second row, not the original second row of A). This is the step that produces zero in row 3, column 2 of U. E_{32} gives the last step of 3 by 3 elimination.

Example 1 $\quad E_{21}$ and E_{31} subtract multiples of row 1 from rows 2 and 3 of A:

$$E_{31}E_{21}A = \begin{bmatrix} 1 & 0 & 0 \\ 0 & 1 & 0 \\ -2 & 0 & 1 \end{bmatrix} \begin{bmatrix} 1 & 0 & 0 \\ 1 & 1 & 0 \\ 0 & 0 & 1 \end{bmatrix} \begin{bmatrix} \mathbf{3} & 1 & 0 \\ \mathbf{-3} & 1 & 1 \\ \mathbf{6} & 8 & 4 \end{bmatrix} = \begin{bmatrix} \mathbf{3} & 1 & 0 \\ \mathbf{0} & 2 & 1 \\ \mathbf{0} & 6 & 4 \end{bmatrix} \quad \begin{matrix}\textbf{two new} \\ \textbf{zeros in} \\ \textbf{column 1}\end{matrix} \qquad (2)$$

To produce a zero in column 2, E_{32} subtracts $\ell_{32} = 3$ times the *new row* 2 from row 3:

$$(E_{32})\,(E_{31}E_{21}A) = \begin{bmatrix} 1 & 0 & 0 \\ 0 & 1 & 0 \\ 0 & -3 & 1 \end{bmatrix} \begin{bmatrix} 3 & 1 & 0 \\ 0 & \mathbf{2} & 1 \\ 0 & \mathbf{6} & 4 \end{bmatrix} = \begin{bmatrix} 3 & 1 & 0 \\ \mathbf{0} & \mathbf{2} & 1 \\ \mathbf{0} & \mathbf{0} & \mathbf{1} \end{bmatrix} = U \quad \begin{matrix}\textbf{\textit{U} has zeros} \\ \textbf{below the} \\ \textbf{main diagonal}\end{matrix} \qquad (3)$$

Notice again: E_{32} is subtracting 3 times the row $0, 2, 1$ and not the original row of A. At the end, the pivots $3, 2, 1$ are on the main diagonal of U: zeros below.

Example 4 will show the "inverse" of each elimination matrix E_{ij}. This leads to the inverse of their product $E = E_{32}E_{31}E_{21}$. That inverse of E is special. *We call it* L.

Inverse Matrices

Suppose A is a square matrix. We look for an "***inverse matrix***" A^{-1} of the same size, such that $\boldsymbol{A^{-1}}$ **times** $\boldsymbol{A}$ **equals** $\boldsymbol{I}$. Whatever A does, A^{-1} undoes. Their product is the identity matrix—which does nothing to a vector, so $A^{-1}A\boldsymbol{x} = \boldsymbol{x}$. *But A^{-1} might not exist.*

The square matrix A needs independent columns to be invertible. Then $A^{-1}A = I$.

What a matrix mostly does is to multiply a vector. Multiplying $A\boldsymbol{x} = \boldsymbol{b}$ by A^{-1} gives $A^{-1}A\boldsymbol{x} = A^{-1}\boldsymbol{b}$. ***This is*** $\boldsymbol{x} = A^{-1}\boldsymbol{b}$. The product $A^{-1}A$ is like multiplying by a number and then dividing by that number. Numbers have inverses if they are not zero. Matrices are more complicated and more interesting. The matrix A^{-1} is called "A inverse."

DEFINITION The matrix A is ***invertible*** if there exists a matrix A^{-1} that "inverts" A:

$$\textbf{Two-sided inverse} \qquad \boldsymbol{A^{-1}A = I} \quad \text{and} \quad \boldsymbol{AA^{-1} = I}. \tag{4}$$

Not all matrices have inverses. This is the first question we ask about a square matrix: Is A invertible? Its columns must be independent. We don't mean that we actually calculate A^{-1}. In most problems we never compute it! Here are seven "notes" about A^{-1}.

Note 1 ***The inverse exists if and only if elimination produces n pivots*** (row exchanges are allowed). Elimination solves $A\boldsymbol{x} = \boldsymbol{b}$ without explicitly using the matrix A^{-1}.

Note 2 The matrix A cannot have two different inverses. Suppose $BA = I$ and also $AC = I$. Then $B = C$, according to this "proof by parentheses":

$$B(AC) = (BA)C \quad \text{gives} \quad BI = IC \quad \text{or} \quad B = C. \tag{5}$$

This shows that a *left-inverse* B (multiplying A from the left) and a *right-inverse* C (multiplying A from the right to give $AC = I$) must be the *same matrix*.

Note 3 If A is invertible, the one and only solution to $A\boldsymbol{x} = \boldsymbol{b}$ is $\boldsymbol{x} = A^{-1}\boldsymbol{b}$:

$$\textbf{\textit{Multiply}} \quad A\boldsymbol{x} = \boldsymbol{b} \quad \textbf{\textit{by}} \quad \boldsymbol{A^{-1}}. \quad \textbf{\textit{Then}} \quad \boldsymbol{x} = A^{-1}A\boldsymbol{x} = \boldsymbol{A^{-1}b}.$$

Note 4 (Important) ***Suppose there is a nonzero vector $\boldsymbol{x}$ such that*** $A\boldsymbol{x} = \boldsymbol{0}$. *Then A has dependent columns. It cannot have an inverse.* No matrix can bring $\boldsymbol{0}$ back to $\boldsymbol{x}$.

If A is invertible, then $A\boldsymbol{x} = \boldsymbol{0}$ can only have the zero solution $\boldsymbol{x} = A^{-1}\boldsymbol{0} = \boldsymbol{0}$.

Note 5 A square matrix is invertible if and only if its columns are independent.

Note 6 A 2 by 2 matrix is invertible if and only if $ad - bc$ is not zero:

$$\textbf{2 by 2 Inverse} \quad \begin{bmatrix} a & b \\ c & d \end{bmatrix}^{-1} = \frac{1}{ad - bc}\begin{bmatrix} d & -b \\ -c & a \end{bmatrix}. \tag{6}$$

This number $ad - bc$ is the *determinant* of A. A matrix is invertible if its determinant is not zero (Chapter 5). The test for n pivots is usually decided before the determinant appears.

Note 7 A triangular matrix has an inverse provided no diagonal entries d_i are zero:

$$\text{If} \quad A = \begin{bmatrix} \boldsymbol{d_1} & \times & \times & \times \\ 0 & \bullet & \times & \times \\ 0 & 0 & \bullet & \times \\ 0 & 0 & 0 & \boldsymbol{d_n} \end{bmatrix} \quad \text{then} \quad A^{-1} = \begin{bmatrix} \boldsymbol{1/d_1} & \times & \times & \times \\ 0 & \bullet & \times & \times \\ 0 & 0 & \bullet & \times \\ 0 & 0 & 0 & \boldsymbol{1/d_n} \end{bmatrix}$$

Example 2 The 2 by 2 matrix $A = \begin{bmatrix} 1 & 2 \\ 1 & 2 \end{bmatrix}$ is not invertible. It fails the test in Note 6, because $ad = bc$. It also fails the test in Note 4, because $A\boldsymbol{x} = \mathbf{0}$ when $\boldsymbol{x} = (2, -1)$. It fails to have two pivots as required by Note 1. Its columns are dependent.

Elimination turns the second row of this matrix A into a zero row. No pivot.

Example 3 Three of these matrices are invertible, and three are singular. Find the inverse when it exists. Give reasons for noninvertibility (zero determinant, too few pivots, nonzero solution to $A\boldsymbol{x} = \mathbf{0}$) for the other three. The matrices are in the order A, B, C, D, S, T:

$$\begin{bmatrix} 4 & 3 \\ 8 & 6 \end{bmatrix} \quad \begin{bmatrix} 4 & 3 \\ 8 & 7 \end{bmatrix} \quad \begin{bmatrix} 6 & 6 \\ 6 & 0 \end{bmatrix} \quad \begin{bmatrix} 6 & 6 \\ 6 & 6 \end{bmatrix} \quad \begin{bmatrix} 1 & 0 & 0 \\ 1 & 1 & 0 \\ 1 & 1 & 1 \end{bmatrix} \quad \begin{bmatrix} 1 & 1 & 1 \\ 1 & 1 & 0 \\ 1 & 1 & 1 \end{bmatrix}$$

Solution

$$B^{-1} = \frac{1}{4}\begin{bmatrix} 7 & -3 \\ -8 & 4 \end{bmatrix} \quad C^{-1} = \frac{1}{36}\begin{bmatrix} 0 & 6 \\ 6 & -6 \end{bmatrix} \quad S^{-1} = \begin{bmatrix} 1 & 0 & 0 \\ -1 & 1 & 0 \\ 0 & -1 & 1 \end{bmatrix}$$

A is not invertible because its determinant is $4 \cdot 6 - 3 \cdot 8 = 24 - 24 = 0$. D is not invertible because there is only one pivot; the second row becomes zero when the first row is subtracted. T has two equal rows (and the second column minus the first column is zero). In other words $T\boldsymbol{x} = \mathbf{0}$ has the nonzero solution $\boldsymbol{x} = (-1, 1, 0)$. *Not invertible.*

The Inverse of a Product AB

For two nonzero numbers a and b, the sum $a + b$ might or might not be invertible. The numbers $a = 3$ and $b = -3$ have inverses $\frac{1}{3}$ and $-\frac{1}{3}$. Their sum $a + b = 0$ has no inverse. But the product $ab = -9$ does have an inverse, which is $\frac{1}{3}$ times $-\frac{1}{3}$.

For matrices A and B, the situation is similar. Their *product* AB has an inverse if and only if A and B are separately invertible (and the same size). The important point is that A^{-1} and B^{-1} come in ***reverse order***:

> If A and B are invertible (same size) then the inverse of AB is $\boldsymbol{B^{-1}A^{-1}}$.
>
> $$(\boldsymbol{AB})^{-1} = \boldsymbol{B^{-1}A^{-1}} \qquad (\boldsymbol{AB})(\boldsymbol{B^{-1}A^{-1}}) = \boldsymbol{AIA^{-1}} = \boldsymbol{AA^{-1}} = \boldsymbol{I} \quad (7)$$

We moved parentheses to multiply BB^{-1} first. Similarly $B^{-1}A^{-1}$ times AB equals I.

$B^{-1}A^{-1}$ illustrates a basic rule of mathematics: Inverses come in reverse order. It is also common sense: If you put on socks and then shoes, the first to be taken off are the _____. The same reverse order applies to three or more matrices:

$$\textbf{Reverse order} \qquad (ABC)^{-1} = C^{-1}B^{-1}A^{-1} \tag{8}$$

Example 4 ***Inverse of an elimination matrix***. If E subtracts 5 times row 1 from row 2, then E^{-1} *adds* 5 times row 1 to row 2:

$$\begin{array}{c}\textbf{\textit{E} subtracts}\\ \boldsymbol{E^{-1}} \textbf{ adds}\end{array} \qquad E = \begin{bmatrix} 1 & 0 & 0 \\ \mathbf{-5} & 1 & 0 \\ 0 & 0 & 1 \end{bmatrix} \quad \text{and} \quad E^{-1} = \begin{bmatrix} 1 & 0 & 0 \\ \mathbf{5} & 1 & 0 \\ 0 & 0 & 1 \end{bmatrix}$$

Multiply EE^{-1} to get the identity matrix I. Also multiply $E^{-1}E$ to get I. We are adding and subtracting the same 5 times row 1. If $AC = I$ then automatically $CA = I$.

For square matrices, an inverse on one side is automatically an inverse on the other side.

Example 5 Suppose F subtracts 4 times row 2 from row 3, and F^{-1} adds it back:

$$F = \begin{bmatrix} 1 & 0 & 0 \\ 0 & 1 & 0 \\ 0 & -4 & 1 \end{bmatrix} \quad \text{and} \quad F^{-1} = \begin{bmatrix} 1 & 0 & 0 \\ 0 & 1 & 0 \\ 0 & 4 & 1 \end{bmatrix}.$$

Now multiply F by the matrix E in Example 4 to find FE. Also multiply E^{-1} times F^{-1} to find $(FE)^{-1}$. Notice the orders FE and $E^{-1}F^{-1}$!

$$FE = \begin{bmatrix} 1 & 0 & 0 \\ \mathbf{-5} & 1 & 0 \\ \mathbf{20} & \mathbf{-4} & 1 \end{bmatrix} \quad \textbf{is inverted by} \quad E^{-1}F^{-1} = \begin{bmatrix} 1 & 0 & 0 \\ \mathbf{5} & 1 & 0 \\ \mathbf{0} & \mathbf{4} & 1 \end{bmatrix}. \tag{9}$$

The result is beautiful and correct. The product FE contains "20" but its inverse doesn't. E subtracts 5 times row 1 from row 2. Then F subtracts 4 times the *new* row 2 (changed by row 1) from row 3. ***In this order FE, row* 3 *feels an effect of size* 20 *from row* 1.**

In the order $E^{-1}F^{-1}$, that effect does not happen. First F^{-1} adds 4 times row 2 to row 3. After that, E^{-1} adds 5 times row 1 to row 2. There is no 20, because row 3 doesn't change again. ***In this order $\boldsymbol{E^{-1}F^{-1}}$, row* 3 *feels no effect from row* 1.**

This is why we choose $A = LU$, to go back from the triangular U to the original A. The multipliers fall into place perfectly in the lower triangular L: Equation (11) below.

The elimination order is FE. The inverse order $L = E^{-1}F^{-1}$ is special. ***The multipliers* 5 *and* 4 *fall into place below the diagonal of* 1*'s in L.***

L is the Inverse of E

E is the product of all the elimination matrices E_{ij}, taking us from A to its upper triangular form $EA = U$. We are assuming for now that no row exchanges are involved (thus $P = I$). The difficulty with E is that multiplying all the separate elimination steps E_{ij} does not produce a good formula. But the inverse matrix E^{-1} becomes beautiful when we multiply the inverse steps E_{ij}^{-1}. Remember that those steps come in the *opposite order*.

With $n = 3$, the complication for $E = E_{32}E_{31}E_{21}$ is in the bottom corner:

$$\boldsymbol{E} = \begin{bmatrix} 1 & & \\ 0 & 1 & \\ 0 & -\ell_{32} & 1 \end{bmatrix} \begin{bmatrix} 1 & & \\ 0 & 1 & \\ -\ell_{31} & 0 & 1 \end{bmatrix} \begin{bmatrix} 1 & & \\ -\ell_{21} & 1 & \\ 0 & 0 & 1 \end{bmatrix} = \begin{bmatrix} 1 & & \\ -\ell_{21} & 1 & \\ (\boldsymbol{\ell_{32}\ell_{21} - \ell_{31}}) & -\ell_{32} & 1 \end{bmatrix}. \quad (10)$$

Watch how that confusion disappears for $\boldsymbol{E^{-1} = L}$. Reverse order is the good way:

$$\boxed{\boldsymbol{E^{-1}} = \begin{bmatrix} 1 & & \\ \ell_{21} & 1 & \\ 0 & 0 & 1 \end{bmatrix} \begin{bmatrix} 1 & & \\ 0 & 1 & \\ \ell_{31} & 0 & 1 \end{bmatrix} \begin{bmatrix} 1 & & \\ 0 & 1 & \\ 0 & \ell_{32} & 1 \end{bmatrix} = \begin{bmatrix} 1 & & \\ \boldsymbol{\ell_{21}} & 1 & \\ \boldsymbol{\ell_{31}} & \boldsymbol{\ell_{32}} & 1 \end{bmatrix} = \boldsymbol{L}} \quad (11)$$

All the multipliers $\boldsymbol{\ell_{ij}}$ appear in their correct positions in $\boldsymbol{L}$. The next section will show that this remains true for all matrix sizes. **Then $\boldsymbol{EA = U}$ becomes $\boldsymbol{A = LU}$.**

Equation (11) is the key to this chapter: each ℓ_{ij} in its place.

Problem Set 2.2

Problems 1–11 are about elimination matrices.

1 Write down the 3 by 3 matrices that produce these elimination steps:

(a) E_{21} subtracts 5 times row 1 from row 2.

(b) E_{32} subtracts -7 times row 2 from row 3.

(c) P exchanges rows 1 and 2, then rows 2 and 3.

2 In Problem 1, applying E_{21} and then E_{32} to $\boldsymbol{b} = (1, 0, 0)$ gives $E_{32}E_{21}\boldsymbol{b} =$ ____. Applying E_{32} before E_{21} gives $E_{21}E_{32}\boldsymbol{b} =$ ____. When E_{32} comes first, row ____ feels no effect from row ____.

3 Which three matrices E_{21}, E_{31}, E_{32} put A into triangular form U?

$$A = \begin{bmatrix} 1 & 1 & 0 \\ 4 & 6 & 1 \\ -2 & 2 & 0 \end{bmatrix} \quad \text{and} \quad E_{32}E_{31}E_{21}A = EA = U.$$

Multiply those E's to get one elimination matrix E. What is $E^{-1} = L$?

4 Include $\boldsymbol{b} = (1, 0, 0)$ as a fourth column in Problem 3 to produce $[\,A \;\; \boldsymbol{b}\,]$. Carry out the elimination steps on this augmented matrix to solve $A\boldsymbol{x} = \boldsymbol{b}$.

5 Suppose $a_{33} = 7$ and the third pivot is 5. If you change a_{33} to 11, the third pivot is _____. If you change a_{33} to _____, there is no third pivot.

6 If every column of A is a multiple of $(1, 1, 1)$, then $A\boldsymbol{x}$ is always a multiple of $(1, 1, 1)$. Do a 3 by 3 example. How many pivots are produced by elimination?

7 Suppose E subtracts 7 times row 1 from row 3.

(a) To *invert* that step you should _____ 7 times row _____ to row _____.

(b) What "inverse matrix" E^{-1} takes that reverse step (so $E^{-1}E = I$)?

(c) If the reverse step is applied first (and then E) show that $EE^{-1} = I$.

8 The ***determinant*** of $M = \left[\begin{smallmatrix} a & b \\ c & d \end{smallmatrix}\right]$ is det $M = ad - bc$. Subtract ℓ times row 1 from row 2 to produce a new M^*. Show that det $M^* =$ det M for every ℓ. When $\ell = c/a$, *the product of pivots equals the determinant*: $(a)(d - \ell b)$ equals $ad - bc$.

9 (a) E_{21} subtracts row 1 from row 2 and then P_{23} exchanges rows 2 and 3. What matrix $M = P_{23}E_{21}$ does both steps at once?

(b) P_{23} exchanges rows 2 and 3 and then E_{31} subtracts row 1 from row 3. What matrix $M = E_{31}P_{23}$ does both steps at once? Explain why the M's are the same but the E's are different.

10 (a) What matrix adds row 1 to row 3 and *at the same time* row 3 to row 1 ?

(b) What matrix adds row 1 to row 3 and *then* adds row 3 to row 1 ?

11 Create a matrix that has $a_{11} = a_{22} = a_{33} = 1$ but elimination produces two negative pivots without row exchanges. (The first pivot is 1.)

12 For these "permutation matrices" find P^{-1} by trial and error (with 1's and 0's):

$$P = \begin{bmatrix} 0 & 0 & 1 \\ 0 & 1 & 0 \\ 1 & 0 & 0 \end{bmatrix} \quad \text{and} \quad P = \begin{bmatrix} 0 & 1 & 0 \\ 0 & 0 & 1 \\ 1 & 0 & 0 \end{bmatrix}.$$

13 Solve for the first column (x, y) and second column (t, z) of A^{-1}. Check AA^{-1}.

$$\begin{bmatrix} 10 & 20 \\ 20 & 50 \end{bmatrix} \begin{bmatrix} x \\ y \end{bmatrix} = \begin{bmatrix} 1 \\ 0 \end{bmatrix} \quad \text{and} \quad \begin{bmatrix} 10 & 20 \\ 20 & 50 \end{bmatrix} \begin{bmatrix} t \\ z \end{bmatrix} = \begin{bmatrix} 0 \\ 1 \end{bmatrix}.$$

14 Find an upper triangular U (not diagonal) with $U^2 = I$. Then $U^{-1} = U$.

15 (a) If A is invertible and $AB = AC$, prove quickly that $B = C$.

(b) If $A = \left[\begin{smallmatrix} 1 & 1 \\ 1 & 1 \end{smallmatrix}\right]$, find two different matrices such that $AB = AC$.

16 (Important) If A has row 1 + row 2 = row 3, show that A is not invertible:

(a) Explain why $A\boldsymbol{x} = (0, 0, 1)$ cannot have a solution. Add eqn 1 + eqn 2.

(b) Which right sides (b_1, b_2, b_3) might allow a solution to $A\boldsymbol{x} = \boldsymbol{b}$?

(c) In the elimination process, what happens to equation 3 ?

17 If A has column 1 + column 2 = column 3, show that A is not invertible:

(a) Find a nonzero solution $\boldsymbol{x}$ to $A\boldsymbol{x} = \boldsymbol{0}$. The matrix is 3 by 3.

(b) Elimination keeps columns $1 + 2 = 3$. Explain why there is no third pivot.

18 Suppose A is invertible and you exchange its first two rows to reach B. Is the new matrix B invertible? How would you find B^{-1} from A^{-1}?

19 (a) Find invertible matrices A and B such that $A + B$ is not invertible.

(b) Find singular matrices A and B such that $A + B$ is invertible.

20 If the product $C = AB$ is invertible (A and B are square), then A itself is invertible. Find a formula for A^{-1} that involves C^{-1} and B.

21 If the product $M = ABC$ of three square matrices is invertible, then B is invertible. (So are A and C.) Find a formula for B^{-1} that involves M^{-1} and A and C.

22 If you add row 1 of A to row 2 to get B, how do you find B^{-1} from A^{-1} ?

23 Prove that a matrix with a column of zeros cannot have an inverse.

24 Multiply $\left[\begin{smallmatrix} a & b \\ c & d \end{smallmatrix}\right]$ times $\left[\begin{smallmatrix} d & -b \\ -c & a \end{smallmatrix}\right]$. What is the inverse of each matrix if $ad \neq bc$?

25 (a) What 3 by 3 matrix E has the same effect as these three steps? Subtract row 1 from row 2, subtract row 1 from row 3, then subtract row 2 from row 3.

(b) What single matrix L has the same effect as these three reverse steps? Add row 2 to row 3, add row 1 to row 3, then add row 1 to row 2.

26 If B is the inverse of A^2, show that AB is the inverse of A.

27 Show that A = 4 ∗ eye (4) − ones (4, 4) is *not* invertible : Multiply A ∗ ones (4, 1).

28 There are sixteen 2 by 2 matrices whose entries are 1's and 0's. How many of them are invertible ?

29 Change I into A^{-1} as elimination reduces A to I (the Gauss-Jordan idea).

$$\begin{bmatrix} A & I \end{bmatrix} = \begin{bmatrix} 1 & 3 & 1 & 0 \\ 2 & 7 & 0 & 1 \end{bmatrix} \quad \text{and} \quad \begin{bmatrix} A & I \end{bmatrix} = \begin{bmatrix} 1 & 4 & 1 & 0 \\ 3 & 9 & 0 & 1 \end{bmatrix}$$

30 Could a 4 by 4 matrix A be invertible if every row contains the numbers $0, 1, 2, 3$ in some order? What if every row of B contains $0, 1, 2, -3$ in some order ?

31 Find A^{-1} and B^{-1} (*if they exist*) by elimination on $[\,A \quad I\,]$ and $[\,B \quad I\,]$:

$$A = \begin{bmatrix} 2 & 1 & 1 \\ 1 & 2 & 1 \\ 1 & 1 & 2 \end{bmatrix} \quad \text{and} \quad B = \begin{bmatrix} 2 & -1 & -1 \\ -1 & 2 & -1 \\ -1 & -1 & 2 \end{bmatrix}.$$

32 Use Gauss-Jordan elimination on $[\,U \quad I\,]$ to find the upper triangular U^{-1}:

$$\boldsymbol{U}\boldsymbol{U}^{-1} = \boldsymbol{I} \qquad \begin{bmatrix} 1 & a & b \\ 0 & 1 & c \\ 0 & 0 & 1 \end{bmatrix} \begin{bmatrix} \\ \boldsymbol{x}_1 & \boldsymbol{x}_2 & \boldsymbol{x}_3 \\ \\ \end{bmatrix} = \begin{bmatrix} 1 & 0 & 0 \\ 0 & 1 & 0 \\ 0 & 0 & 1 \end{bmatrix}.$$

33 True or false (with a counterexample if false and a reason if true):

(a) A 4 by 4 matrix with a row of zeros is not invertible.

(b) Every matrix with 1's down the main diagonal is invertible.

(c) If A is invertible then A^{-1} and A^2 are invertible.

34 (Recommended) Prove that A is invertible if $a \neq 0$ and $a \neq b$ (find the pivots or A^{-1}). Then find three numbers c so that C is not invertible:

$$A = \begin{bmatrix} a & b & b \\ a & a & b \\ a & a & a \end{bmatrix} \qquad C = \begin{bmatrix} 2 & c & c \\ c & c & c \\ 8 & 7 & c \end{bmatrix}.$$

35 This matrix has a remarkable inverse. Find A^{-1} by elimination on $[\,A \quad I\,]$. Extend to a 5 by 5 "alternating matrix" and guess its inverse; then multiply to confirm.

$$\text{Invert } A = \begin{bmatrix} 1 & -1 & 1 & -1 \\ 0 & 1 & -1 & 1 \\ 0 & 0 & 1 & -1 \\ 0 & 0 & 0 & 1 \end{bmatrix} \quad \text{and solve } A\boldsymbol{x} = (1,1,1,1).$$

36 Suppose the matrices P and Q have the same rows as I but in any order. They are "permutation matrices". Show that $P - Q$ is singular by solving $(P - Q)\,\boldsymbol{x} = \boldsymbol{0}$.

37 Find and check the inverses (assuming they exist) of these **block matrices**:

$$\begin{bmatrix} I & 0 \\ C & I \end{bmatrix} \quad \begin{bmatrix} A & 0 \\ C & D \end{bmatrix} \quad \begin{bmatrix} 0 & I \\ I & D \end{bmatrix}.$$

38 How does elimination from A to U on a 3 by 3 matrix tell you if A is invertible?

2.3 Matrix Computations and $A = LU$

1 The elimination steps from A to U cost $\frac{1}{3}n^3$ multiplications and subtractions.

2 Each right side $\boldsymbol{b}$ costs only n^2 : forward to $U\boldsymbol{x} = \boldsymbol{c}$, then back-substitution for $\boldsymbol{x}$.

3 Elimination without row exchanges factors A into LU (two proofs of $A = LU$).

How would you compute the inverse of an n by n matrix A ? Before answering that question I have to ask : Do you really want to know A^{-1} ? It is true that the solution to $A\boldsymbol{x} = \boldsymbol{b}$ (which we do want) is given by $\boldsymbol{x} = A^{-1}\boldsymbol{b}$. Computing A^{-1} and multiplying $A^{-1}\boldsymbol{b}$ is a very slow way to find $\boldsymbol{x}$. We should understand A^{-1} even if we don't use it.

Here is a simple idea for A^{-1}. That matrix is the solution to $AA^{-1} = I$. The identity matrix has n columns $\boldsymbol{e}_1, \boldsymbol{e}_2, \ldots, \boldsymbol{e}_n$. Then $AA^{-1} = I$ is really n equations $A\boldsymbol{x}_k = \boldsymbol{e}_k$ for the n columns $\boldsymbol{x}_k$ of A^{-1}. We have three equations if the matrices are 3 by 3 :

$$\boldsymbol{AA^{-1}=I} \quad A\begin{bmatrix}\boldsymbol{x}_1 \cdots \boldsymbol{x}_n\end{bmatrix} = \begin{bmatrix}\boldsymbol{e}_1 \cdots \boldsymbol{e}_n\end{bmatrix} \quad A\begin{bmatrix} \\ \boldsymbol{x}_1 & \boldsymbol{x}_2 & \boldsymbol{x}_3 \\ \\ \end{bmatrix} = \begin{bmatrix}1 & 0 & 0\\ 0 & 1 & 0\\ 0 & 0 & 1\end{bmatrix}. \quad (1)$$

We are solving n equations and they have the same coefficient matrix A. So we can solve them together. This is called "Gauss-Jordan elimination". Instead of a matrix $\begin{bmatrix} A & \boldsymbol{b}\end{bmatrix}$ augmented by one right hand side $\boldsymbol{b}$, we have a matrix $\begin{bmatrix} A & I\end{bmatrix}$ augmented by n right hand sides (the columns of I). And elimination produces $\begin{bmatrix} I & A^{-1}\end{bmatrix}$.

$$\begin{bmatrix}\boldsymbol{A} & \boldsymbol{I}\end{bmatrix} = \left[\begin{array}{rrr|rrr} 1 & 0 & 0 & 1 & 0 & 0\\ -1 & 1 & 0 & 0 & 1 & 0\\ \mathbf{0} & -1 & 1 & 0 & 0 & 1\end{array}\right] \to \left[\begin{array}{rrr|rrr} 1 & 0 & 0 & 1 & 0 & 0\\ \mathbf{0} & 1 & 0 & 1 & 1 & 0\\ \mathbf{0} & -1 & 1 & 0 & 0 & 1\end{array}\right]$$

$$\textbf{Gauss-Jordan} \qquad \to \left[\begin{array}{rrr|rrr} 1 & 0 & 0 & 1 & 0 & 0\\ \mathbf{0} & 1 & 0 & 1 & 1 & 0\\ \mathbf{0} & \mathbf{0} & 1 & 1 & 1 & 1\end{array}\right] = \begin{bmatrix}\boldsymbol{I} & \boldsymbol{A^{-1}}\end{bmatrix}$$

The key point is that the elimination steps on A only have to be done once. The same steps are applied to the right hand side—but now $AA^{-1} = I$ has n right hand sides. The n solutions $\boldsymbol{x}_i$ to $A\boldsymbol{x}_i = \boldsymbol{e}_i$ go into the n columns of A^{-1}. Then Gauss-Jordan takes $\begin{bmatrix} A & I\end{bmatrix}$ into $\begin{bmatrix} I & A^{-1}\end{bmatrix}$. Here elimination is multiplication by A^{-1}.

In this example A subtracts rows and A^{-1} adds. This is linear algebra's version of the Fundamental Theorem of Calculus : *Derivative of integral of $f(x)$ equals $f(x)$.*

The Cost of Elimination

A very practical question is cost—or computing time. We can solve 1000 equations on a PC. What if $n = 100,000$? (*Is A dense or sparse*?) Large systems come up all the time in scientific computing, where a three-dimensional problem can easily lead to a million unknowns. We can let the calculation run overnight, but we can't leave it for 100 years.

The first stage of elimination produces zeros below the first pivot in column 1. To find each new entry below the pivot row requires one multiplication and one subtraction. *We will count this first stage as n^2 multiplications and n^2 subtractions.* It is actually $n^2 - n$, because row 1 does not change.

The next stage clears out the second column below the second pivot. The working matrix is now of size $n-1$. Estimate this stage by $(n-1)^2$ multiplications and subtractions. The matrices are getting smaller as elimination goes forward. The rough count to reach U is the sum of squares $n^2 + (n-1)^2 + \cdots + 2^2 + 1^2$.

There is an exact formula $\frac{1}{3}n\left(n + \frac{1}{2}\right)(n+1)$ for this sum of squares. When n is large, the $\frac{1}{2}$ and the 1 are not important. *The number that matters is* $\frac{1}{3}\,n^3$. The sum of squares is like the integral of x^2 ! The integral from 0 to n is $\frac{1}{3}\,n^3$:

Elimination on A requires about $\frac{1}{3}\,n^3$ multiplications and $\frac{1}{3}\,n^3$ subtractions.

What about the right side $\boldsymbol{b}$? Going forward, we subtract multiples of b_1 from the lower components $b_2, \ldots, b_n$. This is $n-1$ steps. The second stage takes only $n-2$ steps, because b_1 is not involved. The last stage of forward elimination (**$\boldsymbol{b}$ to $\boldsymbol{c}$**) takes one step.

Now start back substitution. Computing x_n uses one step (divide by the last pivot). The next unknown uses two steps. When we reach x_1 it will require n steps ($n-1$ substitutions of the other unknowns, then division by the first pivot). The total count on the right side, from $\boldsymbol{b}$ to $\boldsymbol{c}$ to $\boldsymbol{x}$—*forward to the bottom and back to the top*—is exactly n^2:

$$[(n-1) + (n-2) + \cdots + 1] \;+\; [1 + 2 + \cdots + (n-1) + n] = n^2. \tag{2}$$

To see that sum, pair off $(n-1)$ with 1 and $(n-2)$ with 2. The pairings leave n terms, each equal to n. That makes n^2. The right side costs a lot less than the left side!

Solve ***Each right side needs n^2 multiplications and n^2 subtractions.***

How long does it take to solve $Ax = b$? For a random matrix of order $n = 1000$, a typical time on a PC is 1 second. The time is multiplied by about 8 when n is multiplied by 2. For professional codes go to **netlib.org**.

According to this n^3 rule, matrices that are 10 times as large (order 10,000) will take a thousand seconds. Matrices of order 100,000 will take a million seconds. This is too expensive without a supercomputer, but remember that these matrices are full. Most large matrices in practice are sparse (many zero entries). In that case $A = LU$ is much faster.

Proving $A = LU$

Elimination is expressed by $EA = U$ and inverted by $A = LU$. Equation (11) in Section 2.2 showed how the multipliers ℓ_{ij} fall exactly into the right positions in E^{-1} which is L.

Why should we want to find a proof? The reason is: A proof means that we have not just seen that pattern and believed it and liked it, but understood it.

The Great Factorization $A = LU$

Let me review the forward steps of elimination. They start with a matrix A and they end with an upper triangular matrix U. Every elimination step E_{ij} produces a lower triangular zero. Those steps E_{ij} subtract ℓ_{ij} times equation j from equation i below it. Row exchanges (permutations) are coming soon but not yet.

To invert one elimination step E_{ij}, we add instead of subtracting:

$$E_{31} = \begin{bmatrix} 1 & & \\ 0 & 1 & \\ -\ell_{31} & 0 & 1 \end{bmatrix} \quad \text{and} \quad L_{31} = E_{31}^{-1} = \begin{bmatrix} 1 & & \\ 0 & 1 & \\ \ell_{31} & 0 & 1 \end{bmatrix}.$$

Equation (10) in Section 2.2 multiplied $E_{32}E_{31}E_{21}$ with a messy result:

$$E = \begin{bmatrix} 1 & & \\ 0 & 1 & \\ 0 & -\ell_{32} & 1 \end{bmatrix} \begin{bmatrix} 1 & & \\ 0 & 1 & \\ -\ell_{31} & 0 & 1 \end{bmatrix} \begin{bmatrix} 1 & & \\ -\ell_{21} & 1 & \\ 0 & 0 & 1 \end{bmatrix} = \begin{bmatrix} 1 & & \\ -\ell_{21} & 1 & \\ (\ell_{32}\ell_{21} - \ell_{31}) & -\ell_{32} & 1 \end{bmatrix}.$$

Equation (11) showed how inverses (in reverse order $E_{21}^{-1}E_{31}^{-1}E_{32}^{-1}$) produced perfection:

$$\boldsymbol{E^{-1}} = \begin{bmatrix} 1 & & \\ \ell_{21} & 1 & \\ 0 & 0 & 1 \end{bmatrix} \begin{bmatrix} 1 & & \\ 0 & 1 & \\ \ell_{31} & 0 & 1 \end{bmatrix} \begin{bmatrix} 1 & & \\ 0 & 1 & \\ 0 & \ell_{32} & 1 \end{bmatrix} = \begin{bmatrix} \mathbf{1} & & \\ \boldsymbol{\ell_{21}} & \mathbf{1} & \\ \boldsymbol{\ell_{31}} & \boldsymbol{\ell_{32}} & \mathbf{1} \end{bmatrix} = \boldsymbol{L}$$

Then elimination $EA = U$ becomes $A = E^{-1}U = LU$ if we run it backward from U to A. These pages aim to show the same result for any matrix size n. The formula $A = LU$ is one of the great matrix factorizations in linear algebra.

Here is one way to understand why L has all the ℓ_{ij} in position, with no mix-up.

The key reason why A equals LU: Ask yourself about the pivot rows that are subtracted from lower rows. Are they the original rows of A? *No*, elimination probably changed them. Are they rows of U? *Yes*, the pivot rows never change again. When computing the third row of U, we subtract multiples of earlier rows of U (*not rows of* A!):

$$\text{Row 3 of } U = (\text{Row 3 of } A) - \ell_{31}(\text{Row 1 of } U) - \ell_{32}(\text{Row 2 of } U). \tag{3}$$

Rewrite this equation to see that the row $[\,\ell_{31} \quad \ell_{32} \quad 1\,]$ is multiplying the matrix U:

$$(\text{Row 3 of } A) = \ell_{31}(\text{Row 1 of } U) + \ell_{32}(\text{Row 2 of } U) + 1(\text{Row 3 of } U). \tag{4}$$

This is exactly row 3 of $A = LU$. That row of L holds ℓ_{31}, ℓ_{32}, 1. All rows look like this, whatever the size of A. With no row exchanges, we have $\boldsymbol{A = LU}$.

Second Proof of $A = LU$: Multiply Columns Times Rows

I would like to present another proof of $A = LU$. The idea is to see elimination as removing *one column of* L *times one row of* U *from* A. The problem becomes one size smaller.

Elimination begins with pivot row = row 1 of A. We multiply that pivot row by the numbers ℓ_{21} and ℓ_{31} and eventually ℓ_{n1}. Then we subtract from row 2 and row 3 and eventually row n of A. By choosing $\ell_{21} = a_{21}/a_{11}$ and $\ell_{31} = a_{31}/a_{11}$ and eventually $\ell_{n1} = a_{n1}/a_{11}$, the subtraction leaves zeros in column 1.

$$\text{Step 1 removes} \begin{bmatrix} 1\,(\text{row }1) \\ \ell_{21}(\text{row }1) \\ \ell_{31}(\text{row }1) \\ \ell_{41}(\text{row }1) \end{bmatrix} \text{ from } A \text{ to leave } \boldsymbol{A_2} = \begin{bmatrix} 0 & 0 & 0 & 0 \\ 0 & \times & \times & \times \\ 0 & \times & \times & \times \\ 0 & \times & \times & \times \end{bmatrix}.$$

Key point: **We removed a rank-one matrix: column times row.** It was the column $\boldsymbol{\ell_1} = (1, \ell_{21}, \ell_{31}, \ell_{41})$ times row 1 of A—the first pivot row $\boldsymbol{u}_1$.

Now we face a similar problem for A_2. And we take a similar step to reach A_3 :

$$\text{Step 2 removes} \begin{bmatrix} 0\,(\text{row 2 of } A_2) \\ 1\,(\text{row 2 of } A_2) \\ \ell_{32}(\text{row 2 of } A_2) \\ \ell_{42}(\text{row 2 of } A_2) \end{bmatrix} \text{ from } A_2 \text{ to leave } \boldsymbol{A_3} = \begin{bmatrix} 0 & 0 & 0 & 0 \\ 0 & 0 & 0 & 0 \\ 0 & 0 & \times & \times \\ 0 & 0 & \times & \times \end{bmatrix}.$$

Row 2 of A_2 was the second pivot row = second row $\boldsymbol{u}_2$ of U. We removed a column $\boldsymbol{\ell}_2 = (0, 1, \ell_{32}, \ell_{42})$ times that second pivot row. Continuing in the same way, every step removes a column $\boldsymbol{\ell}_j$ times a pivot row $\boldsymbol{u}_j$ of U. Now put those pieces back together:

$$\boxed{\boldsymbol{A} = \boldsymbol{\ell}_1\boldsymbol{u}_1 + \boldsymbol{\ell}_2\boldsymbol{u}_2 + \cdots + \boldsymbol{\ell}_n\boldsymbol{u}_n = \begin{bmatrix} \boldsymbol{\ell}_1 \cdots \boldsymbol{\ell}_n \end{bmatrix} \begin{bmatrix} \boldsymbol{u}_1 \\ \vdots \\ \boldsymbol{u}_n \end{bmatrix} = \boldsymbol{LU}} \quad (5)$$

That last crucial step was column-row multiplication of L times U. This was introduced at the very end of Chapter 1. The Problem Set for this section will review this important way to multiply LU—by adding up rank-one matrices (columns of L times rows of U).

Notice that U is upper triangular. The pivot row $\boldsymbol{u}_k$ begins with $k - 1$ zeros. And L is lower triangular with 1's on its main diagonal. Column $\boldsymbol{\ell}_k$ also begins with $k - 1$ zeros.

Elimination Without Row Exchanges

The next section is going to allow row exchanges P_{ij}. They are necessary to move zeros out of the pivot positions. Before we go there, we can answer this basic question: *When is $A = LU$ possible with no row exchanges and no zeros in the pivots?*

Answer **All upper left $\boldsymbol{k}$ by $\boldsymbol{k}$ submatrices of $\boldsymbol{A}$ must be invertible** (sizes $k = 1$ to n).

The reason is that elimination is also factoring every one of those submatrices (k by k corners of A). All those corner matrices A_k agree with L_kU_k (k by k corners of L and U):

$$\begin{bmatrix} \boldsymbol{A_k} & * \\ * & * \end{bmatrix} = \begin{bmatrix} \boldsymbol{L_k} & \mathbf{0} \\ * & * \end{bmatrix} \begin{bmatrix} \boldsymbol{U_k} & * \\ \mathbf{0} & * \end{bmatrix} \quad \text{tells us that } \boldsymbol{A_k = L_kU_k}. \qquad (6)$$

Problem Set 2.3

Problems 1–8 compute the factorization $\boldsymbol{A = LU}$ (and also $A = LDU$).

1 (Important) Forward elimination changes $\left[\begin{smallmatrix} 1 & 1 \\ 1 & 2 \end{smallmatrix}\right]\boldsymbol{x} = \boldsymbol{b}$ to a triangular $\left[\begin{smallmatrix} 1 & 1 \\ 0 & 1 \end{smallmatrix}\right]\boldsymbol{x} = \boldsymbol{c}$:

$$\begin{matrix} x + \ \ y = \mathbf{5} \\ x + 2y = \mathbf{7} \end{matrix} \quad \longrightarrow \quad \begin{matrix} x + y = \mathbf{5} \\ y = \mathbf{2} \end{matrix} \qquad \begin{bmatrix} 1 & 1 & \mathbf{5} \\ 1 & 2 & \mathbf{7} \end{bmatrix} \longrightarrow \begin{bmatrix} 1 & 1 & \mathbf{5} \\ 0 & 1 & \mathbf{2} \end{bmatrix}$$

That step subtracted $\ell_{21} =$ _____ times row 1 from row 2. The reverse step *adds* ℓ_{21} times row 1 to row 2. The matrix for that reverse step is $L =$ _____. Multiply this L times the triangular system $\left[\begin{smallmatrix} 1 & 1 \\ 0 & 1 \end{smallmatrix}\right]\boldsymbol{x}_1 = \left[\begin{smallmatrix} 5 \\ 2 \end{smallmatrix}\right]$ to get _____ $=$ _____. In letters, L multiplies $U\boldsymbol{x} = \boldsymbol{c}$ to give _____.

2 Write down the 2 by 2 triangular systems $L\boldsymbol{c} = \boldsymbol{b}$ and $U\boldsymbol{x} = \boldsymbol{c}$ from Problem 1. Check that $\boldsymbol{c} = (5, 2)$ solves the first one. Find $\boldsymbol{x}$ that solves the second one.

3 What matrix E puts A into triangular form $EA = U$? Multiply by $E^{-1} = L$ to factor A into LU:

$$A = \begin{bmatrix} 2 & 1 & 0 \\ 0 & 4 & 2 \\ 6 & 3 & 5 \end{bmatrix}.$$

4 What two elimination matrices E_{21} and E_{32} put A into upper triangular form $E_{32}E_{21}A = U$? Multiply by E_{32}^{-1} and E_{21}^{-1} to factor A into $LU = E_{21}^{-1}E_{32}^{-1}U$:

$$A = \begin{bmatrix} 1 & 1 & 1 \\ 2 & 4 & 5 \\ 0 & 4 & 0 \end{bmatrix}.$$

5 What three elimination matrices E_{21}, E_{31}, E_{32} put A into its upper triangular form $E_{32}E_{31}E_{21}A = U$? Multiply by E_{32}^{-1}, E_{31}^{-1} and E_{21}^{-1} to factor A into L times U:

$$A = \begin{bmatrix} 1 & 0 & 1 \\ 2 & 2 & 2 \\ 3 & 4 & 5 \end{bmatrix} \qquad L = E_{21}^{-1}E_{31}^{-1}E_{32}^{-1}.$$

6 A and B are symmetric across the diagonal (because $4 = 4$). Find their triple factorizations LDU and say how U is related to L for these symmetric matrices:

Symmetric $$A = \begin{bmatrix} 2 & 4 \\ 4 & 11 \end{bmatrix} \quad \text{and} \quad B = \begin{bmatrix} 1 & 4 & 0 \\ 4 & 12 & 4 \\ 0 & 4 & 0 \end{bmatrix}.$$

7 (*Recommended*) Compute L and U for the symmetric matrix A:

$$A = \begin{bmatrix} a & a & a & a \\ a & b & b & b \\ a & b & c & c \\ a & b & c & d \end{bmatrix}.$$

Find four conditions on a, b, c, d to get $A = LU$ with four pivots.

8 This nonsymmetric matrix will have the same L as in Problem **10**:

Find L and U for $$A = \begin{bmatrix} a & r & r & r \\ a & b & s & s \\ a & b & c & t \\ a & b & c & d \end{bmatrix}.$$

Find the four conditions on a, b, c, d, r, s, t to get $A = LU$ with four pivots.

9 Solve the triangular system $L\boldsymbol{c} = \boldsymbol{b}$ to find $\boldsymbol{c}$. Then solve $U\boldsymbol{x} = \boldsymbol{c}$ to find $\boldsymbol{x}$:

$$L = \begin{bmatrix} 1 & 0 \\ 4 & 1 \end{bmatrix} \quad \text{and} \quad U = \begin{bmatrix} 2 & 4 \\ 0 & 1 \end{bmatrix} \quad \text{and} \quad \boldsymbol{b} = \begin{bmatrix} 2 \\ 11 \end{bmatrix}.$$

For safety multiply LU and solve $A\boldsymbol{x} = \boldsymbol{b}$ as usual. Circle $\boldsymbol{c}$ when you see it.

10 Solve $L\boldsymbol{c} = \boldsymbol{b}$ to find $\boldsymbol{c}$. Then solve $U\boldsymbol{x} = \boldsymbol{c}$ to find $\boldsymbol{x}$. **What was** A?

$$L = \begin{bmatrix} 1 & 0 & 0 \\ 1 & 1 & 0 \\ 1 & 1 & 1 \end{bmatrix} \quad \text{and} \quad U = \begin{bmatrix} 1 & 1 & 1 \\ 0 & 1 & 1 \\ 0 & 0 & 1 \end{bmatrix} \quad \text{and} \quad \boldsymbol{b} = \begin{bmatrix} 4 \\ 5 \\ 6 \end{bmatrix}.$$

11 (a) When you apply the usual elimination steps to L, what matrix do you reach?

$$L = \begin{bmatrix} 1 & 0 & 0 \\ \ell_{21} & 1 & 0 \\ \ell_{31} & \ell_{32} & 1 \end{bmatrix}.$$

(b) When you apply the same steps to I, what matrix do you get ?

(c) When you apply the same steps to LU, what matrix do you get ?

12 If $A = LDU$ and also $A = L_1D_1U_1$ with all factors invertible, then $L = L_1$ and $D = D_1$ and $U = U_1$. *"The three factors are unique."*

Derive the equation $L_1^{-1}LD = D_1U_1U^{-1}$. Are the two sides triangular or diagonal? Deduce $L = L_1$ and $U = U_1$ (they all have diagonal 1's). Then $D = D_1$.

13 *Tridiagonal matrices* have zero entries except on the main diagonal and the two adjacent diagonals. Factor these into $A = LU$ and $A = LDL^{\mathrm{T}}$:

$$A = \begin{bmatrix} 1 & 1 & 0 \\ 1 & 2 & 1 \\ 0 & 1 & 2 \end{bmatrix} \quad \text{and} \quad A = \begin{bmatrix} a & a & 0 \\ a & a+b & b \\ 0 & b & b+c \end{bmatrix}.$$

14 If A and B have nonzeros in the positions marked by x, which zeros (marked by 0) *stay zero* in their factors L and U?

$$A = \begin{bmatrix} x & x & x & x \\ x & x & x & 0 \\ 0 & x & x & x \\ 0 & 0 & x & x \end{bmatrix} \qquad B = \begin{bmatrix} x & x & x & 0 \\ x & x & 0 & x \\ x & 0 & x & x \\ 0 & x & x & x \end{bmatrix}.$$

15 *Easy but important.* If A has pivots $5, 9, 3$ with no row exchanges, what are the pivots for the upper left 2 by 2 submatrix A_2 (without row 3 and column 3 of A) ?

16 Following the second proof of $A = LU$, what three rank 1 matrices add to A ?

$$A = \begin{bmatrix} 1 & 2 & 0 \\ 1 & 5 & 1 \\ 0 & 6 & 4 \end{bmatrix} = \boldsymbol{\ell}_1\boldsymbol{u}_1 + \boldsymbol{\ell}_2\boldsymbol{u}_2 + \boldsymbol{\ell}_3\boldsymbol{u}_3 = LU\,! \quad \text{columns multiply rows}$$

17 Multiply $L^{\mathrm{T}}L$ and LL^{T} by columns times rows when the 3 by 3 lower triangular L has six 1's.

2.4 Permutations and Transposes

1 A **permutation matrix** P has the same rows as I (in any order). There are $n\,!$ different orders.

2 Then Px puts the components $x_1, x_2, \ldots, x_n$ in that new order. And **P^{T} equals P^{-1}**.

3 **Columns of A are rows of A^{T}. The transposes of Ax and AB are $x^{\mathrm{T}}A^{\mathrm{T}}$ and $B^{\mathrm{T}}A^{\mathrm{T}}$.**

4 The idea behind A^{T} is that $Ax \cdot y$ equals $x \cdot A^{\mathrm{T}}y$ because $(Ax)^{\mathrm{T}}y = x^{\mathrm{T}}A^{\mathrm{T}}y = x^{\mathrm{T}}(A^{\mathrm{T}}y)$.

5 A **symmetric matrix** has $S^{\mathrm{T}} = S$. The product $S = A^{\mathrm{T}}A$ is always symmetric.

Permutations

Permutation matrices have a 1 in every row and a 1 in every column. All other entries are zero. When this matrix P multiplies a vector, it changes the order of its components:

Circular shift of x
$1, 2, 3$ to $3, 1, 2$

$$Px = \begin{bmatrix} 0 & 0 & 1 \\ 1 & 0 & 0 \\ 0 & 1 & 0 \end{bmatrix} \begin{bmatrix} x_1 \\ x_2 \\ x_3 \end{bmatrix} = \begin{bmatrix} x_3 \\ x_1 \\ x_2 \end{bmatrix}$$

For this example, P shifts x_1 down to second position. If we repeat, P^2 shifts x_1 to third position. Then $P^3 = I$ and we recover the original order x_1, x_2, x_3.

Shuffling a deck of cards would be a permutation of size $n = 52$. Persi Diaconis proved that a new deck is well shuffled after 7 random permutations.

There are $3! = 6$ permutation matrices of size 3, and $4! = 24$ permutations of size 4. Here are specific tasks for these 4 by 4 permutations, when they multiply a vector x:

Reverse the order to x_4, x_3, x_2, x_1

$$\begin{bmatrix} 0 & 0 & 0 & \mathbf{1} \\ 0 & 0 & \mathbf{1} & 0 \\ 0 & \mathbf{1} & 0 & 0 \\ \mathbf{1} & 0 & 0 & 0 \end{bmatrix}$$

Even x_0, x_2 before odd x_1, x_3 in the Fast Fourier Transform

$$\begin{bmatrix} \mathbf{1} & 0 & 0 & 0 \\ 0 & 0 & \mathbf{1} & 0 \\ 0 & \mathbf{1} & 0 & 0 \\ 0 & 0 & 0 & \mathbf{1} \end{bmatrix}$$

Circular shift Move x_1, x_2, x_3, x_4 to x_4, x_1, x_2, x_3

$$\begin{bmatrix} 0 & 0 & 0 & \mathbf{1} \\ \mathbf{1} & 0 & 0 & 0 \\ 0 & \mathbf{1} & 0 & 0 \\ 0 & 0 & \mathbf{1} & 0 \end{bmatrix}$$

Exchange rows 2 and 3 Exchange again to reset $1, 2, 3, 4$

$$\begin{bmatrix} \mathbf{1} & 0 & 0 & 0 \\ 0 & 0 & \mathbf{1} & 0 \\ 0 & \mathbf{1} & 0 & 0 \\ 0 & 0 & 0 & \mathbf{1} \end{bmatrix}$$

Half of the $n\,!$ permutations of size n are "even" and half are "odd". An even permutation needs an even number of simple row exchanges to reach the matrix I. The last example (1 exchange) was odd. The first example (exchange 1 and 4, exchange 2 and 3) was even.

The rows of P are
the columns of P^{-1}
$P^{-1} =$ transpose of P

$$P^{-1}P = \begin{bmatrix} 0 & \mathbf{1} & 0 \\ 0 & 0 & \mathbf{1} \\ \mathbf{1} & 0 & 0 \end{bmatrix} \begin{bmatrix} 0 & 0 & \mathbf{1} \\ \mathbf{1} & 0 & 0 \\ 0 & \mathbf{1} & 0 \end{bmatrix} = \begin{bmatrix} \mathbf{1} & & \\ & \mathbf{1} & \\ & & \mathbf{1} \end{bmatrix} = I$$

Properties of Permutation Matrices

1. The n 1's appear in n different rows and n different columns of P.
2. The columns of P are *orthogonal*: dot products between columns are all zero.
3. The product P_1P_2 of permutations is also a permutation.
4. If A is invertible, there is a permutation P to order its rows in advance, so that elimination on PA meets **no zeros in the pivot positions**. Then $\boldsymbol{PA = LU}$.

The $PA = LU$ Factorization: Row Exchanges from P

An example will show how elimination can often succeed, even when a zero appears in the pivot position. Suppose elimination starts with 1 as the first pivot. Subtracting 2 times row 1 produces 0 as an unacceptable second pivot:

$$\boldsymbol{A} = \begin{bmatrix} 1 & 2 & a \\ 2 & 4 & b \\ 3 & 7 & c \end{bmatrix} \rightarrow \begin{bmatrix} 1 & 2 & a \\ 0 & \mathbf{0} & b-2a \\ 0 & 1 & c-3a \end{bmatrix} \xrightarrow{\boldsymbol{P}} \begin{bmatrix} 1 & 2 & a \\ 0 & 1 & c-3a \\ 0 & 0 & b-2a \end{bmatrix} = \boldsymbol{U}$$

In spite of this zero, A is probably invertible. To rescue elimination, P will exchange *row* 2 *with row* 3. That brings 1 into the second pivot as shown. So we can continue.

This matrix A is invertible if and only if $b - 2a$ is not zero in the third pivot. Notice that if $b = 2a$, then row 2 of A equals 2 (row 1 of A). In that case, A is surely not invertible.

We can exchange rows 2 and 3 *first* to get PA. Then LU factorization becomes $\boldsymbol{PA = LU}$. The matrix PA sails through elimination without seeing that zero pivot.

$$\underset{\boldsymbol{P}}{\begin{bmatrix} 1 & 0 & 0 \\ 0 & 0 & 1 \\ 0 & 1 & 0 \end{bmatrix}} \underset{\boldsymbol{A}}{\begin{bmatrix} 1 & 2 & a \\ 2 & 4 & b \\ 3 & 7 & c \end{bmatrix}} = \underset{\boldsymbol{PA}}{\begin{bmatrix} 1 & 2 & a \\ 3 & 7 & c \\ 2 & 4 & b \end{bmatrix}} = \underset{\boldsymbol{L}}{\begin{bmatrix} 1 & 0 & 0 \\ 3 & 1 & 0 \\ 2 & 0 & 1 \end{bmatrix}} \underset{\boldsymbol{U}}{\begin{bmatrix} 1 & 2 & a \\ 0 & 1 & c-3a \\ 0 & 0 & b-2a \end{bmatrix}}$$

In principle we might need several row exchanges. Then the overall permutation P includes them all, and still produces $PA = LU$. Daniel Drucker showed me a neat way to keep track of P, by adding a special column to the matrix A. That column tracks the original row numbers, as rows are exchanged. If we do exchanges on that column also, the final permutation P is easy to see. The same example has one row exchange in P.

$$\begin{bmatrix} 1 & 2 & a & \mathbf{1} \\ 2 & 4 & b & \mathbf{2} \\ 3 & 7 & c & \mathbf{3} \end{bmatrix} \rightarrow \begin{bmatrix} 1 & 2 & a & \mathbf{1} \\ 0 & 0 & b-2a & \mathbf{2} \\ 0 & 1 & c-3a & \mathbf{3} \end{bmatrix} \boldsymbol{\rightarrow} \begin{bmatrix} 1 & 2 & a & \mathbf{1} \\ 0 & 1 & c-3a & \mathbf{3} \\ 0 & 0 & b-2a & \mathbf{2} \end{bmatrix}. \ \boldsymbol{P_{132}} \text{ is } \begin{bmatrix} \mathbf{1} & 0 & 0 \\ 0 & 0 & \mathbf{1} \\ 0 & \mathbf{1} & 0 \end{bmatrix}$$

"Partial Pivoting" to Reduce Roundoff Errors

Every good code for elimination allows for extra row exchanges that add safety. **Small pivots are unsafe!** The code does not blindly accept a small pivot when a larger number appears below it (in the same column). The computation is more stable if we exchange those rows, to produce the *largest possible number in the pivot*.

This example had first pivot equal to 1, but column 1 offered larger numbers 2 and 3. **The code will choose the largest number 3 as the first pivot: exchange rows 1 and 3.** The order of rows is tracked by the last column—that column is not part of the matrix.

$$\begin{bmatrix} 1 & 2 & a & \mathbf{1} \\ 2 & 4 & b & \mathbf{2} \\ 3 & 7 & c & \mathbf{3} \end{bmatrix} \xrightarrow{P} \begin{bmatrix} 3 & 7 & c & \mathbf{3} \\ 2 & 4 & b & \mathbf{2} \\ 1 & 2 & a & \mathbf{1} \end{bmatrix} \rightarrow \begin{bmatrix} 3 & 7 & c & \mathbf{3} \\ 0 & -\frac{2}{3} & b-\frac{2}{3}c & \mathbf{2} \\ 0 & -\frac{1}{3} & a-\frac{1}{3}c & \mathbf{1} \end{bmatrix} \rightarrow \begin{bmatrix} 3 & 7 & c & \mathbf{3} \\ 0 & -\frac{2}{3} & b-\frac{2}{3}c & \mathbf{2} \\ 0 & 0 & a-\frac{1}{2}b & \mathbf{1} \end{bmatrix}$$

$$\text{This example has } P = \begin{bmatrix} 0 & 0 & \mathbf{1} \\ 0 & \mathbf{1} & 0 \\ \mathbf{1} & 0 & 0 \end{bmatrix} \text{ and } L = \begin{bmatrix} 1 & 0 & 0 \\ \mathbf{\frac{2}{3}} & 1 & 0 \\ \mathbf{\frac{1}{3}} & \mathbf{\frac{1}{2}} & 1 \end{bmatrix} \quad \text{with } PA = LU.$$

All entries of L are ≤ 1 when each pivot is larger than all the numbers below it.

Fast Fourier Transform

An early example in this section was the "evens before odds permutation". This comes into every step of the Fast Fourier Transform (**FFT**). That step reduces a transform $F\boldsymbol{x}$ to two transforms of half size. The Discrete Fourier Transform is multiplication by F. The FFT may be the most important algorithm in computational science! Step 1 reduces the Fourier matrix F_{1024} with 1024^2 nonzeros to two multiplications by F_{512} (half size):

$$\boldsymbol{F}_{1024} = \begin{bmatrix} \boldsymbol{I} & \boldsymbol{D} \\ \boldsymbol{I} & -\boldsymbol{D} \end{bmatrix} \begin{bmatrix} \boldsymbol{F}_{512} & \mathbf{0} \\ \mathbf{0} & \boldsymbol{F}_{512} \end{bmatrix} \begin{bmatrix} \boldsymbol{P} \end{bmatrix} \qquad P = \begin{bmatrix} \text{rows } 0, 2, 4, 8, \ldots \text{ of } I \\ \text{rows } 1, 3, 5, 7, \ldots \text{ of } I \end{bmatrix}$$

Those zero submatrices cut the computational work in half (plus the small work of the diagonal matrix D and the permutation P). Transform the evens and transform the odds. Then the key idea is **recursion: Reduce $\boldsymbol{F}_{512}$ in the same way to two copies of $\boldsymbol{F}_{256}$.**

As the recursion continues to F_{128} and F_{64} and onwards, the only multiplications will involve the diagonal D's. The permutations combine into one overall P = product of an even-odd permutation at every step. The number of steps from 1024 to 1 is $\log_2 1024 = 10$ (because $2^{10} = 1024$). Every step costs 512 multiplications from the D's. So the total cost of the FFT is proportional to $\frac{1}{2}\boldsymbol{N} \log_2 \boldsymbol{N}$ instead of N^2.

For $N = 1024$, the direct multiplication by F needs $(1024)^2 \approx 1$ million steps. The FFT way reduces 1 million to $\frac{1}{2}(1024)(10) = 5120$ multiplications. That difference makes whole industries successful in dealing with data.

The Transpose of A

We need one more matrix, and fortunately it is much simpler than the inverse. It is the ***"transpose"*** of A, which is denoted by A^{T}. *The columns of A^{T} are the rows of A.*

When A is an m by n matrix, the transpose is n by m. 3 by 2 becomes 2 by 3.

Transpose If $A = \begin{bmatrix} \mathbf{1} & \mathbf{2} & \mathbf{3} \\ 0 & 0 & 4 \end{bmatrix}$ then $A^{\mathrm{T}} = \begin{bmatrix} \mathbf{1} & 0 \\ \mathbf{2} & 0 \\ \mathbf{3} & 4 \end{bmatrix}$.

You can write the rows of A into the columns of A^{T}. Or you can write the columns of A into the rows of A^{T}. The matrix "flips over" its main diagonal. The entry in row i, column j of A^{T} comes from row j, column i of the original A:

Exchange rows and columns $\boxed{(\boldsymbol{A}^{\mathrm{T}})_{ij} = \boldsymbol{A}_{ji}}$

The transpose of a lower triangular matrix is upper triangular. The transpose of A^{T} is A.

Note MATLAB's symbol for A^{T} is A'. Typing $[1 \ \ 2 \ \ 3]$ gives a row vector and the column vector is $\boldsymbol{v} = [1 \ \ 2 \ \ 3]'$. The matrix M with second column $\boldsymbol{w} = [\ 4\ 5\ 6\ ]'$ is $M = [\ \boldsymbol{v}\ \ \boldsymbol{w}\ ]$. Quicker to enter by rows and transpose: $M = [1 \ \ 2 \ \ 3; \ \ 4 \ \ 5 \ \ 6]'$.

The rules for transposes are very direct. We can transpose $A + B$ to get $(A + B)^{\mathrm{T}}$. Or we can transpose A and B separately, and then add $A^{\mathrm{T}} + B^{\mathrm{T}}$—with the same result.

The serious questions are about the transpose of a product AB and an inverse A^{-1}:

Sum	The transpose of	$A + B$	is	$A^{\mathrm{T}} + B^{\mathrm{T}}$.	(1)
Product	The transpose of	AB	is	$(\boldsymbol{AB})^{\mathrm{T}} = \boldsymbol{B}^{\mathrm{T}}\boldsymbol{A}^{\mathrm{T}}$.	(2)
Inverse	The transpose of	A^{-1}	is	$(A^{-1})^{\mathrm{T}} = (A^{\mathrm{T}})^{-1}$.	(3)

Notice especially how $B^{\mathrm{T}}A^{\mathrm{T}}$ comes in reverse order. For inverses, this reverse order was quick to check: $B^{-1}A^{-1}$ times AB produces I because $A^{-1}A = B^{-1}B = I$. To understand $(AB)^{\mathrm{T}} = B^{\mathrm{T}}A^{\mathrm{T}}$, start with $(A\boldsymbol{x})^{\mathrm{T}} = \boldsymbol{x}^{\mathrm{T}}A^{\mathrm{T}}$ when B is just a vector:

$A\boldsymbol{x}$ combines the columns of A while $\boldsymbol{x}^{\mathrm{T}}A^{\mathrm{T}}$ combines the rows of A^{T}.

It is the same combination of the same vectors! In A they are columns, in A^{T} they are rows. So the transpose of the column $A\boldsymbol{x}$ is the row $\boldsymbol{x}^{\mathrm{T}}A^{\mathrm{T}}$. That fits our formula $(A\boldsymbol{x})^{\mathrm{T}} = \boldsymbol{x}^{\mathrm{T}}A^{\mathrm{T}}$. Now we can prove $(AB)^{\mathrm{T}} = B^{\mathrm{T}}A^{\mathrm{T}}$, when B has several columns.

If B has two columns $\boldsymbol{x}_1$ and $\boldsymbol{x}_2$, apply the same idea to each column. The columns of AB are $A\boldsymbol{x}_1$ and $A\boldsymbol{x}_2$. Their transposes appear correctly in the rows of $B^{\mathrm{T}}A^{\mathrm{T}}$:

$$\text{Transposing } AB = \begin{bmatrix} & & \\ A\boldsymbol{x}_1 & A\boldsymbol{x}_2 & \cdots \\ & & \end{bmatrix} \text{ gives } \begin{bmatrix} \boldsymbol{x}_1^{\mathrm{T}}A^{\mathrm{T}} \\ \boldsymbol{x}_2^{\mathrm{T}}A^{\mathrm{T}} \\ \vdots \end{bmatrix} \text{ which is } B^{\mathrm{T}}A^{\mathrm{T}}. \quad (4)$$

The right answer $B^{\mathrm{T}}A^{\mathrm{T}}$ comes out a row at a time. Here are numbers in $(AB)^{\mathrm{T}} = B^{\mathrm{T}}A^{\mathrm{T}}$:

$$\boldsymbol{AB} = \begin{bmatrix} 1 & 0 \\ 1 & 1 \end{bmatrix} \begin{bmatrix} 5 & 0 \\ 4 & 1 \end{bmatrix} = \begin{bmatrix} \mathbf{5} & \mathbf{0} \\ \mathbf{9} & \mathbf{1} \end{bmatrix} \quad \text{and} \quad \boldsymbol{B}^{\mathbf{T}}\boldsymbol{A}^{\mathbf{T}} = \begin{bmatrix} 5 & 4 \\ 0 & 1 \end{bmatrix} \begin{bmatrix} 1 & 1 \\ 0 & 1 \end{bmatrix} = \begin{bmatrix} \mathbf{5} & \mathbf{9} \\ \mathbf{0} & \mathbf{1} \end{bmatrix}.$$

The reverse order rule extends to three or more factors: $(ABC)^{\mathrm{T}}$ equals $C^{\mathrm{T}}B^{\mathrm{T}}A^{\mathrm{T}}$.

Now apply this product rule by transposing both sides of $A^{-1}A = I$. On one side, I^{T} is I. We confirm the rule that $(A^{-1})^{\mathrm{T}}$ *is the inverse of* A^{T}.

$$\textbf{Transpose of inverse} \qquad A^{-1}A = I \quad \text{is transposed to} \quad A^{\mathrm{T}}(A^{-1})^{\mathrm{T}} = I. \qquad (5)$$

Similarly $AA^{-1} = I$ leads to $(A^{-1})^{\mathrm{T}}A^{\mathrm{T}} = I$. We can invert the transpose or we can transpose the inverse. Notice especially: A^{T} ***is invertible exactly when*** A ***is invertible***.

The inverse of $A = \begin{bmatrix} 1 & 0 \\ 6 & 1 \end{bmatrix}$ is $A^{-1} = \begin{bmatrix} 1 & 0 \\ -6 & 1 \end{bmatrix}$. The transpose is $A^{\mathrm{T}} = \begin{bmatrix} 1 & 6 \\ 0 & 1 \end{bmatrix}$.

$$(\boldsymbol{A}^{-1})^{\mathbf{T}} \quad \textbf{\textit{and}} \quad (\boldsymbol{A}^{\mathbf{T}})^{-1} \quad \textbf{\textit{are both equal to}} \quad \begin{bmatrix} 1 & -6 \\ 0 & 1 \end{bmatrix}.$$

The Meaning of Inner Products

The dot product (inner product) of $\boldsymbol{x}$ and $\boldsymbol{y}$ is the sum of numbers $x_i y_i$. Now we have a better way to write $\boldsymbol{x} \cdot \boldsymbol{y}$, without using that unprofessional dot. Use matrix notation $\boldsymbol{x}^{\mathrm{T}}\boldsymbol{y}$:

T is inside ***The dot product or inner product is*** $\boldsymbol{x}^{\mathbf{T}}\boldsymbol{y}$ $\qquad (1 \times n)(n \times 1) = \mathbf{1} \times \mathbf{1}$

T is outside ***The rank one product or outer product is*** $\boldsymbol{x}\boldsymbol{y}^{\mathbf{T}}$ $\quad (n \times 1)(1 \times n) = \boldsymbol{n} \times \boldsymbol{n}$

$\boldsymbol{x}^{\mathrm{T}}\boldsymbol{y}$ is a number, $\boldsymbol{x}\boldsymbol{y}^{\mathrm{T}}$ is a matrix. Quantum mechanics would write those as $< \boldsymbol{x}|\boldsymbol{y} >$ (inner) and $|\boldsymbol{x}><\boldsymbol{y}|$ (outer). Probably our universe is governed by linear algebra? Here are three more examples where the inner product has meaning:

From mechanics	Work = (Movements) (Forces) = $\boldsymbol{x}^{\mathrm{T}}\boldsymbol{f}$
From circuits	Heat loss = (Voltage drops) (Currents) = $\boldsymbol{e}^{\mathrm{T}}\boldsymbol{y}$
From economics	Income = (Quantities) (Prices) = $\boldsymbol{q}^{\mathrm{T}}\boldsymbol{p}$

We are really close to the heart of applied mathematics, and there is one more point to emphasize. It is the deeper connection between inner products and the transpose of A.

We defined A^{T} by flipping the matrix across its main diagonal. That's not mathematics. There is a better way. A^{T} *is the matrix that makes these inner products equal*:

$(\boldsymbol{Ax})^{\mathbf{T}}\boldsymbol{y} = \boldsymbol{x}^{\mathbf{T}}(\boldsymbol{A}^{\mathbf{T}}\boldsymbol{y})$ **Inner product of** Ax **with** y **= Inner product of** x **with** $\boldsymbol{A}^{\mathbf{T}}\boldsymbol{y}$

Example 1 Start with $A = \begin{bmatrix} -1 & 1 & 0 \\ 0 & -1 & 1 \end{bmatrix}$ $\quad \boldsymbol{x} = \begin{bmatrix} x_1 \\ x_2 \\ x_3 \end{bmatrix}$ $\quad \boldsymbol{y} = \begin{bmatrix} y_1 \\ y_2 \end{bmatrix}$

On one side we have $A\boldsymbol{x}$ multiplying $\boldsymbol{y}$ to produce $(x_2 - x_1)\, y_1 + (x_3 - x_2)\, y_2$ That is the same as $x_1\,(-y_1) + x_2\,(y_1 - y_2) + x_3\,(y_2)$. Now $\boldsymbol{x}$ is multiplying $A^{\mathrm{T}}\boldsymbol{y}$.

$$A^{\mathrm{T}}\boldsymbol{y} = \begin{bmatrix} -1 & 0 \\ 1 & -1 \\ 0 & 1 \end{bmatrix} \begin{bmatrix} y_1 \\ y_2 \end{bmatrix} = \begin{bmatrix} -y_1 \\ y_1 - y_2 \\ y_2 \end{bmatrix} \qquad \begin{matrix} (\boldsymbol{Ax})^{\mathrm{T}}\boldsymbol{y} \\ = \\ \boldsymbol{x}^{\mathrm{T}}(\boldsymbol{A}^{\mathrm{T}}\boldsymbol{y}) \end{matrix}$$

Example 2 Will you allow a little calculus? It is important or I wouldn't leave linear algebra. (This is linear algebra for functions.) **Change the matrix to a derivative**: $\boldsymbol{A = d/dt}$. The transpose of d/dt comes from $(\boldsymbol{Ax})^{\mathrm{T}}\boldsymbol{y} = \boldsymbol{x}^{\mathrm{T}}(\boldsymbol{A}^{\mathrm{T}}\boldsymbol{y})$.

First, the dot product $\boldsymbol{x}^{\mathrm{T}}\boldsymbol{y}$ changes from $x_1y_1 + \cdots + x_ny_n$ to an *integral of* $x(t)y(t)$.

Inner product of functions x and y

$$x^{\mathrm{T}}y = (x, y) = \int_{-\infty}^{\infty} x(t)\, y(t)\, dt \quad \text{by definition}$$

Transpose rule for functions $(A\boldsymbol{x})^{\mathrm{T}}\boldsymbol{y} = \boldsymbol{x}^{\mathrm{T}}(A^{\mathrm{T}}\boldsymbol{y})$

$$\int_{-\infty}^{\infty} \frac{dx}{dt}\, y(t)\, dt = \int_{-\infty}^{\infty} x(t) \left(-\frac{dy}{dt}\right) dt \tag{6}$$

I hope you recognize "***integration by parts***". The derivative moves from the first function $x(t)$ to the second function $y(t)$. During that move, a minus sign appears. This tells us that ***the transpose of*** $\boldsymbol{A = d/dt}$ ***is*** $\boldsymbol{A^{\mathrm{T}} = -A = -d/dt}$.

The derivative is *anti-symmetric*. Symmetric matrices have $A^{\mathrm{T}} = A$, anti-symmetric matrices have $A^{\mathrm{T}} = -A$. In some way, the 2 by 3 difference matrix in Example 1 followed this pattern. The 3 by 2 matrix A^{T} was *minus* a difference matrix. It produced $y_1 - y_2$ in the middle component of $A^{\mathrm{T}}\boldsymbol{y}$ instead of the difference $y_2 - y_1$.

Integration by parts is deceptively important and not just a trick.

Symmetric Matrices

For a *symmetric matrix*, transposing A to A^{T} produces no change. In this case A^{T} equals A. Its (j, i) entry across the main diagonal equals its (i, j) entry. In my opinion, these are the most important matrices of all. We give symmetric matrices the special letter S.

A symmetric matrix has $\boldsymbol{S^{\mathrm{T}} = S}$. This means that every $\boldsymbol{s_{ji} = s_{ij}}$.

Symmetric matrices $\quad S = \begin{bmatrix} 1 & 2 \\ 2 & 5 \end{bmatrix} = S^{\mathrm{T}}$ and $D = \begin{bmatrix} 1 & 0 \\ 0 & 10 \end{bmatrix} = D^{\mathrm{T}}$.

The inverse of a symmetric matrix is a symmetric matrix. The transpose of S^{-1} is $(S^{-1})^{\mathrm{T}} = (S^{\mathrm{T}})^{-1} = S^{-1}$. When S is invertible, this says that S^{-1} is symmetric.

Symmetric inverses $\quad S^{-1} = \begin{bmatrix} 5 & -2 \\ -2 & 1 \end{bmatrix}$ and $D^{-1} = \begin{bmatrix} 1 & 0 \\ 0 & 0.1 \end{bmatrix}$.

Now we produce a symmetric matrix S by ***multiplying any matrix*** $\boldsymbol{A}$ ***by*** $\boldsymbol{A^{\mathrm{T}}}$.

Symmetric Products A^TA and AA^T and LDL^T

Choose any matrix A, probably rectangular. Multiply A^T times A. Then the product $S = A^TA$ is automatically a square symmetric matrix:

$$\textit{The transpose of}\quad A^TA \quad \textit{is} \quad A^T(A^T)^T \quad \textit{which is} \quad A^TA \textit{ again.} \tag{7}$$

The matrix AA^T is also symmetric. (The shapes of A and A^T allow multiplication.) But AA^T *is a different matrix from* A^TA. In our experience, most scientific problems that start with a rectangular matrix A end up with A^TA or AA^T or both. As in least squares.

Example 3 Multiply $A = \begin{bmatrix} -1 & 1 & 0 \\ 0 & -1 & 1 \end{bmatrix}$ times $A^T = \begin{bmatrix} -1 & 0 \\ 1 & -1 \\ 0 & 1 \end{bmatrix}$ in both orders.

$$AA^T = \begin{bmatrix} 2 & -1 \\ -1 & 2 \end{bmatrix} \text{ and } A^TA = \begin{bmatrix} 1 & -1 & 0 \\ -1 & 2 & -1 \\ 0 & -1 & 1 \end{bmatrix} \text{ are both symmetric matrices.}$$

The product AA^T is m by m. In the opposite order, A^TA is n by n. Both are symmetric, with positive diagonal (*why*?). But even if $m = n$, it is very likely that $A^TA \neq AA^T$.

Symmetric matrices in elimination $S^T = S$ makes elimination twice as fast, because we can work with half the matrix (plus the diagonal). ***The symmetry is in the triple product*** $\boldsymbol{S = LDL^T}$. The diagonal matrix D of pivots can be divided out, to leave $U = L^T$.

$$\begin{bmatrix} 1 & 2 \\ 2 & 7 \end{bmatrix} = \begin{bmatrix} 1 & 0 \\ 2 & 1 \end{bmatrix} \begin{bmatrix} 1 & 2 \\ 0 & 3 \end{bmatrix} \qquad \begin{matrix} LU \text{ misses the symmetry of } S \\ \text{Divide the pivots } 1,3 \text{ out of } U \end{matrix}$$

$$\begin{bmatrix} 1 & 2 \\ 2 & 7 \end{bmatrix} = \begin{bmatrix} \mathbf{1} & 0 \\ \mathbf{2} & \mathbf{1} \end{bmatrix} \begin{bmatrix} 1 & 0 \\ 0 & 3 \end{bmatrix} \begin{bmatrix} \mathbf{1} & \mathbf{2} \\ 0 & \mathbf{1} \end{bmatrix} \qquad \begin{matrix} \boldsymbol{S = LDL^T} \textbf{ captures the symmetry} \\ \textbf{\textit{Now U is the transpose of L}} \end{matrix}$$

For a rectangular A, this ***saddle-point matrix*** $\boldsymbol{S}$ is symmetric and important:

$$\begin{matrix}\textbf{Block matrix} \\ \textbf{from least squares}\end{matrix} \qquad \boldsymbol{S} = \begin{bmatrix} \boldsymbol{I} & \boldsymbol{A} \\ \boldsymbol{A^T} & \mathbf{0} \end{bmatrix} = \boldsymbol{S^T} \text{ has size } m + n.$$

Apply block elimination to find a **block factorization** $S = LDL^T$. Then test invertibility:

$$\boldsymbol{S} \textbf{\textit{ is invertible}} \iff \boldsymbol{A^TA} \textbf{\textit{ is invertible}} \iff \boldsymbol{Ax \neq 0} \textbf{\textit{ whenever }} \boldsymbol{x \neq 0}$$

$$\begin{matrix}\textbf{Block elimination} \\ \textbf{Subtract } \boldsymbol{A^T}\textbf{(row 1)}\end{matrix} \qquad S = \begin{bmatrix} I & A \\ A^T & 0 \end{bmatrix} \text{ goes to } \begin{bmatrix} I & A \\ 0 & -\boldsymbol{A^TA} \end{bmatrix}. \text{ This is } U.$$

The block pivot matrix D contains I and $-A^TA$. Then L and L^T contain A^T and A:

$$\textbf{Block factorization} \quad \boldsymbol{S = LDL^T} = \begin{bmatrix} I & 0 \\ A^T & I \end{bmatrix} \begin{bmatrix} I & 0 \\ 0 & -A^TA \end{bmatrix} \begin{bmatrix} I & A \\ 0 & I \end{bmatrix}.$$

Problem Set 2.4 1-15

Questions 1–7 are about the rules for transpose matrices.

1 Find A^{T} and A^{-1} and $(A^{-1})^{\mathrm{T}}$ and $(A^{\mathrm{T}})^{-1}$ for $A = \begin{bmatrix} 1 & 0 \\ 9 & 3 \end{bmatrix}$ and $\begin{bmatrix} 1 & c \\ c & 0 \end{bmatrix}$.

2 Verify that $(AB)^{\mathrm{T}}$ equals $B^{\mathrm{T}}A^{\mathrm{T}}$ but those are different from $A^{\mathrm{T}}B^{\mathrm{T}}$:

$$A = \begin{bmatrix} 1 & 0 \\ 2 & 1 \end{bmatrix} \qquad B = \begin{bmatrix} 1 & 3 \\ 0 & 1 \end{bmatrix} \qquad AB = \begin{bmatrix} 1 & 3 \\ 2 & 7 \end{bmatrix}.$$

Show also that AA^{T} is different from $A^{\mathrm{T}}A$. But both of those matrices are ____.

3 (a) The matrix $((AB)^{-1})^{\mathrm{T}}$ comes from $(A^{-1})^{\mathrm{T}}$ and $(B^{-1})^{\mathrm{T}}$. *In what order?*

(b) If U is upper triangular then $(U^{-1})^{\mathrm{T}}$ is ____ triangular.

4 Show that $A^2 = 0$ is possible but $A^{\mathrm{T}}A = 0$ is not possible (unless $A =$ zero matrix).

5 (a) Compute the number $\boldsymbol{x}^{\mathrm{T}}A\boldsymbol{y} = \begin{bmatrix} 0 & 1 \end{bmatrix} \begin{bmatrix} 1 & 2 & 3 \\ 4 & 5 & 6 \end{bmatrix} \begin{bmatrix} 0 \\ 1 \\ 0 \end{bmatrix} =$ ____.

(b) This is the row $\boldsymbol{x}^{\mathrm{T}}A =$ ____ times the column $\boldsymbol{y} = (0, 1, 0)$.

(c) This is the row $\boldsymbol{x}^{\mathrm{T}} = \begin{bmatrix} 0 & 1 \end{bmatrix}$ times the column $A\boldsymbol{y} =$ ____.

6 The transpose of a block matrix $M = \begin{bmatrix} \mathbf{A} & \mathbf{B} \\ \mathbf{C} & \mathbf{D} \end{bmatrix}$ is $M^{\mathrm{T}} =$ ____. Test an example. Under what conditions on A, B, C, D is the block matrix symmetric?

7 True or false:

(a) The block matrix $\begin{bmatrix} \mathbf{0} & \mathbf{A} \\ \mathbf{A} & \mathbf{0} \end{bmatrix}$ is automatically symmetric.

(b) If A and B are symmetric then their product AB is symmetric.

(c) If A is not symmetric then A^{-1} is not symmetric.

(d) When A, B, C are symmetric, the transpose of ABC is CBA.

Questions 8–15 are about permutation matrices.

8 Why are there $n!$ permutation matrices of order n?

9 If P_1 and P_2 are permutation matrices, so is P_1P_2. This still has the rows of I in some order. Give examples with $P_1P_2 \neq P_2P_1$ and $P_3P_4 = P_4P_3$.

P1P2 exchange same pair of rows, P3 P4 exchange different

10 There are 12 "*even*" permutations of $(1, 2, 3, 4)$, with an *even number of exchanges*. Two of them are $(1, 2, 3, 4)$ with zero exchanges and $(4, 3, 2, 1)$ with 2 exchanges. List the other ten. Instead of writing each 4 by 4 matrix, just order the numbers.

11 If P has 1's on the antidiagonal from $(1, n)$ to $(n, 1)$, describe PAP. Note $P = P^{\mathrm{T}}$.

12 Explain why the dot product of $\boldsymbol{x}$ and $\boldsymbol{y}$ equals the dot product of $P\boldsymbol{x}$ and $P\boldsymbol{y}$. Then $(P\boldsymbol{x})^{\mathrm{T}}(P\boldsymbol{y}) = \boldsymbol{x}^{\mathrm{T}}\boldsymbol{y}$ tells us that $P^{\mathrm{T}}P = I$ for any permutation. With $\boldsymbol{x} = (1, 2, 3)$ and $\boldsymbol{y} = (1, 4, 2)$ choose P to show that $P\boldsymbol{x} \cdot \boldsymbol{y}$ is not always $\boldsymbol{x} \cdot P\boldsymbol{y}$.

13 Which permutation makes PA upper triangular? Which permutations make P_1AP_2 lower triangular? ***Multiplying*** A ***on the*** *right* ***by*** P_2 ***exchanges the*** _____ ***of*** A.

$$A = \begin{bmatrix} 0 & 0 & 6 \\ 1 & 2 & 3 \\ 0 & 4 & 5 \end{bmatrix}.$$

14 Find a 3 by 3 permutation matrix with $P^3 = I$ (but not $P = I$). Why can't P be the _____. Find a 4 by 4 permutation $\widehat{P}$ with $\widehat{P}^4 \neq I$.

15 All row exchange matrices are symmetric: $P^{\mathrm{T}} = P$. Then $P^{\mathrm{T}}P = I$ becomes $P^2 = I$. Other permutation matrices may or may not be symmetric.

(a) If P sends row 1 to row 4, then P^{T} sends row _____ to row _____. When $P^{\mathrm{T}} = P$ the row exchanges come in pairs with no overlap.

(b) Find a 4 by 4 example with $P^{\mathrm{T}} = P$ that moves all four rows.

Questions 16–18 are about symmetric matrices and their factorizations.

16 If $A = A^{\mathrm{T}}$ and $B = B^{\mathrm{T}}$, which of these matrices are certainly symmetric?

(a) $A^2 - B^2$ (b) $(A + B)(A - B)$ (c) ABA (d) $ABAB$.

17 (a) How many entries of S can be chosen independently, if $S = S^{\mathrm{T}}$ is 5 by 5 ?

(b) How do L and D (still 5 by 5) give the same number of choices in LDL^{T} ?

(c) How many entries can be chosen if A is *skew-symmetric* ? ($A^{\mathrm{T}} = -A$).

(d) Why does $A^{\mathrm{T}}A$ have no negative numbers on its diagonal ?

18 Factor these symmetric matrices into $S = LDL^{\mathrm{T}}$. The pivot matrix D is diagonal:

$$S = \begin{bmatrix} 1 & 3 \\ 3 & 2 \end{bmatrix} \quad \text{and} \quad S = \begin{bmatrix} 1 & b \\ b & c \end{bmatrix} \quad \text{and} \quad S = \begin{bmatrix} 2 & -1 & 0 \\ -1 & 2 & -1 \\ 0 & -1 & 2 \end{bmatrix}.$$

19 Find the $PA = LU$ factorizations (and check them) for

$$A = \begin{bmatrix} 0 & 1 & 1 \\ 1 & 0 & 1 \\ 2 & 3 & 4 \end{bmatrix} \quad \text{and} \quad A = \begin{bmatrix} 1 & 2 & 0 \\ 2 & 4 & 1 \\ 1 & 1 & 1 \end{bmatrix}.$$

20 Find a 4 by 4 permutation matrix (call it A) that needs 3 row exchanges to reach the end of elimination. For this matrix, what are its factors P, L, and U ?

21 Prove that the identity matrix cannot be the product of three row exchanges (or five). It can be the product of two exchanges (or four).

22 If every row of a 4 by 4 matrix contains the numbers $0, 1, 2, 3$ in some order, can the matrix be symmetric?

23 Start with 9 entries in a 3 by 3 matrix A. Prove that no reordering of rows and reordering of columns can produce A^{T}. (Watch the diagonal entries.)

24 Wires go between Boston, Chicago, and Seattle. Those cities are at voltages x_B, x_C, x_S. With unit resistances between cities, the currents between cities are in $\boldsymbol{y}$:

$$\boldsymbol{y} = A\boldsymbol{x} \quad \text{is} \quad \begin{bmatrix} y_{BC} \\ y_{CS} \\ y_{BS} \end{bmatrix} = \begin{bmatrix} 1 & -1 & 0 \\ 0 & 1 & -1 \\ 1 & 0 & -1 \end{bmatrix} \begin{bmatrix} x_B \\ x_C \\ x_S \end{bmatrix}.$$

(a) Find the total currents $A^{\mathrm{T}}\boldsymbol{y}$ out of the three cities.

(b) Verify that $(A\boldsymbol{x})^{\mathrm{T}}\boldsymbol{y}$ agrees with $\boldsymbol{x}^{\mathrm{T}}(A^{\mathrm{T}}\boldsymbol{y})$—six terms in both.

25 The matrix P that multiplies (x, y, z) to give (z, x, y) is also a rotation matrix. Find P and P^3. The rotation axis $\boldsymbol{a} = (1, 1, 1)$ doesn't move, it equals $P\boldsymbol{a}$. What is the angle of rotation from $\boldsymbol{v} = (2, 3, -5)$ to $P\boldsymbol{v} = (-5, 2, 3)$?

26 Here is a new factorization $A = LS =$ *triangular times symmetric*:

$$\text{Start from } A = LDU. \text{ Then } A \text{ equals } L\,(U^{\mathrm{T}})^{-1} \text{ times } S = U^{\mathrm{T}}DU.$$

Why is $L\,(U^{\mathrm{T}})^{-1}$ triangular? Why is $U^{\mathrm{T}}DU$ symmetric?

27 In algebra, a *group* of matrices includes AB and A^{-1} if it includes A and B. "*Products and inverses stay in the group.*" Which of these sets are groups?
Lower triangular matrices L with 1's on the diagonal, symmetric matrices S, positive matrices M, diagonal invertible matrices D, permutation matrices P, orthogonal matrices with $Q^{\mathrm{T}} = Q^{-1}$. ***Invent two more matrix groups.***

Challenge Problems

28 If you take powers of a permutation matrix, why is some P^k eventually equal to I?
Find a 5 by 5 permutation P so that the smallest power to equal I is P^6.

29 (a) Write down any 3 by 3 matrix M. Split M into $S + A$ where $S = S^{\mathrm{T}}$ is symmetric and $A = -A^{\mathrm{T}}$ is anti-symmetric.

(b) Find formulas for S and A involving M and M^{T}. We want $M = S + A$.

30 Suppose Q^{T} *equals* Q^{-1} (the transpose equals the inverse, so $Q^{\mathrm{T}}Q = I$).

(a) Show that the columns $\boldsymbol{q}_1, \ldots, \boldsymbol{q}_n$ are unit vectors: $\|\boldsymbol{q}_i\|^2 = 1$.

(b) Show that every two columns of Q are perpendicular: $\boldsymbol{q}_1^{\mathrm{T}}\boldsymbol{q}_2 = 0$.

(c) Find a 2 by 2 example with first entry $q_{11} = \cos\theta$.

3 The Four Fundamental Subspaces

Section 3.1 opens with a pure algebra question. *How do we define a "vector space"?* Looking at $\mathbf{R}^3$, the key operations are $\boldsymbol{v} + \boldsymbol{w}$ and $c\boldsymbol{v}$. They are connected by simple laws like $c(\boldsymbol{v} + \boldsymbol{w}) = c\boldsymbol{v} + c\boldsymbol{w}$. We must be able to add $\boldsymbol{v} + \boldsymbol{w}$, and multiply by c. Section 3.1 will give eight rules that the vectors $\boldsymbol{v}$ and the scalars c must satisfy.

Notice that $\boldsymbol{v}$ and $\boldsymbol{w}$ could be matrices! We can add matrices and we can add functions (and multiply by c). So we can have matrix spaces and function spaces. And inside $\mathbf{R}^n$, we could allow only vectors $\boldsymbol{x}$ that satisfy $A\boldsymbol{x} = \mathbf{0}$. That produces the "**nullspace of** $\boldsymbol{A}$". All combinations of those solutions are also solutions: *The nullspace is a subspace.*

Linear algebra gives us a way to solve $A\boldsymbol{x} = \mathbf{0}$. The best system is *elimination*: Simplify the equations to $R\boldsymbol{x} = \mathbf{0}$. Then find a "special solution" for each dependent column. Taking all their combinations is the crucial step to produce the nullspace.

Finally comes the idea of a **basis**: A set of vectors that perfectly describes the space. Their combinations give one and only one way to produce every vector in the space. The r independent columns of A are a basis for $\mathbf{C}(A)$. The $n - r$ special solutions to $A\boldsymbol{x} = \mathbf{0}$ are a basis for $\mathbf{N}(A)$. Those subspaces $\mathbf{C}$ and $\mathbf{N}$ have "*dimensions*" r and $n - r$.

Chapter 2 was about square invertible matrices. All four of the matrices in $PA = LU$ had full rank $r = m = n$. The column space and row space of A were the full space $\mathbf{R}^n$.

Chapter 3 moves to a higher level! It may be the most important chapter in the book. Every m by n matrix is allowed, and there will surely be nonzero solutions to $A\boldsymbol{x} = \mathbf{0}$ if $n > m$. Notice that this nullspace of A is not like the column space and row space. **The nullspace starts with equations $A\boldsymbol{x} = \mathbf{0}$, not with columns or rows from $\boldsymbol{A}$.**

The prime goal of this chapter is the *"Fundamental Theorem of Linear Algebra"*. In Section 3.5 this connects the four subspaces and their dimensions:

Column space of $\boldsymbol{A}$, row space of $\boldsymbol{A}$, nullspace of $\boldsymbol{A}$, nullspace of $\boldsymbol{A}^{\mathrm{T}}$.

I hope the picture that goes with it makes the Fundamental Theorem easy to remember.

3.1 Vector Spaces and Subspaces

1 All linear combinations $c\boldsymbol{v} + d\boldsymbol{w}$ must stay in the vector space.

2 The row space of A is "spanned" by the rows of A. The columns span $\mathbf{C}(A)$.

3 Matrices M_1 to M_N and functions f_1 to f_N span matrix spaces and function spaces.

Start with the vector spaces $\mathbf{R}^1, \mathbf{R}^2, \mathbf{R}^3, \ldots$ The space $\mathbf{R}^n$ contains all column vectors $\boldsymbol{v}$ of length n. The components v_1 to v_n are real numbers. (When complex numbers like $v_1 = 2+3i$ are allowed, the spaces become $\mathbf{C}^1, \mathbf{C}^2, \mathbf{C}^3, \ldots$). We know how to add vectors $\boldsymbol{v}$ and $\boldsymbol{w}$ in $\mathbf{R}^n$. We know how to multiply a vector by a number c or d to get $c\boldsymbol{v}$ or $d\boldsymbol{w}$. So we can find **linear combinations** $c\boldsymbol{v} + d\boldsymbol{w}$ in the vector space $\mathbf{R}^n$.

This operation of "linear combinations" is fundamental for any vector space. It must satisfy eight rules. Those eight rules are listed at the start of Problem Set 3.1 — they start with $\boldsymbol{v}+\boldsymbol{w} = \boldsymbol{w}+\boldsymbol{v}$ and they are easy to check in $\mathbf{R}^n$. They don't need to be memorized!

One important requirement: All linear combinations $c\boldsymbol{v} + d\boldsymbol{w}$ **must stay in the vector space**. The set of positive vectors $(v_1, \ldots, v_n)$ with every $v_i > 0$ is not a vector space. The set of solutions to $A\boldsymbol{x} = (1, 1, \ldots, 1)$ is not a vector space. A line in $\mathbf{R}^n$ is not a vector space unless it goes through $(0, 0, \ldots, 0)$.

If the line does go through $\mathbf{0}$, we can multiply points on the line by any number c and we can add points on the line—without leaving the line. That line in $\mathbf{R}^n$ shows the **idea of a subspace: A vector space inside another vector space**.

Examples of Vector Spaces

This book is mainly about the vector spaces $\mathbf{R}^n$ and their subspaces like lines and planes. The space $\mathbf{Z}$ that only contains the zero vector $\mathbf{0} = (0, 0, \ldots, 0)$ counts as a subspace! Combinations $c\,\mathbf{0} + d\,\mathbf{0}$ are still $\mathbf{0}$ (inside the subspace). $\mathbf{Z}$ is the smallest vector space. We often see $\mathbf{Z}$ as the nullspace of an invertible matrix: If the only solution to $A\boldsymbol{x} = \mathbf{0}$ is the zero vector $\boldsymbol{x} = \mathbf{0}$, then the nullspace of A is $\mathbf{Z}$.

We can certainly accept vector spaces of matrices. The space $\mathbf{R}^{3\times 3}$ contains all 3 by 3 matrices. We can take combinations $cA + dB$ of those matrices. They easily satisfy the eight rules. One subspace would be the 3 by 3 matrices with all 9 entries equal—a "line of matrices". Note that $\mathbf{Z} = \{\text{zero matrix}\}$ and $\mathbf{S}$ = symmetric 3 by 3 matrices are also subspaces: $A+B$ stays symmetric. But the invertible matrices are *not* a subspace.

We can also accept vector spaces of *functions*. The line of functions $\boldsymbol{y} = \boldsymbol{c}\boldsymbol{e}^{\boldsymbol{x}}$ (any c) would be a "line in function space". That line contains all the solutions to the differential equation $dy/dx = y$. Another function space contains all quadratics $\boldsymbol{y} = \boldsymbol{a} + \boldsymbol{b}\boldsymbol{x} + \boldsymbol{c}\boldsymbol{x}^{\mathbf{2}}$. Those are the solutions to $d^3y/dx^3 = 0$. You see how linear differential equations replace linear algebraic equations $A\boldsymbol{x} = \mathbf{0}$ when we move to function space.

In some way the space of 3 by 3 matrices is essentially the same as $\mathbf{R}^9$. The space of functions $f(x) = a + bx + cx^2$ is essentially $\mathbf{R}^3$.

Linear combinations of the matrices and functions are safely in those spaces. This book will stay almost entirely with ordinary column vectors and not functions.

The word "space" means that all linear combinations of the vectors or matrices or functions stay *inside the space*.

Subspaces of Vector Spaces

At different times, we will ask you to think of matrices and functions as vectors. But at all times, the vectors that we need most are ordinary column vectors. They are vectors with n components—but *maybe not all* of the vectors with n components. There are important vector spaces ***inside*** $\mathbf{R}^n$. Those are ***subspaces*** of $\mathbf{R}^n$.

Start with the usual three-dimensional space $\mathbf{R}^3$. Choose a plane through the origin $(0, 0, 0)$. ***That plane is a vector space in its own right.*** If we add two vectors in the plane, their sum is in the plane. If we multiply an in-plane vector by 2 or -5, it stays in the plane. A plane in three-dimensional space is not $\mathbf{R}^2$ (even if it looks like $\mathbf{R}^2$). The vectors have three components and they belong to $\mathbf{R}^3$. The plane is a vector space ***inside*** $\mathbf{R}^3$.

This illustrates one of the most fundamental ideas in linear algebra. The plane going through $(0, 0, 0)$ is a ***subspace*** of the full vector space $\mathbf{R}^3$.

DEFINITION A ***subspace*** of a vector space is a set of vectors (including $\mathbf{0}$) that satisfies two requirements: ***If*** $\boldsymbol{v}$ ***and*** $\boldsymbol{w}$ ***are vectors in the subspace and*** c ***is any scalar, then***

(i) $\boldsymbol{v} + \boldsymbol{w}$ is in the subspace	**(ii)** $c\boldsymbol{v}$ is in the subspace

In other words, the set of vectors is "closed" under addition $\boldsymbol{v} + \boldsymbol{w}$ and multiplication $c\boldsymbol{v}$ (and $d\boldsymbol{w}$). Those operations leave us in the subspace. We can also subtract, because $-\boldsymbol{w}$ is in the subspace and its sum with $\boldsymbol{v}$ is $\boldsymbol{v} - \boldsymbol{w}$. ***All linear combinations stay in the subspace.***

These operations follow the rules of the host space, so the eight required conditions are automatic. We just have to check the linear combinations requirement for a subspace.

First fact: ***Every subspace contains the zero vector.*** The plane in $\mathbf{R}^3$ has to go through $(0, 0, 0)$. We mention this separately, for extra emphasis, but it follows directly from rule **(ii)**. Choose $c = 0$, and the rule requires $0\boldsymbol{v}$ to be in the subspace.

Planes that don't contain the origin fail those tests. Those planes are not subspaces.

Lines through the origin are also subspaces. When we multiply by 5, or add two vectors on the line, we stay on the line. But the line must go through $(0, 0, 0)$.

Another subspace is all of $\mathbf{R}^3$. The whole space is a subspace (*of itself*). Here is a list of all the possible subspaces of $\mathbf{R}^3$:

($\mathbf{L}$) Any line through $(0, 0, 0)$	($\mathbf{R}^3$) The whole space
($\mathbf{P}$) Any plane through $(0, 0, 0)$	($\mathbf{Z}$) The single vector $(0, 0, 0)$

If we try to keep only *part* of a plane or line, the requirements for a subspace don't hold. Look at these examples in $\mathbf{R}^2$—they are not subspaces.

Example 1 Keep only the vectors (x, y) whose components are positive or zero (this is a quarter-plane). The vector $(2, 3)$ is included but $(-2, -3)$ is not. So rule (**ii**) is violated when we try to multiply by $c = -1$. ***The quarter-plane is not a subspace.***

Example 2 Include also the vectors whose components are both negative. Now we have two quarter-planes. Requirement (**ii**) is satisfied; we can multiply by any c. But rule (**i**) now fails. The sum of $\boldsymbol{v} = (2, 3)$ and $\boldsymbol{w} = (-3, -2)$ is $(-1, 1)$, which is outside the quarter-planes. ***Two quarter-planes don't make a subspace.***

Rules (**i**) and (**ii**) involve vector addition $\boldsymbol{v} + \boldsymbol{w}$ and multiplication by scalars c and d. The rules can be combined into a single requirement—*the rule for subspaces*:

A subspace containing $\boldsymbol{v}$ ***and*** $\boldsymbol{w}$ ***must contain all linear combinations*** $c\boldsymbol{v} + d\boldsymbol{w}$.

Example 3 Inside the vector space $\mathbf{M}$ of all 2 by 2 matrices, here are two subspaces:

$$(\mathbf{U}) \text{ All upper triangular matrices } \begin{bmatrix} a & b \\ 0 & d \end{bmatrix} \qquad (\mathbf{D}) \text{ All diagonal matrices } \begin{bmatrix} a & 0 \\ 0 & d \end{bmatrix}.$$

Add any upper triangular matrices in $\mathbf{U}$, and the sum is in $\mathbf{U}$. Add diagonal matrices, and the sum is diagonal. In this case $\mathbf{D}$ is also a subspace of $\mathbf{U}$! Of course the zero matrix is in these subspaces, when a, b, and d all equal zero. $\mathbf{Z}$ is always a subspace.

Multiples of the identity matrix also form a subspace of M. Those matrices cI form a "line of matrices" inside $\mathbf{M}$ and $\mathbf{U}$ and $\mathbf{D}$.

Is the matrix I a subspace by itself? Certainly not. Only the zero matrix is. Your mind will invent more subspaces of 2 by 2 matrices—write them down for Problem 5.

The Column Space of A

The most important subspaces are tied directly to a matrix A. We are trying to solve $A\boldsymbol{x} = \boldsymbol{b}$. If A is not invertible, the system is solvable for some $\boldsymbol{b}$ and not solvable for other $\boldsymbol{b}$. We want to describe the good right sides $\boldsymbol{b}$—the vectors that can be written as A times some vector $\boldsymbol{x}$. Those $\boldsymbol{b}$'s form the "*column space*" of A.

Remember that $A\boldsymbol{x}$ is a combination of the columns of A. To get every possible $\boldsymbol{b}$, we use every possible $\boldsymbol{x}$. Start with the columns of A and ***take all their linear combinations. This produces the column space of A.*** **It is a vector space made up of column vectors.**

DEFINITION The ***column space*** consists of ***all linear combinations of the columns***. Those combinations are all possible vectors $A\boldsymbol{x}$. They fill the column space $\mathbf{C}(A)$.

This column space is crucial to the whole book, and here is why. ***To solve*** $A\boldsymbol{x} = \boldsymbol{b}$ ***is to express*** $\boldsymbol{b}$ ***as a combination of the columns***. The right side $\boldsymbol{b}$ has to be *in the column space* produced by A, or $A\boldsymbol{x} = \boldsymbol{b}$ has no solution!

The equations $A\boldsymbol{x} = \boldsymbol{b}$ ***are solvable if and only if*** $\boldsymbol{b}$ ***is in the column space of*** A.

When $\boldsymbol{b}$ is in the column space $\mathbf{C}(A)$, it is a combination of the columns of A. The coefficients in that combination will solve $A\boldsymbol{x} = \boldsymbol{b}$. The word "space" is justified by taking *all combinations* of the columns. The column space is a subspace of $\mathbf{R}^m$.

Caution: The columns of A do not form a subspace ! The invertible matrices do not form a subspace. The singular matrices do not form a subspace. You have to include all linear combinations. The columns of A **"span"** a subspace when *we take their combinations.*

The Row Space of A

The rows of A are the columns of A^{T}, the n by m transpose matrix. Since we prefer to work with column vectors, we welcome A^{T} :

The row space of A is the column space $\mathbf{C}(A^{\mathrm{T}})$ of the transpose matrix A^{T}

This row space is a subspace of $\mathbf{R}^n$. It contains m column vectors from A^{T} and all their combinations. The equations $A^{\mathrm{T}}\boldsymbol{y} = \boldsymbol{c}$ are solvable exactly when the vector $\boldsymbol{c}$ is in the subspace $\mathbf{C}(A^{\mathrm{T}}) =$ *row space of* A.

Chapter 1 explained why $\mathbf{C}(A)$ and $\mathbf{C}(A^{\mathrm{T}})$ both contain r independent vectors and no more. Then $r =$ rank of $A =$ rank of A^{T}. A new proof is in Section 3.5.

Example The row space of the rank 1 matrix $A = \boldsymbol{u}\boldsymbol{v}^{\mathrm{T}}$ is the line of all column vectors $c\boldsymbol{v}$. This is because every column of $A^{\mathrm{T}} = \boldsymbol{v}\boldsymbol{u}^{\mathrm{T}}$ is a multiple of $\boldsymbol{v}$. One vector $\boldsymbol{v}$ spans the row space, one vector $\boldsymbol{u}$ spans the column space.

The Columns of A Span the Vector Space $\mathbf{C}(A)$

One more new word: *"Span"*. Suppose we start with a set $\boldsymbol{S}$ of vectors in $\mathbf{R}^m$. If $\boldsymbol{S}$ contains only N vectors, it is certainly not a vector space. But if we include *all combinations of the vectors in* $\boldsymbol{S}$, then we have a vector space $\mathbf{V}$. In this case **the set $\boldsymbol{S}$ spans $\mathbf{V}$**. In fact $\mathbf{V}$ is the smallest vector space containing $\boldsymbol{S}$ (because we are forced to include all combinations to produce a vector space).

This is exactly what we did for the columns of A. Those n columns span the column space $\mathbf{C}(A)$ = all combinations of the columns. Independence is not required by the word *span*. In the same way, the m columns of A^{T} span the row space $\mathbf{C}(A^{\mathrm{T}})$.

Test question. Show that the invertible 2 by 2 matrices span $\mathbf{R}^{2\times 2}$.

Examples If the n by n matrix A is invertible, then its columns span $\mathbf{R}^n$.

The invertible 3 by 3 matrices span the matrix space $\mathbf{R}^{3\times 3}$.

The singular (not invertible) 3 by 3 matrices also span $\mathbf{R}^{3\times 3}$.

Next comes the nullspace $\mathbf{N}(A)$ and that requires new ideas. We start with $A\boldsymbol{x} = \mathbf{0}$ (*equations and not vectors*). The solutions $\boldsymbol{x}$ to those equations give the nullspace. It is a vector space because $A\boldsymbol{x} = \mathbf{0}$ and $A\boldsymbol{y} = \mathbf{0}$ lead to $A(c\boldsymbol{x} + d\boldsymbol{y}) = \mathbf{0}$. But we have to work to find those solutions $\boldsymbol{x}$ and $\boldsymbol{y}$.

Test question: When do 10 vectors span $\mathbf{R}^5$? This is very possible.

Problem Set 3.1

The first problems 1–7 are about vector spaces in general. The vectors in those spaces are not necessarily column vectors. In the definition of a *vector space*, vector addition $\boldsymbol{x}+\boldsymbol{y}$ and scalar multiplication $c\boldsymbol{x}$ must obey the following eight rules:

(1) $\boldsymbol{x}+\boldsymbol{y}=\boldsymbol{y}+\boldsymbol{x}$

(2) $\boldsymbol{x}+(\boldsymbol{y}+\boldsymbol{z})=(\boldsymbol{x}+\boldsymbol{y})+\boldsymbol{z}$

(3) There is a unique "zero vector" such that $\boldsymbol{x}+\mathbf{0}=\boldsymbol{x}$ for all $\boldsymbol{x}$

(4) For each $\boldsymbol{x}$ there is a unique vector $-\boldsymbol{x}$ such that $\boldsymbol{x}+(-\boldsymbol{x})=\mathbf{0}$

(5) 1 times $\boldsymbol{x}$ equals $\boldsymbol{x}$

(6) $(c_1c_2)\boldsymbol{x}=c_1(c_2\boldsymbol{x})$ — (1) to (4) about $\boldsymbol{x}+\boldsymbol{y}$

(7) $c(\boldsymbol{x}+\boldsymbol{y})=c\boldsymbol{x}+c\boldsymbol{y}$ — (5) to (6) about $c\boldsymbol{x}$

(8) $(c_1+c_2)\boldsymbol{x}=c_1\boldsymbol{x}+c_2\boldsymbol{x}$. — **(7) to (8) connects them**

---

1 Suppose $(x_1,x_2)+(y_1,y_2)$ is defined to be (x_1+y_2,x_2+y_1). With the usual multiplication $c\boldsymbol{x}=(cx_1,cx_2)$, which of the eight conditions are not satisfied?

2 Suppose the multiplication $c\boldsymbol{x}$ is defined to produce $(cx_1,0)$ instead of (cx_1,cx_2). With the usual addition in $\mathbf{R}^2$, are the eight conditions satisfied?

3 (a) Which rules are broken if we keep only the positive numbers $x>0$ in $\mathbf{R}^1$? Every c must be allowed. The half-line is not a subspace.

(b) The positive numbers with $\boldsymbol{x}+\boldsymbol{y}$ and $c\boldsymbol{x}$ redefined to equal the usual xy and x^c *do satisfy the eight rules*. Test rule 7 when $c=3, x=2, y=1$. (Then $\boldsymbol{x}+\boldsymbol{y}=2$ and $c\boldsymbol{x}=8$.) Which number acts as the "zero vector"?

4 The matrix $A=\begin{bmatrix}2&-2\\2&-2\end{bmatrix}$ is a "vector" in the space $\mathbf{M}$ of all 2 by 2 matrices. Write down the zero vector in this space, the vector $\frac{1}{2}A$, and the vector $-A$. What matrices are in the smallest subspace containing A (the subspace spanned by A)?

5 (a) Describe a subspace of $\mathbf{M}$ that contains $A=\begin{bmatrix}1&0\\0&0\end{bmatrix}$ but not $B=\begin{bmatrix}0&0\\0&-1\end{bmatrix}$.

(b) If a subspace of $\mathbf{M}$ does contain A and B, must it contain I? A - B = I

(c) Describe a subspace of $\mathbf{M}$ that contains no nonzero diagonal matrices.

6 The functions $\boldsymbol{f}(x) = x^2$ and $\boldsymbol{g}(x) = 5x$ are "vectors" in $\mathbf{F}$. This is the vector space of all real functions. (The functions are defined for $-\infty < x < \infty$.) The combination $3\boldsymbol{f}(x) - 4\boldsymbol{g}(x)$ is the function $\boldsymbol{h}(x) =$ _____.

7 Which rule is broken if multiplying $\boldsymbol{f}(x)$ by c gives the function $\boldsymbol{f}(cx)$? Keep the usual addition $\boldsymbol{f}(x) + \boldsymbol{g}(x)$.

Questions 8–15 are about the "subspace requirements": $x + y$ and cx (and then all linear combinations $cx + dy$) stay in the subspace.

8 One subspace requirement can be met while the other fails. Show this by finding

(a) A set of vectors in $\mathbf{R}^2$ for which $\boldsymbol{x} + \boldsymbol{y}$ stays in the set but $\frac{1}{2}\boldsymbol{x}$ may be outside.

(b) A set of vectors in $\mathbf{R}^2$ (other than two quarter-planes) for which every $c\boldsymbol{x}$ stays in the set but $\boldsymbol{x} + \boldsymbol{y}$ may be outside.

9 Which of these subsets of $\mathbf{R}^3$ are actually subspaces ? They all span subspaces !

(a) The plane of vectors (b_1, b_2, b_3) with $b_1 = b_2$.

(b) The plane of vectors with $b_1 = 1$.

(c) The vectors with $b_1 b_2 b_3 = 0$.

(d) All linear combinations of $\boldsymbol{v} = (1, 4, 0)$ and $\boldsymbol{w} = (2, 2, 2)$.

(e) All vectors that satisfy $b_1 + b_2 + b_3 = 0$.

(f) All vectors with $b_1 \le b_2 \le b_3$.

10 Describe the smallest subspace of the matrix space $\mathbf{M}$ that contains

(a) $\begin{bmatrix} 1 & 0 \\ 0 & 0 \end{bmatrix}$ and $\begin{bmatrix} 0 & 1 \\ 0 & 0 \end{bmatrix}$ (b) $\begin{bmatrix} 1 & 1 \\ 0 & 0 \end{bmatrix}$ (c) $\begin{bmatrix} 1 & 1 \\ 1 & 1 \end{bmatrix}$ and $\begin{bmatrix} 1 & 0 \\ 0 & 1 \end{bmatrix}$.

11 Let P be the plane in $\mathbf{R}^3$ with equation $x + y - 2z = 4$. The origin $(0, 0, 0)$ is not in P! Find two vectors in P and check that their sum is not in P.

12 Let $\mathbf{P}_0$ be the plane through $(0, 0, 0)$ parallel to the previous plane P. What is the equation for $\mathbf{P}_0$? Find two vectors in $\mathbf{P}_0$ and check that their sum is in $\mathbf{P}_0$.

13 Suppose $\mathbf{P}$ is a plane through $(0, 0, 0)$ and $\mathbf{L}$ is a line through $(0, 0, 0)$. The smallest vector space containing both $\mathbf{P}$ and $\mathbf{L}$ is either _____ or _____.

14 (a) Show that the set of *invertible* matrices in $\mathbf{M}$ is not a subspace.

(b) Show that the set of *singular* matrices in $\mathbf{M}$ is not a subspace.

15 True or false (check addition in each case by an example):

(a) The symmetric matrices in $\mathbf{M}$ (with $A^{\mathrm{T}} = A$) form a subspace.

(b) The skew-symmetric matrices in $\mathbf{M}$ (with $A^{\mathrm{T}} = -A$) form a subspace.

(c) The unsymmetric matrices in $\mathbf{M}$ (with $A^{\mathrm{T}} \neq A$) **span** a subspace.

Questions 16–26 are about column spaces $\mathbf{C}(A)$ and the equation $Ax = b$.

16 Describe the column spaces (lines or planes) of these particular matrices:

$$A = \begin{bmatrix} 1 & 2 \\ 0 & 0 \\ 0 & 0 \end{bmatrix} \quad \text{and} \quad B = \begin{bmatrix} 1 & 0 \\ 0 & 2 \\ 0 & 0 \end{bmatrix} \quad \text{and} \quad C = \begin{bmatrix} 1 & 0 \\ 2 & 0 \\ 0 & 0 \end{bmatrix}.$$

17 For which right sides (find a condition on b_1, b_2, b_3) are these systems solvable?

$$\text{(a)} \quad \begin{bmatrix} 1 & 4 & 2 \\ 2 & 8 & 4 \\ -1 & -4 & -2 \end{bmatrix} \begin{bmatrix} x_1 \\ x_2 \\ x_3 \end{bmatrix} = \begin{bmatrix} b_1 \\ b_2 \\ b_3 \end{bmatrix} \qquad \text{(b)} \quad \begin{bmatrix} 1 & 4 \\ 2 & 9 \\ -1 & -4 \end{bmatrix} \begin{bmatrix} x_1 \\ x_2 \end{bmatrix} = \begin{bmatrix} b_1 \\ b_2 \\ b_3 \end{bmatrix}.$$

18 Adding row 1 of A to row 2 produces B. Adding column 1 to column 2 produces C. A combination of the columns of (B or C ?) is also a combination of the columns of A. Which two matrices have the same column _____ ?

$$A = \begin{bmatrix} 1 & 2 \\ 2 & 4 \end{bmatrix} \quad \text{and} \quad B = \begin{bmatrix} 1 & 2 \\ 3 & 6 \end{bmatrix} \quad \text{and} \quad C = \begin{bmatrix} 1 & 3 \\ 2 & 6 \end{bmatrix}.$$

19 For which vectors (b_1, b_2, b_3) do these systems have a solution?

$$\begin{bmatrix} 1 & 1 & 1 \\ 0 & 1 & 1 \\ 0 & 0 & 1 \end{bmatrix} \begin{bmatrix} x_1 \\ x_2 \\ x_3 \end{bmatrix} = \begin{bmatrix} b_1 \\ b_2 \\ b_3 \end{bmatrix} \quad \text{and} \quad \begin{bmatrix} 1 & 1 & 1 \\ 0 & 1 & 1 \\ 0 & 0 & 0 \end{bmatrix} \begin{bmatrix} x_1 \\ x_2 \\ x_3 \end{bmatrix} = \begin{bmatrix} b_1 \\ b_2 \\ b_3 \end{bmatrix}$$

$$\text{and} \quad \begin{bmatrix} 1 & 1 & 1 \\ 0 & 0 & 1 \\ 0 & 0 & 1 \end{bmatrix} \begin{bmatrix} x_1 \\ x_2 \\ x_3 \end{bmatrix} = \begin{bmatrix} b_1 \\ b_2 \\ b_3 \end{bmatrix}.$$

20 (Recommended) If we add an extra column $\boldsymbol{b}$ to a matrix A, then the column space gets larger unless _____ . Give an example where the column space gets larger and an example where it doesn't. Why is $A\boldsymbol{x} = \boldsymbol{b}$ solvable exactly when the column space *doesn't* get larger? Then it is the same for A and $\begin{bmatrix} A & \boldsymbol{b} \end{bmatrix}$.

21 The columns of AB are combinations of the columns of A. This means: *The column space of AB is contained in* (possibly equal to) *the column space of A*. Give an example where the column spaces of A and AB are not equal.

22 Suppose $A\boldsymbol{x} = \boldsymbol{b}$ and $A\boldsymbol{y} = \boldsymbol{b}^*$ are both solvable. Then $A\boldsymbol{z} = \boldsymbol{b} + \boldsymbol{b}^*$ is solvable. What is $\boldsymbol{z}$? This translates into: If $\boldsymbol{b}$ and $\boldsymbol{b}^*$ are in the column space $\mathbf{C}(A)$, *then* $\boldsymbol{b} + \boldsymbol{b}^*$ is in $\mathbf{C}(A)$. That is a requirement for a vector space.

23 If A is any 5 by 5 invertible matrix, then its column space is _____. Why?

24 True or false (with a counterexample if false):

(a) The vectors $\boldsymbol{b}$ that are not in the column space $\mathbf{C}(A)$ form a subspace.

(b) If $\mathbf{C}(A)$ contains only the zero vector, then A is the zero matrix.

(c) The column space of $2A$ equals the column space of A.

(d) The column space of $A - I$ equals the column space of A (test this).

25 Construct a 3 by 3 matrix whose column space contains $(1, 1, 0)$ and $(1, 0, 1)$ but not $(1, 1, 1)$. Construct a 3 by 3 matrix whose column space is only a line.

26 If the 9 by 12 system $A\boldsymbol{x} = \boldsymbol{b}$ is solvable for every $\boldsymbol{b}$, then $\mathbf{C}(A) =$ _____ .

Challenge Problems

27 Suppose $\mathbf{S}$ and $\mathbf{T}$ are two subspaces of a vector space $\mathbf{V}$.

(a) **Definition**: The **sum** $\mathbf{S} + \mathbf{T}$ contains all sums $\boldsymbol{s} + \boldsymbol{t}$ of a vector $\boldsymbol{s}$ in $\mathbf{S}$ and a vector $\boldsymbol{t}$ in $\mathbf{T}$. Show that $\mathbf{S} + \mathbf{T}$ satisfies the requirements for a vector space. Addition and scalar multiplication stay inside $\mathbf{S} + \mathbf{T}$.

(b) If $\mathbf{S}$ and $\mathbf{T}$ are lines in $\mathbf{R}^m$, what is the difference between $\mathbf{S} + \mathbf{T}$ and $\mathbf{S} \cup \mathbf{T}$? That union contains all vectors from $\mathbf{S}$ or $\mathbf{T}$ or both. Explain this statement: **The span of $\mathbf{S} \cup \mathbf{T}$ is $\mathbf{S} + \mathbf{T}$.** (Section 3.5 returns to this word "span".)

28 If $\mathbf{S}$ is the column space of A and $\mathbf{T}$ is $\mathbf{C}(B)$, then $\mathbf{S} + \mathbf{T}$ is the column space of what matrix M? The columns of A and B and M are all in $\mathbf{R}^m$. I don't think $A + B$ is always a correct M. We want the columns of M to span $\mathbf{S} + \mathbf{T}$.

29 Show that the matrices A and $\begin{bmatrix} A & AB \end{bmatrix}$ (with extra columns) have the same column space. But find a square matrix with $\mathbf{C}(A^2)$ smaller than $\mathbf{C}(A)$. Important point:

An n by n matrix has $\mathbf{C}(A) = \mathbf{R}^n$ exactly when A is an _____ matrix.

30 Find another independent solution (after $y = e^x$) to the second order differential equation $\boldsymbol{d^2y/dx^2 = y}$. Find two independent solutions to $\boldsymbol{d^2y/dx^2 = -y}$.

Then the 2-dimensional solution space contains all linear combinations $y =$ _____ .

31 Suppose $\boldsymbol{V}$ and $\boldsymbol{W}$ are two subspaces of $\mathbf{R}^n$. Their "intersection" $\boldsymbol{V} \cap \boldsymbol{W}$ contains the vectors that are in both subspaces. (Notice that the zero vector is in $\boldsymbol{V}$ and $\boldsymbol{W}$.) Show that $\boldsymbol{V} \cap \boldsymbol{W}$ is a *subspace* by testing the requirement: If $\boldsymbol{x}$ and $\boldsymbol{y}$ are in $\boldsymbol{V} \cap \boldsymbol{W}$, why is $c\boldsymbol{x} + d\boldsymbol{y}$ in $\boldsymbol{V} \cap \boldsymbol{W}$?

3.2 The Nullspace of A: Solving $Ax = 0$

1 The **nullspace** $\mathbf{N}(A)$ in $\mathbf{R}^n$ contains all solutions $\boldsymbol{x}$ to $A\boldsymbol{x} = \mathbf{0}$. This includes $\boldsymbol{x} = \mathbf{0}$.

2 Elimination from A to U to R_0 does not change the nullspace: $\mathbf{N}(A) = \mathbf{N}(U) = \mathbf{N}(\boldsymbol{R}_0)$.

3 **The reduced row echelon form** $\boldsymbol{R}_0 = \mathbf{rref}(\boldsymbol{A})$ has $\boldsymbol{I}$ in r columns and $\boldsymbol{F}$ in $n - r$ columns.

4 If column j of R_0 is free (no pivot), there is a "*special solution*" to $A\boldsymbol{x} = \mathbf{0}$ with $x_j = 1$.

5 Every short wide matrix with $m < n$ has nonzero solutions to $A\boldsymbol{x} = \mathbf{0}$ in its nullspace.

This section is about the nullspace containing all solutions to $A\boldsymbol{x} = \mathbf{0}$. The m by n matrix A can be square or rectangular. The right hand side is $\boldsymbol{b} = \mathbf{0}$. *One immediate solution is* $\boldsymbol{x} = \mathbf{0}$. For square invertible matrices this is the only solution. For other matrices, we find $n - r$ special solutions to $A\boldsymbol{x} = \mathbf{0}$. *Each solution $\boldsymbol{x}$ belongs to the nullspace of A.*

Elimination will find all solutions and identify this very important subspace.

The nullspace $\mathbf{N}(A)$ ***consists of all solutions to*** $A\boldsymbol{x} = \mathbf{0}$. ***These vectors*** $\boldsymbol{x}$ ***are in*** $\mathbf{R}^n$.

Check that those vectors form a subspace. Suppose $\boldsymbol{x}$ and $\boldsymbol{y}$ are in the nullspace (this means $A\boldsymbol{x} = \mathbf{0}$ and $A\boldsymbol{y} = \mathbf{0}$). The rules of matrix multiplication give $A(\boldsymbol{x} + \boldsymbol{y}) = \mathbf{0} + \mathbf{0}$. The rules also give $A(c\boldsymbol{x}) = c\mathbf{0}$. The right sides are still zero. Therefore $\boldsymbol{x} + \boldsymbol{y}$ and $c\boldsymbol{x}$ are also in the nullspace $\mathbf{N}(A)$, and the test for a subspace is passed.

To repeat: The solution vectors $\boldsymbol{x}$ have n components. They are vectors in $\mathbf{R}^n$, so *the nullspace is a subspace of* $\mathbf{R}^n$. The column space $\mathbf{C}(A)$ is a subspace of $\mathbf{R}^m$.

Example 1 Describe the nullspace of $A = \begin{bmatrix} 1 & 2 \\ 3 & 6 \end{bmatrix}$. This matrix is singular!

Solution Apply elimination to change the linear equations $A\boldsymbol{x} = \mathbf{0}$ to $R_0\boldsymbol{x} = \mathbf{0}$:

$$\begin{matrix} x_1 + 2x_2 = 0 \\ 3x_1 + 6x_2 = 0 \end{matrix} \quad \rightarrow \quad \begin{matrix} x_1 + 2x_2 = 0 \\ \mathbf{0} = \mathbf{0} \end{matrix} \qquad \begin{bmatrix} 1 & 2 \\ 3 & 6 \end{bmatrix} \rightarrow R_0 = \begin{bmatrix} 1 & 2 \\ 0 & 0 \end{bmatrix} = \begin{bmatrix} I & F \\ 0 & 0 \end{bmatrix}$$

There is really only one equation. The second equation is the first equation multiplied by 3. In the row picture, the line $x_1 + 2x_2 = 0$ is the same as the line $3x_1 + 6x_2 = 0$. That line is the nullspace $\mathbf{N}(A)$. It contains all solutions $(x_1, x_2) = (-2c, c) = \boldsymbol{c}\,(\mathbf{-2}, \mathbf{1})$.

To describe the solutions to $A\boldsymbol{x} = \mathbf{0}$, here is an efficient way. Choose one "***special solution***". Then all solutions are multiples of this one. We choose the second component to be $x_2 = 1$ (a special choice). From the equation $x_1 + 2x_2 = 0$, the first component must be $x_1 = -2$. **The special solution is** $\boldsymbol{s} = (\mathbf{-2}, \mathbf{1})$.

Special solution $\quad A\boldsymbol{s} = \mathbf{0} \quad$ The nullspace of $A = \begin{bmatrix} 1 & 2 \\ 3 & 6 \end{bmatrix}$ contains all multiples of $\boldsymbol{s} = \begin{bmatrix} -2 \\ \mathbf{1} \end{bmatrix}$.

This is the best way to describe the nullspace. **The solution** $\boldsymbol{s}$ **is special because the free variable is 1.** Simple formulas for $\boldsymbol{R}$ and $\boldsymbol{s}$ come at the end of this Section 3.2.

If $r < n$, $\mathbf{N}(A)$ consists of all combinations of the $n-r$ special solutions to $A\boldsymbol{x} = \mathbf{0}$.

Example 2 $x + 2y + 3z = 0$ comes from the 1 by 3 matrix $A = \begin{bmatrix} 1 & 2 & 3 \end{bmatrix}$. Then $A\boldsymbol{x} = \mathbf{0}$ produces a plane. All vectors on the plane are perpendicular to $(1, 2, 3)$. *The plane is the nullspace of A.* There are two free variables y and z: Set to 0 and 1.

$$\begin{bmatrix} 1 & 2 & 3 \end{bmatrix}\begin{bmatrix} x \\ y \\ z \end{bmatrix} = 0 \text{ has } \textbf{two special solutions } \boldsymbol{s}_1 = \begin{bmatrix} -2 \\ \mathbf{1} \\ \mathbf{0} \end{bmatrix} \text{ and } \boldsymbol{s}_2 = \begin{bmatrix} -3 \\ \mathbf{0} \\ \mathbf{1} \end{bmatrix}.$$

Those vectors $\boldsymbol{s}_1$ and $\boldsymbol{s}_2$ lie on the plane $x + 2y + 3z = 0$. All vectors on the plane are combinations of $\boldsymbol{s}_1$ and $\boldsymbol{s}_2$. In this example $A = R_0 = \begin{bmatrix} 1 & 2 & 3 \end{bmatrix} = \begin{bmatrix} I & F \end{bmatrix}$.

Notice what is special about $\boldsymbol{s}_1$ and $\boldsymbol{s}_2$. *The last two components are "free" and we choose them specially as* $1, 0$ *and* $0, 1$. Then the first components -2 and -3 are determined by the equation $x + 2y + 3z = 0$.

The solutions to $x + 2y + 3z = \mathbf{6}$ also lie on a plane, but that plane is not a subspace. The vector $\boldsymbol{x} = \mathbf{0}$ is only a solution if $\boldsymbol{b} = \mathbf{0}$. Section 3.3 will show how the solutions to $A\boldsymbol{x} = \boldsymbol{b}$ (if there are any solutions) are shifted away from zero by one particular solution.

Two key steps	(1) **reducing A to its row echelon forms R_0 and R**
in this section	(2) **finding the $n - r$ special solutions $\boldsymbol{s}$ to $R\boldsymbol{x} = \mathbf{0}$**
Section 3.3 has the final step	(3) **finding a particular solution** to $A\boldsymbol{x} = \boldsymbol{b}$

R is connected to A by $A = CR$. As in Chapter 1, C contains r independent columns. Elimination (row operations) will now take us directly from A to R_0 to R, without C.

Example $A = \begin{bmatrix} 1 & 2 & 1 \\ 2 & 4 & 5 \\ 3 & 6 & 9 \end{bmatrix}$. We can see that column 2 is 2 times column 1. Then columns 1 and 3 are independent and the rank is $r = 2$. But we don't want to use this information! We want a systematic way to find dependent columns for *any matrix A*. That systematic way is *a sequence of row operations on A that will lead directly to R.*

The row operations are like the elimination steps in Chapter 2, leading from A to U (upper triangular). But now **we don't stop at U**. We will continue to R_0 and R. We are discovering R before C. **That matrix R will reveal the nullspace of A.**

Step 1 Subtract multiples of row 1 to clear out column 1 below the first pivot. Column 2 is also cleared:

$$\boldsymbol{A} = \begin{bmatrix} \mathbf{1} & 2 & 1 \\ 2 & 4 & 5 \\ 3 & 6 & 9 \end{bmatrix} \rightarrow \begin{bmatrix} 1 & 2 & 1 \\ 0 & 0 & 3 \\ 3 & 6 & 9 \end{bmatrix} \rightarrow \begin{bmatrix} \mathbf{1} & 2 & 1 \\ 0 & 0 & \mathbf{3} \\ 0 & 0 & 6 \end{bmatrix}$$

Step 2 Divide row 2 by 3, to produce *second pivot* $= 1$. Use it to eliminate 6 and 1:

$$\boldsymbol{A} = \begin{bmatrix} 1 & 2 & 1 \\ 0 & 0 & 3 \\ 0 & 0 & 6 \end{bmatrix} \rightarrow \begin{bmatrix} \mathbf{1} & \mathbf{2} & 1 \\ 0 & 0 & \mathbf{1} \\ 0 & 0 & 6 \end{bmatrix} \rightarrow \begin{bmatrix} \mathbf{1} & \mathbf{2} & \mathbf{0} \\ 0 & 0 & \mathbf{1} \\ 0 & 0 & \mathbf{0} \end{bmatrix} = \boldsymbol{R_0}$$

That matrix R_0 is the **reduced row echelon form**. It has the same rank as A (rank 2). The word *echelon* means that the 1's in R_0 go steadily down, left to right. R_0 has the same row space as A (all our row operations were invertible). R_0 has the same nullspace as A. The equations $R_0\boldsymbol{x} = \mathbf{0}$ are linear combinations of the equations $A\boldsymbol{x} = \mathbf{0}$.

Notice the zero row in R_0. We can and will remove it—no change in the row space or nullspace. **$\boldsymbol{R_0}$ becomes $\boldsymbol{R}$ with no zero rows.** *This is the R we wanted in Chapter* 1.

$$\boldsymbol{R_0} = \begin{bmatrix} 1 & 2 & 0 \\ 0 & 0 & 1 \\ \mathbf{0} & \mathbf{0} & \mathbf{0} \end{bmatrix} \qquad \boldsymbol{A} = \boldsymbol{CR} \text{ is } \begin{bmatrix} 1 & 2 & 1 \\ 2 & 4 & 5 \\ 3 & 6 & 9 \end{bmatrix} = \begin{bmatrix} 1 & 1 \\ 2 & 5 \\ 3 & 9 \end{bmatrix} \begin{bmatrix} \mathbf{1} & 2 & 0 \\ 0 & 0 & \mathbf{1} \end{bmatrix}$$

C contains the first r independent columns of A (columns 1 and 3)

R has the identity matrix in columns 1 and 3 and F in column 2 : rank $r = 2$

The special solution to $R\boldsymbol{x} = \mathbf{0}$ is $\boldsymbol{s} = (-2, 1, 0)$ with free variable $= 1$

The nullspace of A and R_0 and R contains all multiples of that solution $\boldsymbol{s}$

This is the same $\boldsymbol{A} = \boldsymbol{CR} = (\boldsymbol{m} \times \boldsymbol{r})\,(\boldsymbol{r} \times \boldsymbol{n})$ that Section 1.4 would produce by looking for independent columns in C. Now we have a good computational system: **Elimination steps from $\boldsymbol{A}$ to $\boldsymbol{R_0}$ and $\boldsymbol{R}$**, then look for the r by r identity matrix inside R.

By creating R, we know the correct columns of A that go into C. Those columns give the identity matrix I in R. Then $A = CR$ is the result of elimination on *any matrix*, going far beyond $A = LU$ to allow every matrix A.

Pivot Columns and Free Columns of $\boldsymbol{R}$ and $\boldsymbol{A}$

If $R\boldsymbol{x} = \mathbf{0}$ then $A\boldsymbol{x} = CR\boldsymbol{x} = \mathbf{0}$. There is a special solution $\boldsymbol{x} = \boldsymbol{s}$ for every column of A without a pivot. The r pivots are the 1's in I, leaving $n - r$ free columns of R. Here is the result of elimination on a 4 by 5 matrix when the rank is $r = 3$ and the independent columns of C are $\boldsymbol{a}_1$ and $\boldsymbol{a}_3$ and $\boldsymbol{a}_5$. The free columns $\boldsymbol{a}_2$ and $\boldsymbol{a}_4$ are not in C.

$$\underset{4 \times 5}{\boldsymbol{A}} = \begin{bmatrix} \\ \boldsymbol{a}_1 & \boldsymbol{a}_2 & \boldsymbol{a}_3 & \boldsymbol{a}_4 & \boldsymbol{a}_5 \\ \\ \end{bmatrix} = \begin{bmatrix} \\ \boldsymbol{a}_1 & \boldsymbol{a}_3 & \boldsymbol{a}_5 \\ \\ \end{bmatrix} \begin{bmatrix} \mathbf{1} & p & 0 & q & 0 \\ 0 & 0 & \mathbf{1} & r & 0 \\ 0 & 0 & 0 & 0 & \mathbf{1} \end{bmatrix} = \underset{4 \times 3}{\boldsymbol{C}} \;\; \underset{3 \times 5}{\boldsymbol{R}}$$

You see the 3 by 3 identity matrix in R. Elimination on the 4 by 5 matrix A led to a 4 by 5 matrix R_0. With rank $r = 3$, the fourth row of R_0 was all zeros. Removing that zero row from R_0 produced the perfect factorization $A = CR$.

Elimination on A is complete and it reached R. The remaining step is to read off the $5 - 3 = 2$ special solutions to $R\boldsymbol{x} = \mathbf{0}$.

What are the $n - r = 5 - 3 = 2$ special solutions $\boldsymbol{s}_1$ and $\boldsymbol{s}_2$? Those vectors solve $R\boldsymbol{s}_1 = \mathbf{0}$ and $R\boldsymbol{s}_2 = \mathbf{0}$. Multiplying by C they also solve $A\boldsymbol{s}_1 = \mathbf{0}$ and $A\boldsymbol{s}_2 = \mathbf{0}$. **The combinations $c_1\boldsymbol{s}_1 + c_2\boldsymbol{s}_2$ fill out the nullspace $\mathbf{N}(A)$.**

To find the special solutions, start with $\boldsymbol{s}_1 = (\_, \mathbf{1}, \_, \mathbf{0}, \_)$ and $\boldsymbol{s}_2 = (\_, \mathbf{0}, \_, \mathbf{1}, \_)$.

We are assigning the values $1, 0$ and $0, 1$ to the $n - r = 5 - 3 = 2$ positions that *don't* correspond to the columns $1, 3, 5$ containing the identity matrix in R. The equations $R\boldsymbol{s}_1 = \mathbf{0}$ and $R\boldsymbol{s}_2 = \mathbf{0}$ tell us how to fill in the rest of those special solutions $\boldsymbol{s}_1$ and $\boldsymbol{s}_2$:

Special solutions to $R\boldsymbol{x} = \mathbf{0}$

$$\boldsymbol{s}_1 = \begin{bmatrix} -p \\ \mathbf{1} \\ 0 \\ \mathbf{0} \\ 0 \end{bmatrix} \quad \text{and} \quad \boldsymbol{s}_2 = \begin{bmatrix} -q \\ \mathbf{0} \\ -r \\ \mathbf{1} \\ 0 \end{bmatrix}$$

The nullspace $\mathbf{N}(A) = \mathbf{N}(R)$ contains all $\boldsymbol{x} = c_1\boldsymbol{s}_1 + c_2\boldsymbol{s}_2$

Those three numbers $-p$ and $-q$ and $-r$ are just negatives of three numbers in R. Elimination has led systematically to $n - r = 2$ independent vectors in the nullspace of R. Those are the two special solutions $\boldsymbol{s}_1$ and $\boldsymbol{s}_2$ to $R\boldsymbol{x} = \mathbf{0}$ and $A\boldsymbol{x} = \mathbf{0}$.

The free components correspond to columns with no pivots. The special choice (one or zero) is only for the free variables in the special solutions.

Example 3 Find the nullspaces of A, B, M and the two special solutions to $M\boldsymbol{x} = \mathbf{0}$.

$$\boldsymbol{A} = \begin{bmatrix} 1 & 2 \\ 3 & 8 \end{bmatrix} \quad \boldsymbol{B} = \begin{bmatrix} A \\ 2A \end{bmatrix} = \begin{bmatrix} 1 & 2 \\ 3 & 8 \\ 2 & 4 \\ 6 & 16 \end{bmatrix} \quad \boldsymbol{M} = \begin{bmatrix} A & 2A \end{bmatrix} = \begin{bmatrix} 1 & 2 & 2 & 4 \\ 3 & 8 & 6 & 16 \end{bmatrix}.$$

Solution The equation $A\boldsymbol{x} = \mathbf{0}$ has only the zero solution $\boldsymbol{x} = \mathbf{0}$. *The nullspace is* $\mathbf{Z}$. It contains only the single point $\boldsymbol{x} = \mathbf{0}$ in $\mathbf{R}^2$. This fact comes from elimination:

$$A\boldsymbol{x} = \begin{bmatrix} 1 & 2 \\ 3 & 8 \end{bmatrix} \to \begin{bmatrix} 1 & 2 \\ 0 & 2 \end{bmatrix} \to \begin{bmatrix} 1 & 0 \\ 0 & 1 \end{bmatrix} = R = I \quad \textbf{No free variables}$$

A is invertible. There are no special solutions. Both columns of this matrix have pivots.

The rectangular matrix B has the same nullspace $\mathbf{Z}$. The first two equations in $B\boldsymbol{x} = \mathbf{0}$ again require $\boldsymbol{x} = \mathbf{0}$. The last two equations would also force $\boldsymbol{x} = \mathbf{0}$. When we add extra equations (giving extra rows), the nullspace certainly cannot become larger. The extra rows impose more conditions on the vectors $\boldsymbol{x}$ in the nullspace.

The rectangular matrix M is different. It has extra columns instead of extra rows. The solution vector $\boldsymbol{x}$ has *four* components. Elimination will produce pivots in the first two columns of M. **The last two columns of M are "free". They don't have pivots.**

$$\boldsymbol{M} = \begin{bmatrix} 1 & 2 & 2 & 4 \\ 3 & 8 & 6 & 16 \end{bmatrix} \quad \boldsymbol{U} = \begin{bmatrix} \mathbf{1} & 2 & 2 & 4 \\ 0 & \mathbf{2} & 0 & 4 \end{bmatrix} \quad \boldsymbol{R} = \begin{bmatrix} \mathbf{1} & 0 & 2 & 0 \\ 0 & \mathbf{1} & 0 & 2 \end{bmatrix} = [\,\boldsymbol{I} \ \ \boldsymbol{F}\,]$$

$\uparrow \ \uparrow$ **pivot columns** $\quad \uparrow \ \uparrow$ **free columns**

For the free variables x_3 and x_4, we make special choices of ones and zeros. First $x_3 = 1$, $x_4 = 0$ and second $x_3 = 0$, $x_4 = 1$. The pivot variables x_1 and x_2 are determined by the equation $U\boldsymbol{x} = \mathbf{0}$ (or $R\boldsymbol{x} = \mathbf{0}$). We get two special solutions in the nullspace of M. This is also the nullspace of U and R: *elimination doesn't change solutions.*

Special solutions

$$\boldsymbol{R} = \begin{bmatrix} 1 & 0 & 2 & 0 \\ 0 & 1 & 0 & 2 \end{bmatrix} \qquad \boldsymbol{s}_1 = \begin{bmatrix} -2 \\ 0 \\ 1 \\ 0 \end{bmatrix} \quad \text{and } \boldsymbol{s}_2 = \begin{bmatrix} 0 \\ -2 \\ 0 \\ 1 \end{bmatrix} \begin{matrix} \leftarrow \textbf{2 pivot} \\ \leftarrow \textbf{variables} \\ \leftarrow \textbf{2 free} \\ \leftarrow \textbf{variables} \end{matrix}$$

$$R\boldsymbol{s}_1 = \mathbf{0} \quad R\boldsymbol{s}_2 = \mathbf{0}$$

The Reduced Row Echelon Form $\boldsymbol{R}$

Summary Elimination will not stop at the upper triangular U. We continue to R_0 and R.

1. ***Produce zeros above the pivots.*** **Use pivot rows to eliminate upward.**
2. ***Produce ones in the pivots.*** **Divide the whole pivot row by its pivot.**

Those steps don't change the zero vector on the right side of the equation. The nullspace stays the same: $\mathbf{N}(A) = \mathbf{N}(U) = \mathbf{N}(R)$. This nullspace becomes easiest to see when we reach the ***reduced row echelon form. The pivot columns of R contain I.***

Reduced form $\boldsymbol{R}$
$$U = \begin{bmatrix} 1 & 2 & 2 & 4 \\ 0 & 2 & 0 & 4 \end{bmatrix} \quad \text{becomes} \quad R = \begin{bmatrix} \mathbf{1} & \mathbf{0} & 2 & 0 \\ \mathbf{0} & \mathbf{1} & 0 & 2 \end{bmatrix}.$$

I subtracted row 2 of U from row 1. Then I multiplied row 2 by $\frac{1}{2}$ to get pivot $= 1$.

Now (**free column 3**) $=$ **2** (**pivot column 1**), so -2 appears in $\boldsymbol{s}_1 = (-2, 0, 1, 0)$. The special solutions are much easier to find from the reduced system $R\boldsymbol{x} = \mathbf{0}$. In each pivot column of R, change all the signs to find $\boldsymbol{s}$. Second special solution $\boldsymbol{s}_2 = (0, -2, 0, 1)$.

Before moving to m by n matrices A and their nullspaces $\mathbf{N}(A)$ and special solutions, allow me to repeat one comment. For many matrices, the only solution to $A\boldsymbol{x} = \mathbf{0}$ is $\boldsymbol{x} = \mathbf{0}$. Their nullspaces $\mathbf{N}(A) = \mathbf{Z}$ contain only that zero vector: *no* special solutions. The only combination of the columns that produces $\boldsymbol{b} = \mathbf{0}$ is then the "zero combination".

This case of a zero nullspace $\mathbf{Z}$ is of the greatest importance. It says that the columns of A are **independent**. No combination of columns gives the zero vector (except $\boldsymbol{x} = \mathbf{0}$). But this can't happen if $n > m$. We can't have n independent columns in $\mathbf{R}^m$.

Important Suppose A has more columns than rows. With $n > m$ there is at least one free variable. The system $A\boldsymbol{x} = \mathbf{0}$ has at least one nonzero solution.

Suppose $A\boldsymbol{x} = \mathbf{0}$ has more unknowns than equations ($\boldsymbol{n} > \boldsymbol{m}$). There must be **at least $\boldsymbol{n} - \boldsymbol{m}$ free columns. $A\boldsymbol{x} = \mathbf{0}$ has nonzero solutions in $\mathbf{N}(A)$.**

The nullspace is a subspace. Its "dimension" is the number of free variables. This central idea—the ***dimension of a subspace***—is explained in Section 3.5 of this chapter.

Pivot Variables and Free Variables in the Echelon Matrix R

$$A=\begin{bmatrix} p & p & f & p & f \\ | & | & | & | & | \\ | & | & | & | & | \\ | & | & | & | & | \end{bmatrix} \quad R=\begin{bmatrix} \mathbf{1} & 0 & a & 0 & c \\ 0 & \mathbf{1} & b & 0 & d \\ 0 & 0 & 0 & \mathbf{1} & e \\ 0 & 0 & 0 & 0 & 0 \end{bmatrix} \quad s_1=\begin{bmatrix} -a \\ -b \\ \mathbf{1} \\ 0 \\ \mathbf{0} \end{bmatrix} \quad s_2=\begin{bmatrix} -c \\ -d \\ \mathbf{0} \\ -e \\ \mathbf{1} \end{bmatrix}$$

3 pivot columns p	I in pivot columns	special $Rs_1=\mathbf{0}$ and $Rs_2=\mathbf{0}$
2 free columns f	F in free columns	$-a$ to $-e$ come from R
to be revealed by R	3 pivots: rank $r=3$	$Rs=\mathbf{0}$ means $As=\mathbf{0}$

R shows clearly: *column* $3 = \boldsymbol{a}\,$(*column* 1) $+ \boldsymbol{b}\,$(*column* 2). The same must be true for A. The special solution $\boldsymbol{s}_1$ repeats that combination so $(-\boldsymbol{a},-\boldsymbol{b},\mathbf{1},0,0)$ has $R\boldsymbol{s}_1=\mathbf{0}$. Nullspace of $A=$ Nullspace of $R=$ all combinations of $\boldsymbol{s}_1$ and $\boldsymbol{s}_2$.

On the next page you will see simple formulas for the echelon matrix R and the $n-r$ special solutions $x=s$ to $Ax=0$ and $Rx=0$.

Example This 4 by 7 reduced row echelon matrix $\boldsymbol{R}_0$ has 3 pivots. Delete row 4 to find R.

$$\boldsymbol{R}_0=\begin{bmatrix} \mathbf{1} & \mathbf{0} & x & x & x & \mathbf{0} & x \\ \mathbf{0} & \mathbf{1} & x & x & x & \mathbf{0} & x \\ \mathbf{0} & \mathbf{0} & 0 & 0 & 0 & \mathbf{1} & x \\ 0 & 0 & 0 & 0 & 0 & 0 & 0 \end{bmatrix}$$

Three pivot variables x_1,x_2,x_6
Four free variables x_3,x_4,x_5,x_7
Four special solutions s in $\mathbf{N}(R_0)=\mathbf{N}(R)$
The pivot rows and columns contain I

$\boldsymbol{R}=\begin{bmatrix}\boldsymbol{I} & \boldsymbol{F}\end{bmatrix}\boldsymbol{P}$ has row 4 removed. The permutation P puts column 3 of I into column 6.

Question What are the column space and the nullspace for this matrix $\boldsymbol{R}$?

Answer The columns of R_0 have four components so they lie in $\mathbf{R}^4$. (Not in $\mathbf{R}^3$!) The fourth component of every column is zero. *The column space of R_0 consists of all vectors of the form* $(b_1,b_2,b_3,0)$. The nullspace $\mathbf{N}(R)=\mathbf{N}(R_0)$ is a subspace of $\mathbf{R}^7$. The solutions to $R_0\boldsymbol{x}=\mathbf{0}$ are combinations of the four special solutions—*one for each free variable*:

1. Columns 3, 4, 5, 7 have no pivots. So the four free variables are x_3,x_4,x_5,x_7.
2. Set one free variable to 1 and set the other three free variables to zero.
3. To find each $\boldsymbol{s}$, solve $R\boldsymbol{s}=\mathbf{0}$ for the pivot variables x_1,x_2,x_6. Four special solutions.

To repeat: *A short wide matrix $(n>m)$ always has nonzero vectors in its nullspace.* There must be at least $n-m$ free variables, since the number of pivots cannot exceed m.

The Echelon Form and Special Solutions in Matrix Language

From the examples you see the steps to R_0 and R. Chapter 2 produced zeros below the pivots in U. Chapter 3 also has zeros *above* the pivots in R. All pivots are 1. We now have a systematic way to identify independent columns in A and to reach $A = CR$.

This row echelon form is famous, but its simple matrix formula is seldom given. This page will give formulas for R_0 and R, along with the special solutions to $A\boldsymbol{s} = \mathbf{0}$. Those $n - r$ special solutions combine to give the nullspace: all solutions to $A\boldsymbol{x} = \mathbf{0}$. R_0 comes from elimination (down and up) on A. Here are the basic formulas.

$$\boldsymbol{R_0} = \begin{bmatrix} \boldsymbol{I} & \boldsymbol{F} \\ 0 & 0 \end{bmatrix} \boldsymbol{P} \qquad \boldsymbol{R} = \begin{bmatrix} \boldsymbol{I} & \boldsymbol{F} \end{bmatrix} \boldsymbol{P} \qquad \boldsymbol{A} = \boldsymbol{CR} = \begin{bmatrix} \boldsymbol{C} & \boldsymbol{CF} \end{bmatrix} \boldsymbol{P} \qquad (1)$$

That column permutation P puts the columns of I and F into their correct positions in R.

F tells how the independent columns in A combine into the dependent columns

Special solutions to $\boldsymbol{Ax} = \mathbf{0}$ Since A has rank r, we expect $n-r$ independent solutions.

$$\boldsymbol{Ax} = \mathbf{0} \quad \text{gives} \quad \boldsymbol{Rx} = \begin{bmatrix} \boldsymbol{I} & \boldsymbol{F} \end{bmatrix} \boldsymbol{Px} = \mathbf{0}$$

Here I is r by r and F is r by $n - r$. Thanks to the simplicity of I, and the fact that $PP^{\mathrm{T}} = I$, we know immediately the matrix $\boldsymbol{S}$ of special solutions $\begin{bmatrix} \boldsymbol{s}_1 & \dots & \boldsymbol{s}_{n-r} \end{bmatrix}$.

$$\boldsymbol{S} = \boldsymbol{P}^{\mathrm{T}} \begin{bmatrix} -\boldsymbol{F} \\ \boldsymbol{I} \end{bmatrix} \quad \text{and} \quad \boldsymbol{RS} = \begin{bmatrix} \boldsymbol{I} & \boldsymbol{F} \end{bmatrix} \boldsymbol{PS} = \begin{bmatrix} \boldsymbol{I} & \boldsymbol{F} \end{bmatrix} \boldsymbol{PP}^{\mathrm{T}} \begin{bmatrix} -\boldsymbol{F} \\ \boldsymbol{I} \end{bmatrix} = 0$$

S has n rows and $n-r$ columns (special solutions). *The identity matrix in S has size $n-r$.* Each column has a 1, as special solutions always do. The other nonzeros in that column come directly from F, with signs reversed to $-F$. The role of P^{T} is to move the 1's into the right positions (free positions) in these special solutions. If the r independent columns of A come first, then P is the identity matrix and S is truly simple: $RS = -F + F = 0$.

Here is a magic factorization that treats rows and columns of A in the same way. C contains the first r independent columns of A as always. **Suppose $\boldsymbol{R}^*$ contains the first $\boldsymbol{r}$ independent rows of $\boldsymbol{A}$.** (We know that row rank = column rank.) The rows of R^* will meet the columns of C in an r by r matrix W. Then $\boldsymbol{A}$ factors into $\boldsymbol{CW}^{-1}\boldsymbol{R}^*$.

$$\boldsymbol{A} = \begin{bmatrix} \\ \boldsymbol{C} \\ \\ \end{bmatrix} \begin{bmatrix} \boldsymbol{W}^{-1} \end{bmatrix} \begin{bmatrix} \quad \boldsymbol{R}^* \quad \end{bmatrix} \text{ as in } \begin{bmatrix} 1 & 4 & 7 \\ 2 & 5 & 8 \\ 3 & 6 & 9 \end{bmatrix} = \begin{bmatrix} 1 & 4 \\ 2 & 5 \\ 3 & 6 \end{bmatrix} \begin{bmatrix} \mathbf{1} & \mathbf{4} \\ \mathbf{2} & \mathbf{5} \end{bmatrix}^{-1} \begin{bmatrix} 1 & 4 & 7 \\ 2 & 5 & 8 \end{bmatrix}$$

The first columns of $W^{-1}R^*$ will be $W^{-1}W = I$. The last column will be the free part F. The permutation is just $P = I$, since the independent rows and columns came first in A.

$\boldsymbol{W}^{-1}\boldsymbol{R}^*$ **is the same matrix as** $\boldsymbol{R} = \begin{bmatrix} \boldsymbol{I} & \boldsymbol{F} \end{bmatrix}$. **The free part is** $\boldsymbol{F} = \begin{bmatrix} \mathbf{1} & \mathbf{4} \\ \mathbf{2} & \mathbf{5} \end{bmatrix}^{-1} \begin{bmatrix} \mathbf{7} \\ \mathbf{8} \end{bmatrix}$.

Three Identical Factorizations $A = CR = C\begin{bmatrix} I & F \end{bmatrix} P = CW^{-1}R^*$

This very optional page completes the presentation of $A = CR$ factorizations—all with the same C and R. C contains the first r independent columns of A, and R^* contains the first r independent rows of A. C and R^* meet in an r by r matrix W. Then $W^{-1} = M$ is the mixing matrix, and a small example from page 32 has grown into the "magic factorization" $\boldsymbol{A} = \boldsymbol{CW^{-1}R^*}$.

$$A = \begin{bmatrix} \mathbf{1} & \mathbf{4} & 7 \\ \mathbf{2} & \mathbf{5} & 8 \\ 3 & 6 & 9 \end{bmatrix} \qquad R = W^{-1}R^* = \frac{1}{3}\begin{bmatrix} -5 & 4 \\ 2 & -1 \end{bmatrix}\begin{bmatrix} 1 & 4 & 7 \\ 2 & 5 & 8 \end{bmatrix} = \begin{bmatrix} \mathbf{1} & \mathbf{0} & \mathbf{-1} \\ \mathbf{0} & \mathbf{1} & \mathbf{2} \end{bmatrix} = \begin{bmatrix} I & F \end{bmatrix}$$

$\boldsymbol{A}$ = any matrix of rank r	$m \times n$	
$\boldsymbol{C}$ = first r independent columns of A	$m \times r$	
$\boldsymbol{R^*}$ = first r independent rows of A	$r \times n$	
$\boldsymbol{W}$ = intersection of C and R^*	$r \times r$	

$$\boldsymbol{A} = \begin{bmatrix} \boldsymbol{W} & \boldsymbol{H} \\ \boldsymbol{J} & \boldsymbol{K} \end{bmatrix} \begin{matrix} r \\ m-r \end{matrix} \quad (\text{columns } r,\ n-r)$$

$$\boldsymbol{C} = \begin{bmatrix} \boldsymbol{W} \\ \boldsymbol{J} \end{bmatrix} \qquad \boldsymbol{R^*} = \begin{bmatrix} \boldsymbol{W} & \boldsymbol{H} \end{bmatrix}$$

Theorem **The r by r matrix $\boldsymbol{W}$ also has rank r and $\boldsymbol{A} = \boldsymbol{CW^{-1}R^*}$.**

1. Combinations V of the rows of R^* must produce the dependent rows in $\begin{bmatrix} \boldsymbol{J} & \boldsymbol{K} \end{bmatrix}$

   Then $\begin{bmatrix} \boldsymbol{J} & \boldsymbol{K} \end{bmatrix} = \boldsymbol{VR^*} = \begin{bmatrix} \boldsymbol{VW} & \boldsymbol{VH} \end{bmatrix}$ for some matrix V and $\boldsymbol{C} = \begin{bmatrix} \boldsymbol{I} \\ \boldsymbol{V} \end{bmatrix}\boldsymbol{W}$

2. Combinations T of the columns of C must produce the dependent columns in $\begin{bmatrix} \boldsymbol{H} \\ \boldsymbol{K} \end{bmatrix}$

   Then $\begin{bmatrix} \boldsymbol{H} \\ \boldsymbol{K} \end{bmatrix} = \boldsymbol{CT} = \begin{bmatrix} \boldsymbol{WT} \\ \boldsymbol{JT} \end{bmatrix}$ for some matrix T and $\boldsymbol{R^*} = \boldsymbol{W}\begin{bmatrix} \boldsymbol{I} & \boldsymbol{T} \end{bmatrix}$

3. $\boldsymbol{A} = \begin{bmatrix} W & H \\ VW & VH \end{bmatrix} = \begin{bmatrix} W & WT \\ VW & VWT \end{bmatrix} = \begin{bmatrix} \boldsymbol{I} \\ \boldsymbol{V} \end{bmatrix}\begin{bmatrix} \boldsymbol{W} \end{bmatrix}\begin{bmatrix} \boldsymbol{I} & \boldsymbol{T} \end{bmatrix} = \boldsymbol{CW^{-1}R^*}$

Since A has rank r, its factors must have rank $\geq r$. From their shapes that means rank r. If C and R^* were not in the first r columns and rows of A, then permutations P_R of the rows and P_C of the columns will give $P_R\,A\,P_C = \begin{bmatrix} W & H \\ J & K \end{bmatrix}$ and the proof goes through.

I. Find C and R^* and W and W^{-1} and $R = W^{-1}R^*$ for the transpose of A above.

II. Explain these statements about the rank of augmented matrices $\begin{bmatrix} A & \boldsymbol{b} \end{bmatrix}$ and $\begin{bmatrix} C & D \end{bmatrix}$.

The rank of $\boldsymbol{A}$ equals the rank of $\begin{bmatrix} \boldsymbol{A} & \boldsymbol{b} \end{bmatrix}$ if and only if $\boldsymbol{Ax} = \boldsymbol{b}$ is solvable.
The rank of $\boldsymbol{C}$ equals the rank of $\begin{bmatrix} \boldsymbol{C} & \boldsymbol{D} \end{bmatrix}$ if and only if $\boldsymbol{CT} = \boldsymbol{D}$ is solvable.

III. If $A = C\,M\,R^*$ has sizes $(m \times r)\,(r \times r)\,(r \times n)$ and rank $A = r$, show that rank $M = r$.

Problem Set 3.2

1 Why do A and $R = EA$ have the same nullspace? We know that E is invertible.

2 Find the row reduced form R and the rank r of A and B (*those depend on* c). Which are the pivot columns? Find the special solutions to $A\boldsymbol{x} = \mathbf{0}$ and $B\boldsymbol{x} = \mathbf{0}$.

Find special solutions $$A = \begin{bmatrix} 1 & 2 & 1 \\ 3 & 6 & 3 \\ 4 & 8 & c \end{bmatrix} \quad \text{and} \quad B = \begin{bmatrix} c & c \\ c & c \end{bmatrix}.$$

3 Create a 2 by 4 matrix R whose special solutions to $R\boldsymbol{x} = \mathbf{0}$ are $\boldsymbol{s}_1$ and $\boldsymbol{s}_2$:

$$\boldsymbol{s}_1 = \begin{bmatrix} -3 \\ \mathbf{1} \\ 0 \\ \mathbf{0} \end{bmatrix} \quad \text{and} \quad \boldsymbol{s}_2 = \begin{bmatrix} -2 \\ \mathbf{0} \\ -6 \\ \mathbf{1} \end{bmatrix}$$

pivot columns 1 and 3
free variables x_2 and x_4
x_2 and x_4 are $1, 0$ and $0, 1$

Describe all 2 by 4 matrices with this nullspace $\mathbf{N}(A)$ spanned by $\boldsymbol{s}_1$ and $\boldsymbol{s}_2$.

4 Reduce A and B to their echelon forms R. Which variables are free?

$$\text{(a)}\ A = \begin{bmatrix} 1 & 2 & 2 & 4 & 6 \\ 1 & 2 & 3 & 6 & 9 \\ 0 & 0 & 1 & 2 & 3 \end{bmatrix} \qquad \text{(b)} \quad B = \begin{bmatrix} 2 & 4 & 2 \\ 0 & 4 & 4 \\ 0 & 8 & 8 \end{bmatrix}.$$

5 For the matrix A in Problem 4, find a special solution to $R\boldsymbol{x} = \mathbf{0}$ for each free variable. Set the free variable to 1. Set the other free variables to zero. Solve $R\boldsymbol{x} = \mathbf{0}$.

6 True or false (with reason if true or example to show it is false):

(a) A square matrix has no free variables.

(b) An invertible matrix has no free variables.

(c) An m by n matrix has no more than n pivot variables.

(d) An m by n matrix has no more than m pivot variables.

7 Put as many 1's as possible in a 4 by 7 echelon matrix U whose pivot columns are

(a) $2, 4, 5$ (b) $1, 3, 6, 7$ (c) 4 and 6.

8 Put as many 1's as possible in a 4 by 8 *reduced* echelon matrix R so that the free columns are (a) $2, 4, 5, 6$ or (b) $1, 3, 6, 7, 8$.

9 Suppose column 4 of a 3 by 5 matrix is all zero. Then x_4 is certainly a ______ variable. The special solution for this variable is the vector $\boldsymbol{x} =$ ______

10 Suppose the first and last columns of a 3 by 5 matrix are the same (not zero). Then ______ is a free variable. Find the special solution for this free variable.

11 The nullspace of a 5 by 5 matrix contains only $\boldsymbol{x} = \mathbf{0}$ when the matrix has ______ pivots. In that case the column space is $\mathbf{R}^5$. Explain why.

12 Suppose an m by n matrix has r pivots. The number of special solutions is _____, by the Counting Theorem. The nullspace contains only $\boldsymbol{x} = \mathbf{0}$ when $r =$ _____. The column space is all of $\mathbf{R}^m$ when the rank is $r =$ _____.

13 (Recommended) The plane $x - 3y - z = 12$ is parallel to $x - 3y - z = 0$. One particular point on this plane is $(12, 0, 0)$. All points on the plane have the form

$$\begin{bmatrix} x \\ y \\ z \end{bmatrix} = \begin{bmatrix} \\ 0 \\ 0 \end{bmatrix} + y \begin{bmatrix} \\ 1 \\ 0 \end{bmatrix} + z \begin{bmatrix} \\ 0 \\ 1 \end{bmatrix}.$$

14 Suppose column 1 + column 3 + column 5 = $\mathbf{0}$ in a 4 by 5 matrix with four pivots. Which column has no pivot? What is the special solution? Describe $\mathbf{N}(A)$.

Questions 15–20 ask for matrices (if possible) with specific properties.

15 Construct a matrix for which $\mathbf{N}(A) =$ all combinations of $(2, 2, 1, 0)$ and $(3, 1, 0, 1)$.

16 Construct A so that $\mathbf{N}(A) =$ all multiples of $(4, 3, 2, 1)$. Its rank is _____.

17 Construct a matrix whose column space contains $(1, 1, 5)$ and $(0, 3, 1)$ and whose nullspace contains $(1, 1, 2)$.

18 Construct a matrix whose column space contains $(1, 1, 0)$ and $(0, 1, 1)$ and whose nullspace contains $(1, 0, 1)$.

19 Construct a 2 by 2 matrix whose nullspace equals its column space. This is possible.

20 Why does no 3 by 3 matrix have a nullspace that equals its column space?

21 If $AB = 0$ then the column space of B is contained in the _____ of A. Why?

22 The reduced form R of a 3 by 3 matrix with randomly chosen entries is almost sure to be _____. What R is virtually certain if the random A is 4 by 3?

23 If $\mathbf{N}(A) =$ all multiples of $\boldsymbol{x} = (2, 1, 0, 1)$, what is R and what is its rank?

24 If the special solutions to $R\boldsymbol{x} = \mathbf{0}$ are in the columns of these nullspace matrices N, go backward to find the nonzero rows of the reduced matrices R:

$$N = \begin{bmatrix} 2 & 3 \\ 1 & 0 \\ 0 & 1 \end{bmatrix} \quad \text{and} \quad N = \begin{bmatrix} 0 \\ 0 \\ 1 \end{bmatrix} \quad \text{and} \quad N = \begin{bmatrix} \\ \\ \end{bmatrix} \quad \text{(empty 3 by 1).}$$

25 (a) What are the five 2 by 2 reduced matrices R whose entries are all 0's and 1's?

(b) What are the eight 1 by 3 matrices containing only 0's and 1's? Are all eight of them reduced echelon matrices R?

26 If A is 4 by 4 and invertible, describe the nullspace of the 4 by 8 matrix $B = [A \;\; A]$. Explain why A and $-A$ always have the same reduced echelon form R.

27 How is the nullspace $\mathbf{N}(C)$ related to the spaces $\mathbf{N}(A)$ and $\mathbf{N}(B)$, if $C = \begin{bmatrix} A \\ B \end{bmatrix}$?

28 Find the reduced R_0 and R for each of these matrices:

$$A = \begin{bmatrix} 0 & 0 & 0 \\ 2 & 4 & 6 \end{bmatrix} \qquad B = \begin{bmatrix} A & A \\ 0 & A \end{bmatrix} \qquad C = \begin{bmatrix} A & A \\ A & 0 \end{bmatrix}$$

29 Suppose the 2 pivot variables come *last* instead of first. Describe the reduced matrix R (3 columns) and the nullspace matrix N containing the special solutions.

30 If A has r pivot columns, how do you know that A^{T} has r pivot columns? Give a 3 by 3 example with different pivot column numbers for A and A^{T}.

31 Fill out these matrices so that they have rank 1:

$$A = \begin{bmatrix} a & b & c \\ d & & \\ g & & \end{bmatrix} \quad \text{and} \quad B = \begin{bmatrix} & 9 & \\ 1 & & \\ 2 & 6 & -3 \end{bmatrix} \quad \text{and} \quad M = \begin{bmatrix} a & b \\ c & \end{bmatrix}.$$

32 If A is a rank one matrix, the second row of R is ______. Do an example.

33 *If A has rank r, then it has an r by r submatrix S that is invertible.* Remove $m - r$ rows and $n - r$ columns to find an invertible submatrix S inside A, B, and C. You could keep the pivot rows and pivot columns:

$$A = \begin{bmatrix} 1 & 2 & 3 \\ 1 & 2 & 4 \end{bmatrix} \qquad B = \begin{bmatrix} 1 & 2 & 3 \\ 2 & 4 & 6 \end{bmatrix} \qquad C = \begin{bmatrix} 0 & 1 & 0 \\ 0 & 0 & 0 \\ 0 & 0 & 1 \end{bmatrix}.$$

34 Suppose A and B have the *same* reduced row echelon form R.

(a) Show that A and B have the same nullspace and the same row space.

(b) We know $E_1 A = R$ and $E_2 B = R$. So A equals an ______ matrix times B.

35 Kirchhoff's Current Law $A^{\mathrm{T}}\boldsymbol{y} = \mathbf{0}$ says that *current in* = *current out* at every node. At node 1 this is $y_3 = y_1 + y_4$ (arrows show the positive direction of each y). Reduce A^{T} to R (3 rows) and find three special solutions in the nullspace of A^{T}.

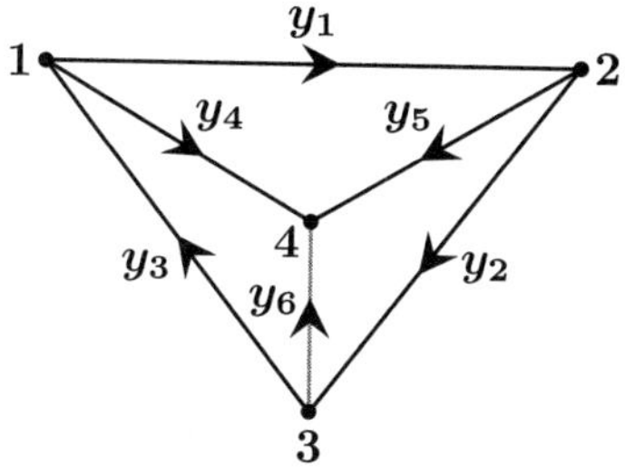

$$A^{\mathrm{T}} = \begin{bmatrix} -1 & 0 & 1 & -1 & 0 & 0 \\ 1 & -1 & 0 & 0 & -1 & 0 \\ 0 & 1 & -1 & 0 & 0 & 1 \\ 0 & 0 & 0 & 1 & 1 & 1 \end{bmatrix}$$

36 C contains the r pivot columns of A. Find the r pivot columns of C^{T} (r by m). Transpose back to find an r by r invertible submatrix S inside A:

$$\text{For } A = \begin{bmatrix} 1 & 2 & 3 \\ 2 & 4 & 6 \\ 2 & 4 & 7 \end{bmatrix} \quad \text{find } C \text{ (3 by 2) and then the invertible } S \text{ (2 by 2).}$$

37 Why is the column space $\mathbf{C}(AB)$ a subspace of $\mathbf{C}(A)$? Then $\operatorname{rank}(AB) \le \operatorname{rank}(A)$.

38 Suppose column j of B is a combination of previous columns of B. Show that column j of AB is the same combination of previous columns of AB. Then AB cannot have new pivot columns, so $\mathbf{rank}(\boldsymbol{AB}) \le \mathbf{rank}(\boldsymbol{B})$.

39 *(Important)* Suppose A and B are n by n matrices, and $AB = I$. Prove from $\operatorname{rank}(AB) \le \operatorname{rank}(A)$ that the rank of A is n. So A is invertible and B must be its inverse. Therefore $BA = I$ *(which is not so obvious!)*.

40 If A is 2 by 3 and B is 3 by 2 and $AB = I$, show from its rank that $BA \neq I$. Give an example of A and B with $AB = I$. For $m < n$, a right inverse is not a left inverse.

41 What is the nullspace matrix N (containing the special solutions) for A, B, C?

$$\text{2 by 2 blocks} \quad A = \begin{bmatrix} I & I \end{bmatrix} \quad \text{and} \quad B = \begin{bmatrix} I & I \\ 0 & 0 \end{bmatrix} \quad \text{and} \quad C = \begin{bmatrix} I & I & I \end{bmatrix}.$$

42 Suppose A is an m by n matrix of rank r. Its reduced echelon form (including any zero rows) is R_0. Describe exactly the matrix Z (its shape and all its entries) that comes from *transposing the reduced row echelon form of* R_0^{T}: $\boldsymbol{Z} = (\mathrm{rref}(\boldsymbol{A}^{\mathrm{T}}))^{\mathrm{T}}$.

43 (Recommended) Suppose $R_0 = \begin{bmatrix} I & F \\ 0 & 0 \end{bmatrix}$ is m by n of rank r. Pivot columns first:

(a) What are the shapes of those four blocks, based on m and n and r?

(b) Find a *right inverse* B with $R_0B = I$ if $r = m$. The zero blocks are gone.

(c) Find a *left inverse* C with $CR_0 = I$ if $r = n$. The F and 0 column is gone.

(d) What is the reduced row echelon form of R_0^{T} (with shapes)?

(e) What is the reduced row echelon form of $R_0^{\mathrm{T}}R_0$ (with shapes)?

44 Suppose you allow elementary *column* operations on A as well as elementary row operations (which get to R_0). What is the "row-and-column reduced form" for an m by n matrix A of rank r?

45 Verify that equation (1) on page 89 is correct: $W = \begin{bmatrix} 1 & 4 \\ 2 & 5 \end{bmatrix}$ is invertible and

$$F = W^{-1}\begin{bmatrix} 7 \\ 8 \end{bmatrix} \quad \text{agrees with } F \text{ in } A = CR = C\begin{bmatrix} I & F \end{bmatrix}.$$

46 The magic factorization is easy if the first r rows and columns of A are independent. What multiple of block row 1 will equal block row 2 of this matrix?

$$\begin{bmatrix} W \\ J \end{bmatrix}\begin{bmatrix} W^{-1} \end{bmatrix}\begin{bmatrix} W & H \end{bmatrix} = \begin{bmatrix} W & H \\ J & JW^{-1}H \end{bmatrix}$$

Elimination: The Big Picture

This page explains elimination at the vector level and subspace level, when A is reduced to R. You know the steps and I won't repeat them. Elimination starts with the first pivot. It moves a column at a time (left to right) and a row at a time (top to bottom) for U. Then upwards elimination produces R_0 and R. Elimination answers two questions:

Question 1 Is this column a combination of previous columns?

If the column contains a pivot, the answer is no. Pivot columns are "independent" of previous columns. If column 4 has no pivot, it is a combination of columns $1, 2, 3$.

Question 2 Is this row a combination of previous rows?

If the row contains a pivot, the answer is no. Pivot rows are independent of previous rows, and their first nonzero is 1 from I. Rows that are all zero in R_0 were and are not independent, and they disappear in R.

It is amazing to me that one pass through the matrix answers both questions 1 and 2. Elimination acts on the rows but the result tells us about the columns! The identity matrix in R locates the first r independent columns in A.Then the free columns F in R tell us *the combinations of those independent columns that produce the dependent columns in* A. This is easy to miss without seeing the factorization $A = CR$.

R tells us the special solutions to $A\boldsymbol{x} = \mathbf{0}$. We could reach R from A by different row exchanges and elimination steps, but it will always be the same R. (This is because the special solutions are decided by A. The formula comes before Problem Set 3.2.) In the language coming soon, R reveals a "basis" for three of the fundamental subspaces:

The **column space** of A—choose the columns of A that produce pivots in R.

The **row space** of A—choose the rows of R as a basis.

The **nullspace** of A—choose the special solutions to $R\boldsymbol{x} = \mathbf{0}$ (and $A\boldsymbol{x} = \mathbf{0}$).

For the left nullspace $\mathbf{N}(A^{\mathrm{T}})$, we look at the elimination step $EA = R_0$. The last $m - r$ rows of R_0 are zero. The last $m - r$ rows of E are a basis for the left nullspace! In reducing $[A \;\; I]$ to $[R_0 \;\; E]$, the matrix E keeps a record of elimination that is otherwise lost.

Suppose we fix C and R^* (m by r and r by n, both rank r). Choose any invertible r by r mixing matrix M. All the matrices CMR^* (and only those) have the **same four fundamental subspaces** as the original A.

3.3 The Complete Solution to $Ax = b$

1 Complete solution to $A\boldsymbol{x} = \boldsymbol{b}$: $\boldsymbol{x} =$ (one particular solution $\boldsymbol{x}_p$) + (any $\boldsymbol{x}_n$ in the nullspace).

2 Elimination on $\begin{bmatrix} A & \boldsymbol{b} \end{bmatrix}$ leads to $\begin{bmatrix} R_0 & \boldsymbol{d} \end{bmatrix}$: Solvable when zero rows of R_0 have zeros in $\boldsymbol{d}$.

3 When $R_0\boldsymbol{x} = \boldsymbol{d}$ is solvable, one very particular solution $\boldsymbol{x}_p$ has all free variables equal to zero.

4 A has **full column rank** $r = n$ when its nullspace $\mathbf{N}(A) =$ zero vector: *no free variables.*

5 A has **full row rank** $r = m$ when its column space $\mathbf{C}(A)$ is $\mathbf{R}^m$: $A\boldsymbol{x} = \boldsymbol{b}$ is always solvable.

The last section totally solved $A\boldsymbol{x} = \mathbf{0}$. Elimination converted the problem to $R_0\boldsymbol{x} = \mathbf{0}$. The free variables were given special values (one and zero). Then the pivot variables were found by back substitution. We paid no attention to the right side $\boldsymbol{b}$ because it stayed at zero. Then zero rows in R_0 were no problem.

Now $\boldsymbol{b}$ is not zero. Row operations on the left side must act also on the right side. $A\boldsymbol{x} = \boldsymbol{b}$ is reduced to a simpler system $R_0\boldsymbol{x} = \boldsymbol{d}$ with the same solutions (if any). One way to organize that is to ***add b as an extra column of the matrix***. I will *"augment" A* with the right side $(b_1, b_2, b_3) = (1, 6, 7)$ to produce the **augmented matrix** $\begin{bmatrix} A & \boldsymbol{b} \end{bmatrix}$:

$$\begin{bmatrix} 1 & 3 & 0 & 2 \\ 0 & 0 & 1 & 4 \\ 1 & 3 & 1 & 6 \end{bmatrix} \begin{bmatrix} x_1 \\ x_2 \\ x_3 \\ x_4 \end{bmatrix} = \begin{bmatrix} \mathbf{1} \\ \mathbf{6} \\ \mathbf{7} \end{bmatrix} \quad \begin{matrix} \text{has the} \\ \text{augmented} \\ \text{matrix} \end{matrix} \quad \begin{bmatrix} 1 & 3 & 0 & 2 & \mathbf{1} \\ 0 & 0 & 1 & 4 & \mathbf{6} \\ 1 & 3 & 1 & 6 & \mathbf{7} \end{bmatrix} = \begin{bmatrix} A & \boldsymbol{b} \end{bmatrix}.$$

When we apply the usual elimination steps to A, reaching R_0, we also apply them to $\boldsymbol{b}$.

In this example we subtract row 1 from row 3. Then we subtract row 2 from row 3. This produces a *row of zeros in* R_0, and it changes $\boldsymbol{b}$ to a new right side $\boldsymbol{d} = (\mathbf{1}, \mathbf{6}, \mathbf{0})$:

$$\begin{bmatrix} 1 & 3 & 0 & 2 \\ 0 & 0 & 1 & 4 \\ 0 & 0 & 0 & 0 \end{bmatrix} \begin{bmatrix} x_1 \\ x_2 \\ x_3 \\ x_4 \end{bmatrix} = \begin{bmatrix} \mathbf{1} \\ \mathbf{6} \\ \mathbf{0} \end{bmatrix} \quad \begin{matrix} \text{has the} \\ \text{augmented} \\ \text{matrix} \end{matrix} \quad \boxed{\begin{bmatrix} 1 & 3 & 0 & 2 & \mathbf{1} \\ 0 & 0 & 1 & 4 & \mathbf{6} \\ 0 & 0 & 0 & 0 & \mathbf{0} \end{bmatrix} = \begin{bmatrix} R_0 & \boldsymbol{d} \end{bmatrix}.}$$

That very last row is crucial. The third equation has become $0 = 0$. So the equations can be solved. In the original matrix A, the first row plus the second row equals the third row. To solve $A\boldsymbol{x} = \boldsymbol{b}$, we need $b_1 + b_2 = b_3$ on the right side too. The all-important property of $\boldsymbol{b}$ was $\mathbf{1} + \mathbf{6} = \mathbf{7}$. That led to $0 = 0$ in the third equation. This was essential.

Here are the same augmented matrices for a general $\boldsymbol{b} = (b_1, b_2, b_3)$:

$$\begin{bmatrix} A & \boldsymbol{b} \end{bmatrix} = \begin{bmatrix} 1 & 3 & 0 & 2 & \boldsymbol{b}_1 \\ 0 & 0 & 1 & 4 & \boldsymbol{b}_2 \\ 1 & 3 & 1 & 6 & \boldsymbol{b}_3 \end{bmatrix} \longrightarrow \begin{bmatrix} 1 & 3 & 0 & 2 & \boldsymbol{b}_1 \\ 0 & 0 & 1 & 4 & \boldsymbol{b}_2 \\ 0 & 0 & 0 & 0 & \boldsymbol{b}_3 - \boldsymbol{b}_1 - \boldsymbol{b}_2 \end{bmatrix} = \begin{bmatrix} R_0 & \boldsymbol{d} \end{bmatrix}$$

Now we get $0 = 0$ in the third equation only if $b_3 - b_1 - b_2 = 0$. This is $b_1 + b_2 = b_3$.

One Particular Solution $Ax_p = b$

For an easy solution $\boldsymbol{x}_p$, *choose the free variables to be zero:* $x_2 = x_4 = 0$. Then the two nonzero equations give the two pivot variables $x_1 = 1$ and $x_3 = 6$. Our particular solution to $A\boldsymbol{x} = \boldsymbol{b}$ (and also $R_0\boldsymbol{x} = \boldsymbol{d}$) is $\boldsymbol{x}_p = (1, 0, 6, 0)$. This particular solution is my favorite: ***free variables = zero, pivot variables from d.***

For a solution to exist, zero rows in R_0 must also be zero in d. Since I is in the pivot rows and pivot columns of R_0, the pivot variables in $x_{\text{particular}}$ come from d:

$$R_0\boldsymbol{x}_p = \begin{bmatrix} \mathbf{1} & 3 & \mathbf{0} & 2 \\ \mathbf{0} & 0 & \mathbf{1} & 4 \\ 0 & 0 & 0 & 0 \end{bmatrix} \begin{bmatrix} \mathbf{1} \\ 0 \\ \mathbf{6} \\ 0 \end{bmatrix} = \begin{bmatrix} \mathbf{1} \\ \mathbf{6} \\ 0 \end{bmatrix} \qquad \begin{matrix} \textbf{Pivot variables } \mathbf{1, 6} \\ \textbf{Free variables } \mathbf{0, 0} \\ \textbf{Solution } \boldsymbol{x_p} = \mathbf{(1, 0, 6, 0)}. \end{matrix}$$

Notice how we *choose* the free variables (as zero) and *solve* for the pivot variables. After the row reduction to R_0, those steps are quick. When the free variables are zero, the pivot variables for $\boldsymbol{x}_p$ are already seen in the right side vector $\boldsymbol{d}$.

$\boldsymbol{x}_{\textbf{particular}}$	***The particular solution solves***	$A\boldsymbol{x}_p = \boldsymbol{b}$
$\boldsymbol{x}_{\textbf{nullspace}}$	***The $n - r$ special solutions solve***	$A\boldsymbol{x}_n = \mathbf{0}$

That particular solution is $(1, 0, 6, 0)$. The two special (nullspace) solutions to $R_0\boldsymbol{x} = \mathbf{0}$ come from the two free columns of R_0, by reversing signs of $3, 2$, and 4. ***Please notice how I write the complete solution $\boldsymbol{x}_p + \boldsymbol{x}_n$ to $A\boldsymbol{x} = \boldsymbol{b}$:***

Complete solution
one $\boldsymbol{x}_p$+many $\boldsymbol{x}_n$
$\boldsymbol{x}_p$ **: free variables**
$\boldsymbol{x}_n$ **: special solutions**

$$\boldsymbol{x} = \boldsymbol{x}_p + \boldsymbol{x}_n = \begin{bmatrix} 1 \\ 0 \\ 6 \\ 0 \end{bmatrix} + x_2 \begin{bmatrix} -3 \\ 1 \\ 0 \\ 0 \end{bmatrix} + x_4 \begin{bmatrix} -2 \\ 0 \\ -4 \\ 1 \end{bmatrix}.$$

Question Suppose A is a square invertible matrix, $m = n = r$. What are $\boldsymbol{x}_p$ and $\boldsymbol{x}_n$?
Answer The particular solution is the one and *only* solution $\boldsymbol{x}_p = A^{-1}\boldsymbol{b}$. There are no special solutions or free variables. $R_0 = I$ has no zero rows. The only vector in the nullspace is $\boldsymbol{x}_n = \mathbf{0}$. The complete solution is $\boldsymbol{x} = \boldsymbol{x}_p + \boldsymbol{x}_n = A^{-1}\boldsymbol{b} + \mathbf{0}$.

We didn't mention the nullspace in Chapter 2, because A was invertible. It was reduced all the way to I. $\begin{bmatrix} A & \boldsymbol{b} \end{bmatrix}$ went to $\begin{bmatrix} I & A^{-1}\boldsymbol{b} \end{bmatrix}$. Then $A\boldsymbol{x} = \boldsymbol{b}$ became $\boldsymbol{x} = A^{-1}\boldsymbol{b}$ which is $\boldsymbol{d}$. This is a special case here, but square invertible matrices are the best. So they got their own chapter before this one.

For small examples we can reduce $\begin{bmatrix} A & \boldsymbol{b} \end{bmatrix}$ to $\begin{bmatrix} R_0 & \boldsymbol{d} \end{bmatrix}$. For a large matrix, MATLAB does it better. One particular solution (not necessarily ours) is $\boldsymbol{x} = A\backslash\boldsymbol{b}$ from backslash. Here is an example with *full column rank.* Both columns have pivots.

Example 1 Find the condition on (b_1, b_2, b_3) for $A\boldsymbol{x} = \boldsymbol{b}$ to be solvable, if

$$A = \begin{bmatrix} 1 & 1 \\ 1 & 2 \\ -2 & -3 \end{bmatrix} \quad \text{and} \quad \boldsymbol{b} = \begin{bmatrix} b_1 \\ b_2 \\ b_3 \end{bmatrix}.$$

Solution Use the augmented matrix, with its extra column $\boldsymbol{b}$. Subtract row 1 of $\begin{bmatrix} A & \boldsymbol{b} \end{bmatrix}$ from row 2. Then add 2 times row 1 to row 3 to reach $\begin{bmatrix} R_0 & \boldsymbol{d} \end{bmatrix}$

$$\begin{bmatrix} 1 & 1 & b_1 \\ 1 & 2 & b_2 \\ -2 & -3 & b_3 \end{bmatrix} \rightarrow \begin{bmatrix} 1 & 1 & b_1 \\ 0 & 1 & b_2 - b_1 \\ 0 & -1 & b_3 + 2b_1 \end{bmatrix} \rightarrow \begin{bmatrix} 1 & 0 & 2b_1 - b_2 \\ 0 & 1 & b_2 - b_1 \\ 0 & 0 & \boldsymbol{b_3 + b_1 + b_2} \end{bmatrix} = \begin{bmatrix} R_0 & \boldsymbol{d} \end{bmatrix}$$

The last equation is $0 = 0$ provided $b_3 + b_1 + b_2 = 0$. This is the condition to put $\boldsymbol{b}$ in the column space. Then $A\boldsymbol{x} = \boldsymbol{b}$ will be solvable. The rows of A add to the zero row. So for consistency (these are equations!) the entries of $\boldsymbol{b}$ must also add to zero.

This example has no free variables since $n - r = 2 - 2$. Therefore no special solutions. The nullspace solution is $\boldsymbol{x}_n = \boldsymbol{0}$. The particular solution to $A\boldsymbol{x} = \boldsymbol{b}$ and $R_0\boldsymbol{x} = \boldsymbol{d}$ is at the top of the final column $\boldsymbol{d}$:

$$\boldsymbol{r = n} \qquad \textbf{One solution to } A\boldsymbol{x} = \boldsymbol{b} \qquad \boldsymbol{x} = \boldsymbol{x}_p + \boldsymbol{x}_n = \begin{bmatrix} 2b_1 - b_2 \\ b_2 - b_1 \end{bmatrix} + \begin{bmatrix} 0 \\ 0 \end{bmatrix}.$$

If $b_3 + b_1 + b_2$ is not zero, there is no solution to $A\boldsymbol{x} = \boldsymbol{b}$ ($\boldsymbol{x}_p$ and $\boldsymbol{x}$ don't exist).

This example is typical of an extremely important case: A has *full column rank*. Every column has a pivot. *The rank is* $r = n$. The matrix is tall and thin ($m \geq n$). Row reduction puts I at the top, when A is reduced to R with rank n:

$$\textbf{Full column rank } \boldsymbol{r = n} \qquad R_0 = \begin{bmatrix} I \\ 0 \end{bmatrix} = \begin{bmatrix} n \text{ by } n \text{ identity matrix} \\ m - n \text{ rows of zeros} \end{bmatrix} \tag{1}$$

There are no free columns or free variables. The nullspace is $\mathbf{Z} = \{\text{zero vector}\}$.

We will collect together the different ways of recognizing this type of matrix.

Every matrix A with **full column rank** ($\boldsymbol{r = n}$) has all these properties:
1. All columns of A are pivot columns. No free variables.
2. The nullspace $\mathbf{N}(A)$ contains only the zero vector $\boldsymbol{x} = \boldsymbol{0}$.
3. If $A\boldsymbol{x} = \boldsymbol{b}$ has a solution (it might not) then it has only *one solution*.

In the essential language of the next section, **this** A **has *independent columns*.** $A\boldsymbol{x} = \boldsymbol{0}$ only happens when $\boldsymbol{x} = \boldsymbol{0}$. Later we will add one more fact to the list above: *The square matrix* $A^{\mathrm{T}}A$ *is invertible when the rank is* n. A may have many rows.

In this case the nullspace of A has shrunk to the zero vector. The solution to $A\boldsymbol{x} = \boldsymbol{b}$ is *unique* (if it exists). There will be $m - n$ zero rows in R_0. So there are $m - n$ conditions on $\boldsymbol{b}$ in order to have $0 = 0$ in those rows, and $\boldsymbol{b}$ in the column space. With full column rank, $A\boldsymbol{x} = \boldsymbol{b}$ has *one solution* or *no solution*.

Full Row Rank and the Complete Solution

The other extreme case is full row rank. Now $A\boldsymbol{x} = \boldsymbol{b}$ has *one or infinitely many* solutions. In this case A must be *short and wide* ($m \le n$). ***A matrix has full row rank if*** $r = m$. "*The rows are independent.*" Every row has a pivot, and here is an example.

Example 2 This system $A\boldsymbol{x} = \boldsymbol{b}$ has $n = 3$ unknowns but only $m = 2$ equations:

Full row rank
$$\begin{aligned} x + \ \ y + z &= 3 \\ x + 2y - z &= 4 \end{aligned} \qquad (\text{rank } \boldsymbol{r} = \boldsymbol{m} = \mathbf{2})$$

These are two planes in xyz space. The planes are not parallel so they intersect in a line. This line of solutions is exactly what elimination will find. ***The particular solution will be one point on the line. Adding the nullspace vectors*** $\boldsymbol{x}_n$ ***will move us along the line in Figure 3.1***. Then $\boldsymbol{x} = \boldsymbol{x}_p +$ all $\boldsymbol{x}_n$ gives the whole line of solutions.

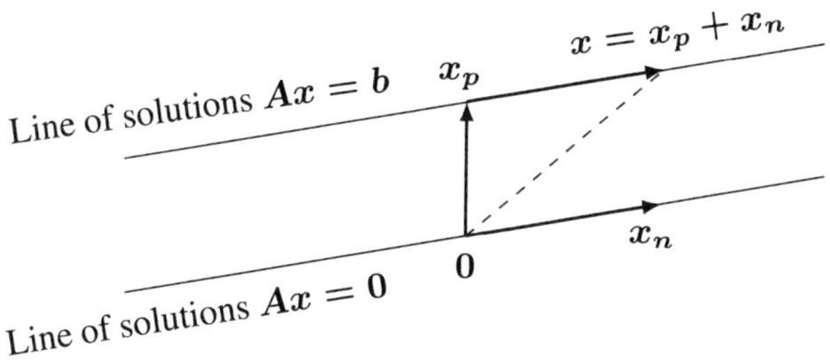

Figure 3.1: Complete solution = *one* particular solution $\boldsymbol{x}_p +$ *all* nullspace solutions $\boldsymbol{x}_n$.

We find $\boldsymbol{x}_p$ and $\boldsymbol{x}_n$ by elimination downwards and then upwards on $\begin{bmatrix} A & \boldsymbol{b} \end{bmatrix}$.

$$\begin{bmatrix} 1 & 1 & 1 & \mathbf{3} \\ 1 & 2 & -1 & \mathbf{4} \end{bmatrix} \rightarrow \begin{bmatrix} 1 & 1 & 1 & \mathbf{3} \\ \mathbf{0} & 1 & -2 & \mathbf{1} \end{bmatrix} \rightarrow \begin{bmatrix} 1 & \mathbf{0} & 3 & \mathbf{2} \\ 0 & 1 & -2 & \mathbf{1} \end{bmatrix} = \begin{bmatrix} R & \boldsymbol{d} \end{bmatrix}.$$

The particular solution $(2, 1, 0)$ *has free variable* $x_3 = 0$. It comes directly from $\boldsymbol{d}$. *The special solution* $\boldsymbol{s}$ *has* $x_3 = 1$. Then $-x_1$ and $-x_2$ come from the free column of R.

It is wise to check that $\boldsymbol{x}_p$ and $\boldsymbol{s}$ satisfy the original equations $A\boldsymbol{x}_p = \boldsymbol{b}$ and $A\boldsymbol{s} = \mathbf{0}$:

$$\begin{aligned} 2 + 1 &= 3 \\ 2 + 2 &= 4 \end{aligned} \qquad \begin{aligned} -3 + 2 + 1 &= 0 \\ -3 + 4 - 1 &= 0 \end{aligned}$$

The nullspace solution $\boldsymbol{x}_n$ is any multiple of $\boldsymbol{s}$. It moves along the line of solutions, starting at $\boldsymbol{x}_{\textbf{particular}}$. ***Please notice again how to write the answer***:

Complete solution
Particular + nullspace
$$\boldsymbol{x} = \boldsymbol{x}_p + \boldsymbol{x}_n = \begin{bmatrix} 2 \\ 1 \\ 0 \end{bmatrix} + x_3 \begin{bmatrix} -3 \\ 2 \\ 1 \end{bmatrix}.$$

This line of solutions is drawn in Figure 3.1. Any point on the line *could* have been chosen as the particular solution. We chose $\boldsymbol{x}_p$ as the point with $x_3 = 0$.

The particular solution is *not* multiplied by an arbitrary constant! The special solution needs that constant, and you understand why—to produce all $\boldsymbol{x}_n$ in the nullspace.

Now we summarize this short wide case of *full row rank.* If $m < n$ the equation $A\boldsymbol{x} = \boldsymbol{b}$ is **underdetermined** (many solutions).

Every matrix A with ***full row rank*** ($\boldsymbol{r}$=$\boldsymbol{m}$) has all these properties:
1. All rows have pivots, and $\boldsymbol{R}_0$ **has no zero rows**: $\boldsymbol{R}_0 = \boldsymbol{R}$.
2. $A\boldsymbol{x} = \boldsymbol{b}$ **has a solution for every right side** $\boldsymbol{b}$.
3. The column space is the whole space $\mathbf{R}^m$.
4. There are $n - r = n - m$ special solutions in the nullspace of A.

In this case with m pivots, ***the rows are linearly independent***. So the columns of A^{T} are linearly independent. The nullspace of A^{T} contains only the zero vector. And that nullspace $\mathbf{N}(A^{\mathrm{T}})$ will be the fourth fundamental subspace.

We are ready for the definition of linear independence, as soon as we summarize the four possibilities—which depend on the rank. Notice how r, m, n are the critical numbers.

The four possibilities for linear equations depend on the rank r

$\boldsymbol{r} = \boldsymbol{m}$	and	$\boldsymbol{r} = \boldsymbol{n}$	*Square and invertible*	$A\boldsymbol{x} = \boldsymbol{b}$	has 1 solution
$\boldsymbol{r} = \boldsymbol{m}$	and	$r < n$	*Short and wide*	$A\boldsymbol{x} = \boldsymbol{b}$	has ∞ solutions
$r < m$	and	$\boldsymbol{r} = \boldsymbol{n}$	*Tall and thin*	$A\boldsymbol{x} = \boldsymbol{b}$	has 0 or 1 solution
$r < m$	and	$r < n$	*Not full rank*	$A\boldsymbol{x} = \boldsymbol{b}$	has 0 or ∞ solutions

The reduced R_0 will fall in the same category as the matrix A. In case the pivot columns happen to come first, we can display these four possibilities. For $R_0\boldsymbol{x} = \boldsymbol{d}$ and $A\boldsymbol{x} = \boldsymbol{b}$ to be solvable, $\boldsymbol{d}$ must end in $m - r$ zeros. F is the free part of R_0.

Four types for $\boldsymbol{R}_0$	$\begin{bmatrix} I \end{bmatrix}$	$\begin{bmatrix} I & F \end{bmatrix}$	$\begin{bmatrix} I \\ 0 \end{bmatrix}$	$\begin{bmatrix} I & F \\ 0 & 0 \end{bmatrix}$
Their ranks	$\boldsymbol{r} = \boldsymbol{m} = \boldsymbol{n}$	$r = \boldsymbol{m} < \boldsymbol{n}$	$\boldsymbol{r} = \boldsymbol{n} < \boldsymbol{m}$	$\boldsymbol{r} < \boldsymbol{m}, \boldsymbol{r} < \boldsymbol{n}$

Cases 1 and 2 have full row rank $r = m$. Cases 1 and 3 have full column rank $r = n$.

■ REVIEW OF THE KEY IDEAS ■

1. The rank r is the number of pivots. The matrix R_0 has $m - r$ zero rows.

2. $A\boldsymbol{x} = \boldsymbol{b}$ is solvable if and only if the last $m - r$ equations reduce to $0 = 0$.

3. One particular solution $\boldsymbol{x}_p$ has all free variables equal to zero.

4. The pivot variables are determined after the free variables are chosen.

5. Full column rank $r = n$ means no free variables: one solution or none.

6. Full row rank $r = m$ means one solution if $m = n$ or infinitely many if $m < n$.

■ WORKED EXAMPLES ■

3.3 A This question connects elimination (**pivot columns and back substitution**) to **column space-nullspace-rank-solvability** (the higher level picture). A has rank 2:

$$A\boldsymbol{x} = \boldsymbol{b} \quad \text{is} \quad \begin{aligned} x_1 + 2x_2 + 3x_3 + \ \ 5x_4 &= b_1 \\ 2x_1 + 4x_2 + 8x_3 + 12x_4 &= b_2 \\ 3x_1 + 6x_2 + 7x_3 + 13x_4 &= b_3 \end{aligned} \qquad A = \begin{bmatrix} 1 & 2 & 3 & 5 \\ 2 & 4 & 8 & 12 \\ 3 & 6 & 7 & 13 \end{bmatrix}$$

1. Reduce $[\,A \ \ \boldsymbol{b}\,]$ to $[\,U \ \ \boldsymbol{c}\,]$, so that $A\boldsymbol{x} = \boldsymbol{b}$ becomes a triangular system $U\boldsymbol{x} = \boldsymbol{c}$.
2. Find the condition on b_1, b_2, b_3 for $A\boldsymbol{x} = \boldsymbol{b}$ to have a solution.
3. Describe the column space of A. Which plane in $\mathbf{R}^3$?
4. Describe the nullspace of A. Which special solutions in $\mathbf{R}^4$?
5. Reduce $[\,U \ \ \boldsymbol{c}\,]$ to $[\,R_0 \ \ \boldsymbol{d}\,]$: Special solutions from R_0, particular solution from $\boldsymbol{d}$.
6. Find a particular solution to $A\boldsymbol{x} = (0, 6, -6)$ and then the complete solution.

Solution

1. The multipliers in elimination are 2 and 3 and -1. They take $[\,A \ \ \boldsymbol{b}\,]$ into $[\,U \ \ \boldsymbol{c}\,]$.

$$\begin{bmatrix} 1 & 2 & 3 & 5 & \mathbf{b_1} \\ 2 & 4 & 8 & 12 & \mathbf{b_2} \\ 3 & 6 & 7 & 13 & \mathbf{b_3} \end{bmatrix} \rightarrow \left[\begin{array}{cccc|c} 1 & 2 & 3 & 5 & \mathbf{b_1} \\ 0 & 0 & 2 & 2 & \mathbf{b_2 - 2b_1} \\ 0 & 0 & -2 & -2 & \mathbf{b_3 - 3b_1} \end{array}\right] \rightarrow \left[\begin{array}{cccc|c} 1 & 2 & 3 & 5 & \mathbf{b_1} \\ 0 & 0 & 2 & 2 & \mathbf{b_2 - 2b_1} \\ 0 & 0 & 0 & 0 & \mathbf{b_3 + b_2 - 5b_1} \end{array}\right]$$

2. The last equation shows the solvability condition $b_3 + b_2 - 5b_1 = 0$. Then $0 = 0$.
3. **First description**: The column space is the plane containing all combinations of the pivot columns $(1, 2, 3)$ and $(3, 8, 7)$. The pivots are in columns 1 and 3. **Second description**: The column space contains all vectors with $b_3 + b_2 - 5b_1 = 0$. That makes $A\boldsymbol{x} = \boldsymbol{b}$ solvable, so $\boldsymbol{b}$ is in the column space. *All columns of A pass this test* $b_3 + b_2 - 5b_1 = 0$. *This is the equation for the plane in the first description*!
4. The special solutions have free variables $x_2 = 1, x_4 = 0$ and then $x_2 = 0, x_4 = 1$:

Special solutions to $A\boldsymbol{x} = \mathbf{0}$
Back substitution in $U\boldsymbol{x} = \mathbf{0}$
or change signs of 2, 2, 1 in R

$$\boldsymbol{s}_1 = \begin{bmatrix} -2 \\ \mathbf{1} \\ 0 \\ \mathbf{0} \end{bmatrix} \qquad \boldsymbol{s}_2 = \begin{bmatrix} -2 \\ \mathbf{0} \\ -1 \\ \mathbf{1} \end{bmatrix}$$

The nullspace $\mathbf{N}(A)$ in $\mathbf{R}^4$ contains all $\boldsymbol{x}_n = c_1\boldsymbol{s}_1 + c_2\boldsymbol{s}_2$.

5. In the reduced form R_0, the third column changes from $(3, 2, 0)$ in U to $(0, 1, 0)$. The right side $\boldsymbol{c} = (0, 6, 0)$ becomes $\boldsymbol{d} = (-9, 3, 0)$ showing -9 and 3 in $\boldsymbol{x}_p$:

$$[\,U \ \ \boldsymbol{c}\,] = \begin{bmatrix} 1 & 2 & 3 & 5 & \mathbf{0} \\ 0 & 0 & 2 & 2 & \mathbf{6} \\ 0 & 0 & 0 & 0 & \mathbf{0} \end{bmatrix} \longrightarrow [\,R_0 \ \ \boldsymbol{d}\,] = \begin{bmatrix} 1 & 2 & 0 & 2 & \mathbf{-9} \\ 0 & 0 & 1 & 1 & \mathbf{3} \\ 0 & 0 & 0 & 0 & \mathbf{0} \end{bmatrix}$$

6. $\boldsymbol{x} = (-9, 0, 3, 0)$ is the very particular solution with free variables = zero.

The complete solution to $A\boldsymbol{x} = (0, 6, -6)$ is $\boldsymbol{x} = \boldsymbol{x}_p + \boldsymbol{x}_n = \boldsymbol{x}_p + c_1\boldsymbol{s}_1 + c_2\boldsymbol{s}_2$.

3.3 B Suppose you have this information about the solutions to $A\boldsymbol{x} = \boldsymbol{b}$ for a specific $\boldsymbol{b}$. What does that tell you about m and n and r (and A itself)? And possibly about $\boldsymbol{b}$.

1. There is exactly one solution.
2. All solutions to $A\boldsymbol{x} = \boldsymbol{b}$ have the form $\boldsymbol{x} = \begin{bmatrix} 2 \\ 1 \end{bmatrix} + c\begin{bmatrix} 1 \\ 1 \end{bmatrix}$.
3. There are no solutions.
4. All solutions to $A\boldsymbol{x} = \boldsymbol{b}$ have the form $\boldsymbol{x} = \begin{bmatrix} 1 \\ 1 \\ 0 \end{bmatrix} + c\begin{bmatrix} 1 \\ 0 \\ 1 \end{bmatrix}$

Solution In case **1**, with exactly one solution, A must have full column rank $r = n$. The nullspace of A contains only the zero vector. Necessarily $m \geq n$.

In case **2**, A must have $n = 2$ columns (and m is arbitrary). With $\begin{bmatrix} 1 \\ 1 \end{bmatrix}$ in the nullspace of A, column 2 is the *negative* of column 1. Also $A \neq 0$: the rank is 1. With $\boldsymbol{x} = \begin{bmatrix} 2 \\ 1 \end{bmatrix}$ as a solution, $\boldsymbol{b} = 2$(column 1) + (column 2). My choice for $\boldsymbol{x}_p$ would be $(1, 0)$.

In case **3** we only know that $\boldsymbol{b}$ is not in the column space of A. The rank of A must be less than m. I guess we know $\boldsymbol{b} \neq \boldsymbol{0}$, otherwise $\boldsymbol{x} = \boldsymbol{0}$ would be a solution.

In case **4**, A must have $n = 3$ columns. With $(1, 0, 1)$ in the nullspace of A, column 3 is $-$(column 1). The rank is $3 - 1 = 2$ and $\boldsymbol{b}$ is column 1 + column 2.

3.3 C Find the complete solution $\boldsymbol{x} = \boldsymbol{x_p} + \boldsymbol{x_n}$ by forward elimination on $[A \ \ \boldsymbol{b}]$:

Solution

$$\begin{bmatrix} 1 & 2 & 1 & 0 & \mathbf{4} \\ 2 & 4 & 4 & 8 & \mathbf{2} \\ 4 & 8 & 6 & 8 & \mathbf{10} \end{bmatrix} \longrightarrow \begin{bmatrix} 1 & 2 & 1 & 0 & \mathbf{4} \\ 0 & 0 & 2 & 8 & \mathbf{-6} \\ 0 & 0 & 0 & 0 & \mathbf{0} \end{bmatrix} \longrightarrow \begin{bmatrix} 1 & 2 & 0 & -4 & \mathbf{7} \\ 0 & 0 & 1 & 4 & \mathbf{-3} \\ 0 & 0 & 0 & 0 & \mathbf{0} \end{bmatrix}.$$

For the nullspace part $\boldsymbol{x_n}$ with $\boldsymbol{b} = \boldsymbol{0}$, set the free variables x_2, x_4 to $1, 0$ and also $0, 1$:

Special solutions $\boldsymbol{s}_1 = (-2, 1, 0, 0)$ and $\boldsymbol{s}_2 = (4, 0, -4, 1)$ **Particular** $\boldsymbol{x_p} = (7, 0, -3, 0)$

Then the complete solution to $A\boldsymbol{x} = \boldsymbol{b}$ is $\boldsymbol{x}_{\text{complete}} = \boldsymbol{x_p} + c_1\boldsymbol{s}_1 + c_2\boldsymbol{s}_2$.

The rows of A produced the zero row from 2(row 1) + (row 2) $-$ (row 3) $= (0, 0, 0, 0)$. Thus $\boldsymbol{y} = (2, 1, -1)$. The same combination for $\boldsymbol{b} = (4, 2, 10)$ gives $2(4) + (2) - (10) = 0$.

If a combination of the rows (on the left side) gives the zero row, then the same combination must give zero on the right side. Of course! *Otherwise no solution.*

Later we will say this again in different words: If every column of A is perpendicular to $\boldsymbol{y} = (2, 1, -1)$, then any combination $\boldsymbol{b}$ of those columns must also be perpendicular to $\boldsymbol{y}$. Otherwise $\boldsymbol{b}$ is not in the column space and $A\boldsymbol{x} = \boldsymbol{b}$ is not solvable.

And again: If $\boldsymbol{y}$ is in the nullspace of A^{T} then $\boldsymbol{y}$ must be perpendicular to every $\boldsymbol{b}$ in the column space of A. Just looking ahead...

Problem Set 3.3

1 (Recommended) Execute the six steps of Worked Example **3.3 A** to describe the column space and nullspace of A and the complete solution to $A\boldsymbol{x} = \boldsymbol{b}$:

$$A = \begin{bmatrix} 2 & 4 & 6 & 4 \\ 2 & 5 & 7 & 6 \\ 2 & 3 & 5 & 2 \end{bmatrix} \qquad \boldsymbol{b} = \begin{bmatrix} b_1 \\ b_2 \\ b_3 \end{bmatrix} = \begin{bmatrix} 4 \\ 3 \\ 5 \end{bmatrix}$$

2 Carry out the same six steps for this matrix A with rank one. You will find *two* conditions on b_1, b_2, b_3 for $A\boldsymbol{x} = \boldsymbol{b}$ to be solvable. Together these two conditions put $\boldsymbol{b}$ into the ______ space (two planes give a line):

$$A = \begin{bmatrix} 1 \\ 3 \\ 2 \end{bmatrix} \begin{bmatrix} 2 & 1 & 3 \end{bmatrix} = \begin{bmatrix} 2 & 1 & 3 \\ 6 & 3 & 9 \\ 4 & 2 & 6 \end{bmatrix} \qquad \boldsymbol{b} = \begin{bmatrix} b_1 \\ b_2 \\ b_3 \end{bmatrix} = \begin{bmatrix} 10 \\ 30 \\ 20 \end{bmatrix}$$

Questions 3–15 are about the solution of $A\boldsymbol{x} = \boldsymbol{b}$. Follow the steps in the text to $\boldsymbol{x}_p$ and $\boldsymbol{x}_n$. Start from the augmented matrix with last column $\boldsymbol{b}$.

3 Write the complete solution as $\boldsymbol{x}_p$ plus any multiple of $\boldsymbol{s}$ in the nullspace:

$$\begin{aligned} x + 3y &= 7 \\ 2x + 6y &= 14 \end{aligned} \qquad\qquad \begin{aligned} x + 3y + 3z &= 1 \\ 2x + 6y + 9z &= 5 \\ -x - 3y + 3z &= 5 \end{aligned}$$

4 Find the complete solution $\boldsymbol{x} = \boldsymbol{x}_p +$ any $\boldsymbol{x}_n$ (also called the *general solution*) to

$$\begin{bmatrix} 1 & 3 & 1 & 2 \\ 2 & 6 & 4 & 8 \\ 0 & 0 & 2 & 4 \end{bmatrix} \begin{bmatrix} x \\ y \\ z \\ t \end{bmatrix} = \begin{bmatrix} 1 \\ 3 \\ 1 \end{bmatrix}.$$

5 Under what conditions on b_1, b_2, b_3 are these systems solvable? Include $\boldsymbol{b}$ as a fourth column in elimination. Find all solutions when that condition holds:

$$\begin{aligned} x + 2y - 2z &= b_1 \\ 2x + 5y - 4z &= b_2 \\ 4x + 9y - 8z &= b_3 \end{aligned} \qquad\qquad \begin{aligned} 2x + 2z &= b_1 \\ 4x + 4y &= b_2 \\ 8x + 8y &= b_3 \end{aligned}$$

6 What conditions on b_1, b_2, b_3, b_4 make each system solvable? Find $\boldsymbol{x}$ in that case:

$$\begin{bmatrix} 1 & 2 \\ 2 & 4 \\ 2 & 5 \\ 3 & 9 \end{bmatrix} \begin{bmatrix} x_1 \\ x_2 \end{bmatrix} = \begin{bmatrix} b_1 \\ b_2 \\ b_3 \\ b_4 \end{bmatrix} \qquad \begin{bmatrix} 1 & 2 & 3 \\ 2 & 4 & 6 \\ 2 & 5 & 7 \\ 3 & 9 & 12 \end{bmatrix} \begin{bmatrix} x_1 \\ x_2 \\ x_3 \end{bmatrix} = \begin{bmatrix} b_1 \\ b_2 \\ b_3 \\ b_4 \end{bmatrix}.$$

7 Show by elimination that (b_1, b_2, b_3) is in the column space if $b_3 - 2b_2 + 4b_1 = 0$. The rank is $r = \mathbf{2}$. What combination of the rows of A gives the zero row?

$$A = \begin{bmatrix} 1 & 3 & 1 \\ 3 & 8 & 2 \\ 2 & 4 & 0 \end{bmatrix}.$$

8 Which vectors (b_1, b_2, b_3) are in the column space of A? Which combinations of the rows of A give zero?

$$\text{(a)}\ A = \begin{bmatrix} 1 & 2 & 1 \\ 2 & 6 & 3 \\ 0 & 2 & 5 \end{bmatrix} \qquad \text{(b)}\ \ A = \begin{bmatrix} 1 & 1 & 1 \\ 1 & 2 & 4 \\ 2 & 4 & 8 \end{bmatrix}.$$

9 Find the complete solution in the form $\boldsymbol{x}_p + \boldsymbol{x}_n$ to these full rank systems:

$$\text{(a)}\ x + y + z = 4 \qquad \text{(b)}\ \begin{aligned} x + y + z &= 4 \\ x - y + z &= 4. \end{aligned}$$

10 Construct a 2 by 3 system $A\boldsymbol{x} = \boldsymbol{b}$ with particular solution $\boldsymbol{x}_p = (2, 4, 0)$ and homogeneous solution $\boldsymbol{x}_n =$ any multiple of $(1, 1, 1)$.

11 Why can't a 1 by 3 system have $\boldsymbol{x}_p = (2, 4, 0)$ and $\boldsymbol{x}_n =$ any multiple of $(1, 1, 1)$?

12 (a) If $A\boldsymbol{x} = \boldsymbol{b}$ has two solutions $\boldsymbol{x}_1$ and $\boldsymbol{x}_2$, find two solutions to $A\boldsymbol{x} = \mathbf{0}$.

(b) Then find another solution to $A\boldsymbol{x} = \mathbf{0}$ and another solution to $A\boldsymbol{x} = \boldsymbol{b}$.

13 Explain why these are all false:

(a) The complete solution is any linear combination of $\boldsymbol{x}_p$ and $\boldsymbol{x}_n$.

(b) A system $A\boldsymbol{x} = \boldsymbol{b}$ has at most one particular solution. This is true if A is ______.

(c) The solution $\boldsymbol{x}_p$ with all free variables zero is the shortest solution (minimum length $\|\boldsymbol{x}\|$). Find a 2 by 2 counterexample.

(d) If A is invertible there is no solution $\boldsymbol{x}_n$ in the nullspace.

14 Suppose column 5 of U has no pivot. Then x_5 is a ______ variable. The zero vector (is) (is not) the only solution to $A\boldsymbol{x} = \mathbf{0}$. If $A\boldsymbol{x} = \boldsymbol{b}$ has a solution, then it has ______ solutions.

15 Suppose row 3 of U has no pivot. Then that row is ______. The equation $U\boldsymbol{x} = \boldsymbol{c}$ is only solvable provided ______. The equation $A\boldsymbol{x} = \boldsymbol{b}$ (*is*) (*is not*) (*might not be*) solvable.

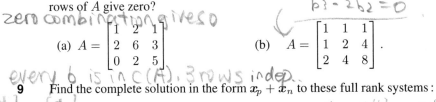

16 The largest possible rank of a 3 by 5 matrix is ______. Then there is a pivot in every ______ of U and R. The solution to $A\boldsymbol{x} = \boldsymbol{b}$ (*always exists*) (*is unique*). The column space of A is ______. An example is $A =$ ______.

17 The largest possible rank of a 6 by 4 matrix is ______. Then there is a pivot in every ______ of U and R. The solution to $A\boldsymbol{x} = \boldsymbol{b}$ (*always exists*) (*is unique*). The nullspace of A is ______. An example is $A =$ ______.

18 Find by elimination the rank of A and also the rank of A^{T}:

$$A = \begin{bmatrix} 1 & 4 & 0 \\ 2 & 11 & 5 \\ -1 & 2 & 10 \end{bmatrix} \quad \text{and} \quad A = \begin{bmatrix} 1 & 0 & 1 \\ 1 & 1 & 2 \\ 1 & 1 & q \end{bmatrix} \quad \text{(rank depends on } q\text{)}.$$

19 If $A\boldsymbol{x} = \boldsymbol{b}$ has infinitely many solutions, why is it impossible for $A\boldsymbol{x} = \boldsymbol{B}$ (new right side) to have only one solution? Could $A\boldsymbol{x} = \boldsymbol{B}$ have no solution?

20 Choose the number q so that (if possible) the ranks are (a) 1 (b) 2 (c) 3:

$$A = \begin{bmatrix} 6 & 4 & 2 \\ -3 & -2 & -1 \\ 9 & 6 & q \end{bmatrix} \quad \text{and} \quad B = \begin{bmatrix} 3 & 1 & 3 \\ q & 2 & q \end{bmatrix}.$$

21 Give examples of matrices A for which the number of solutions to $A\boldsymbol{x} = \boldsymbol{b}$ is

(a) 0 or 1, depending on $\boldsymbol{b}$ (b) ∞, regardless of $\boldsymbol{b}$

(c) 0 or ∞, depending on $\boldsymbol{b}$ (d) 1, regardless of $\boldsymbol{b}$.

22 Write down all known relations between r and m and n if $A\boldsymbol{x} = \boldsymbol{b}$ has

(a) no solution for some $\boldsymbol{b}$ (b) one solution for some $\boldsymbol{b}$, no solution for other $\boldsymbol{b}$

(c) infinitely many solutions for every $\boldsymbol{b}$ (d) exactly one solution for every $\boldsymbol{b}$.

Questions 23–27 are about the reduced echelon matrices R_0 and R.

23 Divide rows by pivots. Then produce zeros *above* those pivots to reach R_0 and R.

$$U = \begin{bmatrix} 2 & 4 & 4 \\ 0 & 3 & 6 \\ 0 & 0 & 0 \end{bmatrix} \quad \text{and} \quad U = \begin{bmatrix} 2 & 4 & 4 \\ 0 & 3 & 6 \\ 0 & 0 & 5 \end{bmatrix} \quad \text{and} \quad U = \begin{bmatrix} 0 & 0 & 4 \\ 0 & 1 & 0 \end{bmatrix}.$$

24 If A is a triangular matrix, when is $R_0 = \text{rref}(A)$ equal to I?

$R_0 = I$
A is square + invertable
diagonal entries are

25 Apply elimination to $U\boldsymbol{x} = \boldsymbol{0}$ and $U\boldsymbol{x} = \boldsymbol{c}$. Reach $R_0\boldsymbol{x} = \boldsymbol{0}$ and $R_0\boldsymbol{x} = \boldsymbol{d}$:

$$\begin{bmatrix} U & \boldsymbol{0} \end{bmatrix} = \begin{bmatrix} 1 & 2 & 3 & \boldsymbol{0} \\ 0 & 0 & 4 & \boldsymbol{0} \end{bmatrix} \quad \text{and} \quad \begin{bmatrix} U & \boldsymbol{c} \end{bmatrix} = \begin{bmatrix} 1 & 2 & 3 & \boldsymbol{5} \\ 0 & 0 & 4 & \boldsymbol{8} \end{bmatrix}.$$

Solve $R_0\boldsymbol{x} = \boldsymbol{0}$ to find $\boldsymbol{x}_n$ with $x_2 = 1$. Solve $R_0\boldsymbol{x} = \boldsymbol{d}$ to find $\boldsymbol{x}_p$ with $x_2 = 0$.

26 Reduce $U\boldsymbol{x} = \mathbf{0}$ and $U\boldsymbol{x} = \boldsymbol{c}$ to $R_0\boldsymbol{x} = \mathbf{0}$ and $R_0\boldsymbol{x} = \boldsymbol{d}$. What are the solutions to $R_0\boldsymbol{x} = \boldsymbol{d}$?

$$\begin{bmatrix} U & 0 \end{bmatrix} = \begin{bmatrix} 3 & 0 & 6 & \mathbf{0} \\ 0 & 0 & 0 & \mathbf{0} \\ 0 & 0 & 2 & \mathbf{0} \end{bmatrix} \quad \text{and} \quad \begin{bmatrix} U & c \end{bmatrix} = \begin{bmatrix} 3 & 0 & 6 & \mathbf{9} \\ 0 & 0 & 0 & \mathbf{4} \\ 0 & 0 & 2 & \mathbf{5} \end{bmatrix}.$$

27 Reduce to $U\boldsymbol{x} = \boldsymbol{c}$ (Gaussian elimination) and then $R_0\boldsymbol{x} = \boldsymbol{d}$. Find a particular solution $\boldsymbol{x}_p$ and all homogeneous solutions $\boldsymbol{x}_n$.

$$A\boldsymbol{x} = \begin{bmatrix} 1 & 0 & 2 & 3 \\ 1 & 3 & 2 & 0 \\ 2 & 0 & 4 & 9 \end{bmatrix} \begin{bmatrix} x_1 \\ x_2 \\ x_3 \\ x_4 \end{bmatrix} = \begin{bmatrix} 2 \\ 5 \\ 10 \end{bmatrix} = \boldsymbol{b}.$$

28 Find matrices A and B with the given property or explain why you can't:

(a) The only solution of $A\boldsymbol{x} = \begin{bmatrix} \mathbf{1} \\ \mathbf{2} \\ \mathbf{3} \end{bmatrix}$ is $\boldsymbol{x} = \begin{bmatrix} \mathbf{0} \\ \mathbf{1} \end{bmatrix}$.

(b) The only solution of $B\boldsymbol{x} = \begin{bmatrix} \mathbf{0} \\ \mathbf{1} \end{bmatrix}$ is $\boldsymbol{x} = \begin{bmatrix} \mathbf{1} \\ \mathbf{2} \\ \mathbf{3} \end{bmatrix}$.

29 Find the LU factorization of A and all solutions to $A\boldsymbol{x} = \boldsymbol{b}$:

$$A = \begin{bmatrix} 1 & 3 & 1 \\ 1 & 2 & 3 \\ 2 & 4 & 6 \\ 1 & 1 & 5 \end{bmatrix} \quad \text{and} \quad \boldsymbol{b} = \begin{bmatrix} 1 \\ 3 \\ 6 \\ 5 \end{bmatrix} \quad \text{and then} \quad \boldsymbol{b} = \begin{bmatrix} 1 \\ 0 \\ 0 \\ 0 \end{bmatrix}.$$

30 The complete solution to $A\boldsymbol{x} = \begin{bmatrix} \mathbf{1} \\ \mathbf{3} \end{bmatrix}$ is $\boldsymbol{x} = \begin{bmatrix} \mathbf{1} \\ \mathbf{0} \end{bmatrix} + c \begin{bmatrix} \mathbf{0} \\ \mathbf{1} \end{bmatrix}$. Find A.

31 (Recommended!) Suppose you know that the 3 by 4 matrix A has the vector $\boldsymbol{s} = (2, 3, 1, 0)$ as the only special solution to $A\boldsymbol{x} = \mathbf{0}$.

(a) What is the *rank* of A and the complete solution to $A\boldsymbol{x} = \mathbf{0}$?

(b) What is the exact row reduced echelon form R_0 of A?

(c) How do you know that $A\boldsymbol{x} = \boldsymbol{b}$ can be solved for all $\boldsymbol{b}$?

32 Suppose $A\boldsymbol{x} = \boldsymbol{b}$ and $C\boldsymbol{x} = \boldsymbol{b}$ have the same (complete) solutions for every $\boldsymbol{b}$. Is it true that A equals C?

33 Describe the column space of a reduced row echelon matrix R_0 with rank r. Removing any zero rows, describe the column space of R.

3.4 Independence, Basis, and Dimension

1 Independent vectors: The only zero combination $c_1\boldsymbol{v}_1 + \cdots + c_k\boldsymbol{v}_k = 0$ has all c's $= 0$.

2 The vectors $\boldsymbol{v}_1, \ldots, \boldsymbol{v}_k$ **span the space S** if $\mathbf{S} =$ all combinations of the $\boldsymbol{v}$'s.

3 The vectors $\boldsymbol{v}_1, \ldots, \boldsymbol{v}_k$ are a **basis for S** if (1) they are independent and (2) they span **S**.

4 The **dimension of a space S** is the number k of vectors in every basis for **S**.

This important section is about the true size of a subspace. There are n columns in an m by n matrix. But the true "dimension" of the column space is not necessarily n. The dimension of $\mathbf{C}(A)$ is measured by counting ***independent columns***. We will see again that *the true dimension of the column space is the rank* r.

The idea of independence applies to any vectors $\boldsymbol{v}_1, \ldots, \boldsymbol{v}_n$ in any vector space. Most of this section concentrates on the subspaces that we know and use—especially the column space and the nullspace of A. In the last part we also study "vectors" that are not column vectors. They can be matrices and functions; they can be linearly independent or dependent. First come the key examples using column vectors.

The goal is to understand a ***basis*** : ***independent vectors that "span the space"***.

Every vector in the space is a unique combination of the basis vectors.

We are at the heart of our subject, and we cannot go on without a basis. The four essential ideas in this section (with first hints at their meaning) are:

1. Independent vectors	(*no extra vectors*)
2. Spanning a space	(*enough vectors to produce the rest*)
3. Basis for a space	(*not too many and not too few*)
4. Dimension of a space	(*the number of vectors in every basis*)

Linear Independence

Our first definition of independence is not so conventional, but you are ready for it.

DEFINITION The columns of A are ***linearly independent*** when the only solution to $A\boldsymbol{x} = \mathbf{0}$ is $\boldsymbol{x} = \mathbf{0}$. ***No other combination*** $A\boldsymbol{x}$ ***of the columns gives the zero vector***.

The columns are independent when the nullspace $\mathbf{N}(A)$ contains only the zero vector. Let me illustrate linear independence (and dependence) with three vectors in $\mathbf{R}^3$.

1. If three vectors in $\mathbf{R}^3$ are *not* in the same plane, they are independent. No combination of $\boldsymbol{v}_1, \boldsymbol{v}_2, \boldsymbol{v}_3$ in Figure 3.2 gives zero except $0\boldsymbol{v}_1 + 0\boldsymbol{v}_2 + 0\boldsymbol{v}_3$.

2. If three vectors $\boldsymbol{w}_1, \boldsymbol{w}_2, \boldsymbol{w}_3$ are *in the same plane* in $\mathbf{R}^3$, they are dependent.

Figure 3.2: Independent: Only $0\boldsymbol{v}_1+0\boldsymbol{v}_2+0\boldsymbol{v}_3$ gives $\mathbf{0}$. Dependent: $\boldsymbol{w}_1-\boldsymbol{w}_2+\boldsymbol{w}_3=\mathbf{0}$.

This idea of independence applies to 7 vectors in 12-dimensional space. If they are the columns of A, and *independent*, the nullspace only contains $\boldsymbol{x}=\mathbf{0}$. None of the vectors is a combination of the other six vectors.

Now we choose different words to express the same idea in any vector space.

DEFINITION The sequence of vectors $\boldsymbol{v}_1, \ldots, \boldsymbol{v}_n$ is ***linearly independent*** if the only combination that gives the zero vector is $0\boldsymbol{v}_1 + 0\boldsymbol{v}_2 + \cdots + 0\boldsymbol{v}_n$.

Linear independence

$$x_1\boldsymbol{v}_1 + x_2\boldsymbol{v}_2 + \cdots + x_n\boldsymbol{v}_n = \mathbf{0} \quad \text{only happens when all } x\text{'s are zero.} \tag{1}$$

If a combination gives $\mathbf{0}$, when the x's are not all zero, the vectors are *dependent.*

Correct language: "The sequence of vectors is linearly independent." *Acceptable shortcut*: "The vectors are independent." *Unacceptable*: "The matrix is independent."

A sequence of vectors is either dependent or independent. They can be combined to give the zero vector (with nonzero x's) or they can't. So the key question is: Which combinations of the vectors give zero? We begin with some small examples in $\mathbf{R}^2$:

(a) The vectors $(1,0)$ and $(1,0.00001)$ are independent.

(b) The vectors $(1,1)$ and $(-1,-1)$ are *dependent.*

(c) The vectors $(1,1)$ and $(0,0)$ are *dependent* because of the zero vector.

(d) In $\mathbf{R}^2$, any three vectors (a,b) and (c,d) and (e,f) are *dependent.*

Dependent columns
$\boldsymbol{x} \neq \mathbf{0}$ in the nullspace

$$\begin{bmatrix} 1 & -1 \\ 1 & -1 \end{bmatrix} \begin{bmatrix} x_1 \\ x_2 \end{bmatrix} = \begin{bmatrix} 0 \\ 0 \end{bmatrix} \quad \text{for } x_1 = 1 \text{ and } x_2 = 1.$$

Three vectors in $\mathbf{R}^2$ cannot be independent! One way to see this: the matrix A with those three columns must have a free variable and then a special solution to $A\boldsymbol{x} = \mathbf{0}$.

Now move to three vectors in $\mathbf{R}^3$. If one of them is a multiple of another one, these vectors are dependent. But the complete test involves all three vectors at once. We put them in a matrix and try to solve $A\boldsymbol{x} = \mathbf{0}$.

Example 1 The columns of this A are dependent. $A\boldsymbol{x} = \mathbf{0}$ has a nonzero solution:

$$A\boldsymbol{x} = \begin{bmatrix} 1 & 0 & 3 \\ 2 & 1 & 5 \\ 1 & 0 & 3 \end{bmatrix} \begin{bmatrix} -3 \\ 1 \\ 1 \end{bmatrix} \quad \text{is} \quad -3 \begin{bmatrix} 1 \\ 2 \\ 1 \end{bmatrix} + 1 \begin{bmatrix} 0 \\ 1 \\ 0 \end{bmatrix} + 1 \begin{bmatrix} 3 \\ 5 \\ 3 \end{bmatrix} = \begin{bmatrix} \mathbf{0} \\ \mathbf{0} \\ \mathbf{0} \end{bmatrix}.$$

The rank is only $r = 2$. *Independent columns produce full column rank* $r = n = 3$. For a *square matrix*, dependent columns imply dependent rows and vice versa.

Question How to find that solution to $A\boldsymbol{x} = \mathbf{0}$? The systematic way is elimination.

$$A = \begin{bmatrix} 1 & 0 & 3 \\ 2 & 1 & 5 \\ 1 & 0 & 3 \end{bmatrix} \text{ reduces to } R_0 = \begin{bmatrix} 1 & 0 & 3 \\ 0 & 1 & -1 \\ 0 & 0 & 0 \end{bmatrix}. \quad \text{Then } F = \begin{bmatrix} 3 \\ -1 \end{bmatrix} \text{ and } \boldsymbol{x} = \begin{bmatrix} -3 \\ 1 \\ 1 \end{bmatrix}.$$

> **Full column rank** The columns of A are independent exactly when the rank is $r = n$. There are n pivots and no free variables and $A = C$. Only $\boldsymbol{x} = \mathbf{0}$ is in the nullspace.

One case is of special importance because it is clear from the start. Suppose seven columns have five components each (**$m = 5$ is less than $n = 7$**). Then the columns *must be dependent*. Any seven vectors from $\mathbf{R}^5$ are dependent. The rank of A cannot be larger than 5. There cannot be more than five pivots in five rows. $A\boldsymbol{x} = \mathbf{0}$ has at least $7 - 5 = 2$ free variables, so it has nonzero solutions—which means that the columns are dependent.

> Any set of n vectors in $\mathbf{R}^m$ must be linearly dependent if $n > m$.

This type of matrix has more columns than rows—it is short and wide. The columns are certainly dependent if $n > m$, because $A\boldsymbol{x} = \mathbf{0}$ has a nonzero solution.

The columns might be dependent or might be independent if $n \leq m$. Elimination will reveal the r pivot columns. *It is those r pivot columns that are independent in C.*

Note Another way to describe linear dependence is this: "*One vector is a combination of the other vectors.*" That sounds clear. Why don't we say this from the start? Our definition was longer: "*Some combination gives the zero vector, other than the trivial combination with every $x = 0$.*" We must rule out the easy way to get the zero vector.

The point is, our definition doesn't pick out one particular vector as guilty. All columns of A are treated the same. We look at $A\boldsymbol{x} = \mathbf{0}$, and it has a nonzero solution or it hasn't.

Vectors that Span a Subspace

The first subspace in this book was the column space. Starting with columns $\boldsymbol{v}_1, \ldots, \boldsymbol{v}_n$, $\mathbf{C}(A)$ includes all combinations $x_1\boldsymbol{v}_1 + \cdots + x_n\boldsymbol{v}_n$. *The column space consists of all combinations $A\boldsymbol{x}$ of the columns.* The single word "span" describes $\mathbf{C}(A)$.

The columns of a matrix span its column space. They might be dependent.

Example 2 Describe the column space and the row space of A.

$$\begin{matrix} m = 3 \\ n = 2 \end{matrix} \qquad A = \begin{bmatrix} 1 & 4 \\ 2 & 7 \\ 3 & 5 \end{bmatrix} \text{ and } A^{\mathrm{T}} = \begin{bmatrix} 1 & 2 & 3 \\ 4 & 7 & 5 \end{bmatrix}.$$

The column space of A is the plane in $\mathbf{R}^3$ spanned by the two columns of A. *The row space of A is spanned by the three rows of A* (which are columns of A^{T}). This row space is all of $\mathbf{R}^2$. Remember: The rows are in $\mathbf{R}^n$ spanning the row space. The columns are in $\mathbf{R}^m$ spanning the column space. Same numbers, different vectors, different spaces.

A Basis for a Vector Space

Two vectors can't span all of $\mathbf{R}^3$, even if they are independent. Four vectors can't be independent, even if they span $\mathbf{R}^3$. We want ***enough independent vectors to span the space*** (and not more). A "***basis***" is just right.

DEFINITION A ***basis*** for a vector space is a sequence of vectors with two properties:

The basis vectors are linearly independent and they span the space.

This combination of properties is fundamental to linear algebra. Every vector $\boldsymbol{v}$ in the space is a combination of the basis vectors, because they span the space. More than that, the combination that produces $\boldsymbol{v}$ is *unique*, because the basis vectors $\boldsymbol{v}_1, \ldots, \boldsymbol{v}_n$ are independent:

There is one and only one way to write $\boldsymbol{v}$ as a combination of the basis vectors.

Reason: Suppose $\boldsymbol{v} = a_1\boldsymbol{v}_1 + \cdots + a_n\boldsymbol{v}_n$ and also $\boldsymbol{v} = b_1\boldsymbol{v}_1 + \cdots + b_n\boldsymbol{v}_n$. By subtraction $(a_1 - b_1)\boldsymbol{v}_1 + \cdots + (a_n - b_n)\boldsymbol{v}_n$ is the zero vector. From the independence of the $\boldsymbol{v}$'s, each $a_i - b_i = 0$. Hence $a_i = b_i$, and there are not two ways to produce $\boldsymbol{v}$.

Example 3 The columns of $I = \begin{bmatrix} 1 & 0 \\ 0 & 1 \end{bmatrix}$ produce the "standard basis" for $\mathbf{R}^2$.

$$\text{The basis vectors } \quad \boldsymbol{i} = \begin{bmatrix} 1 \\ 0 \end{bmatrix} \text{ and } \boldsymbol{j} = \begin{bmatrix} 0 \\ 1 \end{bmatrix} \text{ are independent. They span } \mathbf{R}^2.$$

Everybody thinks of this basis first. The vector $\boldsymbol{i}$ goes across and $\boldsymbol{j}$ goes straight up. The columns of the n by n identity matrix give the "**standard basis**" for $\mathbf{R}^n$.

Now we find many other bases (infinitely many). The basis is not unique!

Example 4 (Important) The columns of *every invertible n by n matrix* give a basis for $\mathbf{R}^n$:

Invertible matrix
Independent columns
Column space is $\mathbf{R}^3$
$$A = \begin{bmatrix} 1 & 0 & 0 \\ 1 & 1 & 0 \\ 1 & 1 & 1 \end{bmatrix}$$

Singular matrix
Dependent columns
Column space $\neq \mathbf{R}^3$
$$B = \begin{bmatrix} 1 & 0 & 1 \\ 1 & 1 & 2 \\ 1 & 1 & 2 \end{bmatrix}.$$

The only solution to $A\boldsymbol{x} = \mathbf{0}$ is $\boldsymbol{x} = A^{-1}\mathbf{0} = \mathbf{0}$. The columns are independent. They span the whole space $\mathbf{R}^n$—because every vector $\boldsymbol{b}$ is a combination of the columns. $A\boldsymbol{x} = \boldsymbol{b}$ can always be solved by $\boldsymbol{x} = A^{-1}\boldsymbol{b}$. Do you see how everything comes together for invertible matrices? Here it is in one sentence :

> The vectors $\boldsymbol{v}_1, \ldots, \boldsymbol{v}_n$ are a ***basis for*** $\mathbf{R}^n$ exactly when they are ***the columns of an n by n invertible matrix***. Thus $\mathbf{R}^n$ has infinitely many different bases.

When the columns are dependent, we keep only the *pivot columns*—the first two columns of B above. Those two columns are independent and they span the column space.

> **Every set of independent vectors can be extended to a basis.**
> **Every spanning set of vectors can be reduced to a basis.**

Example 5 This matrix is not invertible. Its columns are not a basis for anything !

One pivot column
One pivot row $(r = 1)$
$$A = \begin{bmatrix} 2 & 4 \\ 3 & 6 \end{bmatrix} \text{ reduces to } R_0 = \begin{bmatrix} 1 & 2 \\ 0 & 0 \end{bmatrix}.$$

Example 6 Find bases for the column and row spaces of this rank two matrix :

$$R_0 = \begin{bmatrix} 1 & 2 & 0 & 3 \\ 0 & 0 & 1 & 4 \\ 0 & 0 & 0 & 0 \end{bmatrix}.$$

Columns 1 and 3 are the pivot columns. They are a basis for the column space of R_0. The column space is the "xy plane" inside xyz space $\mathbf{R}^3$. That plane is not $\mathbf{R}^2$, it is a subspace of $\mathbf{R}^3$. Columns 2 and 3 are also a basis for the same column space. Which pair of columns of R_0 is *not* a basis for its column space ?

The row space is a subspace of $\mathbf{R}^4$. The simplest basis for that row space is the two nonzero rows of R_0. **The zero vector is never in a basis.**

> **Question** Given five vectors in $\mathbf{R}^7$, ***how do you find a basis for the space they span?***

First answer Make them the **rows of A**, and eliminate to find the nonzero rows in R.

Second answer Put the five vectors into the **columns of A**. Eliminate to find the pivot columns. Those pivot columns in C are a basis for the column space.

Could another basis have more vectors, or fewer? This is a crucial question with a good answer : *No. **All bases for a vector space contain the same** number **of vectors**.*

> ***The number of vectors in any and every basis is the "dimension" of the space.***

Dimension of a Vector Space

We have to prove what was just stated. There are many choices for the basis vectors, but **the number of basis vectors doesn't change**.

> If $\boldsymbol{v}_1, \ldots, \boldsymbol{v}_m$ and $\boldsymbol{w}_1, \ldots, \boldsymbol{w}_n$ are both bases for the same vector space, then $m = n$.

Proof Suppose that there are more $\boldsymbol{w}$'s than $\boldsymbol{v}$'s. From $n > m$ we want to reach a contradiction. The $\boldsymbol{v}$'s are a basis, so $\boldsymbol{w}_1$ must be a combination of the $\boldsymbol{v}$'s. If $\boldsymbol{w}_1$ equals $a_{11}\boldsymbol{v}_1 + \cdots + a_{m1}\boldsymbol{v}_m$, this is the first column of a matrix multiplication $VA = W$:

Each $\boldsymbol{w}$ is a combination of the $\boldsymbol{v}$'s

$$W = \begin{bmatrix} & & & \\ \boldsymbol{w}_1 & \boldsymbol{w}_2 & \ldots & \boldsymbol{w}_n \\ & & & \end{bmatrix} = \begin{bmatrix} & & \\ \boldsymbol{v}_1 & \ldots & \boldsymbol{v}_m \\ & & \end{bmatrix} \begin{bmatrix} a_{11} & & a_{1n} \\ \vdots & & \vdots \\ a_{m1} & & a_{mn} \end{bmatrix} = VA.$$

We don't know each a_{ij}, but we know the shape of A (it is m by n). The second vector $\boldsymbol{w}_2$ is also a combination of the $\boldsymbol{v}$'s. The coefficients in that combination fill the second column of A. The key is that A has a row for every $\boldsymbol{v}$ and a column for every $\boldsymbol{w}$. A is a *short wide matrix*, since we assumed $n > m$. So $A\boldsymbol{x} = \mathbf{0}$ ***has a nonzero solution***.

$A\boldsymbol{x} = 0$ gives $VA\boldsymbol{x} = \mathbf{0}$ which is $W\boldsymbol{x} = \mathbf{0}$. *A combination of the $\boldsymbol{w}$'s gives zero!* Then the $\boldsymbol{w}$'s could not be a basis—our assumption $n > m$ is **not possible** for two bases.

If $m > n$ we exchange the $\boldsymbol{v}$'s and $\boldsymbol{w}$'s and repeat the same steps. The only way to avoid a contradiction is to have $m = n$. This completes the proof that $m = n$.

The number of basis vectors is the dimension. So the dimension of $\mathbf{R}^n$ is n. We now define the important word ***dimension***.

> **DEFINITION** The ***dimension of a space*** is the ***number of vectors in every basis***.

The dimension matches our intuition. The line through $\boldsymbol{v} = (1, 5, 2)$ has dimension one. It is a subspace with this one vector $\boldsymbol{v}$ in its basis. Perpendicular to that line is the plane $x + 5y + 2z = 0$. This plane has dimension 2. To prove it, we find a basis $(-5, 1, 0)$ and $(-2, 0, 1)$. The dimension is 2 because the basis contains two vectors.

The plane is the nullspace of the matrix $A = \begin{bmatrix} 1 & 5 & 2 \end{bmatrix}$, which has two free variables. Our basis vectors $(-5, 1, 0)$ and $(-2, 0, 1)$ are the "special solutions" to $A\boldsymbol{x} = \mathbf{0}$. The $n - r$ special solutions always give *a basis for the nullspace*: dimension $n - r$.

Note about the language of linear algebra We never say "the rank of a space" or "the dimension of a basis" or "the basis of a matrix". Those terms have no meaning. It is the ***dimension of the column space*** that equals the ***rank of the matrix***.

Bases for Matrix Spaces and Function Spaces

The words "independence" and "basis" and "dimension" are not limited to column vectors. We can ask whether three matrices A_1, A_2, A_3 are independent. When they are 3 by 4 matrices, some combination might give the zero matrix. We can also ask the dimension of the full 3 by 4 matrix space. (It is 12.)

In differential equations, $d^2y/dx^2 = y$ has a space of solutions. One basis is $y = e^x$ and $y = e^{-x}$. Counting the basis functions gives the dimension 2 for this solution space. (The dimension is 2 because of the second derivative.)

Matrix spaces The vector space $\mathbf{M}$ contains all 2 by 2 matrices. Its dimension is 4.

$$\textbf{One basis is}\quad A_1, A_2, A_3, A_4 = \begin{bmatrix} \mathbf{1} & 0 \\ 0 & 0 \end{bmatrix}, \begin{bmatrix} 0 & \mathbf{1} \\ 0 & 0 \end{bmatrix}, \begin{bmatrix} 0 & 0 \\ \mathbf{1} & 0 \end{bmatrix}, \begin{bmatrix} 0 & 0 \\ 0 & \mathbf{1} \end{bmatrix}.$$

Those matrices are linearly independent. We are not looking at their columns, but at the whole matrix. Combinations of those four matrices can produce any matrix in $\mathbf{M}$.

$$\begin{array}{l}\textbf{Every } A \textbf{ combines} \\ \textbf{the basis matrices}\end{array} \qquad c_1A_1 + c_2A_2 + c_3A_3 + c_4A_4 = \begin{bmatrix} c_1 & c_2 \\ c_3 & c_4 \end{bmatrix} = A.$$

A is zero only if the c's are all zero—this proves independence of A_1, A_2, A_3, A_4.

The three matrices A_1, A_2, A_4 are a basis for a subspace—the upper triangular matrices. Its dimension is 3. A_1 and A_4 are a basis for the diagonal matrices. What is a basis for the symmetric matrices? Keep A_1 and A_4, and throw in $A_2 + A_3$.

The dimension of the whole n by n matrix space is n^2.

The dimension of the subspace of *upper triangular* matrices is $\frac{1}{2}n^2 + \frac{1}{2}n$.

The dimension of the subspace of *diagonal* matrices is n.

The dimension of the subspace of *symmetric* matrices is $\frac{1}{2}n^2 + \frac{1}{2}n$ (why ?).

Function spaces The equations $d^2y/dx^2 = 0$ and $d^2y/dx^2 = -y$ and $d^2y/dx^2 = y$ involve the second derivative. In calculus we solve to find the functions $y(x)$:

$$\begin{array}{ll} y'' = 0 & \text{is solved by any linear function } y = cx + d \\ y'' = -y & \text{is solved by any combination } y = c\sin x + d\cos x \\ y'' = y & \text{is solved by any combination } y = ce^x + de^{-x}. \end{array}$$

That solution space for $y'' = -y$ has two basis functions: $\sin x$ and $\cos x$. The space for $y'' = 0$ has x and 1. It is the "nullspace" of the second derivative! The dimension is 2 in each case (these are second-order equations).

The solutions of $y'' = 2$ don't form a subspace—the right side $b = 2$ is not zero. A particular solution is $y(x) = x^2$. The complete solution is $y(x) = x^2 + cx + d$. All those functions satisfy $y'' = 2$. Notice the *particular solution plus any function $cx + d$ in the nullspace*. A linear differential equation is like a linear matrix equation $A\boldsymbol{x} = \boldsymbol{b}$.

We end here with the space $\mathbf{Z}$ that contains only the zero vector. The dimension of this space is *zero*. ***The empty set*** (containing no vectors) ***is a basis for*** $\mathbf{Z}$. We can never allow the zero vector into a basis, because then linear independence is lost.

The key words in this section were **independence**, **span**, **basis**, **dimension**.

1. The columns of A are ***independent*** if $\boldsymbol{x} = \mathbf{0}$ is the only solution to $A\boldsymbol{x} = \mathbf{0}$.
2. The vectors $\boldsymbol{v}_1, \ldots, \boldsymbol{v}_r$ ***span*** a space if their combinations fill that space.
3. ***A basis consists of linearly independent vectors that span the space.*** Every vector in the space is a *unique* combination of the basis vectors.
4. All bases for a space have the same number of vectors. This number of vectors in a basis is the ***dimension*** of the space.
5. The pivot columns are one basis for the column space. The dimension is r.

■ WORKED EXAMPLES ■

3.4 A (***Important example***) Suppose $\boldsymbol{v}_1, \ldots, \boldsymbol{v}_n$ is a basis for $\mathbf{R}^n$ and the n by n matrix A is invertible. **Show that $A\boldsymbol{v}_1, \ldots, A\boldsymbol{v}_n$ is also a basis for $\mathbf{R}^n$.**

Solution In *matrix language*: Put the basis vectors $\boldsymbol{v}_1, \ldots, \boldsymbol{v}_n$ in the columns of an invertible (!) matrix V. Then $A\boldsymbol{v}_1, \ldots, A\boldsymbol{v}_n$ are the columns of AV. Since A is invertible, so is AV. Its columns give a basis.

In *vector language*: Suppose $c_1A\boldsymbol{v}_1 + \cdots + c_nA\boldsymbol{v}_n = \mathbf{0}$. This is $A\boldsymbol{v} = \mathbf{0}$ with $\boldsymbol{v} = c_1\boldsymbol{v}_1 + \cdots + c_n\boldsymbol{v}_n$. Multiply by A^{-1} to reach $\boldsymbol{v} = \mathbf{0}$. By linear independence of the $\boldsymbol{v}$'s, all $c_i = 0$. This shows that the $A\boldsymbol{v}$'s are independent.

To show that the $A\boldsymbol{v}$'s span $\mathbf{R}^n$, solve $c_1A\boldsymbol{v}_1 + \cdots + c_nA\boldsymbol{v}_n = \boldsymbol{b}$ which is the same as $c_1\boldsymbol{v}_1 + \cdots + c_n\boldsymbol{v}_n = A^{-1}\boldsymbol{b}$. Since the $\boldsymbol{v}$'s are a basis, this must be solvable.

3.4 B Start with the vectors $\boldsymbol{v}_1 = (1, 2, 0)$ and $\boldsymbol{v}_2 = (2, 3, 0)$. **(a)** Are they linearly independent? **(b)** Are they a basis for any space? **(c)** What is the dimension of $\mathbf{V}$? **(d)** Which matrices A have $\mathbf{V}$ as their column space? **(e)** Which matrices have $\mathbf{V}$ as their nullspace? **(f)** Describe all vectors $\boldsymbol{v}_3$ that complete a basis $\boldsymbol{v}_1, \boldsymbol{v}_2, \boldsymbol{v}_3$ for $\mathbf{R}^3$.

Solution

(a) $\boldsymbol{v}_1$ and $\boldsymbol{v}_2$ are independent—the only combination to give $\mathbf{0}$ is $0\boldsymbol{v}_1 + 0\boldsymbol{v}_2$.

(b) Yes, they are a basis for the space $\mathbf{V}$ they span: All vectors $(x, y, 0)$.

(c) The dimension of $\mathbf{V}$ is 2 since the basis contains two vectors.

(d) This $\mathbf{V}$ is the column space of any 3 by n matrix A of rank 2, if every column is a combination of $\boldsymbol{v}_1$ and $\boldsymbol{v}_2$. In particular A could just have columns $\boldsymbol{v}_1$ and $\boldsymbol{v}_2$.

(e) This $\mathbf{V}$ is the nullspace of any m by 3 matrix B of rank 1, if every row is a multiple of $(0, 0, 1)$. In particular take $B = [0\ \ 0\ \ 1]$. Then $B\boldsymbol{v}_1 = \mathbf{0}$ and $B\boldsymbol{v}_2 = \mathbf{0}$.

(f) Any third vector $\boldsymbol{v}_3 = (a, b, c)$ will complete a basis for $\mathbf{R}^3$ provided $c \neq 0$.

3.4 C Start with three independent vectors $\boldsymbol{w}_1, \boldsymbol{w}_2, \boldsymbol{w}_3$. Take combinations of those vectors to produce $\boldsymbol{v}_1, \boldsymbol{v}_2, \boldsymbol{v}_3$. Write the combinations in matrix form as $V = WB$:

$$\begin{aligned} \boldsymbol{v}_1 &= \boldsymbol{w}_1 + \boldsymbol{w}_2 \\ \boldsymbol{v}_2 &= \boldsymbol{w}_1 + 2\boldsymbol{w}_2 + \boldsymbol{w}_3 \\ \boldsymbol{v}_3 &= \qquad\; \boldsymbol{w}_2 + c\boldsymbol{w}_3 \end{aligned} \quad \text{which is} \quad \begin{bmatrix} \boldsymbol{v}_1 & \boldsymbol{v}_2 & \boldsymbol{v}_3 \end{bmatrix} = \begin{bmatrix} \boldsymbol{w}_1 & \boldsymbol{w}_2 & \boldsymbol{w}_3 \end{bmatrix} \begin{bmatrix} 1 & 1 & 0 \\ 1 & 2 & 1 \\ 0 & 1 & \boldsymbol{c} \end{bmatrix}$$

What is the test on B to see if $V = WB$ has independent columns? If $\boldsymbol{c} \neq 1$ show that $\boldsymbol{v}_1, \boldsymbol{v}_2, \boldsymbol{v}_3$ are linearly independent. If $\boldsymbol{c} = 1$ show that the $\boldsymbol{v}$'s are linearly *dependent.*

Solution For independent columns, *the nullspace of V must contain only the zero vector.* $V\boldsymbol{x} = 0$ requires $\boldsymbol{x} = (0, 0, 0)$.

If $\boldsymbol{c} = \mathbf{1}$ in our problem, we can see *dependence* in two ways. First, $\boldsymbol{v}_1 + \boldsymbol{v}_3$ will be the same as $\boldsymbol{v}_2$. Then $\boldsymbol{v}_1 - \boldsymbol{v}_2 + \boldsymbol{v}_3 = \mathbf{0}$—which says that the $\boldsymbol{v}$'s are dependent.

The other way is to look at the nullspace of B. If $c = 1$, the vector $\boldsymbol{x} = (1, -1, 1)$ is in that nullspace, and $B\boldsymbol{x} = \mathbf{0}$. Then certainly $WB\boldsymbol{x} = \mathbf{0}$ which is the same as $V\boldsymbol{x} = 0$. So the $\boldsymbol{v}$'s are dependent: $\boldsymbol{v}_1 - \boldsymbol{v}_2 + \boldsymbol{v}_3 = \mathbf{0}$.

Now suppose $\boldsymbol{c} \neq \mathbf{1}$. Then the matrix B is invertible. So if $\boldsymbol{x}$ is *any nonzero vector* we know that $B\boldsymbol{x}$ is nonzero. Since the $\boldsymbol{w}$'s are given as independent, we further know that $WB\boldsymbol{x}$ is nonzero. Since $V = WB$, this says that $\boldsymbol{x}$ is *not* in the nullspace of V. In other words $\boldsymbol{v}_1, \boldsymbol{v}_2, \boldsymbol{v}_3$ are independent.

The general rule is "independent $\boldsymbol{v}$'s from independent $\boldsymbol{w}$'s when B is invertible". And if these vectors are in $\mathbf{R}^3$, they are not only independent—they are a basis for $\mathbf{R}^3$. ***"Basis of v's from basis of w's when the change of basis matrix B is invertible."***

Problem Set 3.4

Questions 1–10 are about linear independence and linear dependence.

1 Show that $\boldsymbol{v}_1, \boldsymbol{v}_2, \boldsymbol{v}_3$ are independent but $\boldsymbol{v}_1, \boldsymbol{v}_2, \boldsymbol{v}_3, \boldsymbol{v}_4$ are dependent:

$$\boldsymbol{v}_1 = \begin{bmatrix} 1 \\ 0 \\ 0 \end{bmatrix} \quad \boldsymbol{v}_2 = \begin{bmatrix} 1 \\ 1 \\ 0 \end{bmatrix} \quad \boldsymbol{v}_3 = \begin{bmatrix} 1 \\ 1 \\ 1 \end{bmatrix} \quad \boldsymbol{v}_4 = \begin{bmatrix} 2 \\ 3 \\ 4 \end{bmatrix}$$

Solve $c_1\boldsymbol{v}_1 + c_2\boldsymbol{v}_2 + c_3\boldsymbol{v}_3 + c_4\boldsymbol{v}_4 = \mathbf{0}$ or $A\boldsymbol{x} = \mathbf{0}$. The $\boldsymbol{v}$'s go in the columns of A.

2 (Recommended) Find the largest possible number of independent vectors among

$$\boldsymbol{v}_1 = \begin{bmatrix} 1 \\ -1 \\ 0 \\ 0 \end{bmatrix} \quad \boldsymbol{v}_2 = \begin{bmatrix} 1 \\ 0 \\ -1 \\ 0 \end{bmatrix} \quad \boldsymbol{v}_3 = \begin{bmatrix} 1 \\ 0 \\ 0 \\ -1 \end{bmatrix} \quad \boldsymbol{v}_4 = \begin{bmatrix} 0 \\ 1 \\ -1 \\ 0 \end{bmatrix} \quad \boldsymbol{v}_5 = \begin{bmatrix} 0 \\ 1 \\ 0 \\ -1 \end{bmatrix} \quad \boldsymbol{v}_6 = \begin{bmatrix} 0 \\ 0 \\ 1 \\ -1 \end{bmatrix}$$

3 Prove that if $a = 0$ or $d = 0$ or $f = 0$ (3 cases), the columns of U are dependent:

$$U = \begin{bmatrix} a & b & c \\ 0 & d & e \\ 0 & 0 & f \end{bmatrix}.$$

4 If a, d, f in Question 3 are nonzero, show that the only solution to $U\boldsymbol{x} = \mathbf{0}$ is $\boldsymbol{x} = \mathbf{0}$. An upper triangular U with no diagonal zeros has independent columns.

5 Decide the dependence or independence of

(a) the vectors $(1, 3, 2)$ and $(2, 1, 3)$ and $(3, 2, 1)$

(b) the vectors $(1, -3, 2)$ and $(2, 1, -3)$ and $(-3, 2, 1)$.

6 Choose three independent columns of U. Then make two other choices. Do the same for A.

$$U = \begin{bmatrix} 2 & 3 & 4 & 1 \\ 0 & 6 & 7 & 0 \\ 0 & 0 & 0 & 9 \\ 0 & 0 & 0 & 0 \end{bmatrix} \quad \text{and} \quad A = \begin{bmatrix} 2 & 3 & 4 & 1 \\ 0 & 6 & 7 & 0 \\ 0 & 0 & 0 & 9 \\ 4 & 6 & 8 & 2 \end{bmatrix}.$$

7 If $\boldsymbol{w}_1, \boldsymbol{w}_2, \boldsymbol{w}_3$ are independent vectors, show that the differences $\boldsymbol{v}_1 = \boldsymbol{w}_2 - \boldsymbol{w}_3$ and $\boldsymbol{v}_2 = \boldsymbol{w}_1 - \boldsymbol{w}_3$ and $\boldsymbol{v}_3 = \boldsymbol{w}_1 - \boldsymbol{w}_2$ are *dependent*. Find a combination of the $\boldsymbol{v}$'s that gives zero. Which matrix A in $[\,\boldsymbol{v}_1\ \boldsymbol{v}_2\ \boldsymbol{v}_3\,] = [\,\boldsymbol{w}_1\ \boldsymbol{w}_2\ \boldsymbol{w}_3\,]\,A$ is singular?

8 If $\boldsymbol{w}_1, \boldsymbol{w}_2, \boldsymbol{w}_3$ are independent vectors, show that the sums $\boldsymbol{v}_1 = \boldsymbol{w}_2 + \boldsymbol{w}_3$ and $\boldsymbol{v}_2 = \boldsymbol{w}_1 + \boldsymbol{w}_3$ and $\boldsymbol{v}_3 = \boldsymbol{w}_1 + \boldsymbol{w}_2$ are *independent*. (Write $c_1\boldsymbol{v}_1 + c_2\boldsymbol{v}_2 + c_3\boldsymbol{v}_3 = \mathbf{0}$ in terms of the $\boldsymbol{w}$'s. Find and solve equations for the c's, to show they are zero.)

9 Suppose $\boldsymbol{v}_1, \boldsymbol{v}_2, \boldsymbol{v}_3, \boldsymbol{v}_4$ are vectors in $\mathbf{R}^3$.

(a) These four vectors are dependent because ______.

(b) The two vectors $\boldsymbol{v}_1$ and $\boldsymbol{v}_2$ will be dependent if ______.

(c) The vectors $\boldsymbol{v}_1$ and $(0,0,0)$ are dependent because ______.

10 Find two independent vectors on the plane $x+2y-3z-t=0$ in $\mathbf{R}^4$. Then find three independent vectors. Why not four? This plane is the nullspace of what matrix?

Questions 11–14 are about the space *spanned* by a set of vectors. Take all linear combinations of the vectors.

11 Describe the subspace of $\mathbf{R}^3$ (is it a line or plane or $\mathbf{R}^3$?) spanned by

(a) the two vectors $(1,1,-1)$ and $(-1,-1,1)$

(b) the three vectors $(0,1,1)$ and $(1,1,0)$ and $(0,0,0)$

(c) all vectors in $\mathbf{R}^3$ with whole number components

(d) all vectors with positive components.

12 The vector $\boldsymbol{b}$ is in the subspace spanned by the columns of A when ______ has a solution. The vector $\boldsymbol{c}$ is in the row space of A when ______ has a solution.

True or false: If the zero vector is in the row space, the rows are dependent.

13 Find the dimensions of these 4 spaces. Which two of the spaces are the same? (a) column space of A, (b) column space of U, (c) row space of A, (d) row space of U:

$$A = \begin{bmatrix} 1 & 1 & 0 \\ 1 & 3 & 1 \\ 3 & 1 & -1 \end{bmatrix} \quad \text{and} \quad U = \begin{bmatrix} 1 & 1 & 0 \\ 0 & 2 & 1 \\ 0 & 0 & 0 \end{bmatrix}.$$

14 $\boldsymbol{v}+\boldsymbol{w}$ and $\boldsymbol{v}-\boldsymbol{w}$ are combinations of $\boldsymbol{v}$ and $\boldsymbol{w}$. Write $\boldsymbol{v}$ and $\boldsymbol{w}$ as combinations of $\boldsymbol{v}+\boldsymbol{w}$ and $\boldsymbol{v}-\boldsymbol{w}$. The two pairs of vectors ______ the same space. When are they a basis for the same space?

Questions 15–25 are about the requirements for a basis.

15 If $\boldsymbol{v}_1, \ldots, \boldsymbol{v}_n$ are linearly independent, the space they span has dimension ______. These vectors are a ______ for that space. If the vectors are the columns of an m by n matrix, then m is ______ than n. If $m=n$, that matrix is ______.

16 Find a basis for each of these subspaces of $\mathbf{R}^4$:

(a) All vectors whose components are equal.

(b) All vectors whose components add to zero.

(c) All vectors that are perpendicular to $(1,1,0,0)$ and $(1,0,1,1)$.

(d) The column space and the nullspace of I (4 by 4).

17 Find three different bases for the column space of $U = \left[\begin{smallmatrix} 1 & 0 & 1 & 0 & 1 \\ 0 & 1 & 0 & 1 & 0 \end{smallmatrix}\right]$. Then find two different bases for the row space of U.

18 Suppose $\boldsymbol{v}_1, \boldsymbol{v}_2, \ldots, \boldsymbol{v}_6$ are six vectors in $\mathbf{R}^4$.

(a) Those vectors (do)(do not)(might not) span $\mathbf{R}^4$.

(b) Those vectors (are)(are not)(might be) linearly independent.

(c) Any four of those vectors (are)(are not)(might be) a basis for $\mathbf{R}^4$.

19 The columns of A are n vectors from $\mathbf{R}^m$. If they are linearly independent, what is the rank of A? If they span $\mathbf{R}^m$, what is the rank? If they are a basis for $\mathbf{R}^m$, what then? *Looking ahead*: The rank r counts the number of ______ columns.

20 Find a basis for the plane $x - 2y + 3z = 0$ in $\mathbf{R}^3$. Then find a basis for the intersection of that plane with the xy plane. Then find a basis for all vectors perpendicular to the plane.

21 Suppose the columns of a 5 by 5 matrix A are a basis for $\mathbf{R}^5$.

(a) The equation $A\boldsymbol{x} = \mathbf{0}$ has only the solution $\boldsymbol{x} = \mathbf{0}$ because ______.

(b) If $\boldsymbol{b}$ is in $\mathbf{R}^5$ then $A\boldsymbol{x} = \boldsymbol{b}$ is solvable because the basis vectors ______ $\mathbf{R}^5$.

Conclusion: A is invertible. Its rank is 5. Its rows are also a basis for $\mathbf{R}^5$.

22 Suppose $\mathbf{S}$ is a 5-dimensional subspace of $\mathbf{R}^6$. True or false (example if false):

(a) Every basis for $\mathbf{S}$ can be extended to a basis for $\mathbf{R}^6$ by adding one more vector.

(b) Every basis for $\mathbf{R}^6$ can be reduced to a basis for $\mathbf{S}$ by removing one vector.

23 U comes from A by subtracting row 1 from row 3:

$$A = \begin{bmatrix} 1 & 3 & 2 \\ 0 & 1 & 1 \\ 1 & 3 & 2 \end{bmatrix} \quad \text{and} \quad U = \begin{bmatrix} 1 & 3 & 2 \\ 0 & 1 & 1 \\ 0 & 0 & 0 \end{bmatrix}.$$

Find bases for the two column spaces. Find bases for the two row spaces. Find bases for the two nullspaces. Which spaces stay fixed in elimination?

24 True or false (give a good reason):

(a) If the columns of a matrix are dependent, so are the rows.

(b) The column space of a 2 by 2 matrix is the same as its row space.

(c) The column space of a 2 by 2 matrix has the same dimension as its row space.

(d) The columns of a matrix are a basis for the column space.

25 Suppose $\boldsymbol{v}_1, \ldots, \boldsymbol{v}_k$ span $\mathbf{R}^n$. How would you reduce this set to a basis? Suppose $\boldsymbol{v}_1, \ldots, \boldsymbol{v}_j$ are independent in $\mathbf{R}^n$. How would you add vectors to reach a basis?

26 For which numbers c and d do these matrices have rank 2?

$$A = \begin{bmatrix} 1 & 2 & 5 & 0 & 5 \\ 0 & 0 & c & 2 & 2 \\ 0 & 0 & 0 & d & 2 \end{bmatrix} \quad \text{and} \quad B = \begin{bmatrix} c & d \\ d & c \end{bmatrix}.$$

Questions 27–31 are about spaces where the "vectors" are matrices.

27 Find a basis (and the dimension) for each of these subspaces of 3 by 3 matrices:

(a) All diagonal matrices.

(b) All symmetric matrices ($A^{\mathrm{T}} = A$).

(c) All skew-symmetric matrices ($A^{\mathrm{T}} = -A$).

28 Construct six linearly independent 3 by 3 echelon matrices $U_1, \ldots, U_6$.

29 Find a basis for the space of all 2 by 3 matrices whose columns add to zero. Find a basis for the subspace whose rows also add to zero.

30 What subspace of 3 by 3 matrices is spanned (take all combinations) by

(a) the invertible matrices?

(b) the rank one matrices?

(c) the identity matrix?

31 Find a basis for the space of 2 by 3 matrices whose nullspace contains $(2, 1, 1)$.

Questions 32–36 are about spaces where the "vectors" are functions.

32 (a) Find all functions that satisfy $\frac{dy}{dx} = 0$.

(b) Choose a particular function that satisfies $\frac{dy}{dx} = 3$.

(c) Find all functions that satisfy $\frac{dy}{dx} = 3$.

33 The cosine space $\mathbf{F}_3$ contains all combinations $y(x) = A\cos x + B\cos 2x + C\cos 3x$. Find a basis for the subspace with $y(0) = 0$.

34 Find a basis for the space of functions that satisfy

(a) $\frac{dy}{dx} - 2y = 0$

(b) $\frac{dy}{dx} - \frac{y}{x} = 0$.

35 Suppose $y_1(x), y_2(x), y_3(x)$ are three different functions of x. The vector space they span could have dimension 1, 2, or 3. Give an example of y_1, y_2, y_3 to show each possibility.

36 Find a basis for the space of polynomials $p(x)$ of degree ≤ 3. Find a basis for the subspace with $p(1) = 0$.

37 Find a basis for the space $\mathbf{S}$ of vectors (a, b, c, d) with $a + c + d = 0$ and also for the space $\mathbf{T}$ with $a + b = 0$ and $c = 2d$. What is the dimension of the intersection $\mathbf{S} \cap \mathbf{T}$?

38 Suppose A is 5 by 4 with rank 4. Show that $A\boldsymbol{x} = \boldsymbol{b}$ has no solution when the 5 by 5 matrix $[\, A \;\; \boldsymbol{b} \,]$ is invertible. Show that $A\boldsymbol{x} = \boldsymbol{b}$ is solvable when $[\, A \;\; \boldsymbol{b} \,]$ is singular.

39 Find bases for all solutions to $d^2y/dx^2 = y(x)$ and then $d^2y/dx^2 = -y(x)$.

Challenge Problems

40 Write the 3 by 3 identity matrix as a combination of the other five permutation matrices! Then show that those five matrices are linearly independent. This is a basis for the subspace of 3 by 3 matrices with row and column sums all equal.

41 Choose $\boldsymbol{x} = (x_1, x_2, x_3, x_4)$ in $\mathbf{R}^4$. It has 24 rearrangements like (x_2, x_1, x_3, x_4) and (x_4, x_3, x_1, x_2). Those 24 vectors, including $\boldsymbol{x}$ itself, span a subspace $\mathbf{S}$. Find specific vectors $\boldsymbol{x}$ so that the dimension of $\mathbf{S}$ is: (a) zero, (b) one, (c) three, (d) four.

42 Intersections and sums have $\dim(\mathbf{V}) + \dim(\mathbf{W}) = \dim(\mathbf{V} \cap \mathbf{W}) + \dim(\mathbf{V} + \mathbf{W})$. Start with a basis $\boldsymbol{u}_1, \ldots, \boldsymbol{u}_r$ for the intersection $\mathbf{V} \cap \mathbf{W}$. Extend with $\boldsymbol{v}_1, \ldots, \boldsymbol{v}_s$ to a basis for $\mathbf{V}$, and separately $\boldsymbol{w}_1, \ldots, \boldsymbol{w}_t$ to a basis for $\mathbf{W}$. Prove that the $\boldsymbol{u}$'s, $\boldsymbol{v}$'s and $\boldsymbol{w}$'s together are ***independent***: dimensions $(\boldsymbol{r}+\boldsymbol{s})+(\boldsymbol{r}+\boldsymbol{t}) = (\boldsymbol{r})+(\boldsymbol{r}+\boldsymbol{s}+\boldsymbol{t})$.

43 Inside $\mathbf{R}^n$, suppose dimension $(\mathbf{V})$ + dimension $(\mathbf{W}) > n$. Show that some nonzero vector is in both $\mathbf{V}$ and $\mathbf{W}$.

44 Suppose A is 10 by 10 and $A^2 = 0$ (zero matrix). So A multiplies each column of A to give the zero vector. Then the column space of A is contained in the _____ . If A has rank r, those subspaces have dimension $r \leq 10 - r$. So the rank is $r \leq 5$.

3.5 Dimensions of the Four Subspaces

1 The column space $\mathbf{C}(A)$ and the row space $\mathbf{C}(A^{\mathrm{T}})$ both have *dimension* r (the rank of A).

2 The nullspace $\mathbf{N}(A)$ has *dimension* $n-r$. The left nullspace $\mathbf{N}(A^{\mathrm{T}})$ has *dimension* $m-r$.

3 Elimination often changes $\mathbf{C}(A)$ and $\mathbf{N}(A^{\mathrm{T}})$ (but their dimensions don't change).

The main theorem in this chapter connects ***rank*** and ***dimension***. The ***rank*** of a matrix counts independent columns. The ***dimension*** of a subspace is the number of vectors in a basis. We can count pivots or basis vectors. *The rank of A reveals the dimensions of all four fundamental subspaces.* Here are the subspaces, including the new one.

Two subspaces come directly from A, and the other two come from A^{T}.

	Four Fundamental Subspaces	**Dimensions**
1.	The ***row space*** is $\mathbf{C}(A^{\mathrm{T}})$, a subspace of $\mathbf{R}^n$.	r
2.	The ***column space*** is $\mathbf{C}(A)$, a subspace of $\mathbf{R}^m$.	r
3.	The ***nullspace*** is $\mathbf{N}(A)$, a subspace of $\mathbf{R}^n$.	$n-r$
4.	The ***left nullspace*** is $\mathbf{N}(A^{\mathrm{T}})$, a subspace of $\mathbf{R}^m$.	$m-r$

We know $\mathbf{C}(A)$ and $\mathbf{N}(A)$ pretty well. Now $\mathbf{C}(A^{\mathrm{T}})$ and $\mathbf{N}(A^{\mathrm{T}})$ come forward. The row space contains all combinations of the rows. *This row space is the column space of* A^{T}.

For the left nullspace we solve $A^{\mathrm{T}}\boldsymbol{y}=\mathbf{0}$—that system is n by m. In Example 2 this produces one of the great equations of applied mathematics—Kirchhoff's Current Law. The currents flow around a network, and they can't pile up at the nodes. The four subspaces come from nodes and edges and loops and trees. Those subspaces are connected in an absolutely beautiful way.

Part 1 of the Fundamental Theorem finds the dimensions of the four subspaces. One fact stands out: ***The row space and column space have the same dimension*** r. This number r is the rank of A (Chapter 1). The other important fact involves the two nullspaces:

$\mathbf{N}(A)$ ***and*** $\mathbf{N}(A^{\mathrm{T}})$ ***have dimensions*** $n-r$ ***and*** $m-r$***, to make up the full*** n ***and*** m.

Part 2 of the Fundamental Theorem will describe how the four subspaces fit together: Nullspace perpendicular to row space, and $\mathbf{N}(A^{\mathrm{T}})$ perpendicular to $\mathbf{C}(A)$. That completes the "right way" to understand $A\boldsymbol{x}=\boldsymbol{b}$. Stay with it—you are doing real mathematics.

The Four Subspaces for R_0

Suppose A is reduced to its row echelon form R_0. For that special form, the four subspaces are easy to identify. We will find a basis for each subspace and check its dimension. Then we watch how the subspaces change (two of them don't change!) as we look back at A. The main point is that *the four dimensions are the same for A and R_0.*

For A and R, one of the four subspaces can have different dimensions. **Which one?**

As a specific 3 by 5 example, look at the four subspaces for this echelon matrix R_0:

$$\begin{matrix} m=3 \\ n=5 \\ r=2 \end{matrix} \qquad \boldsymbol{R_0} = \begin{bmatrix} \mathbf{1} & \mathbf{3} & \mathbf{5} & 0 & \mathbf{7} \\ 0 & 0 & 0 & \mathbf{1} & \mathbf{2} \\ 0 & 0 & 0 & 0 & 0 \end{bmatrix} \qquad \begin{matrix} \textbf{pivot rows 1 and 2} \\ \\ \textbf{pivot columns 1 and 4} \end{matrix}$$

The rank of this matrix is $\boldsymbol{r=2}$ (*two pivots*). Take the four subspaces in order.

1. The ***row space*** has dimension 2, matching the rank.

Reason: The first two rows are a basis. The row space contains combinations of all three rows, but the third row (the zero row) adds nothing to the row space.

The pivot rows 1 and 2 are independent. That is obvious for this example, and it is always true. If we look only at the pivot columns, we see the r by r identity matrix. There is no way to combine its rows to give the zero row (except by the combination with all coefficients zero). So the r pivot rows (the rows of R) are a basis for the row space.

The dimension of the row space is the rank r. The nonzero rows of R_0 form a basis.

2. The ***column space*** of R_0 also has dimension $r=2$.

Reason: The pivot columns 1 and 4 form a basis. They are independent because they contain the r by r identity matrix. No combination of those pivot columns can give the zero column (except the combination with all coefficients zero). And they also span the column space. Every other (free) column is a combination of the pivot columns. Actually the combinations we need are the three special solutions!

Column 2 is 3 (column 1). The special solution is $(-3,1,0,0,0)$.

Column 3 is 5 (column 1). The special solution is $(-5,0,1,0,0,)$.

Column 5 is 7 (column 1) $+$ 2 (column 4). That solution is $(-7,0,0,-2,1)$.

The pivot columns are independent, and they span $\mathbf{C}(R_0)$, so they are a basis for $\mathbf{C}(R_0)$.

The dimension of the column space is the rank r. The pivot columns form a basis.

3. The ***nullspace*** of R_0 has dimension $\boldsymbol{n - r = 5 - 2}$. The 3 free variables give **3 special solutions** to $R_0\boldsymbol{x} = \boldsymbol{0}$. Set the free variables to 1 and 0 and 0.

$$s_2 = \begin{bmatrix} \mathbf{-3} \\ \mathbf{1} \\ \mathbf{0} \\ 0 \\ \mathbf{0} \end{bmatrix} \quad s_3 = \begin{bmatrix} \mathbf{-5} \\ \mathbf{0} \\ \mathbf{1} \\ 0 \\ \mathbf{0} \end{bmatrix} \quad s_5 = \begin{bmatrix} \mathbf{-7} \\ \mathbf{0} \\ \mathbf{0} \\ \mathbf{-2} \\ \mathbf{1} \end{bmatrix}$$

$R_0\boldsymbol{x} = \boldsymbol{0}$ has the complete solution
$\boldsymbol{x} = x_2\boldsymbol{s}_2 + x_3\boldsymbol{s}_3 + x_5\boldsymbol{s}_5$
The nullspace has dimension 3.

Reason : There is a special solution for each free variable. With n variables and r pivots, that leaves $n-r$ free variables and special solutions. The special solutions are independent, because they contain the identity matrix in rows 2, 3, 5.

The nullspace $\mathbf{N}(A)$ ***has dimension*** $\boldsymbol{n - r}$. ***The special solutions form a basis.***

4. The ***nullspace of*** $\boldsymbol{R_0^{\mathrm{T}}}$ ***(left nullspace of*** $\boldsymbol{R_0}$***) has dimension*** $\boldsymbol{m - r = 3 - 2}$.

Reason : R_0 has r independent rows and $m - r$ zero rows. Then R_0^{T} has r independent columns and $\boldsymbol{m - r}$ **zero columns**. So $\boldsymbol{y}$ in the nullspace of R_0^{T} can have nonzeros in its last $m - r$ entries. The example has $m - r = 1$ zero columns in R_0^{T} and 1 nonzero in $\boldsymbol{y}$.

$$R_0^{\mathrm{T}}\boldsymbol{y} = \begin{bmatrix} 1 & 0 & 0 \\ 3 & 0 & 0 \\ 5 & 0 & 0 \\ 0 & 1 & 0 \\ 7 & 2 & 0 \end{bmatrix} \begin{bmatrix} y_1 \\ y_2 \\ y_3 \end{bmatrix} = \begin{bmatrix} 0 \\ 0 \\ 0 \\ 0 \\ 0 \end{bmatrix} \quad \text{is solved by } \boldsymbol{y} = \begin{bmatrix} 0 \\ 0 \\ y_3 \end{bmatrix}. \tag{1}$$

Because of zero rows in R_0 and zero columns in R_0^{T}, it is easy to see the dimension (and even a basis) for this fourth fundamental subspace :

If $\boldsymbol{R_0}$ **has** $\boldsymbol{m - r}$ **zero rows, its left nullspace has dimension** $\boldsymbol{m - r}$.

Why is this a "*left* nullspace"? Because we can transpose $R_0^{\mathrm{T}}\boldsymbol{y} = \boldsymbol{0}$ to $\boldsymbol{y}^{\mathrm{T}}R_0 = \boldsymbol{0}^{\mathrm{T}}$. Now $\boldsymbol{y}^{\mathrm{T}}$ is a row vector to the *left* of R. This subspace came fourth, and some linear algebra books omit it—but that misses the beauty of the whole subject.

In $\mathbf{R}^n$ ***the row space and nullspace have dimensions*** $\boldsymbol{r}$ and $\boldsymbol{n - r}$ (adding to n).
In $\mathbf{R}^m$ ***the column space and left nullspace have dimensions*** $\boldsymbol{r}$ and $\boldsymbol{m - r}$ (total m).

We have a job still to do. ***The four subspace dimensions for*** $\boldsymbol{A}$ ***are the same as for*** $\boldsymbol{R_0}$. The job is to explain why. A is now any matrix that reduces to $R_0 = \mathsf{rref}(A)$.

This $\boldsymbol{A}$ **reduces to** $\boldsymbol{R_0}$ $$A = \begin{bmatrix} 1 & 3 & 5 & 0 & 7 \\ 0 & 0 & 0 & 1 & 2 \\ \mathbf{1} & \mathbf{3} & \mathbf{5} & \mathbf{1} & \mathbf{9} \end{bmatrix}$$ **Same row space as** $\boldsymbol{R_0}$
Different column space
But same dimension !

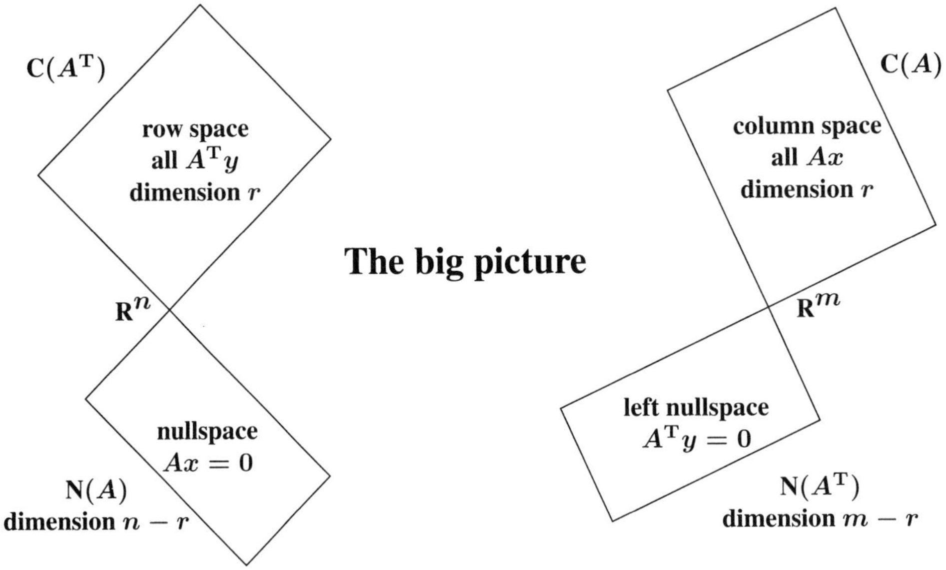

Figure 3.3: The dimensions of the Four Fundamental Subspaces (for R_0 and for A).

The Four Subspaces for A

1 ***A has the same row space as R_0 and R.* Same dimension r and same basis.**

Reason: Every row of A is a combination of the rows of R_0. Also every row of R_0 is a combination of the rows of A. Elimination changes *rows*, but not *row spaces*.

Since A has the same row space as R_0, the first r rows of R_0 are still a basis. Or we could choose r suitable rows of the original A. They might not always be the *first* r rows of A, because those could be dependent. The good r rows of A are the ones that end up as pivot rows in R_0 and R.

2 ***The column space of A has dimension r. The column rank equals the row rank.***

The number of independent columns = the number of independent rows.

Wrong reason : "A and R_0 have the same column space." This is false. *The columns of R_0 often end in zeros.* The columns of A don't often end in zeros. Then $\mathbf{C}(A)$ is not $\mathbf{C}(R_0)$.

Right reason : The ***same combinations*** of the columns are zero (or not) for A and R_0. Dependent in $A \Leftrightarrow$ dependent in R_0. Say that another way: $A\boldsymbol{x} = \mathbf{0}$ *exactly when* $R_0\boldsymbol{x} = \mathbf{0}$. The column spaces are different, but their *dimensions* are the same—equal to the rank r.

Conclusion The r pivot columns of A are a basis for *its* column space $\mathbf{C}(A)$.

3 ***A has the same nullspace as R_0.*** **Same dimension $n - r$ and same basis.**

Reason: The elimination steps don't change the solutions. The special solutions are a basis for this nullspace (as we always knew). There are $n-r$ free variables, so the dimension of the nullspace is $n-r$. This is the **Counting Theorem**: $\boldsymbol{r + (n - r)}$ **equals** $\boldsymbol{n}$.

$$\textbf{(dimension of column space)} + \textbf{(dimension of nullspace)} = \textbf{dimension of } \mathbf{R}^n.$$

4 ***The left nullspace of A*** (the nullspace of A^{T}) ***has dimension $m - r$.***

Reason: A^{T} is just as good a matrix as A. When we know the dimensions for every A, we also know them for A^{T}. Its column space was proved to have dimension r. Since A^{T} is n by m, the "whole space" is now $\mathbf{R}^m$. The counting rule for A was $r+(n-r)=n$. **The counting rule for $\boldsymbol{A^{\mathrm{T}}}$ is $\boldsymbol{r + (m - r) = m}$.** We have all details of a big theorem:

***Fundamental Theorem of Linear Algebra*, Part 1**

The column space and row space both have dimension r.

The nullspaces have dimensions $n - r$ and $m - r$.

By concentrating on *spaces* of vectors, not on individual numbers or vectors, we get these clean rules. You will soon take them for granted—eventually they begin to look obvious. But if you write down an 11 by 17 matrix with 187 nonzero entries, I don't think most people would see why these facts are true:

Two key facts

$$\text{dimension of } \mathbf{C}(A) = \text{dimension of } \mathbf{C}(A^{\mathrm{T}}) = \text{rank of } A$$
$$\text{dimension of } \mathbf{C}(A) + \text{dimension of } \mathbf{N}(A) = 17.$$

Every vector $A\boldsymbol{x} = \boldsymbol{b}$ in the column space comes from exactly one $\boldsymbol{x}$ in the row space! (If we also have $A\boldsymbol{y} = \boldsymbol{b}$ then $A(\boldsymbol{x}-\boldsymbol{y}) = \boldsymbol{b}-\boldsymbol{b} = \mathbf{0}$. So $\boldsymbol{x}-\boldsymbol{y}$ is in the nullspace as well as the row space, which forces $\boldsymbol{x} = \boldsymbol{y}$.) From its row space to its column space, *A is like an r by r invertible matrix.*

It is the nullspaces that will force us to define a "pseudoinverse of A": page 133.

Example 1 $A = \begin{bmatrix} 1 & 2 & 3 \\ 2 & 4 & 6 \end{bmatrix}$ has $m = 2$ with $n = 3$. The rank is $r = 1$.

The row space is the line through $(1,2,3)$. The nullspace is the plane $x_1 + 2x_2 + 3x_3 = 0$. The line and plane dimensions still add to $1 + 2 = 3$. The column space and left nullspace are **perpendicular lines in $\mathbf{R}^2$**. Dimensions $1 + 1 = 2$.

$$\text{Column space} = \text{line through } \begin{bmatrix} 1 \\ 2 \end{bmatrix} \qquad \text{Left nullspace} = \text{line through } \begin{bmatrix} 2 \\ -1 \end{bmatrix}.$$

Final point: The y's in the left nullspace combine the rows of A to give the zero row.

Example 2 You have nearly finished three chapters with made-up equations, and this can't continue forever. Here is a better example of five equations (one equation for every edge in Figure 3.4). The five equations have four unknowns (one for every node). The matrix in $A\boldsymbol{x} = \boldsymbol{b}$ is an **incidence matrix**. This matrix A has 1 and -1 on every row.

Differences $A\boldsymbol{x} = \boldsymbol{b}$
across edges $1, 2, 3, 4, 5$
between nodes $1, 2, 3, 4$
$m = 5$ and $n = 4$

$$\boxed{\begin{array}{rrrrl} -x_1 & +x_2 & & & = b_1 \\ -x_1 & & +x_3 & & = b_2 \\ & -x_2 & +x_3 & & = b_3 \\ & -x_2 & & +x_4 & = b_4 \\ & & -x_3 & +x_4 & = b_5 \end{array}} \qquad (2)$$

If you understand the four fundamental subspaces for this matrix (*the column spaces and the nullspaces for A and A^{T}*) you have captured a central idea of linear algebra.

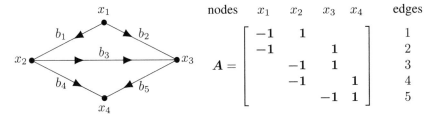

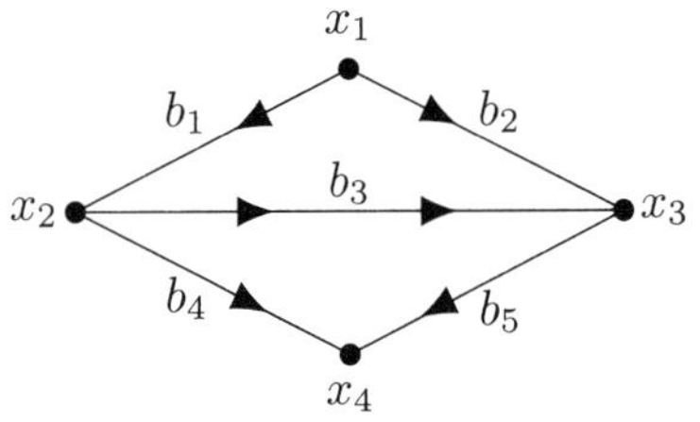

$$\begin{array}{c} \text{nodes} \quad x_1 \quad x_2 \quad x_3 \quad x_4 \qquad \text{edges} \\ \boldsymbol{A} = \begin{bmatrix} -1 & 1 & & \\ -1 & & 1 & \\ & -1 & 1 & \\ & -1 & & 1 \\ & & -1 & 1 \end{bmatrix} \begin{array}{c} 1 \\ 2 \\ 3 \\ 4 \\ 5 \end{array} \end{array}$$

Figure 3.4: A **"graph"** with 5 edges and 4 nodes. A is its 5 by 4 **incidence matrix**.

The nullspace $\mathbf{N}(A)$ To find the nullspace we set $\boldsymbol{b} = \mathbf{0}$. Then the first equation says $x_1 = x_2$. The second equation is $x_3 = x_1$. Equation 4 is $x_2 = x_4$. *All four unknowns x_1, x_2, x_3, x_4 have the same value c.* The vectors $\boldsymbol{x} = (c, c, c, c)$ fill the nullspace of A.

That nullspace is a line in $\mathbf{R}^4$. The special solution $\boldsymbol{x} = (1, 1, 1, 1)$ is a basis for $\mathbf{N}(A)$. The dimension of $\mathbf{N}(A)$ is 1 (one vector in the basis). *The rank of A must be* 3, *since $n - r = 4 - 3 = 1$.* We now know the dimensions of all four subspaces.

The column space $\mathbf{C}(A)$ There must be $r = 3$ independent columns. The fast way is to look at the first 3 columns of A. The systematic way is to find $R_0 = \mathsf{rref}(A)$.

$$\textbf{Columns } \mathbf{1, 2, 3} \textbf{ of } \boldsymbol{A} \quad \begin{array}{rrr} -1 & 1 & 0 \\ -1 & 0 & 1 \\ 0 & -1 & 1 \\ 0 & -1 & 0 \\ 0 & 0 & -1 \end{array} \qquad \boldsymbol{R_0} = \begin{array}{c} \textbf{reduced row} \\ \textbf{echelon form} \end{array} = \begin{bmatrix} 1 & 0 & 0 & -1 \\ 0 & 1 & 0 & -1 \\ 0 & 0 & 1 & -1 \\ 0 & 0 & 0 & 0 \\ 0 & 0 & 0 & 0 \end{bmatrix}$$

From R_0 we see again the special solution $\boldsymbol{x} = (1, 1, 1, 1)$. The first 3 columns are basic, the fourth column is free. To produce a basis for $\mathbf{C}(A)$ and not $\mathbf{C}(R_0)$, we must go back to *columns* $1, 2, 3$ *of* A. The column space has dimension $r = 3$.

The row space $\mathbf{C}(A^{\mathrm{T}})$ The dimension must again be $r = 3$. But the first 3 rows of A are *not independent*: row 3 = row 2 − row 1. So row 3 became zero in elimination, and row 3 was exchanged with row 4. *The first three independent rows are rows* $1, 2, 4$. Those three rows are a basis (one possible basis) for the row space.

Edges $1, 2, 3$ form a **loop** in the graph:	Dependent rows $1, 2, 3$.
Edges $1, 2, 4$ form a **tree**. **Trees have no loops**!	Independent rows $1, 2, 4$.

The left nullspace $\mathbf{N}(A^{\mathrm{T}})$ Now we solve $A^{\mathrm{T}}\boldsymbol{y} = \mathbf{0}$. Combinations of the rows give zero. We already noticed that row 3 = row 2 − row 1, so one solution is $\boldsymbol{y} = (1, -1, 1, 0, 0)$. I would say: *That* $\boldsymbol{y}$ *comes from following the upper loop in the graph.* Another $\boldsymbol{y}$ comes from going around the lower loop and it is $\boldsymbol{y} = (0, 0, \mathbf{-1}, \mathbf{1}, \mathbf{-1})$: *row* 3 = *row* 4 − *row* 5. Those two $\boldsymbol{y}$'s are independent, they solve $A^{\mathrm{T}}\boldsymbol{y} = \mathbf{0}$, and the dimension of $\mathbf{N}(A^{\mathrm{T}})$ is $\boldsymbol{m} - \boldsymbol{r} = 5 - 3 = \mathbf{2}$. So we have a basis for the left nullspace.

You may ask how "loops" and "trees" got into this problem. That didn't have to happen. We could have used elimination to solve $A^{\mathrm{T}}\boldsymbol{y} = \mathbf{0}$. The 4 by 5 matrix A^{T} would have three pivot columns $1, 2, 4$ and two free columns $3, 5$. There are two special solutions and the nullspace of A^{T} has dimension two: $m - r = 5 - 3 = 2$. But *loops* and *trees* identify *dependent rows* and *independent rows* in a beautiful way for every incidence matrix.

The equations $A\boldsymbol{x} = \boldsymbol{b}$ give "voltages" x_1, x_2, x_3, x_4 at the four nodes. The equations $A^{\mathrm{T}}\boldsymbol{y} = \mathbf{0}$ give "currents" y_1, y_2, y_3, y_4, y_5 on the five edges. These two equations are **Kirchhoff's Voltage Law** and **Kirchhoff's Current Law**. Those laws apply to an electrical network. But the ideas behind the words apply all over engineering and science and economics and business. Linear algebra connects the laws to the four subspaces.

Graphs are *the most important model in discrete applied mathematics.* You see graphs everywhere: roads, pipelines, blood flow, the brain, the Web, the economy of a country or the world. We can understand their matrices A and A^{T}. Here is a summary.

The incidence matrix A comes from a connected graph with n nodes and m edges. The row space and column space have dimensions $r = n - 1$. The nullspaces of A and A^{T} have dimensions 1 and $m - n + 1$:

$\mathbf{N}(A)$ The constant vectors $(c, c, \ldots, c)$ make up the nullspace of A : dim $= 1$.

$\mathbf{C}(A^{\mathrm{T}})$ The edges of any tree give r independent rows of A : $r = n - 1$.

$\mathbf{C}(A)$ ***Voltage Law***: The components of $A\boldsymbol{x}$ add to zero around all loops: dim $= n - 1$.

$\mathbf{N}(A^{\mathrm{T}})$ ***Current Law***: $A^{\mathrm{T}}\boldsymbol{y} =$ (**flow in**)−(**flow out**) $= \mathbf{0}$ is solved by loop currents.

There are $\boldsymbol{m} - \boldsymbol{r} = \boldsymbol{m} - \boldsymbol{n} + \mathbf{1}$ ***independent small loops in the graph.***

For every graph in a plane, linear algebra yields ***Euler's formula***: Theorem 1 in topology!

$$(\textit{nodes}) - (\textit{edges}) + (\textit{small loops}) = (n) - (m) + (m - n + 1) = 1$$

Rank Two Matrices = Rank One plus Rank One

Rank one matrices have the form $\boldsymbol{u}\boldsymbol{v}^{\mathrm{T}}$. Here is a matrix A of rank $r = 2$. We can't see r immediately from A. So we reduce the matrix by row operations to R_0. $\boldsymbol{R_0}$ **has the same row space as** $\boldsymbol{A}$. Throw away its zero row to find R—also with the same row space.

$$\textbf{Rank two} \qquad A = \begin{bmatrix} 1 & 0 & 3 \\ 1 & 1 & 7 \\ 4 & 2 & 20 \end{bmatrix} = \begin{bmatrix} \mathbf{1} & \mathbf{0} \\ \mathbf{1} & \mathbf{1} \\ \mathbf{4} & \mathbf{2} \end{bmatrix} \begin{bmatrix} \mathbf{1} & \mathbf{0} & \mathbf{3} \\ \mathbf{0} & \mathbf{1} & \mathbf{4} \end{bmatrix} = \boldsymbol{CR} \qquad (3)$$

Now look at columns. The pivot columns of R are clearly $(1, 0)$ and $(0, 1)$. Then the pivot columns of A are also in columns 1 and 2: $\boldsymbol{u}_1 = (1, 1, 4)$ and $\boldsymbol{u}_2 = (0, 1, 2)$. Notice that C has those same first two columns! That was guaranteed since multiplying by two columns of the identity matrix (in R) won't change the pivot columns $\boldsymbol{u}_1$ and $\boldsymbol{u}_2$.

When you put in letters for the columns and rows, you see **rank 2 = rank 1 + rank 1**.

$$\begin{array}{l}\textbf{Matrix } \boldsymbol{A} \\ \textbf{Rank two}\end{array} \qquad A = \begin{bmatrix} & & \\ \boldsymbol{u}_1 & \boldsymbol{u}_2 & \boldsymbol{u}_3 \\ & & \end{bmatrix} \begin{bmatrix} \boldsymbol{v}_1^{\mathrm{T}} \\ \boldsymbol{v}_2^{\mathrm{T}} \\ \textbf{zero row} \end{bmatrix} = \boldsymbol{u}_1\boldsymbol{v}_1^{\mathrm{T}} + \boldsymbol{u}_2\boldsymbol{v}_2^{\mathrm{T}}$$

Columns of $\boldsymbol{C}$ **times rows of** $\boldsymbol{R}$. *Every rank r matrix is a sum of r rank one matrices*

■ WORKED EXAMPLES ■

3.5 A **Put four 1's into a 5 by 6 matrix of zeros**, keeping the dimension of its *row space* as small as possible. Describe all the ways to make the dimension of its *column space* as small as possible. Then describe all the ways to make the dimension of its *nullspace* as small as possible. How to make the *sum of the dimensions of all four subspaces small*?

Solution The rank is 1 if the four 1's go into the same row, or into the same column. They can also go into *two rows and two columns* (so $a_{ii} = a_{ij} = a_{ji} = a_{jj} = 1$). Since the column space and row space always have the same dimensions, this answers the first two questions: Dimension 1.

The nullspace has its smallest possible dimension $6 - 4 = 2$ when the rank is $r = 4$. To achieve rank 4, the 1's must go into four different rows and four different columns.

You can't do anything about the sum $r + (n - r) + r + (m - r) = \boldsymbol{n} + \boldsymbol{m}$. It will be $6 + 5 = 11$ no matter how the 1's are placed. The sum is 11 even if there aren't any 1's...

If all the other entries of A are 2's instead of 0's, how do these answers change ?

3.5 B All the rows of AB are combinations of the rows of B. So the row space of AB is contained in (possibly equal to) the row space of B. **Rank $(AB) \leq$ rank (B).**

All columns of AB are combinations of the columns of A. So the column space of AB is contained in (possibly equal to) the column space of A. **Rank $(AB) \leq$ rank (A).**

If we multiply A by an *invertible* matrix B, the rank will not change. The rank can't drop, because when we multiply by the inverse matrix the rank can't jump back up.

Appendix 1 collects the key facts about the ranks of matrices.

Problem Set 3.5

1 (a) If a 7 by 9 matrix has rank 5, what are the dimensions of the four subspaces? What is the sum of all four dimensions?

(b) If a 3 by 4 matrix has rank 3, what are its column space and left nullspace?

2 Find bases and dimensions for the four subspaces associated with A and B:

$$A = \begin{bmatrix} 1 & 2 & 4 \\ 2 & 4 & 8 \end{bmatrix} \quad \text{and} \quad B = \begin{bmatrix} 1 & 2 & 4 \\ 2 & 5 & 8 \end{bmatrix}.$$

3 Find a basis for each of the four subspaces associated with A:

$$A = \begin{bmatrix} 0 & 1 & 2 & 3 & 4 \\ 0 & 1 & 2 & 4 & 6 \\ 0 & 0 & 0 & 1 & 2 \end{bmatrix} = \begin{bmatrix} 1 & 0 & 0 \\ 1 & 1 & 0 \\ 0 & 1 & 1 \end{bmatrix} \begin{bmatrix} 0 & 1 & 2 & 3 & 4 \\ 0 & 0 & 0 & 1 & 2 \\ 0 & 0 & 0 & 0 & 0 \end{bmatrix}.$$

4 Construct a matrix with the required property or explain why this is impossible:

(a) Column space contains $\begin{bmatrix} 1 \\ 1 \\ 0 \end{bmatrix}, \begin{bmatrix} 0 \\ 0 \\ 1 \end{bmatrix}$, row space contains $\begin{bmatrix} 1 \\ 2 \end{bmatrix}, \begin{bmatrix} 2 \\ 5 \end{bmatrix}$.

(b) Column space has basis $\begin{bmatrix} 1 \\ 1 \\ 3 \end{bmatrix}$, nullspace has basis $\begin{bmatrix} 3 \\ 1 \\ 1 \end{bmatrix}$.

(c) Dimension of nullspace $= 1 +$ dimension of left nullspace.

(d) Nullspace contains $\begin{bmatrix} 1 \\ 3 \end{bmatrix}$, column space contains $\begin{bmatrix} 3 \\ 1 \end{bmatrix}$.

(e) Row space $=$ column space, nullspace $\neq$ left nullspace.

5 If $\mathbf{V}$ is the subspace spanned by $(1,1,1)$ and $(2,1,0)$, find a matrix A that has $\mathbf{V}$ as its row space. Find a matrix B that has $\mathbf{V}$ as its nullspace. Multiply AB.

6 Without using elimination, find dimensions and bases for the four subspaces for

$$A = \begin{bmatrix} 0 & 3 & 3 & 3 \\ 0 & 0 & 0 & 0 \\ 0 & 1 & 0 & 1 \end{bmatrix} \quad \text{and} \quad B = \begin{bmatrix} 1 \\ 4 \\ 5 \end{bmatrix}.$$

7 Suppose the 3 by 3 matrix A is invertible. Write down bases for the four subspaces for A, and also for the 3 by 6 matrix $B = [\,A \quad A\,]$. (*The basis for $\mathbf{Z}$ is empty.*)

8 What are the dimensions of the four subspaces for A, B, and C, if I is the 3 by 3 identity matrix and 0 is the 3 by 2 zero matrix ?

$$A = \begin{bmatrix} I & 0 \end{bmatrix} \quad \text{and} \quad B = \begin{bmatrix} I & I \\ 0^{\mathrm{T}} & 0^{\mathrm{T}} \end{bmatrix} \quad \text{and} \quad C = \begin{bmatrix} 0 \end{bmatrix}.$$

9 Which subspaces are the same for these matrices of different sizes?

$$\text{(a) } \begin{bmatrix} A \end{bmatrix} \text{ and } \begin{bmatrix} A \\ A \end{bmatrix} \qquad \text{(b) } \begin{bmatrix} A \\ A \end{bmatrix} \text{ and } \begin{bmatrix} A & A \\ A & A \end{bmatrix}.$$

Prove that all three of those matrices have the *same rank* r.

10 If the entries of a 3 by 3 matrix are chosen randomly between 0 and 1, what are the most likely dimensions of the four subspaces? What if the random matrix is 3 by 5?

11 (Important) A is an m by n matrix of rank r. Suppose there are right sides $\boldsymbol{b}$ for which $A\boldsymbol{x} = \boldsymbol{b}$ has *no solution.*

(a) What are all inequalities ($<$ or $\leq$) that must be true between m, n, and r?

(b) How do you know that $A^{\mathrm{T}}\boldsymbol{y} = \mathbf{0}$ has solutions other than $\boldsymbol{y} = \mathbf{0}$?

12 Construct a matrix with $(1, 0, 1)$ and $(1, 2, 0)$ as a basis for its row space and its column space. Why can't this be a basis for the row space and nullspace ?

13 True or false (with a reason or a counterexample):

(a) If $m = n$ then the row space of A equals the column space.

(b) The matrices A and $-A$ share the same four subspaces.

(c) If A and B share the same four subspaces then A is a multiple of B.

14 Without computing A, find bases for its four fundamental subspaces :

$$A = \begin{bmatrix} 1 & 0 & 0 \\ 6 & 1 & 0 \\ 9 & 8 & 1 \end{bmatrix} \begin{bmatrix} 1 & 2 & 3 & 4 \\ 0 & 1 & 2 & 3 \\ 0 & 0 & 1 & 2 \end{bmatrix}.$$

15 If you exchange the first two rows of A, which of the four subspaces stay the same? If $\boldsymbol{v} = (1, 2, 3, 4)$ is in the left nullspace of A, write down a vector in the left nullspace of the new matrix after the row exchange.

16 *Explain why* $\boldsymbol{v} = (1, 0, -1)$ *cannot be a row of A and also in the nullspace.*

17 Describe the four subspaces of $\mathbf{R}^3$ associated with

$$A = \begin{bmatrix} 0 & 1 & 0 \\ 0 & 0 & 1 \\ 0 & 0 & 0 \end{bmatrix} \quad \text{and} \quad I + A = \begin{bmatrix} 1 & 1 & 0 \\ 0 & 1 & 1 \\ 0 & 0 & 1 \end{bmatrix}.$$

18 Can tic-tac-toe be completed (5 ones and 4 zeros in A) so that rank $(A) = 2$ but neither side passed up a winning move ?

19 (Left nullspace) Add the extra column $\boldsymbol{b}$ and reduce A to echelon form:

$$\begin{bmatrix} A & \boldsymbol{b} \end{bmatrix} = \begin{bmatrix} 1 & 2 & 3 & b_1 \\ 4 & 5 & 6 & b_2 \\ 7 & 8 & 9 & b_3 \end{bmatrix} \quad \rightarrow \quad \begin{bmatrix} 1 & 2 & 3 & b_1 \\ 0 & -3 & -6 & b_2 - 4b_1 \\ 0 & 0 & 0 & b_3 - 2b_2 + b_1 \end{bmatrix}.$$

A combination of the rows of A has produced the zero row. What combination is it? (Look at $b_3 - 2b_2 + b_1$ on the right side.) Which vectors are in the nullspace of A^{T} and which vectors are in the nullspace of A?

20 (a) Check that the solutions to $A\boldsymbol{x} = \boldsymbol{0}$ are perpendicular to the rows of A:

$$\boldsymbol{A} = \begin{bmatrix} 1 & 0 & 0 \\ 2 & 1 & 0 \\ 3 & 4 & 1 \end{bmatrix} \begin{bmatrix} 4 & 2 & 0 & 1 \\ 0 & 0 & 1 & 3 \\ 0 & 0 & 0 & 0 \end{bmatrix} = \boldsymbol{E}\boldsymbol{R}_0 = \boldsymbol{C}\boldsymbol{R}.$$

(b) How many independent solutions to $A^{\mathrm{T}}\boldsymbol{y} = \boldsymbol{0}$? Why does $\boldsymbol{y}^{\mathrm{T}} =$ row 3 of E^{-1}?

21 Suppose A is the sum of two matrices of rank one: $A = \boldsymbol{u}\boldsymbol{v}^{\mathrm{T}} + \boldsymbol{w}\boldsymbol{z}^{\mathrm{T}}$.

(a) Which vectors span the column space of A?

(b) Which vectors span the row space of A?

(c) The rank is less than 2 if ______ or if ______.

(d) Compute A and its rank if $\boldsymbol{u} = \boldsymbol{z} = (1,0,0)$ and $\boldsymbol{v} = \boldsymbol{w} = (0,0,1)$.

22 Construct $A = \boldsymbol{u}\boldsymbol{v}^{\mathrm{T}} + \boldsymbol{w}\boldsymbol{z}^{\mathrm{T}}$ whose column space has basis $(1,2,4), (2,2,1)$ and whose row space has basis $(1,0), (1,1)$. Write A as (3 by 2) times (2 by 2).

23 Without multiplying matrices, find bases for the row and column spaces of A:

$$A = \begin{bmatrix} 1 & 2 \\ 4 & 5 \\ 2 & 7 \end{bmatrix} \begin{bmatrix} 3 & 0 & 3 \\ 1 & 1 & 2 \end{bmatrix}.$$

How do you know from these shapes that A cannot be invertible?

24 (Important) $A^{\mathrm{T}}\boldsymbol{y} = \boldsymbol{d}$ is solvable when $\boldsymbol{d}$ is in which of the four subspaces? The solution $\boldsymbol{y}$ is unique when the ______ contains only the zero vector.

25 True or false (with a reason or a counterexample):

(a) A and A^{T} have the same number of pivots.

(b) A and A^{T} have the same left nullspace.

(c) If the row space equals the column space then $A^{\mathrm{T}} = A$.

(d) If $A^{\mathrm{T}} = -A$ then the row space of A equals the column space.

26 If a, b, c are given with $a \neq 0$, how would you choose d so that $\begin{bmatrix} a & b \\ c & d \end{bmatrix}$ has rank 1? Find a basis for the row space and nullspace. Show they are perpendicular!

27 Find the ranks of the 8 by 8 checkerboard matrix B and the chess matrix C:

$$B = \begin{bmatrix} 1 & 0 & 1 & 0 & 1 & 0 & 1 & 0 \\ 0 & 1 & 0 & 1 & 0 & 1 & 0 & 1 \\ 1 & 0 & 1 & 0 & 1 & 0 & 1 & 0 \\ \cdot & \cdot & \cdot & \cdot & \cdot & \cdot & \cdot & \cdot \\ 0 & 1 & 0 & 1 & 0 & 1 & 0 & 1 \end{bmatrix} \quad \text{and} \quad C = \begin{bmatrix} r & n & b & q & k & b & n & r \\ p & p & p & p & p & p & p & p \\ & & & \text{four zero rows} & & & & \\ p & p & p & p & p & p & p & p \\ r & n & b & q & k & b & n & r \end{bmatrix}$$

The numbers r, n, b, q, k, p are all different. Find bases for the row space and left nullspace of B and C. Challenge problem: Find a basis for the nullspace of C.

Challenge Problems

28 If $A = \boldsymbol{u}\boldsymbol{v}^{\mathrm{T}}$ is a 2 by 2 matrix of rank 1, redraw Figure 3.5 to show clearly the Four Fundamental Subspaces. If B produces those same four subspaces, what is the exact relation of B to A?

29 $\mathbf{M}$ is the space of 3 by 3 matrices. Multiply every matrix X in $\mathbf{M}$ by

$$A = \begin{bmatrix} 1 & 0 & -1 \\ -1 & 1 & 0 \\ 0 & -1 & 1 \end{bmatrix}. \quad \text{Notice: } A \begin{bmatrix} 1 \\ 1 \\ 1 \end{bmatrix} = \begin{bmatrix} 0 \\ 0 \\ 0 \end{bmatrix}.$$

(a) Which matrices X lead to $AX =$ zero matrix?

(b) Which matrices have the form AX for some matrix X?

(a) finds the "nullspace" of that operation AX and (b) finds the "column space". What are the dimensions of those two subspaces of $\mathbf{M}$? Why do they add to 9?

30 Suppose the m by n matrices A and B have *the same four subspaces*. If they are both in row reduced echelon form, prove that $\boldsymbol{F}$ **must equal** $\boldsymbol{G}$:

$$A = \begin{bmatrix} I & F \\ 0 & 0 \end{bmatrix} \qquad B = \begin{bmatrix} I & G \\ 0 & 0 \end{bmatrix}.$$

Every Matrix A has a Pseudoinverse A^+

If the columns of A are independent, then $A^+ = (A^{\mathrm{T}}A)^{-1}A^{\mathrm{T}}$ is a **left-inverse**: $A^+A = I$.
If the rows of A are independent, then $A^+ = A^{\mathrm{T}}(AA^{\mathrm{T}})^{-1}$ is a **right-inverse**: $AA^+ = I$.
This page allows dependent columns and dependent rows, and creates the **pseudoinverse** $\boldsymbol{A^+}$.

Here is the key idea for $A^+\boldsymbol{b}$. **Split** $\boldsymbol{b}$ **into** $\boldsymbol{p}$ **and** $\boldsymbol{e}$. The part $\boldsymbol{p}$ in the column space equals $A\boldsymbol{x}^+$ for one vector $\boldsymbol{x}^+$ in the row space (see page 125). The part $\boldsymbol{e}$ is in the nullspace of A^{T} (the fourth subspace). Then the best possible inverse A^+ has $A^+\boldsymbol{p} = \boldsymbol{x}^+$ and $A^+\boldsymbol{e} = \mathbf{0}$ on the two parts. By linearity $A^+\boldsymbol{b} = A^+(\boldsymbol{p}+\boldsymbol{e}) = \boldsymbol{x}^+$.

In short, A takes its row space to its column space. A^+ inverts that invertible part.

$$A = \begin{bmatrix} 3 & 0 & 0 \\ 0 & 0 & 0 \end{bmatrix} \text{ has } A^+ = \begin{bmatrix} \frac{1}{3} & 0 \\ 0 & 0 \\ 0 & 0 \end{bmatrix} \text{ and } AA^+ = \begin{bmatrix} 1 & 0 \\ 0 & 0 \end{bmatrix} \text{ and } A^+A = \begin{bmatrix} 1 & 0 & 0 \\ 0 & 0 & 0 \\ 0 & 0 & 0 \end{bmatrix}.$$

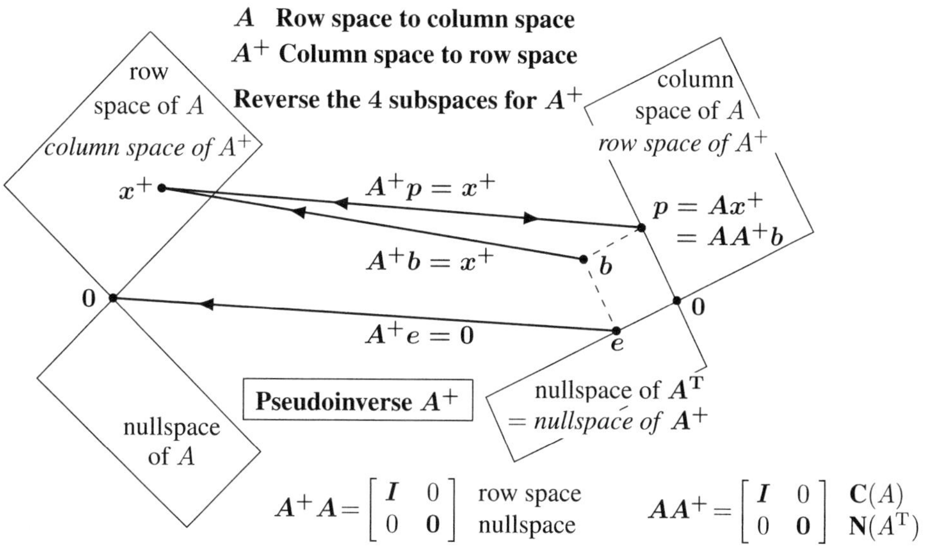

Figure 3.5: Vectors $\boldsymbol{p} = A\boldsymbol{x}^+$ in the column space of A go back to $\boldsymbol{x}^+$ in the row space.

Notice ! Suppose you start from that vector $\boldsymbol{b}$ (not in the column space). Then $A^+\boldsymbol{b} = \boldsymbol{x}^+$ is in the row space and $AA^+\boldsymbol{b} = A\boldsymbol{x}^+ = \boldsymbol{p}$ is in the column space. So $AA^+ \neq I$. *But* $\boldsymbol{p}$ *is as close to* $\boldsymbol{b}$ *as possible.* Actually $\boldsymbol{p}$ is the "**projection**" of $\boldsymbol{b}$ onto the column space: the subject of Chapter 4. Symmetrically, $A^+A\boldsymbol{x}$ is the **projection** of $\boldsymbol{x}$ onto the row space.

Examples $\begin{bmatrix} 1 & 1 \\ 1 & 1 \end{bmatrix}^+ = \frac{1}{4}\begin{bmatrix} 1 & 1 \\ 1 & 1 \end{bmatrix}$ If $A = CR = (m\times r)\,(r\times n)$ then $A^+ = R^+C^+$

4 Orthogonality

Two vectors are orthogonal when their dot product is zero: $\boldsymbol{v} \cdot \boldsymbol{w} = \boldsymbol{v}^{\mathrm{T}}\boldsymbol{w} = 0$. This chapter moves to **orthogonal subspaces** and **orthogonal bases** and **orthogonal matrices**. The vectors in two subspaces, and the vectors in a basis, and the column vectors in Q, all pairs will be orthogonal. Think of $a^2 + b^2 = c^2$ for a *right triangle* with sides $\boldsymbol{v}$ and $\boldsymbol{w}$.

Orthogonal vectors	$\boldsymbol{v}^{\mathrm{T}}\boldsymbol{w} = 0$	and	$\|\|\boldsymbol{v}\|\|^2 + \|\|\boldsymbol{w}\|\|^2 = \|\|\boldsymbol{v} + \boldsymbol{w}\|\|^2$.

The right side is $(\boldsymbol{v} + \boldsymbol{w})^{\mathrm{T}}(\boldsymbol{v} + \boldsymbol{w})$. This equals $\boldsymbol{v}^{\mathrm{T}}\boldsymbol{v} + \boldsymbol{w}^{\mathrm{T}}\boldsymbol{w}$ when $\boldsymbol{v}^{\mathrm{T}}\boldsymbol{w} = \boldsymbol{w}^{\mathrm{T}}\boldsymbol{v} = 0$.

Subspaces entered Chapter 3 to throw light on $A\boldsymbol{x} = \boldsymbol{b}$. Right away we needed the column space and the nullspace. Then the light turned onto A^{T}, uncovering two more subspaces. Those four fundamental subspaces reveal what a matrix really does.

A matrix multiplies a vector: *A times* $\boldsymbol{x}$. At the first level this is only numbers. At the second level $A\boldsymbol{x}$ is a combination of column vectors. The third level shows subspaces. But I don't think you have seen the whole picture until you study Figure 4.2. **Those fundamental subspaces are orthogonal:**

The nullspace $\mathbf{N}(A)$ contains all vectors orthogonal to the row space $\mathbf{C}(A^{\mathrm{T}})$.

The nullspace $\mathbf{N}(A^{\mathrm{T}})$ contains all vectors orthogonal to the column space $\mathbf{C}(A)$.

$A\boldsymbol{x} = \mathbf{0}$ makes $\boldsymbol{x}$ orthogonal to each row. $A^{\mathrm{T}}\boldsymbol{y} = \mathbf{0}$ makes $\boldsymbol{y}$ orthogonal to each column.

A key idea in this chapter is **projection**: If $\boldsymbol{b}$ is outside the column space of A, find the closest point $\boldsymbol{p}$ that is inside. The line from $\boldsymbol{b}$ to $\boldsymbol{p}$ shows the error $\boldsymbol{e}$. That line is perpendicular to the column space. The *least squares equation* $A^{\mathrm{T}}A\widehat{\boldsymbol{x}} = A^{\mathrm{T}}\boldsymbol{b}$ produces the closest $\boldsymbol{p} = A\widehat{\boldsymbol{x}}$ and smallest possible $\boldsymbol{e}$. It gives the best $\widehat{\boldsymbol{x}}$ when $A\boldsymbol{x} = \boldsymbol{b}$ is unsolvable. That best $\widehat{\boldsymbol{x}}$ makes $||A\widehat{\boldsymbol{x}} - \boldsymbol{b}||^2 = ||\boldsymbol{e}||^2$ as small as possible.

The equation $A^{\mathrm{T}}A\widehat{\boldsymbol{x}} = A^{\mathrm{T}}\boldsymbol{b}$ is easy when $A^{\mathrm{T}}A = I$. Then A has orthonormal columns: *perpendicular unit vectors.* That won't happen by accident but we can make it happen. Orthogonalizing the columns $\boldsymbol{a}_1$ to $\boldsymbol{a}_n$ produces columns $\boldsymbol{q}_1$ to $\boldsymbol{q}_n$ with $\boldsymbol{q}_i^{\mathrm{T}}\boldsymbol{q}_j = 0$ and $\boldsymbol{q}_i^{\mathrm{T}}\boldsymbol{q}_i = 1$. The orthogonal matrix Q has $Q^{\mathrm{T}}Q = I$ and it connects to A by $\boldsymbol{A} = \boldsymbol{QR}$.

Orthogonal matrices are perfect for computations. The whole of Section 7.4 will highlight Q. In many ways $A = QR$ is better than $A = LU$, and this chapter shows why.

4.1 Orthogonality of the Four Subspaces

1 Orthogonal vectors have $\boldsymbol{v}^{\mathrm{T}}\boldsymbol{w} = 0$. Then $||\boldsymbol{v}||^2 + ||\boldsymbol{w}||^2 = ||\boldsymbol{v}+\boldsymbol{w}||^2$ as in $a^2 + b^2 = c^2$.

2 Subspaces $\mathbf{V}$ and $\mathbf{W}$ are orthogonal when $\boldsymbol{v}^{\mathrm{T}}\boldsymbol{w} = 0$ for every $\boldsymbol{v}$ in $\mathbf{V}$ and every $\boldsymbol{w}$ in $\mathbf{W}$.

3 The row space of A is orthogonal to the nullspace. The column space is orthogonal to $\mathbf{N}(A^{\mathrm{T}})$.

4 The dimensions add to $r + (n - r) = n$ and $r + (m - r) = m$: Orthogonal complements.

5 If n vectors in $\mathbf{R}^n$ are independent, they span $\mathbf{R}^n$. If n vectors span $\mathbf{R}^n$, they are independent.

Chapter 1 connected dot products $\boldsymbol{v}^{\mathrm{T}}\boldsymbol{w}$ to the angle between $\boldsymbol{v}$ and $\boldsymbol{w}$. For 90° angles we have $\boldsymbol{v}^{\mathrm{T}}\boldsymbol{w} = 0$. This chapter moves up from orthogonal vectors $\boldsymbol{v}$ and $\boldsymbol{w}$ to orthogonal subspaces $\mathbf{V}$ and $\mathbf{W}$.

The subspaces fit together to show the hidden reality of A times $\boldsymbol{x}$. The 90° angles between subspaces are new—and we can say now what those right angles mean.

The row space is perpendicular to the nullspace. Every row of A is perpendicular to every solution of $A\boldsymbol{x} = \mathbf{0}$. That gives the 90° angle on the left side of the figure. This perpendicularity of subspaces is Part 2 of the Fundamental Theorem of Linear Algebra.

The column space is perpendicular to the nullspace of A^{T}. When we want to solve $A\boldsymbol{x} = \boldsymbol{b}$ and can't do it, this nullspace of A^{T} contains the error $\boldsymbol{e} = \boldsymbol{b} - A\widehat{\boldsymbol{x}}$ in the "least-squares" solution $\widehat{\boldsymbol{x}}$. A key application of linear algebra.

DEFINITION Two subspaces $\mathbf{V}$ and $\mathbf{W}$ of a vector space are ***orthogonal*** if every vector $\boldsymbol{v}$ in $\mathbf{V}$ is perpendicular to every vector $\boldsymbol{w}$ in $\mathbf{W}$:

Orthogonal subspaces	$\boldsymbol{v}^{\mathrm{T}}\boldsymbol{w} = 0$ *for all* $\boldsymbol{v}$ *in* $\mathbf{V}$ *and all* $\boldsymbol{w}$ *in* $\mathbf{W}$.

Example 1 Two walls of a room look perpendicular but those two subspaces are **not orthogonal!** The meeting line is in both $\mathbf{V}$ and $\mathbf{W}$—and this line is not perpendicular to itself. Two planes in $\mathbf{R}^3$ (dimensions 2 and 2) cannot be orthogonal subspaces.

Example 2 The floor and a vertical line do give orthogonal subspaces. They have dimensions $2 + 1 = 3$. Perpendicular lines through $\mathbf{0}$ also give orthogonal subspaces.

When a vector is in two orthogonal subspaces, it *must* be zero. It is perpendicular to itself. It is $\boldsymbol{v}$ and it is $\boldsymbol{w}$, so $\boldsymbol{v}^{\mathrm{T}}\boldsymbol{v} = 0$. This has to be the zero vector.

The crucial examples for linear algebra come from the four fundamental subspaces. Zero is the only point where the nullspace meets the row space. More than that, the **nullspace and row space of A meet at 90°**. This key fact comes directly from $A\boldsymbol{x} = \mathbf{0}$:

> Every vector $\boldsymbol{x}$ in the nullspace is perpendicular to every row of A, because $A\boldsymbol{x} = \mathbf{0}$.
> ***The nullspace*** $\mathbf{N}(A)$ ***and the row space*** $\mathbf{C}(A^{\mathrm{T}})$ ***are orthogonal subspaces of*** $\mathbf{R}^n$.

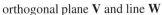

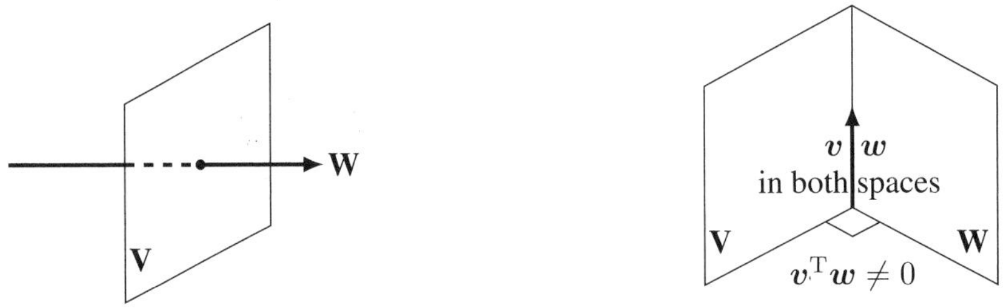

orthogonal plane **V** and line **W** two planes can't be orthogonal in $\mathbf{R}^3$

Figure 4.1: Orthogonality is impossible when $\dim \mathbf{V} + \dim \boldsymbol{W} > \dim$ (whole space).

To see why $\boldsymbol{x}$ is perpendicular to the rows, look at $A\boldsymbol{x} = \mathbf{0}$. Each row multiplies $\boldsymbol{x}$:

$$A\boldsymbol{x} = \begin{bmatrix} \textbf{row 1} \\ \vdots \\ \textbf{row } m \end{bmatrix} \begin{bmatrix} \\ \boldsymbol{x} \\ \\ \end{bmatrix} = \begin{bmatrix} 0 \\ \vdots \\ 0 \end{bmatrix} \quad \begin{matrix} \longleftarrow \quad (\textbf{row 1}) \cdot \boldsymbol{x} \textbf{ is zero} \\ \\ \longleftarrow \quad (\textbf{row } m) \cdot \boldsymbol{x} \textbf{ is zero} \end{matrix} \tag{1}$$

The first equation says that row 1 is perpendicular to $\boldsymbol{x}$. The last equation says that row m is perpendicular to $\boldsymbol{x}$. *Every row has a zero dot product with* $\boldsymbol{x}$. Then $\boldsymbol{x}$ is also perpendicular to every *combination* of the rows. The whole row space $\mathbf{C}(A^T)$ is orthogonal to $\mathbf{N}(A)$.

Here is a second proof of that orthogonality for readers who like matrix shorthand. The vectors in the row space of A are combinations $A^T\boldsymbol{y}$ of the rows.

Nullspace orthogonal to row space $\quad \boldsymbol{x}^T(A^T\boldsymbol{y}) = (A\boldsymbol{x})^T\boldsymbol{y} = \mathbf{0}^T\boldsymbol{y} = 0. \quad (2)$

We like the first proof. You can see those rows of A multiplying $\boldsymbol{x}$ to produce zeros in equation (1). The second proof shows why A and A^T are both in the Fundamental Theorem.

Part 1 of the Fundamental Theorem gave the dimensions of the four subspaces. The row and column spaces have the same dimension r (they are drawn the same size). The two nullspaces have the remaining dimensions $n - r$ and $m - r$. Now we know that ***the row space and nullspace are orthogonal subspaces inside*** $\mathbf{R}^n$.

Example 3 The rows of A are perpendicular to $\boldsymbol{x} = (1, 1, -1)$ in the nullspace:

$$A\boldsymbol{x} = \begin{bmatrix} 1 & 3 & 4 \\ 5 & 2 & 7 \end{bmatrix} \begin{bmatrix} 1 \\ 1 \\ -1 \end{bmatrix} = \begin{bmatrix} 0 \\ 0 \end{bmatrix} \quad \text{gives the dot products} \quad \begin{matrix} 1 + 3 - 4 = 0 \\ 5 + 2 - 7 = 0 \end{matrix}$$

Now turn to the column space of A and the nullspace of A^T : the other pair.

> Every vector $\boldsymbol{y}$ in the nullspace of A^T is perpendicular to every column of A.
> ***The left nullspace*** $\mathbf{N}(\boldsymbol{A}^T)$ ***and the column space*** $\mathbf{C}(\boldsymbol{A})$ ***are orthogonal in*** $\mathbf{R}^m$.

Apply the original proof to A^T. The nullspace of A^T is orthogonal to the row space of A^T—and the row space of A^T is the column space of A. Q.E.D.

For a visual proof, look at $A^{\mathrm{T}}\boldsymbol{y} = \mathbf{0}$. Each column of A multiplies $\boldsymbol{y}$ to give 0: column of A multiplies $\boldsymbol{y}$ to give 0:

$$\mathbf{C}(\boldsymbol{A}) \perp \mathbf{N}(\boldsymbol{A}^{\mathrm{T}}) \qquad A^{\mathrm{T}}\boldsymbol{y} = \begin{bmatrix} (\textbf{column } \mathbf{1})^{\mathrm{T}} \\ \cdots \\ (\textbf{column } \boldsymbol{n})^{\mathrm{T}} \end{bmatrix} \begin{bmatrix} \\ \boldsymbol{y} \\ \\ \end{bmatrix} = \begin{bmatrix} 0 \\ \cdot \\ 0 \end{bmatrix}. \tag{3}$$

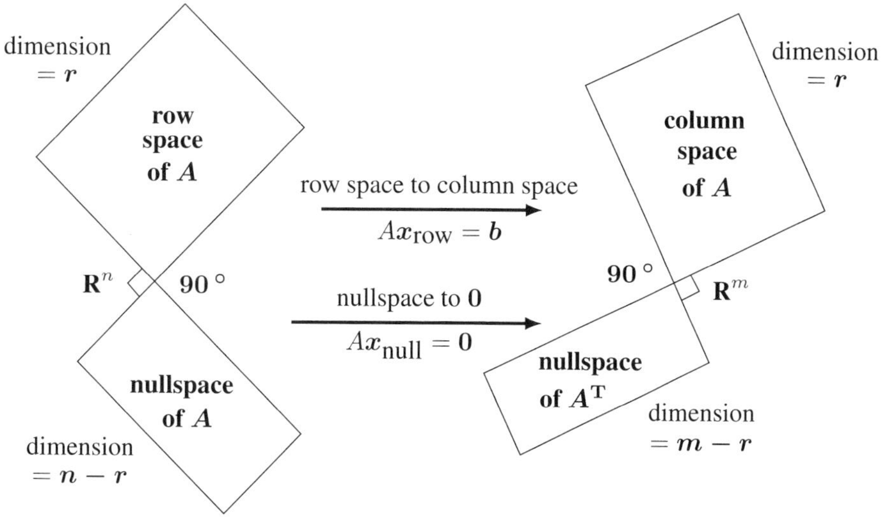

Figure 4.2: **Two pairs of orthogonal subspaces**. The dimensions add to n and add to m. **This is the Big Picture**—two subspaces in $\mathbf{R}^n$ and two subspaces in $\mathbf{R}^m$.

Orthogonal Complements

Important The fundamental subspaces are more than just orthogonal (in pairs). Their dimensions are also right. Two lines could be perpendicular in $\mathbf{R}^3$, **but those lines *could not be* the row space and nullspace of a 3 by 3 matrix**. The lines have dimensions 1 and 1, adding to 2. But the correct dimensions r and $n - r$ must add to $n = 3$.

Figure 4.1 showed two walls of a room. Dimensions $2 + 2 \neq 3$ must fail.

The fundamental subspaces of a 3 by 3 matrix have dimensions 2 and 1, or 3 and 0. Those pairs of subspaces are not only orthogonal, they are *orthogonal complements*.

DEFINITION The ***orthogonal complement*** $\mathbf{V}^{\perp}$ of a subspace $\mathbf{V}$ contains ***every vector that is perpendicular to*** $\mathbf{V}$. The dimensions of $\mathbf{V}$ and $\mathbf{V}^{\perp}$ add to (dimension of the whole space).

By this definition, the nullspace is the orthogonal complement of the row space. ***Every*** $\boldsymbol{x}$ that is perpendicular to the rows satisfies $A\boldsymbol{x} = \mathbf{0}$, and lies in the nullspace. The reverse is also true. ***If v is orthogonal to the nullspace, it must be in the row space.***

In the same way, the left nullspace and column space are orthogonal in $\mathbf{R}^m$, and they are orthogonal complements. Their dimensions r and $m - r$ add to the full dimension m.

> ***Fundamental Theorem of Linear Algebra*, Part 2**
>
> $\mathbf{N}(A)$ ***is the orthogonal complement of the row space*** $\mathbf{C}(A^{\mathrm{T}})$ **(in** $\mathbf{R}^n$**).**
>
> $\mathbf{N}(A^{\mathrm{T}})$ ***is the orthogonal complement of the column space*** $\mathbf{C}(A)$ **(in** $\mathbf{R}^m$**).**

Part 1 gave the dimensions of the subspaces. Part 2 gives the 90° angles between them. Every $\boldsymbol{x}$ can be split into a *row space component* $\boldsymbol{x}_r$ and a *nullspace component* $\boldsymbol{x}_n$. When A multiplies $\boldsymbol{x} = \boldsymbol{x}_r + \boldsymbol{x}_n$, Figure 4.3 shows what happens: $A\boldsymbol{x}_n = 0$ and $A\boldsymbol{x}_r = A\boldsymbol{x}$ is in the column space.

Every vector $A\boldsymbol{x}$ goes to the column space ! Multiplying by A cannot do anything else. More than that: *Every vector* $\boldsymbol{b}$ *in the column space comes from exactly one vector* $\boldsymbol{x}_r$ *in the row space.* Proof: If $A\boldsymbol{x}_r = A\boldsymbol{x}'_r$, the difference $\boldsymbol{x}_r - \boldsymbol{x}'_r$ is in the nullspace. It is also in the row space, where $\boldsymbol{x}_r$ and $\boldsymbol{x}'_r$ came from. This difference must be the zero vector, because the nullspace and row space are perpendicular. Therefore $\boldsymbol{x}_r = \boldsymbol{x}'_r$.

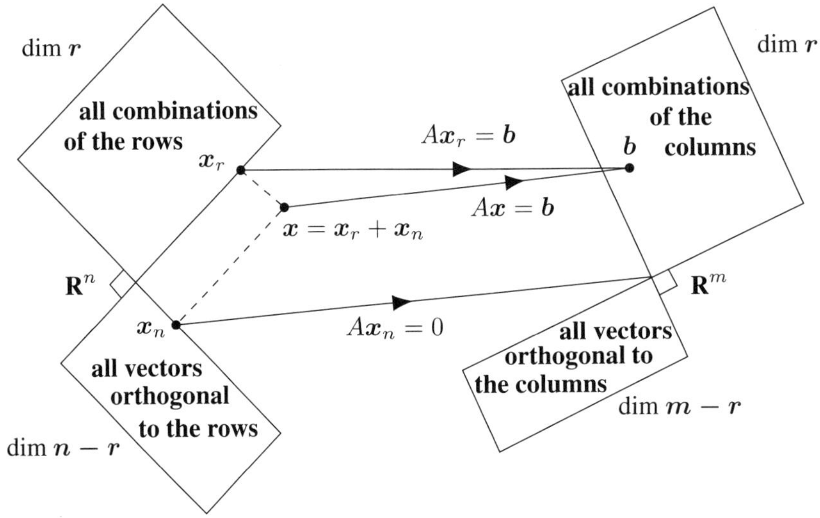

Figure 4.3: This update of Figure 4.2 shows the true action of A on $\boldsymbol{x} = \boldsymbol{x}_r + \boldsymbol{x}_n$. A times $\boldsymbol{x}_r$ is in the column space. A times $\boldsymbol{x}_n$ is the zero vector.

There is an r by r invertible matrix hiding inside A, if we throw away the two nullspaces. ***From row space to column space, A is invertible*** (page 127). The **pseudoinverse** $\boldsymbol{A^+}$ will invert that part of A (page 133).

Example 4 **Every matrix of rank r has an r by r invertible submatrix.** B has rank 2:

$$B = \begin{bmatrix} 1 & 2 & 3 & 4 & 5 \\ 1 & 2 & 4 & 5 & 6 \\ 1 & 2 & 4 & 5 & 6 \end{bmatrix} \quad \text{contains} \quad \begin{bmatrix} 1 & 3 \\ 1 & 4 \end{bmatrix} \quad \text{in the pivot rows and columns.}$$

More than this: Every matrix can be *diagonalized*, when we choose the right orthogonal bases. I hope you reach that amazing fact: the ***Singular Value Decomposition of A***.

Let me repeat: The only vector in two orthogonal subspaces is the zero vector.

Combining Bases from Subspaces

A basis contains *linearly independent* vectors *that span the space.* Normally we have to check both properties of a basis. When the count is right, one property implies the other: Every vector is a combination of the basis vectors **in exactly one way**.

Any n independent vectors in $\mathbf{R}^n$ must span $\mathbf{R}^n$. So they are a basis.
Any n vectors that span $\mathbf{R}^n$ must be independent. So they are a basis.

Starting with the correct number of vectors, one property of a basis produces the other. This is true in any vector space, but we care most about $\mathbf{R}^n$. When the vectors go into the columns of an n by n *square* matrix A, here are the same two facts:

If the n columns of A are independent, they span $\mathbf{R}^n$. So $A\boldsymbol{x} = \boldsymbol{b}$ is solvable.
If the n columns span $\mathbf{R}^n$, they are independent. So $A\boldsymbol{x} = \boldsymbol{b}$ has only one solution.
If $AB = I$ for square matrices, then $BA = I$.

Uniqueness implies existence and existence implies uniqueness. ***Then A is invertible.*** If there are no free variables, the solution $\boldsymbol{x}$ is unique. There must be n pivot columns. Then back substitution solves $A\boldsymbol{x} = \boldsymbol{b}$ (the solution exists).

Starting in the opposite direction, suppose that $A\boldsymbol{x} = \boldsymbol{b}$ can be solved for every $\boldsymbol{b}$ (*existence of solutions*). Then elimination produced no zero rows. There are n pivots and no free variables. The nullspace contains only $\boldsymbol{x} = \mathbf{0}$ (*uniqueness of solutions*).

With bases for the row space and the nullspace, we have $r + (n - r) = n$ vectors. This is the right number. Those n vectors are independent.[2] *Therefore they span* $\mathbf{R}^n$.

Each $\boldsymbol{x}$ is the sum $\boldsymbol{x}_r + \boldsymbol{x}_n$ of a row space vector $\boldsymbol{x}_r$ and a nullspace vector $\boldsymbol{x}_n$.

The splitting $\boldsymbol{x}_r + \boldsymbol{x}_n$ in Figure 4.3 shows the key point of orthogonal complements—the dimensions add to n and all vectors are fully accounted for.

Example 5 For $A = \begin{bmatrix} 1 & 2 \\ 3 & 6 \end{bmatrix}$ split $\boldsymbol{x} = \begin{bmatrix} 4 \\ 3 \end{bmatrix}$ into $\boldsymbol{x}_r + \boldsymbol{x}_n = \begin{bmatrix} 2 \\ 4 \end{bmatrix} + \begin{bmatrix} 2 \\ -1 \end{bmatrix}$.

The vector $(2, 4)$ is in the row space. The orthogonal vector $(2, -1)$ is in the nullspace. The next section will compute this splitting by a projection matrix P.

Example 6 Suppose S is a six-dimensional subspace of nine-dimensional space R^9.

(a) What are the possible dimensions of subspaces orthogonal to $\mathbf{S}$? **0, 1, 2, 3**

(b) What are the possible dimensions of the orthogonal complement $\mathbf{S}^\perp$ of $\mathbf{S}$? **3**

(c) What is the smallest possible size of a matrix A that has row space $\mathbf{S}$? **6 by 9**

(d) What is the smallest possible size of a matrix B that has nullspace $\mathbf{S}^\perp$? **6 by 9**

[2] If a combination of all n vectors gives $\boldsymbol{x}_r + \boldsymbol{x}_n = \mathbf{0}$, then $\boldsymbol{x}_r = -\boldsymbol{x}_n$ is in both subspaces. So $\boldsymbol{x}_r = \boldsymbol{x}_n = \mathbf{0}$. All coefficients of the row space basis and of the nullspace basis must be zero. This proves independence of the n vectors together.

Problem Set 4.1

1 Construct any 2 by 3 matrix of rank one. Copy Figure 4.2 and put one vector in each subspace (and put two in the nullspace). Which vectors are orthogonal?

2 Redraw Figure 4.3 for a 3 by 2 matrix of rank $r = 2$. Which subspace is $\boldsymbol{Z}$ (zero vector only)? The nullspace part of any vector $\boldsymbol{x}$ in $\mathbf{R}^2$ is $\boldsymbol{x}_n =$ ______ .

3 Construct a matrix with the required property or say why that is impossible:

(a) Column space contains $\begin{bmatrix} 1 \\ 2 \\ -3 \end{bmatrix}$ and $\begin{bmatrix} 2 \\ -3 \\ 5 \end{bmatrix}$, nullspace contains $\begin{bmatrix} 1 \\ 1 \\ 1 \end{bmatrix}$

(b) Row space contains $\begin{bmatrix} 1 \\ 2 \\ -3 \end{bmatrix}$ and $\begin{bmatrix} 2 \\ -3 \\ 5 \end{bmatrix}$, nullspace contains $\begin{bmatrix} 1 \\ 1 \\ 1 \end{bmatrix}$

(c) $A\boldsymbol{x} = \begin{bmatrix} 1 \\ 1 \\ 1 \end{bmatrix}$ has a solution and $A^{\mathrm{T}} \begin{bmatrix} 1 \\ 0 \\ 0 \end{bmatrix} = \begin{bmatrix} 0 \\ 0 \\ 0 \end{bmatrix}$

(d) Every row is orthogonal to every column (A is not the zero matrix)

(e) Columns add up to a column of zeros, rows add to a row of 1's.

4 If $AB = 0$ then the columns of B are in the ______ of A. The rows of A are in the ______ of B. With $AB = 0$, why can't A and B be 3 by 3 matrices of rank 2?

5 (a) If $A\boldsymbol{x} = \boldsymbol{b}$ has a solution and $A^{\mathrm{T}}\boldsymbol{y} = \mathbf{0}$, *is* $(\boldsymbol{y}^{\mathrm{T}}\boldsymbol{x} = 0)$ *or* $(\boldsymbol{y}^{\mathrm{T}}\boldsymbol{b} = 0)$?

(b) If $A^{\mathrm{T}}\boldsymbol{y} = (1, 1, 1)$ has a solution and $A\boldsymbol{x} = \mathbf{0}$, then ______ .

6 This system of equations $A\boldsymbol{x} = \boldsymbol{b}$ has *no solution* (they lead to $0 = 1$):

$$\begin{aligned} x + 2y + 2z &= 5 \\ 2x + 2y + 3z &= 5 \\ 3x + 4y + 5z &= 9 \end{aligned}$$

Find numbers y_1, y_2, y_3 to multiply the equations so they add to $0 = 1$. You have found a vector $\boldsymbol{y}$ in which subspace? Its dot product $\boldsymbol{y}^{\mathrm{T}}\boldsymbol{b}$ is 1, so no solution $\boldsymbol{x}$.

7 Every system $A\boldsymbol{x} = \boldsymbol{b}$ with no solution is like the one in Problem 6. There are numbers $y_1, \ldots, y_m$ that multiply the m equations so they add up to $0 = 1$. This is called **Fredholm's Alternative: If $\boldsymbol{b}$ is not in $\mathbf{C}(A)$, then part of $\boldsymbol{b}$ is in $\mathbf{N}(A^{\mathrm{T}})$.**

Exactly one problem has a solution: $A\boldsymbol{x} = \boldsymbol{b}$ **OR** $A^{\mathrm{T}}\boldsymbol{y} = \mathbf{0}$ with $\boldsymbol{y}^{\mathrm{T}}\boldsymbol{b} = 1$.

Multiply the equations $x_1 - x_2 = 1$ and $x_2 - x_3 = 1$ and $x_1 - x_3 = 1$ by numbers y_1, y_2, y_3 chosen so that the equations add up to $0 = 1$.

8 In Figure 4.3, how do we know that $A\boldsymbol{x}_r$ is equal to $A\boldsymbol{x}$? How do we know that this vector is in the column space? If $A = \begin{bmatrix} 1 & 1 \\ 1 & 1 \end{bmatrix}$ and $\boldsymbol{x} = \begin{bmatrix} 1 \\ 0 \end{bmatrix}$ what is $\boldsymbol{x}_r$?

9 **If $A^TAx = 0$ then $Ax = 0$.** Reason: Ax is in the nullspace of A^T and also in the ____ of A and those spaces are ____. *Conclusion*: $Ax = \mathbf{0}$ and therefore *A^TA has the same nullspace as A. This key fact is repeated in the next section.*

10 Suppose A is a symmetric matrix ($A^T = A$).

(a) Why is its column space perpendicular to its nullspace? N(A) ⊥ C(A^T)

(b) If $Ax = \mathbf{0}$ and $Az = 5z$, which subspaces contain these "eigenvectors" x and z? **Symmetric matrices have perpendicular eigenvectors $x^Tz = 0$.** √ x ⊥ z

11 Draw Figure 4.2 to show each subspace correctly for $A = \begin{bmatrix} 1 & 2 \\ 3 & 6 \end{bmatrix}$ and $B = \begin{bmatrix} 1 & 0 \\ 3 & 0 \end{bmatrix}$.

12 Find x_r and x_n and draw Figure 4.3 properly if $A = \begin{bmatrix} 1 & -1 \\ 0 & 0 \end{bmatrix}$ and $x = \begin{bmatrix} 2 \\ 0 \end{bmatrix}$.

Questions 13–23 are about orthogonal subspaces.

13 Put bases for the subspaces **V** and **W** into the columns of matrices V and W. Explain why the test for orthogonal subspaces can be written V^TW = zero matrix. This matches $v^Tw = 0$ for orthogonal vectors.

14 The floor **V** and the wall **W** are not orthogonal subspaces, because they share a nonzero vector (along the line where they meet). No planes **V** and **W** in $\mathbf{R}^3$ can be orthogonal! Find a vector in the column spaces of both matrices:

$$A = \begin{bmatrix} 1 & 2 \\ 1 & 3 \\ 1 & 2 \end{bmatrix} \quad \text{and} \quad B = \begin{bmatrix} 5 & 4 \\ 6 & 3 \\ 5 & 1 \end{bmatrix}$$

This will be a vector Ax and also $B\widehat{x}$. Think 3 by 4 with the matrix $[\,A \;\; B\,]$.

15 Extend Problem 14 to a p-dimensional subspace **V** and a q-dimensional subspace **W** of $\mathbf{R}^n$. What inequality on $p+q$ guarantees that **V** intersects **W** in a nonzero vector? These subspaces cannot be orthogonal.

16 Prove that every y in $\mathbf{N}(A^T)$ is perpendicular to every Ax in the column space, using the matrix shorthand of equation (2). Start from $A^Ty = \mathbf{0}$.

17 If **S** is the subspace of $\mathbf{R}^3$ containing only the zero vector, what is $\mathbf{S}^\perp$? If **S** is spanned by $(1, 1, 1)$, what is $\mathbf{S}^\perp$? If **S** is spanned by $(1, 1, 1)$ and $(1, 1, -1)$, what is a basis for $\mathbf{S}^\perp$?

18 Suppose **S** only contains two vectors $(1, 5, 1)$ and $(2, 2, 2)$ (not a subspace). Then $\mathbf{S}^\perp$ is the nullspace of the matrix $A =$ ____. $\mathbf{S}^\perp$ is a subspace even if **S** is not.

19 Suppose **L** is a one-dimensional subspace (a line) in $\mathbf{R}^3$. Its orthogonal complement $\mathbf{L}^\perp$ is the ____ perpendicular to **L**. Then $(\mathbf{L}^\perp)^\perp$ is a ____ perpendicular to $\mathbf{L}^\perp$. In fact $(\mathbf{L}^\perp)^\perp$ is the same as ____.

20 Suppose **V** is the whole space $\mathbf{R}^4$. Then $\mathbf{V}^\perp$ contains only the vector ____. Then $(\mathbf{V}^\perp)^\perp$ is ____. So $(\mathbf{V}^\perp)^\perp$ is the same as ____.

21 Suppose **S** is spanned by the vectors $(1,2,2,3)$ and $(1,3,3,2)$. Find two vectors that span $\mathbf{S}^\perp$. This is the same as solving $A\boldsymbol{x} = \mathbf{0}$ for which A?

22 If **P** is the plane of vectors in $\mathbf{R}^4$ satisfying $x_1 + x_2 + x_3 + x_4 = 0$, write a basis for $\mathbf{P}^\perp$. Construct a matrix that has **P** as its nullspace.

23 If a subspace **S** is contained in a subspace **V**, explain why $\mathbf{S}^\perp$ contains $\mathbf{V}^\perp$.

Questions 24-28 are about perpendicular columns and rows.

24 Suppose an n by n matrix is invertible: $AA^{-1} = I$. Then the first column of A^{-1} is orthogonal to the space spanned by which rows of A?

25 Find $A^{\mathrm{T}}A$ if the columns of A are unit vectors, perpendicular to each other.

26 Construct a 3 by 3 matrix A with no zero entries whose columns are mutually perpendicular. Compute $A^{\mathrm{T}}A$. Why is it a diagonal matrix?

27 The lines $3x + y = b_1$ and $6x + 2y = b_2$ are ____. They are the same line if ____. In that case (b_1, b_2) is perpendicular to the vector ____. The nullspace of the matrix is the line $3x + y =$ ____. One particular vector in that nullspace is ____.

28 Why is each of these statements false?

(a) $(1,1,1)$ is perpendicular to $(1,1,-2)$ so the planes $x + y + z = 0$ and $x + y - 2z = 0$ are orthogonal subspaces.

(b) The subspace spanned by $(1,1,0,0,0)$ and $(0,0,0,1,1)$ is the orthogonal complement of the subspace spanned by $(1,-1,0,0,0)$ and $(2,-2,3,4,-4)$.

(c) Two subspaces that meet only in the zero vector are orthogonal.

29 Find a matrix A with $\boldsymbol{v} = (1,2,3)$ in the row space and column space. Find B with $\boldsymbol{v}$ in the nullspace and column space. Which pairs of subspaces can't share $\boldsymbol{v}$?

30 Suppose A is 3 by 4 and B is 4 by 5 and $AB = 0$. So $\mathbf{N}(A)$ contains $\mathbf{C}(B)$. Prove from the dimensions of $\mathbf{N}(A)$ and $\mathbf{C}(B)$ that $\operatorname{rank}(A) + \operatorname{rank}(B) \leq 4$.

31 The command $N = \mathsf{null}(A)$ will produce a basis for the nullspace of A. Then the command $B = \mathsf{null}(N')$ will produce a basis for the ____ of A.

32 What are the conditions for nonzero vectors $\boldsymbol{r}, \boldsymbol{n}, \boldsymbol{c}, \boldsymbol{l}$ in $\mathbf{R}^2$ to be bases for the four fundamental subspaces $\mathbf{C}(A^{\mathrm{T}}), \mathbf{N}(A), \mathbf{C}(A), \mathbf{N}(A^{\mathrm{T}})$ of a 2 by 2 matrix?

33 When can the vectors $\boldsymbol{r}_1, \boldsymbol{r}_2, \boldsymbol{n}_1, \boldsymbol{n}_2, \boldsymbol{c}_1, \boldsymbol{c}_2, \boldsymbol{l}_1, \boldsymbol{l}_2$ in $\mathbf{R}^4$ be bases for the four fundamental subspaces of a 4 by 4 matrix? What is one possible A?

4.2 Projections onto Subspaces

1 The projection of a vector $\boldsymbol{b}$ onto the line through $\boldsymbol{a}$ is the closest point $\boldsymbol{p} = \boldsymbol{a}(\boldsymbol{a}^{\mathrm{T}}\boldsymbol{b} \,/\, \boldsymbol{a}^{\mathrm{T}}\boldsymbol{a})$.

2 The error $\boldsymbol{e} = \boldsymbol{b} - \boldsymbol{p}$ is perpendicular to $\boldsymbol{a}$: Right triangle $\boldsymbol{b}\,\boldsymbol{p}\,\boldsymbol{e}$ has $||\boldsymbol{p}||^2 + ||\boldsymbol{e}||^2 = ||\boldsymbol{b}||^2$.

3 The **projection** of $\boldsymbol{b}$ onto a subspace **S** is the closest vector $\boldsymbol{p}$ in **S**; $\boldsymbol{b} - \boldsymbol{p}$ is orthogonal to **S**.

4 Then the projection of $\boldsymbol{b}$ onto the column space of A is the vector $\boldsymbol{p} = A(A^{\mathrm{T}}A)^{-1}A^{\mathrm{T}}\boldsymbol{b}$.

5 The **projection matrix** onto $\mathbf{C}(A)$ is $\boxed{P = A(A^{\mathrm{T}}A)^{-1}A^{\mathrm{T}}.}$ Then $\boldsymbol{p} = P\boldsymbol{b}$ and $P^2 = P$.

This section of the book is about **closest points**. We have a point $\boldsymbol{b}$ that is not in a subspace **S** (both are in m dimensions). What point $\boldsymbol{p}$ in the subspace is closest to $\boldsymbol{b}$? A picture of the problem suggests the key to the solution: **The line from $\boldsymbol{b}$ to $\boldsymbol{p}$ is perpendicular to the subspace**. That line in Figure 4.5 shows us the error $\boldsymbol{e} = \boldsymbol{b} - \boldsymbol{p}$.

Our first examples are "projecting" $\boldsymbol{b}$ onto special subspaces like the xy plane. There is a **projection matrix P** that multiplies $\boldsymbol{b}$ and produces its projection $\boldsymbol{p} = P\boldsymbol{b}$.

1 What are the projections of $\boldsymbol{b} = (2, 3, 4)$ onto the z axis and onto the xy plane?

2 What matrices P_1 and P_2 produce those projections $P\boldsymbol{b}$ onto a line and a plane?

When $\boldsymbol{b}$ is projected onto a line, ***its projection p is the part of b along that line***. If $\boldsymbol{b}$ is projected onto a plane, $\boldsymbol{p}$ is the part in that plane. ***The projection p is** $P\boldsymbol{b}$.*

The projection onto the z axis we call $\boldsymbol{p}_1$. The second projection drops straight down to the xy plane. The picture in your mind should be Figure 4.4. Start with $\boldsymbol{b} = (2, 3, 4)$. The z-projection gives $\boldsymbol{p}_1 = (0, 0, 4)$. The projection down gives $\boldsymbol{p}_2 = (2, 3, 0)$. Those are the parts of $\boldsymbol{b}$ along the z axis and in the xy plane.

The projection matrices P_1 and P_2 are 3 by 3. They multiply $\boldsymbol{b}$ with 3 components to produce $\boldsymbol{p}$ with 3 components. Projection onto a line comes from a rank one matrix. Projection onto a plane comes from a rank two matrix:

Projection matrix
Onto the z axis: $\quad P_1 = \begin{bmatrix} 0 & 0 & 0 \\ 0 & 0 & 0 \\ 0 & 0 & \mathbf{1} \end{bmatrix} \quad$ Onto the xy plane: $\quad P_2 = \begin{bmatrix} \mathbf{1} & 0 & 0 \\ 0 & \mathbf{1} & 0 \\ 0 & 0 & 0 \end{bmatrix}.$

P_1 picks out the z component of every vector. P_2 picks out the x and y components. To find the projections $\boldsymbol{p}_1$ and $\boldsymbol{p}_2$ of $\boldsymbol{b}$, multiply $\boldsymbol{b}$ by P_1 and P_2 (small $\boldsymbol{p}$ for the vector, capital P for the matrix that multiplies $\boldsymbol{b}$ to produce $\boldsymbol{p}$):

$$\boldsymbol{p}_1 = P_1\boldsymbol{b} = \begin{bmatrix} 0 & 0 & 0 \\ 0 & 0 & 0 \\ 0 & 0 & 1 \end{bmatrix} \begin{bmatrix} x \\ y \\ z \end{bmatrix} = \begin{bmatrix} \mathbf{0} \\ \mathbf{0} \\ \boldsymbol{z} \end{bmatrix} \qquad \boldsymbol{p}_2 = P_2\boldsymbol{b} = \begin{bmatrix} 1 & 0 & 0 \\ 0 & 1 & 0 \\ 0 & 0 & 0 \end{bmatrix} \begin{bmatrix} x \\ y \\ z \end{bmatrix} = \begin{bmatrix} \boldsymbol{x} \\ \boldsymbol{y} \\ \mathbf{0} \end{bmatrix}.$$

In this case $\boldsymbol{p}_1$ is perpendicular to $\boldsymbol{p}_2$. The xy plane and the z axis are ***orthogonal subspaces***, like the floor of a room and the line between two walls. Then $P_2P_1 = 0$.

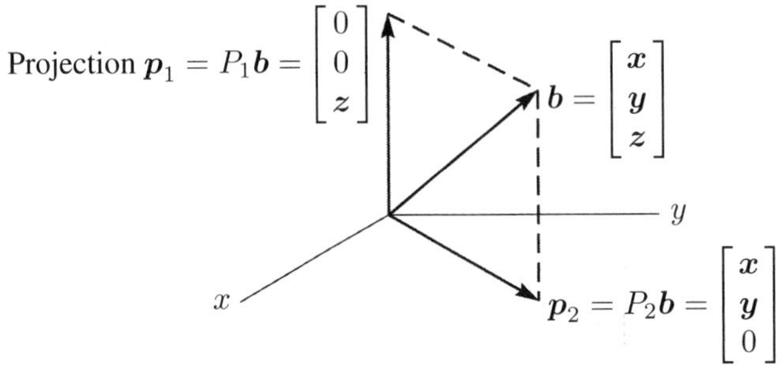

Figure 4.4: The projections $\boldsymbol{p}_1 = P_1\boldsymbol{b}$ and $\boldsymbol{p}_2 = P_2\boldsymbol{b}$ onto the z axis and the xy plane.

More than just orthogonal, the line and plane are orthogonal ***complements***. Their dimensions add to $1 + 2 = 3$. Every vector $\boldsymbol{b}$ in the whole space is the sum of its parts in the two subspaces. The projections $\boldsymbol{p}_1$ and $\boldsymbol{p}_2$ are exactly those two parts of $\boldsymbol{b}$:

$$\text{The vectors give } \boldsymbol{p}_1 + \boldsymbol{p}_2 = \boldsymbol{b}. \qquad \text{The matrices give } P_1 + P_2 = I. \tag{1}$$

This is perfect. Our goal is reached—for this example. We have the same goal for any line and any plane and any n-dimensional subspace of $\mathbf{R}^m$. The object is to find the part $\boldsymbol{p}$ in each subspace, and also the projection matrix P that produces that part $\boldsymbol{p} = P\boldsymbol{b}$. Every subspace of $\mathbf{R}^m$ has its own m by m projection matrix P.

The best description of a subspace is a basis. We put the basis vectors into the columns of A. ***Now we are projecting onto the column space of A***! Certainly the z axis is the column space of the 3 by 1 matrix A_1. The xy plane is the column space of A_2. That plane is *also* the column space of A_3 (a subspace has many bases). So $\boldsymbol{p}_2 = \boldsymbol{p}_3$ and $P_2 = P_3$.

$$A_1 = \begin{bmatrix} 0 \\ 0 \\ 1 \end{bmatrix} \quad \text{and} \quad A_2 = \begin{bmatrix} 1 & 0 \\ 0 & 1 \\ 0 & 0 \end{bmatrix} \quad \text{and} \quad A_3 = \begin{bmatrix} 1 & 2 \\ 2 & 3 \\ 0 & 0 \end{bmatrix} \quad \begin{matrix} A_3 \text{ has the} \\ \text{same column} \\ \text{space as } A_2 \end{matrix}$$

Our problem is *to project any* $\boldsymbol{b}$ *onto the column space of an m by n matrix A*. Start with a line (dimension $n = 1$). The matrix A will have only one column. Call it $\boldsymbol{a}$.

Projection Onto a Line

A line goes through the origin in the direction of $\boldsymbol{a} = (a_1, \ldots, a_m)$. Along that line, we want the point $\boldsymbol{p}$ closest to $\boldsymbol{b} = (b_1, \ldots, b_m)$. The key to projection is orthogonality: ***The line from b to p is perpendicular to the vector a***. This is the dotted line marked $\boldsymbol{e} = \boldsymbol{b} - \boldsymbol{p}$ for the error on the left side of Figure 4.5. We now compute $\boldsymbol{p}$ by algebra.

The projection $\boldsymbol{p}$ will be some multiple of $\boldsymbol{a}$. Call it $\boldsymbol{p} = \widehat{x}\boldsymbol{a}$ = "$\boldsymbol{x}$ **hat**" times $\boldsymbol{a}$. Computing this number $\widehat{x}$ will give the vector $\boldsymbol{p}$. Then from the formula for $\boldsymbol{p}$, we will read off the projection matrix P. These three steps will lead to all projection matrices: **find $\widehat{x}$, then find the vector $\boldsymbol{p} = A\boldsymbol{x}$, then find the matrix $\boldsymbol{P}$.**

The dotted line $\boldsymbol{b} - \boldsymbol{p}$ is the "error" $\boldsymbol{e} = \boldsymbol{b} - \widehat{x}\boldsymbol{a}$. It is perpendicular to $\boldsymbol{a}$—this will determine $\widehat{x}$. Use the fact that $\boldsymbol{b} - \widehat{x}\boldsymbol{a}$ **is perpendicular to** $\boldsymbol{a}$ when their dot product is zero:

Projecting $\boldsymbol{b}$ onto $\boldsymbol{a}$ with error $\boldsymbol{e} = \boldsymbol{b} - \widehat{x}\,\boldsymbol{a}$ $\boldsymbol{a}\cdot(\boldsymbol{b} - \widehat{x}\,\boldsymbol{a}) = 0$ or $\boldsymbol{a}\cdot\boldsymbol{b} = \widehat{x}\,\boldsymbol{a}\cdot\boldsymbol{a}$	$\widehat{x} = \dfrac{\boldsymbol{a}\cdot\boldsymbol{b}}{\boldsymbol{a}\cdot\boldsymbol{a}} = \dfrac{\boldsymbol{a}^{\mathrm{T}}\boldsymbol{b}}{\boldsymbol{a}^{\mathrm{T}}\boldsymbol{a}}$	(2)

The multiplication $\boldsymbol{a}^{\mathrm{T}}\boldsymbol{b}$ is the same as $\boldsymbol{a}\cdot\boldsymbol{b}$. Using the transpose is better, because it applies also to matrices. Our formula $\widehat{x} = \boldsymbol{a}^{\mathrm{T}}\boldsymbol{b}/\boldsymbol{a}^{\mathrm{T}}\boldsymbol{a}$ gives the projection $\boldsymbol{p} = \widehat{x}\boldsymbol{a}$.

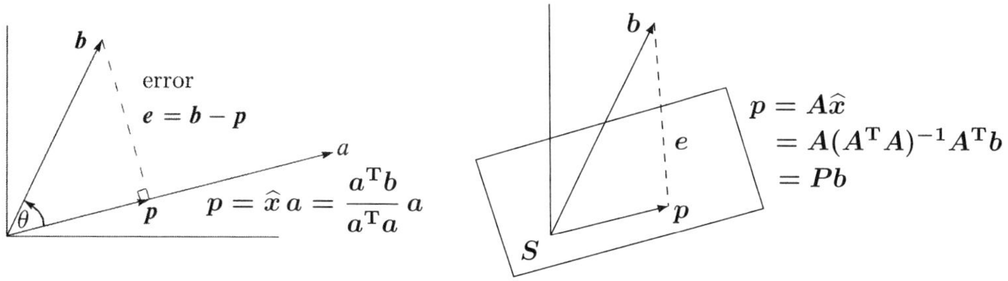

Figure 4.5: The projection $\boldsymbol{p}$ of $\boldsymbol{b}$ onto a line (left) and onto $\boldsymbol{S}$ = column space of A.

> **The projection of b onto the line through a is the vector $p = \widehat{x}\boldsymbol{a} = \dfrac{\boldsymbol{a}^{\mathrm{T}}\boldsymbol{b}}{\boldsymbol{a}^{\mathrm{T}}\boldsymbol{a}}\,\boldsymbol{a}$.**
>
> Special case 1: If $\boldsymbol{b} = \boldsymbol{a}$ then $\widehat{x} = 1$. The projection of $\boldsymbol{a}$ onto $\boldsymbol{a}$ is itself. $P\boldsymbol{a} = \boldsymbol{a}$.
>
> Special case 2: If $\boldsymbol{b}$ is perpendicular to $\boldsymbol{a}$ then $\boldsymbol{a}^{\mathrm{T}}\boldsymbol{b} = 0$. The projection is $\boldsymbol{p} = \boldsymbol{0}$.

Example 1 Project $b = \begin{bmatrix}1\\1\\1\end{bmatrix}$ onto $\boldsymbol{a} = \begin{bmatrix}1\\2\\2\end{bmatrix}$ to find $\boldsymbol{p} = \widehat{x}\boldsymbol{a} = \dfrac{5}{9}\,\boldsymbol{a}$ in Figure 4.5

Solution The number $\widehat{x}$ is the ratio of $\boldsymbol{a}^{\mathrm{T}}\boldsymbol{b} = 5$ to $\boldsymbol{a}^{\mathrm{T}}\boldsymbol{a} = 9$. So the projection is $\boldsymbol{p} = \frac{5}{9}\boldsymbol{a}$. The error vector between $\boldsymbol{b}$ and $\boldsymbol{p}$ is $\boldsymbol{e} = \boldsymbol{b} - \boldsymbol{p}$. Those vectors $\boldsymbol{p}$ and $\boldsymbol{e}$ will add to $\boldsymbol{b}$.

$$\boldsymbol{p} = \frac{5}{9}\boldsymbol{a} = \left(\frac{5}{9}, \frac{10}{9}, \frac{10}{9}\right) \quad \text{and} \quad \boldsymbol{e} = \Big(1, 1, 1\Big) - \boldsymbol{p} = \left(\frac{4}{9}, -\frac{1}{9}, -\frac{1}{9}\right).$$

The error $\boldsymbol{e}$ should be perpendicular to $\boldsymbol{a} = (1, 2, 2)$ and it is: $\boldsymbol{e}^{\mathrm{T}}\boldsymbol{a} = \frac{4}{9} - \frac{2}{9} - \frac{2}{9} = 0$.

Look at the right triangle of $\boldsymbol{b}$, $\boldsymbol{p}$, and $\boldsymbol{e}$. The vector $\boldsymbol{b}$ is split into two parts—its component along the line is $\boldsymbol{p}$, its perpendicular part is $\boldsymbol{e}$. Those two sides $\boldsymbol{p}$ and $\boldsymbol{e}$ have length $||\boldsymbol{p}|| = ||\boldsymbol{b}|| \cos\theta$ and $||\boldsymbol{e}|| = ||\boldsymbol{b}|| \sin\theta$. Trigonometry matches the dot product:

$$\boldsymbol{p} = \frac{\boldsymbol{a}^{\mathrm{T}}\boldsymbol{b}}{\boldsymbol{a}^{\mathrm{T}}\boldsymbol{a}}\boldsymbol{a} \quad \text{has length} \quad ||\boldsymbol{p}|| = \frac{||\boldsymbol{a}||\,||\boldsymbol{b}|| \cos\theta}{||\boldsymbol{a}||^2}||\boldsymbol{a}|| = ||\boldsymbol{b}|| \cos\theta. \tag{3}$$

The dot product is a lot simpler than getting involved with $\cos\theta$ and the length of $\boldsymbol{b}$. The example has square roots in $\cos\theta = 5/3\sqrt{3}$ and $||\boldsymbol{b}|| = \sqrt{3}$. There are no square roots in the projection $\boldsymbol{p} = 5\boldsymbol{a}/9$. The good way to $5/9$ is $\boldsymbol{a}^{\mathrm{T}}\boldsymbol{b}/\boldsymbol{a}^{\mathrm{T}}\boldsymbol{a}$.

Now comes the ***projection matrix***. In the formula for $\boldsymbol{p}$, what matrix is multiplying $\boldsymbol{b}$? You can see the matrix better if the number $\widehat{\boldsymbol{x}}$ is on the right side of $\boldsymbol{a}$:

Projection matrix P	$\boldsymbol{p} = \boldsymbol{a}\widehat{\boldsymbol{x}} = \boldsymbol{a}\dfrac{\boldsymbol{a}^{\mathrm{T}}\boldsymbol{b}}{\boldsymbol{a}^{\mathrm{T}}\boldsymbol{a}} = P\boldsymbol{b}$ ***when the matrix is*** $P = \dfrac{\boldsymbol{a}\boldsymbol{a}^{\mathrm{T}}}{\boldsymbol{a}^{\mathrm{T}}\boldsymbol{a}}$.

P is a column times a row! The column is $\boldsymbol{a}$, the row is $\boldsymbol{a}^{\mathrm{T}}$. Then divide by the number $\boldsymbol{a}^{\mathrm{T}}\boldsymbol{a}$. The projection matrix P is m by m, but ***its rank is one***. We are projecting onto a one-dimensional subspace, the line through $\boldsymbol{a}$. *That line is the column space of* P.

Example 2 Find the projection matrix $P = \dfrac{\boldsymbol{a}\boldsymbol{a}^{\mathrm{T}}}{\boldsymbol{a}^{\mathrm{T}}\boldsymbol{a}}$ onto the line through $\boldsymbol{a} = \begin{bmatrix}1\\2\\2\end{bmatrix}$.

Solution Multiply column $\boldsymbol{a}$ times row $\boldsymbol{a}^{\mathrm{T}}$ and divide by $\boldsymbol{a}^{\mathrm{T}}\boldsymbol{a} = 9$:

$$\textbf{Projection matrix} \qquad P = \frac{\boldsymbol{a}\boldsymbol{a}^{\mathrm{T}}}{\boldsymbol{a}^{\mathrm{T}}\boldsymbol{a}} = \frac{1}{9}\begin{bmatrix}1\\2\\2\end{bmatrix}\begin{bmatrix}1 & 2 & 2\end{bmatrix} = \frac{1}{9}\begin{bmatrix}1 & 2 & 2\\2 & 4 & 4\\2 & 4 & 4\end{bmatrix}.$$

This matrix projects *any* vector $\boldsymbol{b}$ onto $\boldsymbol{a}$. Check $\boldsymbol{p} = P\boldsymbol{b}$ for $\boldsymbol{b} = (1, 1, 1)$ in Example 1:

$$\boldsymbol{p} = P\boldsymbol{b} = \frac{1}{9}\begin{bmatrix}1 & 2 & 2\\2 & 4 & 4\\2 & 4 & 4\end{bmatrix}\begin{bmatrix}1\\1\\1\end{bmatrix} = \frac{1}{9}\begin{bmatrix}5\\10\\10\end{bmatrix} \qquad \text{which is correct.}$$

If the vector $\boldsymbol{a}$ is doubled, the matrix P stays the same! It still projects onto the same line. If P is squared, P^2 equals P. ***Projecting a second time doesn't change anything***, **so $\boldsymbol{P^2 = P}$**. The diagonal entries of P add to $\frac{1}{9}(1 + 4 + 4) = 1 =$ dimension of line.

$$P^2 = \frac{\boldsymbol{a}\boldsymbol{a}^{\mathrm{T}}}{\boldsymbol{a}^{\mathrm{T}}\boldsymbol{a}}\frac{\boldsymbol{a}\boldsymbol{a}^{\mathrm{T}}}{\boldsymbol{a}^{\mathrm{T}}\boldsymbol{a}} = P \text{ when you cancel the number } \frac{\boldsymbol{a}^{\mathrm{T}}\boldsymbol{a}}{\boldsymbol{a}^{\mathrm{T}}\boldsymbol{a}}$$

The matrix $I - P$ should be a projection too. It produces the other side $\boldsymbol{e}$ of the triangle—the perpendicular part of $\boldsymbol{b}$. Note that $(I - P)\boldsymbol{b}$ equals $\boldsymbol{b} - \boldsymbol{p}$ which is $\boldsymbol{e}$ in the left nullspace.

When P projects onto one subspace, $I - P$ projects onto the perpendicular subspace.

Now we move beyond lines and planes in $\mathbf{R}^3$. Projecting onto an n-dimensional subspace of $\mathbf{R}^m$ takes more effort. The crucial formulas will be collected in equations (5)–(6)–(7). Basically you need to remember those three equations.

Projection Onto a Subspace

Start with n vectors $\boldsymbol{a}_1, \ldots, \boldsymbol{a}_n$ in $\mathbf{R}^m$. Assume that these $\boldsymbol{a}$'s are linearly independent.

Problem: Find the combination $\boldsymbol{p} = \widehat{x}_1 \boldsymbol{a}_1 + \cdots + \widehat{x}_n \boldsymbol{a}_n$ closest to a given vector $\boldsymbol{b}$. We are projecting each $\boldsymbol{b}$ in $\mathbf{R}^m$ onto the n-dimensional subspace spanned by the $\boldsymbol{a}$'s.

With $n = 1$ (one vector $\boldsymbol{a}_1$) this is projection onto a line. The line is the column space of A, which has just one column. In general the matrix A has n columns $\boldsymbol{a}_1, \ldots, \boldsymbol{a}_n$.

The combinations in $\mathbf{R}^m$ are the vectors $A\boldsymbol{x}$ in the column space. We are looking for the particular combination $\boldsymbol{p} = A\widehat{\boldsymbol{x}}$ (***the projection***) that is closest to $\boldsymbol{b}$. The hat over $\widehat{\boldsymbol{x}}$ indicates the *best* choice $\widehat{\boldsymbol{x}}$, to give the closest vector in the column space. That choice is $\widehat{x} = \boldsymbol{a}^{\mathrm{T}}\boldsymbol{b}/\boldsymbol{a}^{\mathrm{T}}\boldsymbol{a}$ when $n = 1$. For $n > 1$, the best $\widehat{\boldsymbol{x}} = (\widehat{x}_1, \ldots, \widehat{x}_n)$ is to be found now.

We compute projections onto n-dimensional subspaces in three steps as before:

Find the vector $\widehat{\boldsymbol{x}}$. Find the projection $\boldsymbol{p} = A\widehat{\boldsymbol{x}}$. Find the projection matrix P.

The key is in the geometry! The dotted line in Figure 4.5 goes from $\boldsymbol{b}$ to the nearest point $A\widehat{\boldsymbol{x}}$ in the subspace. ***This error vector $\boldsymbol{b} - A\widehat{\boldsymbol{x}}$ is perpendicular to the subspace.*** The error $\boldsymbol{b} - A\widehat{\boldsymbol{x}}$ makes a right angle with all the vectors $\boldsymbol{a}_1, \ldots, \boldsymbol{a}_n$ in the base. Those n right angles give the n equations for $\widehat{\boldsymbol{x}}$:

$$\begin{array}{c} \boldsymbol{a}_1^{\mathrm{T}}(\boldsymbol{b} - A\widehat{\boldsymbol{x}}) = 0 \\ \vdots \\ \boldsymbol{a}_n^{\mathrm{T}}(\boldsymbol{b} - A\widehat{\boldsymbol{x}}) = 0 \end{array} \quad \text{or} \quad \begin{bmatrix} - \boldsymbol{a}_1^{\mathrm{T}} - \\ \vdots \\ - \boldsymbol{a}_n^{\mathrm{T}} - \end{bmatrix} \begin{bmatrix} \\ \boldsymbol{b} - A\widehat{\boldsymbol{x}} \\ \\ \end{bmatrix} = \begin{bmatrix} \\ \mathbf{0} \\ \\ \end{bmatrix}. \tag{4}$$

The matrix with those rows $\boldsymbol{a}_i^{\mathrm{T}}$ is A^{T}. The n equations are exactly $A^{\mathrm{T}}(\boldsymbol{b} - A\widehat{\boldsymbol{x}}) = \mathbf{0}$.

Rewrite $A^{\mathrm{T}}(\boldsymbol{b} - A\widehat{\boldsymbol{x}}) = \mathbf{0}$ in its famous form $\boldsymbol{A^{\mathrm{T}}A\widehat{x} = A^{\mathrm{T}}b}$. This is the equation for $\widehat{\boldsymbol{x}}$, and the coefficient matrix is $A^{\mathrm{T}}A$. Now we can find $\widehat{\boldsymbol{x}}$ and $\boldsymbol{p}$ and P, in that order.

The combination $\boldsymbol{p} = \widehat{x}_1\boldsymbol{a}_1 + \cdots + \widehat{x}_n\boldsymbol{a}_n$ that is closest to $\boldsymbol{b}$ is $\boldsymbol{p} = A\widehat{\boldsymbol{x}}$:

$$\textbf{Find } \widehat{\boldsymbol{x}}\ (n \times 1) \quad A^{\mathrm{T}}(\boldsymbol{b} - A\widehat{\boldsymbol{x}}) = \mathbf{0} \quad \text{or} \quad A^{\mathrm{T}}A\widehat{\boldsymbol{x}} = A^{\mathrm{T}}\boldsymbol{b}. \tag{5}$$

This symmetric matrix $A^{\mathrm{T}}A$ is n by n. It is invertible if the $\boldsymbol{a}$'s are independent. The solution is $\widehat{\boldsymbol{x}} = (A^{\mathrm{T}}A)^{-1}A^{\mathrm{T}}\boldsymbol{b}$. The ***projection*** of $\boldsymbol{b}$ onto the subspace is $\boldsymbol{p}$:

$$\textbf{Find } \boldsymbol{p}\ (m \times 1) \qquad \boldsymbol{p} = A\widehat{\boldsymbol{x}} = A(A^{\mathrm{T}}A)^{-1}A^{\mathrm{T}}\boldsymbol{b}. \tag{6}$$

The next formula picks out the ***projection matrix*** that is multiplying $\boldsymbol{b}$ in (6):

$$\textbf{Find } \boldsymbol{P}\ (m \times m) \qquad P = A(A^{\mathrm{T}}A)^{-1}A^{\mathrm{T}}. \tag{7}$$

Compare with projection onto a line. A has one column and $A^{\mathrm{T}}A = \boldsymbol{a}^{\mathrm{T}}\boldsymbol{a}$ (1 by 1).

$$\textbf{For } n = 1 \qquad \widehat{x} = \frac{\boldsymbol{a}^{\mathrm{T}}\boldsymbol{b}}{\boldsymbol{a}^{\mathrm{T}}\boldsymbol{a}} \quad \textbf{and} \quad p = \boldsymbol{a}\frac{\boldsymbol{a}^{\mathrm{T}}\boldsymbol{b}}{\boldsymbol{a}^{\mathrm{T}}\boldsymbol{a}} \quad \textbf{and} \quad P = \frac{\boldsymbol{a}\,\boldsymbol{a}^{\mathrm{T}}}{\boldsymbol{a}^{\mathrm{T}}\boldsymbol{a}}$$

Those formulas are identical with (5) and (6) and (7). The number $\boldsymbol{a}^{\mathrm{T}}\boldsymbol{a}$ becomes the matrix $A^{\mathrm{T}}A$. When it is a number, we divide by it. When it is a matrix, we invert it. The linear independence of the columns $\boldsymbol{a}_1, \ldots, \boldsymbol{a}_n$ guarantees that $A^{\mathrm{T}}A$ is invertible.

The key step was $A^{\mathrm{T}}(\boldsymbol{b} - A\widehat{\boldsymbol{x}}) = \mathbf{0}$. We used geometry ($\boldsymbol{e}$ is orthogonal to each $\boldsymbol{a}$). *Linear algebra gives this "normal equation" too*, in a very quick and beautiful way:

1. Our subspace is the column space of A.
2. The error vector $\boldsymbol{e} = \boldsymbol{b} - A\widehat{\boldsymbol{x}}$ is perpendicular to that column space.
3. Now $\boldsymbol{e}$ is in the nullspace of A^{T} ! Then $A^{\mathrm{T}}(\boldsymbol{b} - A\widehat{\boldsymbol{x}}) = \mathbf{0}$ and $A^{\mathrm{T}}A\widehat{\boldsymbol{x}} = A^{\mathrm{T}}\boldsymbol{b}$.

The left nullspace is important in projections. That nullspace of A^{T} contains the error vector $\boldsymbol{e} = \boldsymbol{b} - A\widehat{\boldsymbol{x}}$. The vector $\boldsymbol{b}$ is split into the projection $\boldsymbol{p}$ and the error $\boldsymbol{e} = \boldsymbol{b} - \boldsymbol{p}$. Projection produces a right triangle with sides $\boldsymbol{p}$, $\boldsymbol{e}$, and $\boldsymbol{b}$.

Example 3 If $A = \begin{bmatrix} 1 & 0 \\ 1 & 1 \\ 1 & 2 \end{bmatrix}$ and $\boldsymbol{b} = \begin{bmatrix} 6 \\ 0 \\ 0 \end{bmatrix}$ find $\widehat{\boldsymbol{x}}$ and $\boldsymbol{p}$ and P.

Solution Compute the square matrix $A^{\mathrm{T}}A$ and the vector $A^{\mathrm{T}}\boldsymbol{b}$. Solve $A^{\mathrm{T}}A\widehat{\boldsymbol{x}} = A^{\mathrm{T}}\boldsymbol{b}$:

$$A^{\mathrm{T}}A = \begin{bmatrix} 1 & 1 & 1 \\ 0 & 1 & 2 \end{bmatrix} \begin{bmatrix} 1 & 0 \\ 1 & 1 \\ 1 & 2 \end{bmatrix} = \begin{bmatrix} 3 & 3 \\ 3 & 5 \end{bmatrix} \quad \text{and} \quad A^{\mathrm{T}}\boldsymbol{b} = \begin{bmatrix} 1 & 1 & 1 \\ 0 & 1 & 2 \end{bmatrix} \begin{bmatrix} 6 \\ 0 \\ 0 \end{bmatrix} = \begin{bmatrix} 6 \\ 0 \end{bmatrix}.$$

$$\text{Equations } A^{\mathrm{T}}A\widehat{\boldsymbol{x}} = A^{\mathrm{T}}\boldsymbol{b}: \begin{bmatrix} 3 & 3 \\ 3 & 5 \end{bmatrix} \begin{bmatrix} \widehat{x}_1 \\ \widehat{x}_2 \end{bmatrix} = \begin{bmatrix} 6 \\ 0 \end{bmatrix} \quad \text{gives} \quad \widehat{\boldsymbol{x}} = \begin{bmatrix} \widehat{x}_1 \\ \widehat{x}_2 \end{bmatrix} = \begin{bmatrix} \mathbf{5} \\ \mathbf{-3} \end{bmatrix}. \tag{8}$$

The combination $\boldsymbol{p} = A\widehat{\boldsymbol{x}}$ is the projection of $\boldsymbol{b}$ onto the column space of A:

$$\boldsymbol{p} = \mathbf{5}\begin{bmatrix} 1 \\ 1 \\ 1 \end{bmatrix} - \mathbf{3}\begin{bmatrix} 0 \\ 1 \\ 2 \end{bmatrix} = \begin{bmatrix} 5 \\ 2 \\ -1 \end{bmatrix}. \quad \text{The error is} \quad \boldsymbol{e} = \boldsymbol{b} - \boldsymbol{p} = \begin{bmatrix} 1 \\ -2 \\ 1 \end{bmatrix}. \tag{9}$$

Two checks on the calculation. First, the error $\boldsymbol{e} = (1, -2, 1)$ is perpendicular to both columns $(1, 1, 1)$ and $(0, 1, 2)$ of A. Second, the matrix P times $\boldsymbol{b} = (6, 0, 0)$ correctly gives $\boldsymbol{p} = (5, 2, -1)$. That solves the problem for one particular $\boldsymbol{b}$.

The projection matrix is $P = A(A^{\mathrm{T}}A)^{-1}A^{\mathrm{T}}$. The determinant of $A^{\mathrm{T}}A$ is $15 - 9 = 6$. Then multiply A times $(A^{\mathrm{T}}A)^{-1}$ times A^{T} to reach P:

$$(A^{\mathrm{T}}A)^{-1} = \frac{1}{6}\begin{bmatrix} 5 & -3 \\ -3 & 3 \end{bmatrix} \quad \text{and} \quad P = \frac{1}{6}\begin{bmatrix} 5 & 2 & -1 \\ 2 & 2 & 2 \\ -1 & 2 & 5 \end{bmatrix}. \tag{10}$$

We must have $P^2 = P$, because a second projection doesn't change the first projection!

Warning The matrix $P = A(A^{\mathrm{T}}A)^{-1}A^{\mathrm{T}}$ is deceptive. You might try to split $(A^{\mathrm{T}}A)^{-1}$ into A^{-1} times $(A^{\mathrm{T}})^{-1}$. If you make that mistake, and substitute it into P, you will find $P = AA^{-1}(A^{\mathrm{T}})^{-1}A^{\mathrm{T}}$. Apparently everything cancels. This looks like $P = I$, the identity matrix. We want to say why this is wrong.

The matrix A is rectangular. It has no inverse matrix. We cannot split $(A^{\mathrm{T}}A)^{-1}$ into A^{-1} times $(A^{\mathrm{T}})^{-1}$ because there is no A^{-1} in the first place.

In our experience, a problem that involves a rectangular matrix almost always leads to $A^{\mathrm{T}}A$. When A has independent columns, $A^{\mathrm{T}}A$ is invertible. This fact is so crucial that we state it clearly and give a proof.

$A^{\mathrm{T}}A$ is invertible if and only if A has linearly independent columns.

Proof $A^{\mathrm{T}}A$ is a square matrix (n by n). For every matrix A, we will now show that $A^{\mathrm{T}}A$ ***has the same nullspace as*** A. When the columns of A are linearly independent, its nullspace contains only the zero vector. Then $A^{\mathrm{T}}A$, with this same nullspace, is invertible.

Let A be any matrix. If $\boldsymbol{x}$ is in its nullspace, then $A\boldsymbol{x} = \mathbf{0}$. Multiplying by A^{T} gives $A^{\mathrm{T}}A\boldsymbol{x} = \mathbf{0}$. So $\boldsymbol{x}$ is also in the nullspace of $A^{\mathrm{T}}A$.

Now start with the nullspace of $A^{\mathrm{T}}A$. **From** $A^{\mathrm{T}}A\boldsymbol{x} = \mathbf{0}$ **we must prove** $A\boldsymbol{x} = \mathbf{0}$. We can't multiply by $(A^{\mathrm{T}})^{-1}$, which generally doesn't exist. Just multiply by $\boldsymbol{x}^{\mathrm{T}}$:

$$(\boldsymbol{x}^{\mathrm{T}})A^{\mathrm{T}}A\boldsymbol{x} = 0 \quad \text{or} \quad (A\boldsymbol{x})^{\mathrm{T}}(A\boldsymbol{x}) = 0 \quad \text{or} \quad \|A\boldsymbol{x}\|^2 = 0. \tag{11}$$

We have shown: If $A^{\mathrm{T}}A\boldsymbol{x} = \mathbf{0}$ then $A\boldsymbol{x}$ has length zero. Therefore $A\boldsymbol{x} = \mathbf{0}$. Every vector $\boldsymbol{x}$ in one nullspace is in the other nullspace. If $A^{\mathrm{T}}A$ has dependent columns, so has A. If $A^{\mathrm{T}}A$ has independent columns, so has A. This is the good case: $A^{\mathrm{T}}A$ is invertible.

When A has independent columns, $A^{\mathrm{T}}A$ is square and symmetric and invertible.

$$\underset{A^{\mathrm{T}}}{\begin{bmatrix} 1 & 1 & 0 \\ 2 & 2 & 0 \end{bmatrix}} \underset{\text{dependent}}{\overset{A}{\begin{bmatrix} 1 & 2 \\ 1 & 2 \\ 0 & 0 \end{bmatrix}}} = \underset{\textbf{singular}}{\overset{A^{\mathrm{T}}A}{\begin{bmatrix} 2 & 4 \\ 4 & 8 \end{bmatrix}}} \qquad \underset{A^{\mathrm{T}}}{\begin{bmatrix} 1 & 1 & 0 \\ 2 & 2 & 1 \end{bmatrix}} \underset{\text{indep.}}{\overset{A}{\begin{bmatrix} 1 & 2 \\ 1 & 2 \\ 0 & 1 \end{bmatrix}}} = \underset{\textbf{invertible}}{\overset{A^{\mathrm{T}}A}{\begin{bmatrix} 2 & 4 \\ 4 & 9 \end{bmatrix}}}$$

Very brief summary To find the projection $\boldsymbol{p} = \widehat{x}_1\boldsymbol{a}_1 + \cdots + \widehat{x}_n\boldsymbol{a}_n$, solve $A^{\mathrm{T}}A\widehat{\boldsymbol{x}} = A^{\mathrm{T}}\boldsymbol{b}$. This gives $\widehat{\boldsymbol{x}}$. The projection of $\boldsymbol{b}$ is $\boldsymbol{p} = A\widehat{\boldsymbol{x}}$ and the error is $\boldsymbol{e} = \boldsymbol{b} - \boldsymbol{p} = \boldsymbol{b} - A\widehat{\boldsymbol{x}}$. The projection matrix $P = A(A^{\mathrm{T}}A)^{-1}A^{\mathrm{T}}$ gives $\boldsymbol{p} = P\boldsymbol{b}$. **This matrix satisfies** $P^2 = P$. ***The distance from b to the subspace*** $\mathbf{C}(A)$ ***is*** $\|\boldsymbol{e}\| = \|\boldsymbol{b} - \boldsymbol{p}\|$ ***(p = closest point).***

Example Suppose your pulse is measured at $x = 70$ beats per minute, then $x = 80$, then $x = 120$. Those three equations $Ax = \boldsymbol{b}$ in one unknown have $A^{\mathrm{T}} = [1\ 1\ 1]$ and $\boldsymbol{b} = (70, 80, 120)$. ***Then*** $\widehat{x} = \mathbf{90}^\circ$ ***is the average of*** $\mathbf{70, 80, 120}$. Use calculus or algebra:

1. Minimize $E = (x - 70)^2 + (x - 80)^2 + (x - 120)^2$ by solving $dE/dx = 6x - 540 = 0$.

2. Project $\boldsymbol{b} = (70, 80, 120)$ onto $\boldsymbol{a} = (1, 1, 1)$ to find $\widehat{x} = \dfrac{\boldsymbol{a}^{\mathrm{T}}\boldsymbol{b}}{\boldsymbol{a}^{\mathrm{T}}\boldsymbol{a}} = \dfrac{70 + 80 + 120}{3} = 90$.

Problem Set 4.2

Questions 1–9 ask for projections $\boldsymbol{p}$ onto lines. Also errors $\boldsymbol{e} = \boldsymbol{b} - \boldsymbol{p}$ and matrices P.

1 Project the vector $\boldsymbol{b}$ onto the line through $\boldsymbol{a}$. Check that $\boldsymbol{e}$ is perpendicular to $\boldsymbol{a}$:

$$\text{(a)}\ \boldsymbol{b} = \begin{bmatrix}1\\2\\2\end{bmatrix} \quad\text{and}\quad \boldsymbol{a} = \begin{bmatrix}1\\1\\1\end{bmatrix} \qquad \text{(b)}\quad \boldsymbol{b} = \begin{bmatrix}1\\3\\1\end{bmatrix} \quad\text{and}\quad \boldsymbol{a} = \begin{bmatrix}-1\\-3\\-1\end{bmatrix}.$$

2 *Draw* the projection of $\boldsymbol{b}$ onto $\boldsymbol{a}$ and also compute it from $\boldsymbol{p} = \widehat{x}\boldsymbol{a}$:

$$\text{(a)}\ \boldsymbol{b} = \begin{bmatrix}\cos\theta\\ \sin\theta\end{bmatrix} \quad\text{and}\quad \boldsymbol{a} = \begin{bmatrix}1\\0\end{bmatrix} \qquad \text{(b)}\quad \boldsymbol{b} = \begin{bmatrix}1\\1\end{bmatrix} \quad\text{and}\quad \boldsymbol{a} = \begin{bmatrix}1\\-1\end{bmatrix}.$$

3 In Problem 1, find the projection matrix $P = \boldsymbol{a}\boldsymbol{a}^{\mathrm{T}}/\boldsymbol{a}^{\mathrm{T}}\boldsymbol{a}$ onto the line through each vector $\boldsymbol{a}$. Verify in both cases that $P^2 = P$. Multiply $P\boldsymbol{b}$ in each case to compute the projection $\boldsymbol{p}$. **Projection matrices onto lines have rank 1.**

4 Construct the projection matrices P_1 and P_2 onto the lines through the $\boldsymbol{a}$'s in Problem 2. Is it true that $(P_1 + P_2)^2 = P_1 + P_2$? This *would* be true if $P_1 P_2 = 0$.

5 Compute the projection matrices $\boldsymbol{a}\boldsymbol{a}^{\mathrm{T}}/\boldsymbol{a}^{\mathrm{T}}\boldsymbol{a}$ onto the lines through $\boldsymbol{a}_1 = (-1, 2, 2)$ and $\boldsymbol{a}_2 = (2, 2, -1)$. Multiply those projection matrices and explain why their product $P_1 P_2$ is what it is.

6 Project $\boldsymbol{b} = (1, 0, 0)$ onto the lines through $\boldsymbol{a}_1$ and $\boldsymbol{a}_2$ in Problem 5 and also onto $\boldsymbol{a}_3 = (2, -1, 2)$. Add up the three projections $\boldsymbol{p}_1 + \boldsymbol{p}_2 + \boldsymbol{p}_3$.

7 Continuing Problems 5–6, find the projection matrix P_3 onto $\boldsymbol{a}_3 = (2, -1, 2)$. Verify that $P_1 + P_2 + P_3 = I$. This is because the basis $\boldsymbol{a}_1, \boldsymbol{a}_2, \boldsymbol{a}_3$ is orthogonal!

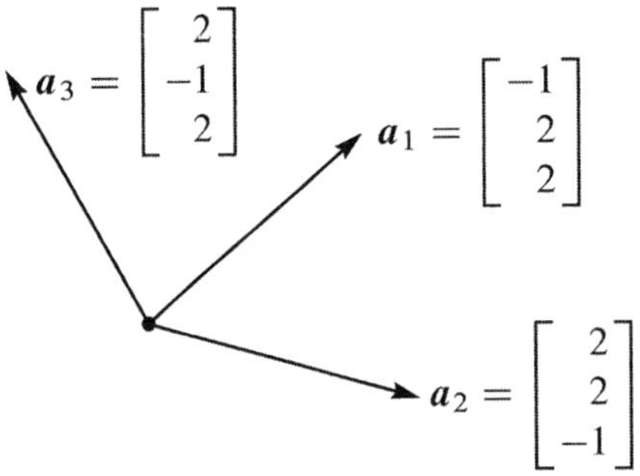

Questions 5–6–7: orthogonal

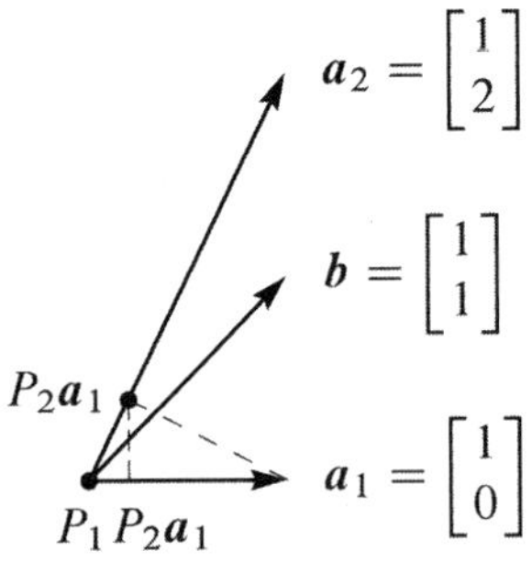

Questions 8–9–10: not orthogonal

8 Project the vector $\boldsymbol{b} = (1,1)$ onto the lines through $\boldsymbol{a}_1 = (1,0)$ and $\boldsymbol{a}_2 = (1,2)$. Draw the projections $\boldsymbol{p}_1$ and $\boldsymbol{p}_2$ and add $\boldsymbol{p}_1 + \boldsymbol{p}_2$. The projections do not add to $\boldsymbol{b}$ because the $\boldsymbol{a}$'s are not orthogonal.

9 In Problem 8, the projection of $\boldsymbol{b}$ onto the *plane* of $\boldsymbol{a}_1$ and $\boldsymbol{a}_2$ will equal $\boldsymbol{b}$. Find $P = A(A^{\mathrm{T}}A)^{-1}A^{\mathrm{T}}$ for $A = \begin{bmatrix} \boldsymbol{a}_1 & \boldsymbol{a}_2 \end{bmatrix} = \left[\begin{smallmatrix} 1 & 1 \\ 0 & 2 \end{smallmatrix}\right]$ = invertible matrix.

10 Project $\boldsymbol{a}_1 = (1,0)$ onto $\boldsymbol{a}_2 = (1,2)$. Then project the result back onto $\boldsymbol{a}_1$. Draw these projections and multiply the projection matrices P_1P_2: Is this a projection?

Questions 11–20 ask for projections, and projection matrices, onto subspaces.

11 Project $\boldsymbol{b}$ onto the column space of A by solving $A^{\mathrm{T}}A\widehat{\boldsymbol{x}} = A^{\mathrm{T}}\boldsymbol{b}$ and $\boldsymbol{p} = A\widehat{\boldsymbol{x}}$:

$$\text{(a)}\ A = \begin{bmatrix} 1 & 1 \\ 0 & 1 \\ 0 & 0 \end{bmatrix} \text{ and } \boldsymbol{b} = \begin{bmatrix} 2 \\ 3 \\ 4 \end{bmatrix} \qquad \text{(b)}\quad A = \begin{bmatrix} 1 & 1 \\ 1 & 1 \\ 0 & 1 \end{bmatrix} \text{ and } \boldsymbol{b} = \begin{bmatrix} 4 \\ 4 \\ 6 \end{bmatrix}.$$

Find $\boldsymbol{e} = \boldsymbol{b} - \boldsymbol{p}$. It should be perpendicular to the columns of A.

12 Compute the projection matrices P_1 and P_2 onto the column spaces in Problem 11. Verify that $P_1\boldsymbol{b}$ gives the first projection $\boldsymbol{p}_1$. Also verify $P_2^2 = P_2$.

13 (Quick and Recommended) Suppose A is the 4 by 4 identity matrix with its last column removed. A is 4 by 3. Project $\boldsymbol{b} = (1,2,3,4)$ onto the column space of A. What shape is the projection matrix P and what is P?

14 Suppose $\boldsymbol{b}$ equals 2 times the first column of A. What is the projection of $\boldsymbol{b}$ onto the column space of A? Is $P = I$ for sure in this case? Compute $\boldsymbol{p}$ and P when $\boldsymbol{b} = (0,2,4)$ and the columns of A are $(0,1,2)$ and $(1,2,0)$.

15 If A is doubled, then $P = \mathbf{2}A(\mathbf{4}A^{\mathrm{T}}A)^{-1}\mathbf{2}A^{\mathrm{T}}$. This is the same as $A(A^{\mathrm{T}}A)^{-1}A^{\mathrm{T}}$. The column space of $2A$ is the same as _____. Is $\widehat{\boldsymbol{x}}$ the same for A and $2A$?

16 What linear combination of $(1,2,-1)$ and $(1,0,1)$ is closest to $\boldsymbol{b} = (2,1,1)$?

17 (*Important*) If $P^2 = P$ show that $(I-P)^2 = I - P$. When P projects onto the column space of A, $I - P$ projects onto the _____.

18 (a) If P is the 2 by 2 projection matrix onto the line through $(1,1)$, then $I - P$ is the projection matrix onto _____.

(b) If P is the 3 by 3 projection matrix onto the line through $(1,1,1)$, then $I - P$ is the projection matrix onto _____.

19 To find the projection matrix onto the plane $x - y - 2z = 0$, choose two vectors in that plane and make them the columns of A. The plane will be the column space of A! Then compute $P = A(A^{\mathrm{T}}A)^{-1}A^{\mathrm{T}}$.

20 To find the projection matrix P onto the same plane $x - y - 2z = 0$, write down a vector $\boldsymbol{e}$ that is perpendicular to that plane. Compute the projection $Q = \boldsymbol{e}\boldsymbol{e}^{\mathrm{T}}/\boldsymbol{e}^{\mathrm{T}}\boldsymbol{e}$ and then $P = I - Q$.

21 Multiply the matrix $P = A(A^{\mathrm{T}}A)^{-1}A^{\mathrm{T}}$ by itself. Cancel to prove that $\boldsymbol{P^2 = P}$. Explain why $P(P\boldsymbol{b})$ always equals $P\boldsymbol{b}$: The vector $P\boldsymbol{b}$ is in the column space of A so its projection onto that column space is ____ .

22 Prove that $P = A(A^{\mathrm{T}}A)^{-1}A^{\mathrm{T}}$ is symmetric by computing P^{T}. Remember that the inverse of a symmetric matrix is symmetric.

23 If A is square and invertible, the warning against splitting $(A^{\mathrm{T}}A)^{-1}$ does not apply. Then $AA^{-1}(A^{\mathrm{T}})^{-1}A^{\mathrm{T}} = I$. *When A is invertible, why is $P = I$?* ***What is* $\boldsymbol{e}$*?***

24 The nullspace of A^{T} is ____ to the column space $\boldsymbol{C}(A)$. So if $A^{\mathrm{T}}\boldsymbol{b} = \boldsymbol{0}$, the projection of $\boldsymbol{b}$ onto $\boldsymbol{C}(A)$ should be $\boldsymbol{p} =$ ____ . Check that $P = A(A^{\mathrm{T}}A)^{-1}A^{\mathrm{T}}$ gives this answer.

25 The projection matrix P onto an n-dimensional subspace of $\mathbf{R}^m$ has rank $r = n$. ***Reason:*** The projections $P\boldsymbol{b}$ fill the subspace $\mathbf{S}$. So $\mathbf{S}$ is the ____ of P.

26 If an m by m matrix has $A^2 = A$ and its rank is m, prove that $A = I$.

27 The important fact that ends the section is this: ***If*** $\boldsymbol{A^{\mathrm{T}}Ax = 0}$ ***then*** $\boldsymbol{Ax = 0}$. *New Proof*: The vector $A\boldsymbol{x}$ is in the nullspace of ____ . $A\boldsymbol{x}$ is always in the column space of ____ . **To be in both of those perpendicular spaces, $\boldsymbol{Ax}$ must be zero.**

28 Use $P^{\mathrm{T}} = P$ and $P^2 = P$ to prove that the length squared of column 2 always equals the diagonal entry P_{22}. This number is $\frac{2}{6} = \frac{4}{36} + \frac{4}{36} + \frac{4}{36}$ for

$$P = \frac{1}{6}\begin{bmatrix} 5 & 2 & -1 \\ 2 & 2 & 2 \\ -1 & 2 & 5 \end{bmatrix}.$$

29 If B has rank m (full row rank, independent rows) show that BB^{T} is invertible.

30 (a) Find the 2 by 2 projection matrix P_C onto the column space of A (after looking closely at the matrix!)

$$A = \begin{bmatrix} 3 & 6 & 6 \\ 4 & 8 & 8 \end{bmatrix}$$

(b) Find the 3 by 3 projection matrix P_R onto the row space of A. Multiply $B = P_C A P_R$. Your answer B should be a little surprising—can you explain it?

31 In $\mathbf{R}^m$, suppose I give you $\boldsymbol{b}$ and also a combination $\boldsymbol{p}$ of $\boldsymbol{a}_1, \ldots, \boldsymbol{a}_n$. How would you test to see if $\boldsymbol{p}$ is the projection of $\boldsymbol{b}$ onto the subspace spanned by the $\boldsymbol{a}$'s?

32 Suppose you know the average $\widehat{x}_{\text{old}}$ of $b_1, b_2, \ldots, b_{999}$. When b_{1000} arrives, check that $\widehat{x}_{\text{new}} = \widehat{x}_{\text{old}} + (b_{1000} - \widehat{x}_{\text{old}})/1000$. That step updates $\widehat{x}_{\text{old}}$ to $\widehat{x}_{\text{new}}$.

33 Suppose P_1 and P_2 are projection matrices ($P_i^2 = P_i = P_i^{\mathrm{T}}$). Prove this fact:

P_1P_2 is a projection matrix if and only if $P_1P_2 = P_2P_1$.

4.3 Least Squares Approximations

1 Solving $\boxed{A^{\mathrm{T}}A\,\widehat{\boldsymbol{x}} = A^{\mathrm{T}}\boldsymbol{b}}$ gives the projection $\boldsymbol{p} = A\widehat{\boldsymbol{x}}$ of $\boldsymbol{b}$ onto the column space of A.

2 When $A\boldsymbol{x} = \boldsymbol{b}$ has no solution, $\widehat{\boldsymbol{x}}$ is the "least-squares solution": $||\boldsymbol{b} - A\widehat{\boldsymbol{x}}||^2 =$ minimum.

3 Setting partial derivatives of $E = ||A\boldsymbol{x} - \boldsymbol{b}||^2$ to zero $\left(\frac{\partial E}{\partial x_i} = 0\right)$ also produces $A^{\mathrm{T}}A\widehat{\boldsymbol{x}} = A^{\mathrm{T}}\boldsymbol{b}$.

4 To fit points $(t_1, b_1), \ldots, (t_m, b_m)$ by a straight line, A has columns $(1, \ldots, 1)$ and $(t_1, \ldots, t_m)$.

It often happens that $A\boldsymbol{x} = \boldsymbol{b}$ has no solution. The usual reason is: *too many equations.* The matrix A has more rows than columns. There are more equations than unknowns (m is greater than n). The n columns span a small part of m-dimensional space. Unless all measurements are perfect, $\boldsymbol{b}$ is outside that column space of A. Elimination reaches an impossible equation and stops. But we can't stop just because measurements include noise !

To repeat: We cannot always get the error $\boldsymbol{e} = \boldsymbol{b} - A\boldsymbol{x}$ down to zero. When $\boldsymbol{e}$ is zero, $\boldsymbol{x}$ is an exact solution to $A\boldsymbol{x} = \boldsymbol{b}$. *When the error $\boldsymbol{e}$ is as small as possible, $\widehat{\boldsymbol{x}}$ is a least squares solution.* The words "least squares" mean that $||\boldsymbol{b} - A\widehat{\boldsymbol{x}}||^2$ is a *minimum.* Our goal in this section is to compute $\widehat{\boldsymbol{x}}$ and use it. These are real problems that need answers.

Note In statistics this problem is **linear regression**: $\boldsymbol{x}$ and $\boldsymbol{b}$ often become $\boldsymbol{Y}$ and $\boldsymbol{X}$.

The previous section emphasized $\boldsymbol{p}$ (the projection). This section emphasizes $\widehat{\boldsymbol{x}}$ (the least squares solution). They are connected by $\boldsymbol{p} = A\widehat{\boldsymbol{x}}$. The fundamental equation is still $A^{\mathrm{T}}A\widehat{\boldsymbol{x}} = A^{\mathrm{T}}\boldsymbol{b}$. Here is a short unofficial way to reach this "*normal equation*":

When $A\boldsymbol{x} = \boldsymbol{b}$ has no solution, multiply by A^{T} and solve $A^{\mathrm{T}}A\widehat{\boldsymbol{x}} = A^{\mathrm{T}}\boldsymbol{b}$.

Example 1 A crucial application of least squares is fitting a straight line to m points. Start with three points: *Find the closest line to the points* $(0, 6), (1, 0)$, *and* $(2, 0)$.

No straight line $b = C + Dt$ goes through those three points. We are asking for two numbers C and D that satisfy three equations: $n = 2$ and $m = 3$ and $m > n$. Here are the three equations at $t = 0, 1, 2$ to match the given values $b = 6, 0, 0$:

$\boldsymbol{t = 0}$	The first point is on the line $b = C + Dt$ if	$C + D \cdot 0 = 6$
$\boldsymbol{t = 1}$	The second point is on the line $b = C + Dt$ if	$C + D \cdot 1 = 0$
$\boldsymbol{t = 2}$	The third point is on the line $b = C + Dt$ if	$C + D \cdot 2 = 0$

This 3 by 2 system has *no solution*: $\boldsymbol{b} = (6, 0, 0)$ is not a combination of the columns $(1, 1, 1)$ and $(0, 1, 2)$. Read off A and $\boldsymbol{x}$ and $\boldsymbol{b}$ from those equations:

$$A = \begin{bmatrix} 1 & 0 \\ 1 & 1 \\ 1 & 2 \end{bmatrix} \quad \boldsymbol{x} = \begin{bmatrix} C \\ D \end{bmatrix} \quad \boldsymbol{b} = \begin{bmatrix} 6 \\ 0 \\ 0 \end{bmatrix} \quad \begin{array}{c} A\boldsymbol{x} = \boldsymbol{b} \text{ is } \textit{not} \text{ solvable} \\ \boldsymbol{x} \text{ is overdetermined} \end{array}$$

The same numbers were in Example 3 in the last section. We computed $\widehat{\boldsymbol{x}} = (5, -3)$. **Those numbers are the best C and D, so $5 - 3t$ will be the best line for the 3 points**. We must connect projections to least squares, by explaining why $A^{\mathrm{T}}A\widehat{\boldsymbol{x}} = A^{\mathrm{T}}\boldsymbol{b}$.

In practical problems, there could easily be $m = 100$ points instead of $m = 3$. They don't exactly match any straight line $C + Dt$. Our numbers $6, 0, 0$ exaggerate the error so you can see e_1, e_2, and e_3 in Figure 4.6.

Minimizing the Error

How do we make the error $\boldsymbol{e} = \boldsymbol{b} - A\boldsymbol{x}$ as small as possible? This is an important question with a beautiful answer. The best $\boldsymbol{x}$ (called $\widehat{\boldsymbol{x}}$) can be found by geometry (the error $\boldsymbol{e}$ meets the column space of A at $90°$). It can be found by algebra: $A^{\mathrm{T}}A\widehat{\boldsymbol{x}} = A^{\mathrm{T}}\boldsymbol{b}$. Calculus gives the same solution $\widehat{\boldsymbol{x}}$: the derivative of the error $||A\boldsymbol{x} - \boldsymbol{b}||^2$ is zero at $\widehat{\boldsymbol{x}}$.

By geometry Every $A\boldsymbol{x}$ lies in the plane of the columns $(1, 1, 1)$ and $(0, 1, 2)$. In that plane, we look for the point closest to $\boldsymbol{b}$. *The nearest point is the projection $\boldsymbol{p}$.*

The best choice for $A\widehat{\boldsymbol{x}}$ is the projection $\boldsymbol{p}$. The smallest possible error is $\boldsymbol{e} = \boldsymbol{b} - \boldsymbol{p}$, perpendicular to the columns. *The three points at heights (p_1, p_2, p_3) do lie on a line*, because $\boldsymbol{p}$ is in the column space of A. The best line $C + Dt$ comes from $\widehat{\boldsymbol{x}} = (C, D)$.

By algebra Every vector $\boldsymbol{b}$ splits into two parts. The part in the column space is $\boldsymbol{p}$. The perpendicular part is $\boldsymbol{e}$. There is an equation we cannot solve $(A\boldsymbol{x} = \boldsymbol{b})$. There is an equation $A\widehat{\boldsymbol{x}} = \boldsymbol{p}$ we can and do solve (by removing $\boldsymbol{e}$ and solving $A^{\mathrm{T}}A\widehat{\boldsymbol{x}} = A^{\mathrm{T}}\boldsymbol{b}$):

$$A\boldsymbol{x} = \boldsymbol{b} = \boldsymbol{p} + \boldsymbol{e} \text{ is impossible} \qquad A\widehat{\boldsymbol{x}} = \boldsymbol{p} \text{ is solvable} \qquad \widehat{\boldsymbol{x}} \text{ is } (A^{\mathrm{T}}A)^{-1}A^{\mathrm{T}}\boldsymbol{b}. \quad (1)$$

The solution to $A\widehat{\boldsymbol{x}} = \boldsymbol{p}$ leaves the least possible error (which is $\boldsymbol{e}$):

$$\textbf{Squared length for any } x \qquad ||A\boldsymbol{x} - \boldsymbol{b}||^2 = ||A\boldsymbol{x} - \boldsymbol{p}||^2 + ||\boldsymbol{e}||^2. \quad (2)$$

This is the law $c^2 = a^2 + b^2$ for a right triangle. The vector $A\boldsymbol{x} - \boldsymbol{p}$ in the column space is perpendicular to $\boldsymbol{e}$ in the left nullspace. We reduce $A\boldsymbol{x} - \boldsymbol{p}$ to **zero** by choosing $\boldsymbol{x} = \widehat{\boldsymbol{x}}$. That leaves thc smallest possible error $\boldsymbol{e} = (e_1, e_2, e_3)$ which we can't reduce.

Notice what "smallest" means. The *squared length* of $A\boldsymbol{x} - \boldsymbol{b}$ is minimized:

> ***The least squares solution $\widehat{x}$ makes $E = ||A\boldsymbol{x} - \boldsymbol{b}||^2$ as small as possible.***

Figure 4.6a shows the closest line. It misses by distances $e_1, e_2, e_3 = 1, -2, 1$. *Those are vertical distances*. The least squares line minimizes $E = e_1^2 + e_2^2 + e_3^2$.

Figure 4.6b shows the same problem in 3-dimensional space ($\boldsymbol{b}\,\boldsymbol{p}\,\boldsymbol{e}$ space). The vector $\boldsymbol{b}$ is not in the column space of A. That is why we could not solve $A\boldsymbol{x} = \boldsymbol{b}$. No line goes through the three points. The smallest possible error is the perpendicular vector $\boldsymbol{e}$. This is $\boldsymbol{e} = \boldsymbol{b} - A\widehat{\boldsymbol{x}}$, the vector of errors $(1, -2, 1)$ in the three equations. Those are the distances from the best line. Behind both figures is the fundamental equation $A^{\mathrm{T}}A\widehat{\boldsymbol{x}} = A^{\mathrm{T}}\boldsymbol{b}$.

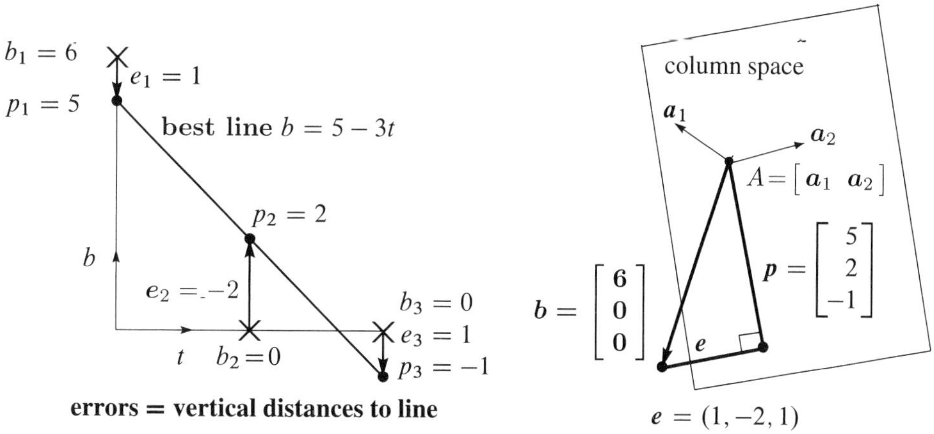

Figure 4.6: **Best line and projection: Two pictures, same problem.** The line has heights $\boldsymbol{p} = (5, 2, -1)$ with errors $\boldsymbol{e} = (1, -2, 1)$. The equations $A^{\mathrm{T}} A\widehat{\boldsymbol{x}} = A^{\mathrm{T}}\boldsymbol{b}$ give $\widehat{\boldsymbol{x}} = (5, -3)$. The best line is $b = 5 - 3t$ and the closest point is $\boldsymbol{p} = 5\boldsymbol{a}_1 - 3\boldsymbol{a}_2$. Same answer!

Notice that the errors $1, -2, 1$ add to zero. *Reason*: The error $\boldsymbol{e} = (e_1, e_2, e_3)$ is perpendicular to the first column $(1, 1, 1)$ in A. The dot product gives $e_1 + e_2 + e_3 = 0$.

By calculus Most functions are minimized by calculus! The graph of E bottoms out and the derivative in every direction is zero. Here the error function E to be minimized is a *sum of squares* $e_1^2 + e_2^2 + e_3^2$ (the square of the error in each equation):

$$E = \|A\boldsymbol{x} - \boldsymbol{b}\|^2 = (C + D \cdot 0 - 6)^2 + (C + D \cdot 1)^2 + (C + D \cdot 2)^2. \qquad (3)$$

The unknowns C and D tell us the closest line $C + Dt$. With two unknowns there are *two derivatives*—both zero at the minimum. They are "partial derivatives" because $\partial E/\partial C$ treats D as constant and $\partial E/\partial D$ treats C as constant:

$$\partial E/\partial C = 2(C + D \cdot 0 - 6) \quad + 2(C + D \cdot 1) \quad + 2(C + D \cdot 2) \quad = 0$$

$$\partial E/\partial D = 2(C + D \cdot 0 - 6)(\mathbf{0}) + 2(C + D \cdot 1)(\mathbf{1}) + 2(C + D \cdot 2)(\mathbf{2}) = 0.$$

$\partial E/\partial D$ contains the extra factors $\mathbf{0}, \mathbf{1}, \mathbf{2}$ from the chain rule. (The last derivative from $(C + 2D)^2$ was 2 times $C + 2D$ times that extra 2.) Those factors are $\mathbf{1}, \mathbf{1}, \mathbf{1}$ in $\partial E/\partial C$.

It is no accident that those factors 1, 1, 1 and 0, 1, 2 in the derivatives of $\|A\boldsymbol{x} - \boldsymbol{b}\|^2$ are the columns of A. Now cancel 2 from every term and collect all C's and all D's:

$$\begin{array}{ll} \text{The } C \text{ derivative is zero:} & 3C + 3D = 6 \\ \text{The } D \text{ derivative is zero:} & 3C + 5D = 0 \end{array} \qquad \textbf{This matrix} \begin{bmatrix} 3 & 3 \\ 3 & 5 \end{bmatrix} \textbf{ is } \boldsymbol{A}^{\mathrm{T}}\boldsymbol{A}\textbf{!} \qquad (4)$$

These equations are identical with $A^{\mathrm{T}}A\widehat{x} = A^{\mathrm{T}}\boldsymbol{b}$. The best C and D are the components of $\widehat{x}$. The equations from calculus are the same as the "normal equations" from linear algebra. These are the key equations of least squares = linear regression.

The partial derivatives of $\|\boldsymbol{A}x - \boldsymbol{b}\|^2$ ***are zero at*** $\widehat{x}$ ***when*** $\boldsymbol{A}^{\mathrm{T}}\boldsymbol{A}\widehat{x} = \boldsymbol{A}^{\mathrm{T}}\boldsymbol{b}$.

The solution is $C = 5$ and $D = -3$. Therefore $b = 5 - 3t$ is the best line—it comes closest to the three points. At $t = 0, 1, 2$ this line goes through $\boldsymbol{p} = 5, 2, -1$. It could not go through $\boldsymbol{b} = 6, 0, 0$. The errors are $1, -2, 1$. This is the vector $\boldsymbol{e}$!

The Big Picture for Least Squares

The key figure of this book shows the four subspaces and the true action of a matrix. The vector $\boldsymbol{x}$ on the left side of Figure 4.3 went to $\boldsymbol{b} = A\boldsymbol{x}$ on the right side. In that figure $\boldsymbol{x}$ was split into $\boldsymbol{x}_r + \boldsymbol{x}_n$. There were *many* solutions to $A\boldsymbol{x} = \boldsymbol{b}$.

In this section the situation is just the opposite. There are *no* solutions to $A\boldsymbol{x} = \boldsymbol{b}$. *Instead of splitting up* $\boldsymbol{x}$ *we are splitting up* $\boldsymbol{b} = \boldsymbol{p} + \boldsymbol{e}$. Figure 4.7 shows the big picture for least squares. Instead of $A\boldsymbol{x} = \boldsymbol{b}$ we solve $A\widehat{\boldsymbol{x}} = \boldsymbol{p}$. The error $\boldsymbol{e} = \boldsymbol{b} - \boldsymbol{p}$ is unavoidable.

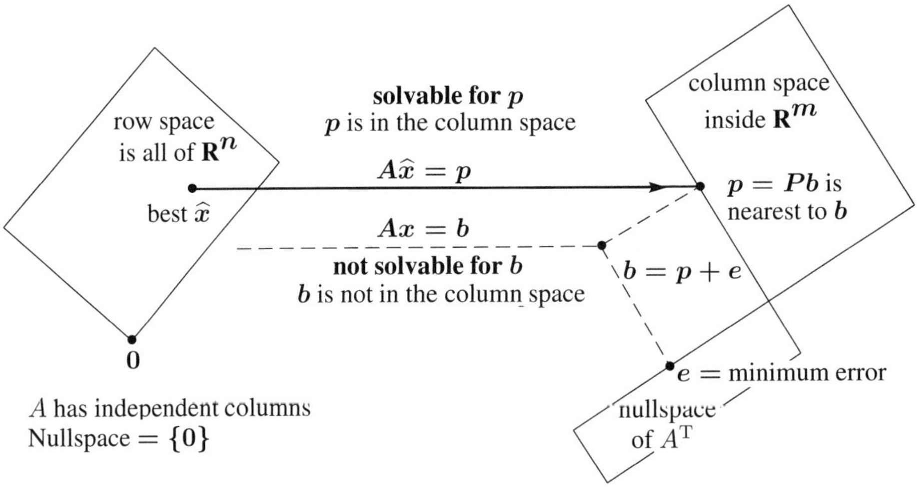

Figure 4.7: The projection $\boldsymbol{p} = A\widehat{\boldsymbol{x}}$ is closest to $\boldsymbol{b}$, so $\widehat{\boldsymbol{x}}$ minimizes $E = \|\boldsymbol{b} - A\boldsymbol{x}\|^2$.

Notice how the nullspace $\mathbf{N}(A)$ is very small—just one point. With independent columns, the only solution to $A\boldsymbol{x} = \mathbf{0}$ is $\boldsymbol{x} = \mathbf{0}$. Then $A^{\mathrm{T}}A$ is invertible. The equation $A^{\mathrm{T}}A\widehat{\boldsymbol{x}} = A^{\mathrm{T}}\boldsymbol{b}$ fully determines the best vector $\widehat{\boldsymbol{x}}$. The error has $A^{\mathrm{T}}\boldsymbol{e} = \mathbf{0}$.

Fitting a Straight Line

Fitting a line is the clearest application of least squares. It starts with $m > 2$ points, hopefully near a straight line. At times $t_1, \ldots, t_m$ those points (t_i, b_i) are at heights $b_1, \ldots, b_m$. The best line $C + Dt$ misses the points by vertical distances $e_1, \ldots, e_m$. No line is perfect, and the least squares line minimizes $E = e_1^2 + \cdots + e_m^2$.

The first example in this section had three points in Figure 4.6. Now we allow m points (and m can be large). The two components of $\widehat{\boldsymbol{x}}$ are still C and D.

A line goes through the m points when we exactly solve $A\boldsymbol{x} = \boldsymbol{b}$. Generally we can't do it. Two unknowns C and D determine a line, so A has only $n = 2$ columns. To fit the m points, we are trying to solve m equations (and we only have two unknowns !).

$$A\boldsymbol{x} = \boldsymbol{b} \quad \text{is} \quad \begin{matrix} \boldsymbol{C + Dt_1 = b_1} \\ \boldsymbol{C + Dt_2 = b_2} \\ \vdots \\ \boldsymbol{C + Dt_m = b_m} \end{matrix} \quad \text{with} \quad A = \begin{bmatrix} 1 & t_1 \\ 1 & t_2 \\ \vdots & \vdots \\ 1 & t_m \end{bmatrix} \quad \text{and} \quad \widehat{\boldsymbol{x}} = \begin{bmatrix} C \\ D \end{bmatrix} \tag{5}$$

The column space of A is so thin that almost certainly the vector $\boldsymbol{b}$ is outside of it. When $\boldsymbol{b}$ happens to lie in the column space, the points happen to lie on a line. That case $\boldsymbol{b} = \boldsymbol{p}$ is very unusual. Then $A\boldsymbol{x} = \boldsymbol{b}$ is solvable and $\boldsymbol{e} = (0, \ldots, 0)$.

The closest line $C + Dt$ has heights $p_1, \ldots, p_m$ with errors $e_1, \ldots, e_m$.

Solve $A^{\mathrm{T}}A\widehat{x} = A^{\mathrm{T}}b$ for $\widehat{x} = (C, D)$. The errors are $e_i = b_i - C - Dt_i$.

Fitting points by a straight line is so important that we now find the two equations $A^{\mathrm{T}}A\widehat{\boldsymbol{x}} = A^{\mathrm{T}}\boldsymbol{b}$, once and for all. The two columns of A are independent (unless all of the times t_i are the same). So we turn to least squares and solve $A^{\mathrm{T}}A\widehat{\boldsymbol{x}} = A^{\mathrm{T}}\boldsymbol{b}$.

$$\textbf{Dot-product matrix} \quad A^{\mathrm{T}}A = \begin{bmatrix} 1 & \cdots & 1 \\ t_1 & \cdots & t_m \end{bmatrix} \begin{bmatrix} 1 & t_1 \\ \vdots & \vdots \\ 1 & t_m \end{bmatrix} = \begin{bmatrix} m & \sum t_i \\ \sum t_i & \sum t_i^2 \end{bmatrix}. \tag{6}$$

On the right side of the normal equation is the 2 by 1 vector $A^{\mathrm{T}}\boldsymbol{b}$:

$$A^{\mathrm{T}}\boldsymbol{b} = \begin{bmatrix} 1 & \cdots & 1 \\ t_1 & \cdots & t_m \end{bmatrix} \begin{bmatrix} b_1 \\ \vdots \\ b_m \end{bmatrix} = \begin{bmatrix} \sum b_i \\ \sum t_i b_i \end{bmatrix}. \tag{7}$$

In a specific problem, the t's and b's are given. The best $\widehat{\boldsymbol{x}} = (C, D)$ is $(A^{\mathrm{T}}A)^{-1}A^{\mathrm{T}}\boldsymbol{b}$.

The line $C + Dt$ minimizes $e_1^2 + \cdots + e_m^2 = \|A\boldsymbol{x} - \boldsymbol{b}\|^2$ when $A^{\mathrm{T}}A\widehat{\boldsymbol{x}} = A^{\mathrm{T}}\boldsymbol{b}$:

$$A^{\mathrm{T}}A\widehat{\boldsymbol{x}} = A^{\mathrm{T}}\boldsymbol{b} \qquad \begin{bmatrix} m & \sum t_i \\ \sum t_i & \sum t_i^2 \end{bmatrix} \begin{bmatrix} C \\ D \end{bmatrix} = \begin{bmatrix} \sum b_i \\ \sum t_i b_i \end{bmatrix}. \tag{8}$$

The vertical errors at the m points on the line are the components of $\boldsymbol{e} = \boldsymbol{b} - \boldsymbol{p}$. This error vector (the *residual*) $\boldsymbol{b} - A\widehat{\boldsymbol{x}}$ is perpendicular to the columns of A (geometry). The error is in the nullspace of A^{T} (linear algebra). The best $\widehat{\boldsymbol{x}} = (C, D)$ minimizes the total error E, the sum of squares (calculus):

$$E(\boldsymbol{x}) = \|A\boldsymbol{x} - \boldsymbol{b}\|^2 = (C + Dt_1 - b_1)^2 + \cdots + (C + Dt_m - b_m)^2.$$

Calculus sets the derivatives $\partial E/\partial C$ and $\partial E/\partial D$ to zero, and produces $A^{\mathrm{T}}A\widehat{\boldsymbol{x}} = A^{\mathrm{T}}\boldsymbol{b}$.

Other least squares problems have more than two unknowns. Fitting by the best parabola has $n = 3$ coefficients C, D, E (see below). In general we are fitting m data points by n parameters $x_1, \ldots, x_n$. The matrix A has n columns and $n < m$. The derivatives of $\|A\boldsymbol{x} - \boldsymbol{b}\|^2$ give the n equations $A^{\mathrm{T}}A\widehat{\boldsymbol{x}} = A^{\mathrm{T}}\boldsymbol{b}$. **The derivative of a square is linear.** This is why the method of least squares is so popular.

Example 2 A has *orthogonal columns* when the measurement times t_i add to zero.

Suppose $b = 1, 2, 4$ at times $t = -2, 0, 2$. *Those times add to zero.* The columns of A have *zero dot product*: $(1, 1, 1)$ is orthogonal to $(-2, 0, 2)$:

$$\begin{aligned} C + D(-2) &= 1 \\ C + \quad D(0) &= 2 \\ C + \quad D(2) &= 4 \end{aligned} \qquad \text{or} \qquad A\boldsymbol{x} = \begin{bmatrix} 1 & -2 \\ 1 & 0 \\ 1 & 2 \end{bmatrix} \begin{bmatrix} C \\ D \end{bmatrix} = \begin{bmatrix} 1 \\ 2 \\ 4 \end{bmatrix}.$$

When the columns of A are orthogonal, $A^{\mathrm{T}}A$ will be a diagonal matrix (this is good):

$$A^{\mathrm{T}}A\widehat{\boldsymbol{x}} = A^{\mathrm{T}}\boldsymbol{b} \qquad \text{is} \qquad \begin{bmatrix} 3 & \mathbf{0} \\ \mathbf{0} & 8 \end{bmatrix} \begin{bmatrix} C \\ D \end{bmatrix} = \begin{bmatrix} 7 \\ 6 \end{bmatrix}. \tag{9}$$

Main point: Since $A^{\mathrm{T}}A$ *is diagonal*, we quickly find $C = \frac{7}{3}$ and $D = \frac{6}{8}$. The zeros in $A^{\mathrm{T}}A$ are dot products of perpendicular columns in A. The diagonal matrix $A^{\mathrm{T}}A$, with entries $m = 3$ and $t_1^2 + t_2^2 + t_3^2 = 8$, is almost as simple as the identity matrix.

Orthogonal columns are so helpful that it is worth *shifting the times by subtracting the average time* $\widehat{t} = (t_1 + \cdots + t_m)/m$. If the original times were $1, 3, 5$ then their average is $\widehat{t} = 3$. The shifted times $T = t - \widehat{t} = t - 3$ are $-2, 0, 2$. Those times add to zero!

$$\begin{aligned} T_1 &= 1 - 3 = -2 \\ T_2 &= 3 - 3 = \quad 0 \\ T_3 &= 5 - 3 = \quad 2 \end{aligned} \qquad A_{\text{new}} = \begin{bmatrix} 1 & T_1 \\ 1 & T_2 \\ 1 & T_3 \end{bmatrix} \qquad A_{\text{new}}^{\mathrm{T}}A_{\text{new}} = \begin{bmatrix} 3 & \mathbf{0} \\ \mathbf{0} & 8 \end{bmatrix}.$$

Now C and D come from the easy equation (9). Then the best straight line uses $C + DT$ which is $C + D(t - \widehat{t}\,) = C + D(t - 3)$. Problem 30 even gives a formula for C and D.

That was a perfect example of the "Gram-Schmidt idea" coming in the next section: *Make the columns orthogonal in advance.* Then $A_{\text{new}}^{\mathrm{T}}A_{\text{new}}$ is diagonal and $\widehat{\boldsymbol{x}}_{\text{new}}$ is easy.

Dependent Columns in A : Which $\widehat{x}$ is best ?

From the start, this chapter assumed independent columns in A. Then $A^{\mathrm{T}}A$ is invertible. $A^{\mathrm{T}}A\widehat{\boldsymbol{x}} = A^{\mathrm{T}}\boldsymbol{b}$ produces the only least squares solution to $A\boldsymbol{x} = \boldsymbol{b}$. **Which $\widehat{x}$ is best if A has dependent columns ?** All the dashed lines have the same errors $\boldsymbol{e} = (1, -1)$.

$$\begin{bmatrix} 1 & 1 \\ 1 & 1 \end{bmatrix}\begin{bmatrix} x_1 \\ x_2 \end{bmatrix} = \begin{bmatrix} 3 \\ 1 \end{bmatrix} = \boldsymbol{b} \qquad \begin{bmatrix} 1 & 1 \\ 1 & 1 \end{bmatrix}\begin{bmatrix} \widehat{x}_1 \\ \widehat{x}_2 \end{bmatrix} = \begin{bmatrix} 2 \\ 2 \end{bmatrix} = \boldsymbol{p}$$

$$A\boldsymbol{x} = \boldsymbol{b} \qquad \begin{bmatrix} 1 & 1 \\ 1 & 1 \end{bmatrix} \text{ is singular} \qquad A\widehat{\boldsymbol{x}} = \boldsymbol{p}$$

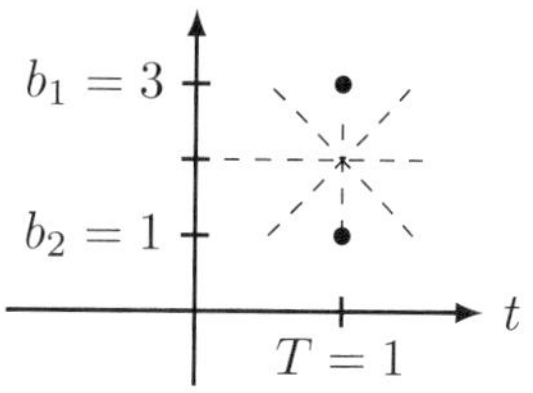

The measurements $b_1 = 3$ and $b_2 = 1$ are at the same time T! A straight line $C + Dt$ cannot go through both points. I think we are right to project $\boldsymbol{b} = (3, 1)$ to $\boldsymbol{p} = (2, 2)$ in the column space of A. That changes the equation $A\boldsymbol{x} = \boldsymbol{b}$ to the equation $A\widehat{\boldsymbol{x}} = \boldsymbol{p}$. An equation with no solution has become an equation with infinitely many solutions. The problem is that A has dependent columns and $\widehat{x}_1 + \widehat{x}_2 = 2$ has many solutions.

Which solution $\widehat{\boldsymbol{x}}$ should we choose? All the dashed lines in the figure have the same two errors 1 and -1 at time T. Those errors $(1, -1) = \boldsymbol{e} = \boldsymbol{b} - \boldsymbol{p}$ are as small as possible. But this doesn't tell us which dashed line is best. My instinct is to go for the horizontal line at height 2.

The *"pseudoinverse" of* A will choose the **shortest solution $x^+ = A^+b$ to** $A\widehat{\boldsymbol{x}} = \boldsymbol{p}$. Here, that shortest solution will be $\boldsymbol{x}^+ = (1, 1)$. This is the particular solution in the row space of A, and $\boldsymbol{x}^+$ has length $\sqrt{2}$. (Both solutions $\widehat{\boldsymbol{x}} = (2, 0)$ and $(0, 2)$ have length 2.) We are choosing the nullspace component of the solution $\boldsymbol{x}^+$ to be zero.

When A has independent columns, the nullspace only contains the zero vector and the pseudoinverse is our usual left inverse $L = (A^{\mathrm{T}}A)^{-1}A^{\mathrm{T}}$. When I write it that way, the pseudoinverse sounds like the best way to choose $\boldsymbol{x}$. The shortest solution $\boldsymbol{x}^+$ is often called the **minimum norm solution** : its nullspace component is zero.

Comment MATLAB experiments with singular matrices produced either **Inf** or **NaN** (Not a Number) or $\mathbf{10^{16}}$ (a bad number). There is a warning in every case! I believe that **Inf** and **NaN** and $\mathbf{10^{16}}$ come from the possibilities $\mathbf{0}\boldsymbol{x} = \boldsymbol{b}$ and $\mathbf{0}\boldsymbol{x} = \mathbf{0}$ and $\mathbf{10^{-16}}\,\boldsymbol{x} = \mathbf{1}$.

Those are three small examples of three big difficulties: singular with no solution, singular with many solutions, and very very close to singular. Try more experiments...

Fitting by a Parabola

If we throw a ball, it would be crazy to fit the path by a straight line. A parabola $b = C + Dt + Et^2$ allows the ball to go up and come down again (b is the height at time t). The actual path is not a perfect parabola, but the whole theory of projectiles starts there.

When Galileo dropped a stone from the Leaning Tower of Pisa, it accelerated. The distance contains a quadratic term $\frac{1}{2}gt^2$. (Galileo's point was that the stone's mass is not involved.) Without that t^2 term we could never send a satellite into its orbit. But even with a nonlinear function like t^2, the unknowns C, D, E still appear linearly! Fitting points by the best parabola is still a problem in linear algebra.

Problem Fit heights $b_1, \ldots, b_m$ at times $t_1, \ldots, t_m$ by a parabola $C + Dt + Et^2$.

Solution With $m > 3$ points, the m equations for an exact fit are generally unsolvable:

$$\begin{aligned} C + Dt_1 + Et_1^2 &= b_1 \\ &\vdots \\ C + Dt_m + Et_m^2 &= b_m \end{aligned} \qquad \text{is } A\boldsymbol{x} = \boldsymbol{b} \text{ with this } m \text{ by 3 matrix} \qquad A = \begin{bmatrix} 1 & t_1 & t_1^2 \\ \vdots & \vdots & \vdots \\ 1 & t_m & t_m^2 \end{bmatrix}. \tag{10}$$

Least squares The closest parabola $C + Dt + Et^2$ chooses $\widehat{\boldsymbol{x}} = (C, D, E)$ to solve the three normal equations $A^{\mathrm{T}}A\widehat{\boldsymbol{x}} = A^{\mathrm{T}}\boldsymbol{b}$.

May I ask you to convert this to a problem of projection? The column space of A has dimension ____. The projection of $\boldsymbol{b}$ is $\boldsymbol{p} = A\widehat{\boldsymbol{x}}$, which combines the three columns using the coefficients C, D, E. The error at the first data point is $e_1 = b_1 - C - Dt_1 - Et_1^2$. The total squared error is $e_1^2 +$ ____. If you prefer to minimize by calculus, take the partial derivatives of E with respect to ____, ____, ____. These three derivatives will be zero when $\widehat{\boldsymbol{x}} = (C, D, E)$ solves the 3 by 3 system of equations $A^{\mathrm{T}}A\widehat{\boldsymbol{x}} = A^{\mathrm{T}}\boldsymbol{b}$.

Example 3 For a parabola $b = C + Dt + Et^2$ to go through the three heights $b = 6, 0, 0$ when $t = 0, 1, 2$, the equations for C, D, E are

$$\begin{aligned} C + D \cdot 0 + E \cdot 0^2 &= 6 \\ C + D \cdot 1 + E \cdot 1^2 &= 0 \\ C + D \cdot 2 + E \cdot 2^2 &= 0. \end{aligned} \tag{11}$$

This is $A\boldsymbol{x} = \boldsymbol{b}$. We can solve it exactly. Three data points give three equations and a square matrix. The solution is $\boldsymbol{x} = (C, D, E) = (\mathbf{6}, \mathbf{-9}, \mathbf{3})$. The parabola through the three points is $b = \mathbf{6 - 9}\boldsymbol{t} + \mathbf{3}\boldsymbol{t}^{\mathbf{2}}$.

What does this mean for projection? The matrix has three columns, which span the whole space $\mathbf{R}^3$. The projection matrix is the identity. The projection of $\boldsymbol{b}$ is $\boldsymbol{b}$. The error is zero. We didn't need $A^{\mathrm{T}}A\widehat{\boldsymbol{x}} = A^{\mathrm{T}}\boldsymbol{b}$, because we solved $A\boldsymbol{x} = \boldsymbol{b}$.

If there are $m = 4$ data points, then we need $A^{\mathrm{T}}A$ and least squares.

Three Ways to Measure Error

Start with nine measurements b_1 to b_9, *all zero*, at times $t = 1, \ldots, 9$. The tenth measurement $b_{10} = 40$ is an outlier. Find the **best horizontal line $y = C$** to fit the ten points $(1,0), (2,0), \ldots, (9,0), (10,40)$ using three options for the error E:

(1) Least squares $E_2 = e_1^2 + \cdots + e_{10}^2$ (then the normal equation for C is linear)

(2) Least maximum error $E_\infty = |e_{\max}|$ **(3) Least sum of errors $E_1 = |e_1| + \cdots + |e_{10}|$.**

Solution (1) The least squares fit to $0, 0, \ldots, 0, 40$ by a horizontal line is $C = 4$:

$$A = \text{column of 1's} \quad A^{\mathrm{T}}A = 10 \quad A^{\mathrm{T}}\boldsymbol{b} = \text{sum of } b_i = 40. \quad \text{So } 10\,C = 40.$$

(2) The least maximum error requires $\boldsymbol{C = 20}$, halfway between 0 and 40.

(3) The least sum requires $\boldsymbol{C = 0}$ (!!). The sum of errors $9|C| + |40 - C|$ would increase if C moves up from zero.

The least sum comes from the *median* measurement (the median of $0, \ldots, 0, 40$ is zero). Many statisticians feel that the least squares solution is too heavily influenced by outliers like $b_{10} = 40$, and they prefer least sum. But the equations become *nonlinear*.

Now find the least squares line $C + Dt$ through those ten points $(1,0)$ to $(10,40)$:

$$A^{\mathrm{T}}A = \begin{bmatrix} 10 & \sum t_i \\ \sum t_i & \sum t_i^2 \end{bmatrix} = \begin{bmatrix} 10 & 55 \\ 55 & 385 \end{bmatrix} \qquad A^{\mathrm{T}}\boldsymbol{b} = \begin{bmatrix} \sum b_i \\ \sum t_i b_i \end{bmatrix} = \begin{bmatrix} 40 \\ 400 \end{bmatrix}$$

Those come from equation (8). Then $A^{\mathrm{T}}A\widehat{\boldsymbol{x}} = A^{\mathrm{T}}\boldsymbol{b}$ gives $C = -8$ and $D = 24/11$.

Problem Set 4.3

Problems 1–11 use four data points $\boldsymbol{b} = (0, 8, 8, 20)$ to bring out the key ideas.

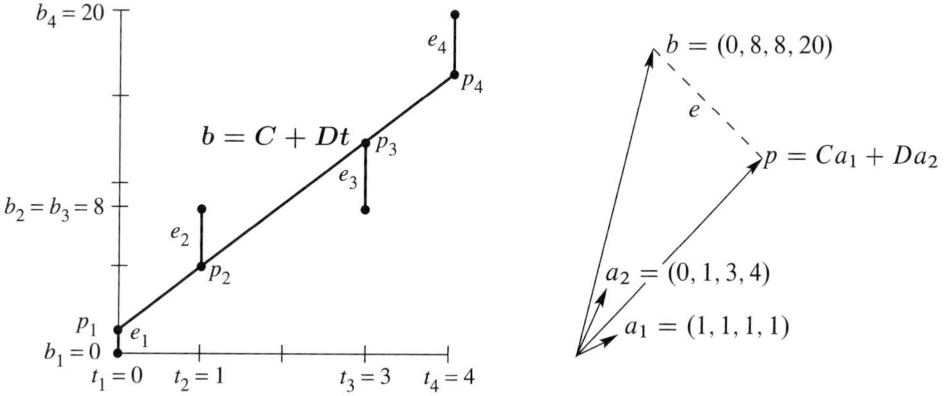

Figure 4.8: **Problems 1–11**: The closest line $C + Dt$ matches $C\boldsymbol{a}_1 + D\boldsymbol{a}_2$ in $\mathbf{R}^4$.

1 With $b = 0, 8, 8, 20$ at $t = 0, 1, 3, 4$, set up and solve the normal equations $A^{\mathrm{T}}A\widehat{\boldsymbol{x}} = A^{\mathrm{T}}\boldsymbol{b}$. For the best straight line in Figure 4.8a, find its four heights p_i and four errors e_i. What is the minimum value $E = e_1^2 + e_2^2 + e_3^2 + e_4^2$?

2 (Line $C + Dt$ does go through p's) With $b = 0, 8, 8, 20$ at times $t = 0, 1, 3, 4$, write down the four equations $A\boldsymbol{x} = \boldsymbol{b}$ (unsolvable). Change the measurements to $p = 1, 5, 13, 17$ and find an exact solution to $A\widehat{\boldsymbol{x}} = \boldsymbol{p}$.

3 Check that $\boldsymbol{e} = \boldsymbol{b} - \boldsymbol{p} = (-1, 3, -5, 3)$ is perpendicular to both columns of the same matrix A. What is the shortest distance $\|e\|$ from $\boldsymbol{b}$ to the column space of A?

4 (By calculus) Write down $E = \|A\boldsymbol{x} - \boldsymbol{b}\|^2$ as a sum of four squares—the last one is $(C + 4D - 20)^2$. Find the derivative equations $\partial E/\partial C = 0$ and $\partial E/\partial D = 0$. Divide by 2 to obtain the normal equations $A^{\mathrm{T}}A\widehat{\boldsymbol{x}} = A^{\mathrm{T}}\boldsymbol{b}$.

5 Find the height C of the best *horizontal line* to fit $\boldsymbol{b} = (0, 8, 8, 20)$. An exact fit would solve the unsolvable equations $C = 0$, $C = 8$, $C = 8$, $C = 20$. Find the 4 by 1 matrix A in these equations and solve $A^{\mathrm{T}}A\widehat{x} = A^{\mathrm{T}}\boldsymbol{b}$. Draw the horizontal line at height $\widehat{x} = C$ and the four errors in $\boldsymbol{e}$.

6 Project $\boldsymbol{b} = (0, 8, 8, 20)$ onto the line through $\boldsymbol{a} = (1, 1, 1, 1)$. Find $\widehat{x} = \boldsymbol{a}^{\mathrm{T}}\boldsymbol{b}/\boldsymbol{a}^{\mathrm{T}}\boldsymbol{a}$ and the projection $\boldsymbol{p} = \widehat{x}\boldsymbol{a}$. Check that $\boldsymbol{e} = \boldsymbol{b} - \boldsymbol{p}$ is perpendicular to $\boldsymbol{a}$, and find the shortest distance $\|\boldsymbol{e}\|$ from $\boldsymbol{b}$ to the line through $\boldsymbol{a}$.

7 Find the closest line $b = Dt$, *through the origin*, to the same four points. An exact fit would solve $D \cdot 0 = 0$, $D \cdot 1 = 8$, $D \cdot 3 = 8$, $D \cdot 4 = 20$. Find the 4 by 1 matrix and solve $A^{\mathrm{T}}A\widehat{x} = A^{\mathrm{T}}\boldsymbol{b}$. Redraw Figure 4.8a showing the best line $b = Dt$ and the e's.

8 Project $\boldsymbol{b} = (0, 8, 8, 20)$ onto the line through $\boldsymbol{a} = (0, 1, 3, 4)$. Find $\widehat{x} = D$ and $\boldsymbol{p} = \widehat{x}\boldsymbol{a}$. The best C in Problems 5–6 and the best D in Problems 7–8 do *not* agree with the best (C, D) in Problems 1–4. That is because $(1, 1, 1, 1)$ and $(0, 1, 3, 4)$ are _____ perpendicular.

9 For the closest parabola $b = C + Dt + Et^2$ to the same four points, write down the unsolvable equations $A\boldsymbol{x} = \boldsymbol{b}$ in three unknowns $\boldsymbol{x} = (C, D, E)$. Set up the three normal equations $A^{\mathrm{T}}A\widehat{\boldsymbol{x}} = A^{\mathrm{T}}\boldsymbol{b}$ (solution not required). In Figure 4.8a you are now fitting a parabola to 4 points—what is happening in Figure 4.8b?

10 For the closest cubic $b = C + Dt + Et^2 + Ft^3$ to the same four points, write down the four equations $A\boldsymbol{x} = \boldsymbol{b}$. Solve them by elimination. In Figure 4.8a this cubic now goes exactly through the points. What are $\boldsymbol{p}$ and $\boldsymbol{e}$?

11 The average of the four times is $\widehat{t} = \frac{1}{4}(0 + 1 + 3 + 4) = 2$. The average of the four b's is $\widehat{b} = \frac{1}{4}(0 + 8 + 8 + 20) = 9$.

(a) Verify that the best line goes through the center point $(\widehat{t}, \widehat{b}) = (2, 9)$.

(b) Explain why $C + D\widehat{t} = \widehat{b}$ comes from the first equation in $A^{\mathrm{T}}A\widehat{\boldsymbol{x}} = A^{\mathrm{T}}\boldsymbol{b}$.

Questions 12–16 introduce basic ideas of statistics—the foundation for least squares.

12 (Recommended) This problem projects $\boldsymbol{b} = (b_1, \ldots, b_m)$ onto the line through $\boldsymbol{a} = (1, \ldots, 1)$. We solve m equations $\boldsymbol{a}x = \boldsymbol{b}$ in one unknown x (by least squares).

(a) Solve $\boldsymbol{a}^{\mathrm{T}}\boldsymbol{a}\widehat{x} = \boldsymbol{a}^{\mathrm{T}}\boldsymbol{b}$ to show that $\widehat{x}$ is the *mean* (the average) of the $\boldsymbol{b}$'s.

(b) Find $\boldsymbol{e} = \boldsymbol{b} - \boldsymbol{a}\widehat{x}$ and the *variance* $\|\boldsymbol{e}\|^2$ and the *standard deviation* $\|\boldsymbol{e}\|$.

(c) The horizontal line $\widehat{b} = 3$ is closest to $\boldsymbol{b} = (1, 2, 6)$. Check that $\boldsymbol{p} = (3, 3, 3)$ is perpendicular to $\boldsymbol{e}$ and find the 3 by 3 projection matrix P.

13 First assumption behind least squares: $A\boldsymbol{x} = \boldsymbol{b}$ – ***(noise e with mean zero)***. Multiply the error vector $\boldsymbol{e} = \boldsymbol{b} - A\boldsymbol{x}$ by $(A^{\mathrm{T}}A)^{-1}A^{\mathrm{T}}$ to get $\widehat{\boldsymbol{x}} - \boldsymbol{x}$ on the right. The estimation errors $\widehat{\boldsymbol{x}} - \boldsymbol{x}$ also average to zero. The estimate $\widehat{\boldsymbol{x}}$ is *unbiased.*

14 Second assumption behind least squares: The m errors $\boldsymbol{e}_i$ are independent with variance σ^2, so the average of $(\boldsymbol{b} - A\boldsymbol{x})(\boldsymbol{b} - A\boldsymbol{x})^{\mathrm{T}}$ is $\sigma^2 I$. Multiply on the left by $(A^{\mathrm{T}}A)^{-1}A^{\mathrm{T}}$ and on the right by $A(A^{\mathrm{T}}A)^{-1}$ to show that the average matrix $(\widehat{\boldsymbol{x}} - \boldsymbol{x})(\widehat{\boldsymbol{x}} - \boldsymbol{x})^{\mathrm{T}}$ is $\sigma^2(A^{\mathrm{T}}A)^{-1}$. ***This is the covariance matrix in Section 8.4.***

15 A doctor takes 4 readings of your heart rate. The best solution to $x = b_1, \ldots, x = b_4$ is the average $\widehat{x}$ of $b_1, \ldots, b_4$. The matrix A is a column of 1's. Problem 14 gives the expected error $(\widehat{x} - x)^2$ as $\sigma^2(A^{\mathrm{T}}A)^{-1} =$ _____. ***By averaging, the variance drops from σ^2 to $\sigma^2/4$.***

16 If you know the average $\widehat{x}_9$ of 9 numbers $b_1, \ldots, b_9$, how can you quickly find the average $\widehat{x}_{10}$ with one more number b_{10}? The idea of *recursive* least squares is to avoid adding 10 numbers. What number multiplies $\widehat{x}_9$ in computing $\widehat{x}_{10}$?

$$\widehat{x}_{10} = \tfrac{1}{10}b_{10} + \underline{\qquad}\,\widehat{x}_9 = \tfrac{1}{10}(b_1 + \cdots + b_{10})$$

Questions 17–24 give more practice with $\widehat{x}$ and p and e.

17 Write down three equations for the line $b = C + Dt$ to go through $b = 7$ at $t = -1$, $b = 7$ at $t = 1$, and $b = 21$ at $t = 2$. Find the least squares solution $\widehat{\boldsymbol{x}} = (C, D)$ and draw the closest line.

18 Find the projection $\boldsymbol{p} = A\widehat{\boldsymbol{x}}$ in Problem 17. This gives the three heights of the closest line. Show that the error vector is $\boldsymbol{e} = (2, -6, 4)$. Why is $P\boldsymbol{e} = \boldsymbol{0}$?

19 Suppose the measurements at $t = -1, 1, 2$ are the errors $2, -6, 4$ in Problem 18. Compute $\widehat{\boldsymbol{x}}$ and the closest line to these new measurements. Explain the answer: $\boldsymbol{b} = (2, -6, 4)$ is perpendicular to _____ so the projection is $\boldsymbol{p} = \boldsymbol{0}$.

20 Suppose the measurements at $t = -1, 1, 2$ are $\boldsymbol{b} = (5, 13, 17)$. Compute $\widehat{\boldsymbol{x}}$ and the closest line and $\boldsymbol{e}$. The error is $\boldsymbol{e} = \boldsymbol{0}$ because this $\boldsymbol{b}$ is _____.

21 Which of the four subspaces contains the error vector $\boldsymbol{e}$? Which contains $\boldsymbol{p}$? Which contains $\widehat{\boldsymbol{x}}$? What is the nullspace of A?

22 Find the best line $C + Dt$ to fit $b = 4, 2, -1, 0, 0$ at times $t = -2, -1, 0, 1, 2$.

23 Is the error vector $\boldsymbol{e}$ orthogonal to $\boldsymbol{b}$ or $\boldsymbol{p}$ or $\boldsymbol{e}$ or $\widehat{\boldsymbol{x}}$? Show that $\|\boldsymbol{e}\|^2$ equals $\boldsymbol{e}^{\mathrm{T}}\boldsymbol{b}$ which equals $\boldsymbol{b}^{\mathrm{T}}\boldsymbol{b} - \boldsymbol{p}^{\mathrm{T}}\boldsymbol{b}$. This is the smallest total error E.

24 The partial derivatives of $\|A\boldsymbol{x}\|^2$ with respect to $x_1, \ldots, x_n$ fill the vector $2A^{\mathrm{T}}A\boldsymbol{x}$. The derivatives of $2\boldsymbol{b}^{\mathrm{T}}A\boldsymbol{x}$ fill the vector $2A^{\mathrm{T}}\boldsymbol{b}$. So the derivatives of $\|A\boldsymbol{x} - \boldsymbol{b}\|^2$ are zero when ______ .

Challenge Problems

25 *What condition on* $(t_1, b_1), (t_2, b_2), (t_3, b_3)$ *puts those three points onto a straight line?* A column space answer is: (b_1, b_2, b_3) must be a combination of $(1,\ 1,\ 1)$ and $(t_1,\ t_2,\ t_3)$. Try to reach a specific equation connecting the t's and b's. I should have thought of this question sooner!

26 Find the *plane* that gives the best fit to the 4 values $\boldsymbol{b} = (0, 1, 3, 4)$ at the corners $(1, 0)$ and $(0, 1)$ and $(-1, 0)$ and $(0, -1)$ of a square. The equations $C + Dx + Ey = b$ at those 4 points are $A\boldsymbol{x} = \boldsymbol{b}$ with 3 unknowns $\boldsymbol{x} = (C, D, E)$. What is A? At the center $(0, 0)$ of the square, show that $C + Dx + Ey =$ average of the b's.

27 (Distance between lines) The points $P = (x, x, x)$ and $Q = (y, 3y, -1)$ are on two lines in space that don't meet. Choose x and y to minimize the squared distance $\|P - Q\|^2$. The line connecting the closest P and Q is perpendicular to ______ .

28 Suppose the columns of A are not independent. How could you find a matrix B so that $P = B(B^{\mathrm{T}}B)^{-1}B^{\mathrm{T}}$ does give the projection onto the column space of A? (The usual formula will fail when $A^{\mathrm{T}}A$ is not invertible.)

29 Usually there will be exactly one hyperplane in $\mathbf{R}^n$ that contains the n given points $\boldsymbol{x} = \mathbf{0}, \boldsymbol{a}_1, \ldots, \boldsymbol{a}_{n-1}$. (Example for $n = 3$: There will be one plane containing $0, \boldsymbol{a}_1, \boldsymbol{a}_2$ unless ______.) What is the test to have exactly one plane in $\mathbf{R}^n$?

30 Example 2 shifted the times t_i to make them add to zero. We subtracted away the average time $\widehat{t} = (t_1 + \cdots + t_m)/m$ to get $T_i = t_i - \widehat{t}$. Those T_i add to zero.

With the columns $(1, \ldots, 1)$ and $(T_1, \ldots, T_m)$ now orthogonal, $A^{\mathrm{T}}A$ is diagonal. Its entries are m and $T_1^2 + \cdots + T_m^2$. Show that the best C and D have direct formulas:

$$\boldsymbol{T} \textbf{ is } \boldsymbol{t} - \widehat{\boldsymbol{t}} \qquad C = \frac{b_1 + \cdots + b_m}{m} \qquad \text{and} \qquad D = \frac{b_1T_1 + \cdots + b_mT_m}{T_1^2 + \cdots + T_m^2}.$$

The best line is $C + DT$ ***or*** $C + D(t - \widehat{t})$. The time shift that makes $A^{\mathrm{T}}A$ diagonal is an example of the Gram-Schmidt process: *orthogonalize the columns of* A *in advance. This is in Section* 4.4.

4.4 Orthogonal Matrices and Gram-Schmidt

1 The columns $\boldsymbol{q_1}, \ldots, \boldsymbol{q_n}$ are orthonormal if $\boldsymbol{q}_i^{\mathrm{T}}\boldsymbol{q}_j = \left\{ \begin{array}{c} 0 \text{ for } i \neq j \\ 1 \text{ for } i = j \end{array} \right\}$. Then $\boxed{Q^{\mathrm{T}}Q = I\,.}$

2 The least squares solution to $Q\boldsymbol{x} = \boldsymbol{b}$ is $\widehat{\boldsymbol{x}} = Q^{\mathrm{T}}\boldsymbol{b}$. Projection of $\boldsymbol{b}$ is $\boldsymbol{p} = QQ^{\mathrm{T}}\boldsymbol{b} = P\boldsymbol{b}$.

3 **Gram-Schmidt** produces orthonormal $\boldsymbol{q}_i$ from independent $\boldsymbol{a}_i$. Start with $\boldsymbol{q}_1 = \boldsymbol{a}_1 / \, ||\boldsymbol{a}_1||$.

4 $\boldsymbol{q}_i = (\boldsymbol{a}_i -$ projection $\boldsymbol{p}_i) \,/\, ||\boldsymbol{a}_i - \boldsymbol{p}_i||$; subtract $\boldsymbol{p}_i = (\boldsymbol{a}_i^{\mathrm{T}}\boldsymbol{q}_1)\boldsymbol{q}_1 + \cdots + (\boldsymbol{a}_i^{\mathrm{T}}\boldsymbol{q}_{i-1})\boldsymbol{q}_{i-1}$.

5 $\boldsymbol{a}_i$ will be a combination of $\boldsymbol{q}_1$ to $\boldsymbol{q}_i$. Then $\boldsymbol{A} = \boldsymbol{QR}$: **orthogonal Q and triangular R.**

This section has two goals, **why** and **how**. The first is to see why orthogonality is good. If Q has orthonormal columns, then $Q^{\mathrm{T}}Q = I$. *Least squares becomes easy.* The second goal is to convert independent vectors in A to orthonormal vectors in Q. You will see how Gram-Schmidt combines the columns of A to produce right angles between columns of Q.

From Chapter 3, a basis consists of independent vectors that span the space. The basis vectors could meet at any angle (except $0°$ and $180°$). But every time we visualize axes, they are perpendicular. *In our imagination, the coordinate axes are practically always orthogonal.* This simplifies the picture and it greatly simplifies the computations.

The vectors $\boldsymbol{q}_1, \ldots, \boldsymbol{q}_n$ are ***orthogonal*** when their dot products $\boldsymbol{q}_i \cdot \boldsymbol{q}_j$ are zero. More exactly $\boldsymbol{q}_i^{\mathrm{T}}\boldsymbol{q}_j = 0$ whenever $i \neq j$. With one more step—just *divide each vector by its length*—the vectors become ***orthogonal unit vectors***. Their lengths are all 1 (normal). Then the basis is ***orthonormal***.

DEFINITION The vectors $\boldsymbol{q}_1, \ldots, \boldsymbol{q}_n$ are ***orthonormal*** if their dot products are 0 or 1 :

$$\boldsymbol{q}_i^{\mathrm{T}}\boldsymbol{q}_j = \begin{cases} 0 & \text{when } i \neq j \quad (\textbf{\textit{orthogonal}} \text{ vectors}) \\ 1 & \text{when } i = j \quad (\textbf{\textit{unit}} \text{ vectors: } ||\boldsymbol{q}_i|| = 1) \end{cases}$$

A matrix with orthonormal columns is assigned the special letter Q.

The matrix Q is easy to work with because $Q^{\mathrm{T}}Q = I$. This repeats in matrix language that the columns $\boldsymbol{q}_1, \ldots, \boldsymbol{q}_n$ are orthonormal. Q is not always required to be square.

A matrix Q with orthonormal columns satisfies $Q^{\mathrm{T}}Q = I$:

$$\boldsymbol{Q}^{\mathrm{T}}\boldsymbol{Q} = \begin{bmatrix} -\, \boldsymbol{q}_1^{\mathrm{T}} \,- \\ -\, \boldsymbol{q}_n^{\mathrm{T}} \,- \end{bmatrix} \begin{bmatrix} | & | \\ \boldsymbol{q}_1 & \boldsymbol{q}_n \\ | & | \end{bmatrix} = \begin{bmatrix} 1 & \cdot & 0 \\ \cdot & \cdot & \cdot \\ 0 & \cdot & 1 \end{bmatrix} = \boldsymbol{I}$$

When row i of Q^{T} multiplies column j of Q, the dot product is $\boldsymbol{q}_i^{\mathrm{T}}\boldsymbol{q}_j$. Off the diagonal $(i \neq j)$ that dot product is zero by orthogonality. On the diagonal $(i = j)$ the unit vectors give $\boldsymbol{q}_i^{\mathrm{T}}\boldsymbol{q}_i = \|\boldsymbol{q}_i\|^2 = 1$. Q can be rectangular $(m > n)$ or square $(m = n)$.

When Q is square, $Q^{\mathrm{T}}Q = I$ means that $Q^{\mathrm{T}} = Q^{-1}$: transpose = inverse.

When Q is not square, its rank is $n < m$. So QQ^{T} cannot equal $I_{m \times m}$.

If the columns are only orthogonal (not unit vectors), dot products still give a diagonal matrix $Q^{\mathrm{T}}Q$ (but not the identity matrix). This diagonal matrix is almost as good as I. The important thing is orthogonality—then it is easy to produce unit vectors.

To repeat: $Q^{\mathrm{T}}Q = I$ even when Q is rectangular. In that case Q^{T} is only an inverse from the left. For square matrices we also have $QQ^{\mathrm{T}} = I$, so Q^{T} is the two-sided inverse of Q. The rows of a square Q are orthonormal like the columns. ***The inverse is the transpose***. In this square case we call Q an ***orthogonal matrix***.[1]

Here are three examples of orthogonal matrices—rotation and permutation and reflection. The quickest test is $Q^{\mathrm{T}}Q = I$.

Example 1 **(Rotation)** Q rotates every vector in the plane by the angle θ and Q^{-1} by $-\theta$:

$$Q = \begin{bmatrix} \cos\theta & -\sin\theta \\ \sin\theta & \cos\theta \end{bmatrix} \quad \text{and} \quad Q^{\mathrm{T}} = Q^{-1} = \begin{bmatrix} \cos\theta & \sin\theta \\ -\sin\theta & \cos\theta \end{bmatrix}.$$

The columns of Q are orthogonal (take their dot product). They are unit vectors because $\sin^2\theta + \cos^2\theta = 1$. Those columns give an ***orthonormal basis*** for the plane $\mathbf{R}^2$.

Example 2 **(Permutation)** These matrices change the order to (y, z, x) and (y, x):

$$\begin{bmatrix} 0 & 1 & 0 \\ 0 & 0 & 1 \\ 1 & 0 & 0 \end{bmatrix} \begin{bmatrix} x \\ y \\ z \end{bmatrix} = \begin{bmatrix} y \\ z \\ x \end{bmatrix} \quad \text{and} \quad \begin{bmatrix} 0 & 1 \\ 1 & 0 \end{bmatrix} \begin{bmatrix} x \\ y \end{bmatrix} = \begin{bmatrix} y \\ x \end{bmatrix}.$$

All columns of these Q's are unit vectors (their lengths are obviously 1). They are also orthogonal (the 1's appear in different places). *The inverse of a permutation matrix is its transpose*: $Q^{-1} = Q^{\mathrm{T}}$. The inverse puts the components back into their original order:

$$\textbf{Inverse = transpose:} \quad \begin{bmatrix} 0 & 0 & 1 \\ 1 & 0 & 0 \\ 0 & 1 & 0 \end{bmatrix} \begin{bmatrix} \boldsymbol{y} \\ \boldsymbol{z} \\ \boldsymbol{x} \end{bmatrix} = \begin{bmatrix} \boldsymbol{x} \\ \boldsymbol{y} \\ \boldsymbol{z} \end{bmatrix} \quad \text{and} \quad \begin{bmatrix} 0 & 1 \\ 1 & 0 \end{bmatrix} \begin{bmatrix} \boldsymbol{y} \\ \boldsymbol{x} \end{bmatrix} = \begin{bmatrix} \boldsymbol{x} \\ \boldsymbol{y} \end{bmatrix}.$$

Every permutation matrix is an orthogonal matrix.

Example 3 **(Reflection)** If $\boldsymbol{u}$ is any unit vector, set $Q = I - 2\boldsymbol{u}\boldsymbol{u}^{\mathrm{T}}$. Notice that $\boldsymbol{u}\boldsymbol{u}^{\mathrm{T}}$ is a matrix while $\boldsymbol{u}^{\mathrm{T}}\boldsymbol{u}$ is the number $\|\boldsymbol{u}\|^2 = 1$. Then Q^{T} and Q^{-1} both equal Q:

$$Q^{\mathrm{T}} = I - 2\boldsymbol{u}\boldsymbol{u}^{\mathrm{T}} = Q \quad \text{and} \quad Q^{\mathrm{T}}Q = I - 4\boldsymbol{u}\boldsymbol{u}^{\mathrm{T}} + 4\boldsymbol{u}\boldsymbol{u}^{\mathrm{T}}\boldsymbol{u}\boldsymbol{u}^{\mathrm{T}} = I. \qquad (1)$$

[1] "Orthonormal matrix" would have been a better name for Q, but it's not used. Any matrix with orthonormal columns has the letter Q. We only call it an **orthogonal matrix when it is square.**

Reflection matrices $I - 2\boldsymbol{u}\boldsymbol{u}^{\mathrm{T}}$ are symmetric and also orthogonal. If you square them, you get the identity matrix: $Q^2 = Q^{\mathrm{T}}Q = I$. Reflecting twice through a mirror brings back the original, like $(-1)^2 = 1$. Notice $\boldsymbol{u}^{\mathrm{T}}\boldsymbol{u} = 1$ inside $4\boldsymbol{u}\boldsymbol{u}^{\mathrm{T}}\boldsymbol{u}\boldsymbol{u}^{\mathrm{T}}$ in equation (1).

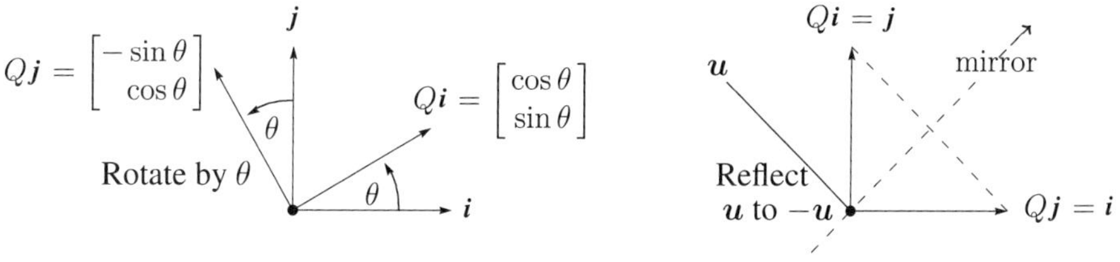

Figure 4.9: Rotation by $Q = \begin{bmatrix} c & -s \\ s & c \end{bmatrix}$ and reflection across 45° by $Q = \begin{bmatrix} 0 & 1 \\ 1 & 0 \end{bmatrix}$.

As an example choose the direction $\boldsymbol{u} = (-1/\sqrt{2}, 1/\sqrt{2})$. Compute $2\boldsymbol{u}\boldsymbol{u}^{\mathrm{T}}$ (column times row) and subtract from I to get the reflection matrix Q across the 45° line.

$$\textbf{Reflection} \quad Q = I - 2\begin{bmatrix} .5 & -.5 \\ -.5 & .5 \end{bmatrix} = \begin{bmatrix} \mathbf{0} & \mathbf{1} \\ \mathbf{1} & \mathbf{0} \end{bmatrix} \quad \text{and} \quad \begin{bmatrix} 0 & 1 \\ 1 & 0 \end{bmatrix}\begin{bmatrix} x \\ y \end{bmatrix} = \begin{bmatrix} \boldsymbol{y} \\ \boldsymbol{x} \end{bmatrix}.$$

When (x, y) goes to (y, x), a vector like $(3, 3)$ doesn't move. It is on the mirror line.

Rotations preserve the length of every vector. So do reflections. So do permutations.

If Q has orthonormal columns ($Q^{\mathrm{T}}Q = I$), it leaves lengths unchanged: $||Q\boldsymbol{x}|| = ||\boldsymbol{x}||$

Same length for $Q\boldsymbol{x}$ $\quad (Q\boldsymbol{x})^{\mathrm{T}}(Q\boldsymbol{x}) = \boldsymbol{x}^{\mathrm{T}}Q^{\mathrm{T}}Q\boldsymbol{x} = \boldsymbol{x}^{\mathrm{T}}I\boldsymbol{x} = \boldsymbol{x}^{\mathrm{T}}\boldsymbol{x}$ (2)

Same dot product: $\quad (Q\boldsymbol{x})^{\mathrm{T}}(Q\boldsymbol{y}) = \boldsymbol{x}^{\mathrm{T}}Q^{\mathrm{T}}Q\boldsymbol{y} = \boldsymbol{x}^{\mathrm{T}}\boldsymbol{y}$. Just use $Q^{\mathrm{T}}Q = I$

Projections Using Orthonormal Bases: Q Replaces A

Orthogonal matrices are excellent for computations—numbers can never grow too large when lengths of vectors are fixed. Stable computer codes use Q's as much as possible.

For projections onto subspaces, all formulas involve $A^{\mathrm{T}}A$. The entries of $A^{\mathrm{T}}A$ are the dot products between the columns of A. Usually $A^{\mathrm{T}}A$ is not diagonal.

Suppose those columns are actually orthonormal. The $\boldsymbol{a}$'s become the $\boldsymbol{q}$'s. Then $A^{\mathrm{T}}A$ *simplifies to* $Q^{\mathrm{T}}Q = I$. Look at the improvements in $\widehat{\boldsymbol{x}}$ and $\boldsymbol{p}$ and the projection matrix $P = QQ^{\mathrm{T}}$. Instead of $Q^{\mathrm{T}}Q$ we print a blank for the identity matrix:

$$\underline{\qquad}\,\widehat{\boldsymbol{x}} = Q^{\mathrm{T}}\boldsymbol{b} \quad \text{and} \quad \boldsymbol{p} = Q\widehat{\boldsymbol{x}} \quad \text{and} \quad P = Q\,\underline{\qquad}\,Q^{\mathrm{T}}. \tag{3}$$

The least squares solution of $Q\boldsymbol{x} = \boldsymbol{b}$ ***is*** $\widehat{\boldsymbol{x}} = Q^{\mathrm{T}}\boldsymbol{b}$. There are no matrices to invert. This is the point of an orthonormal basis. The best $\widehat{\boldsymbol{x}} = Q^{\mathrm{T}}\boldsymbol{b}$ just has dot products of $\boldsymbol{q}_1, \ldots, \boldsymbol{q}_n$ with $\boldsymbol{b}$. We have 1-dimensional projections! $A^{\mathrm{T}}A$ is now $Q^{\mathrm{T}}Q = I$. There is no coupling. When A is Q, with orthonormal columns, here is $\boldsymbol{p} = Q\widehat{\boldsymbol{x}} = QQ^{\mathrm{T}}\boldsymbol{b}$:

$$\textbf{Projection onto } q\text{'s} \qquad p = \begin{bmatrix} | & & | \\ q_1 & \cdots & q_n \\ | & & | \end{bmatrix} \begin{bmatrix} q_1^{\mathrm{T}} b \\ \vdots \\ q_n^{\mathrm{T}} b \end{bmatrix} = q_1(q_1^{\mathrm{T}} b) + \cdots + q_n(q_n^{\mathrm{T}} b). \tag{4}$$

Important case when Q is square: If $m = n$, the subspace is the whole space. Then $Q^{\mathrm{T}} = Q^{-1}$ and $\widehat{x} = Q^{\mathrm{T}} b$ is the same as $x = Q^{-1} b$. The solution is exact! The projection of b onto the whole space is b itself. In this case $p = b$ and $P = QQ^{\mathrm{T}} = I$.

You may think that projection onto the whole space is not worth mentioning. But when Q is square and $p = b$, our formula assembles b out of its 1-dimensional projections. If $q_1, \ldots, q_n$ is an orthonormal basis for the whole space, then b is equal to $QQ^{\mathrm{T}} b$. **Every b is the sum of its components along the q's:**

$$b = q_1(q_1^{\mathrm{T}} b) + q_2(q_2^{\mathrm{T}} b) + \cdots + q_n(q_n^{\mathrm{T}} b). \tag{5}$$

Example 4 The columns of this orthogonal Q are orthonormal vectors q_1, q_2, q_3:

$$m = n = 3 \qquad Q = \frac{1}{3}\begin{bmatrix} -1 & 2 & 2 \\ 2 & -1 & 2 \\ 2 & 2 & -1 \end{bmatrix} \quad \text{has} \quad Q^{\mathrm{T}} Q = QQ^{\mathrm{T}} = I.$$

The separate projections of $b = (0, 0, 1)$ onto q_1 and q_2 and q_3 are p_1 and p_2 and p_3:

$$q_1(q_1^{\mathrm{T}} b) = \tfrac{2}{3} q_1 \quad \text{and} \quad q_2(q_2^{\mathrm{T}} b) = \tfrac{2}{3} q_2 \quad \text{and} \quad q_3(q_3^{\mathrm{T}} b) = -\tfrac{1}{3} q_3.$$

The sum of the first two is the projection of b onto the *plane* of q_1 and q_2. The sum of all three is the projection of b onto the *whole space*—which is b itself:

$$\begin{array}{l}\textbf{Reconstruct } b \\ b = p_1 + p_2 + p_3\end{array} \qquad \frac{2}{3} q_1 + \frac{2}{3} q_2 - \frac{1}{3} q_3 = \frac{1}{9}\begin{bmatrix} -2+4-2 \\ 4-2-2 \\ 4+4+1 \end{bmatrix} = \begin{bmatrix} 0 \\ 0 \\ 1 \end{bmatrix} = b.$$

Transforms $QQ^{\mathrm{T}} = I$ is the foundation of Fourier series and all the great "transforms" of applied mathematics. They break vectors b or functions $f(x)$ into perpendicular pieces. Then by adding the pieces in (5), the inverse transform puts b and $f(x)$ back together.

$$\textbf{Fouricr series} \qquad f(x) = a_0 + a_1 \cos x + b_1 \sin x + a_2 \cos 2x + b_2 \sin 2x + \cdots$$

Only two differences. Those are functions. The sine-cosine basis is infinite: $m = n = \infty$.

The Gram-Schmidt Process

The point of this section is that "orthogonal is good". Projections and least squares always involve $A^{\mathrm{T}}A$. When this matrix becomes $Q^{\mathrm{T}}Q = I$, the inverse is no problem. The one-dimensional projections are uncoupled. The best $\widehat{\boldsymbol{x}}$ is $Q^{\mathrm{T}}\boldsymbol{b}$ (just n separate dot products). For this to be true, we had to say "*If* the vectors are orthonormal". ***Now we explain the "Gram-Schmidt way" to create orthonormal vectors.***

Start with three independent vectors $\boldsymbol{a}, \boldsymbol{b}, \boldsymbol{c}$. We intend to construct three orthogonal vectors $\boldsymbol{A}, \boldsymbol{B}, \boldsymbol{C}$. Sooner or later we will divide $\boldsymbol{A}, \boldsymbol{B}, \boldsymbol{C}$ by their lengths. That produces three orthonormal vectors $\boldsymbol{q}_1 = \boldsymbol{A}/\|\boldsymbol{A}\|$, $\boldsymbol{q}_2 = \boldsymbol{B}/\|\boldsymbol{B}\|$, $\boldsymbol{q}_3 = \boldsymbol{C}/\|\boldsymbol{C}\|$.

Gram-Schmidt Begin by choosing $\boldsymbol{A} = \boldsymbol{a}$. This first direction is accepted as it comes. The next direction $\boldsymbol{B}$ must be perpendicular to $\boldsymbol{A}$. ***Start with b and subtract its projection along A.*** This leaves the perpendicular part, which is the orthogonal vector $\boldsymbol{B}$:

First Gram-Schmidt step $$\boldsymbol{B} = \boldsymbol{b} - \frac{\boldsymbol{A}^{\mathrm{T}}\boldsymbol{b}}{\boldsymbol{A}^{\mathrm{T}}\boldsymbol{A}}\boldsymbol{A}. \qquad (6)$$

$\boldsymbol{A}$ and $\boldsymbol{B}$ are orthogonal in Figure 4.10. Multiply equation (6) by $\boldsymbol{A}^{\mathrm{T}}$ to verify that $\boldsymbol{A}^{\mathrm{T}}\boldsymbol{B} = \boldsymbol{A}^{\mathrm{T}}\boldsymbol{b} - \boldsymbol{A}^{\mathrm{T}}\boldsymbol{b} = 0$. This vector $\boldsymbol{B}$ is what we have called the error vector $\boldsymbol{e}$, perpendicular to $\boldsymbol{A}$. Notice that $\boldsymbol{B}$ in equation (6) is not zero (otherwise $\boldsymbol{a}$ and $\boldsymbol{b}$ would be dependent). The directions $\boldsymbol{A}$ and $\boldsymbol{B}$ are now set.

The third direction starts with $\boldsymbol{c}$. This is not a combination of $\boldsymbol{A}$ and $\boldsymbol{B}$ (because $\boldsymbol{c}$ is not a combination of $\boldsymbol{a}$ and $\boldsymbol{b}$). But most likely $\boldsymbol{c}$ is not perpendicular to $\boldsymbol{A}$ and $\boldsymbol{B}$. So subtract off its components in those two directions to get a perpendicular direction $\boldsymbol{C}$:

Next Gram-Schmidt step $$\boldsymbol{C} = \boldsymbol{c} - \frac{\boldsymbol{A}^{\mathrm{T}}\boldsymbol{c}}{\boldsymbol{A}^{\mathrm{T}}\boldsymbol{A}}\boldsymbol{A} - \frac{\boldsymbol{B}^{\mathrm{T}}\boldsymbol{c}}{\boldsymbol{B}^{\mathrm{T}}\boldsymbol{B}}\boldsymbol{B}. \qquad (7)$$

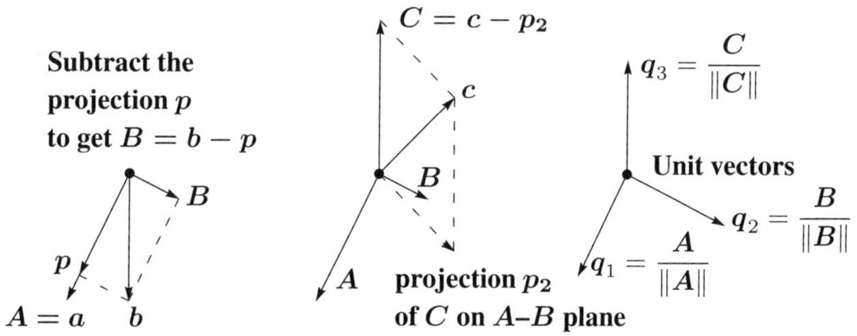

Figure 4.10: First project $\boldsymbol{b}$ onto the line through $\boldsymbol{a}$ and find the orthogonal $\boldsymbol{B}$ as $\boldsymbol{b} - \boldsymbol{p}$. Then project $\boldsymbol{c}$ onto the $\boldsymbol{AB}$ plane and find $\boldsymbol{C}$ as $\boldsymbol{c} - \boldsymbol{p}_2$. Divide by $\|\boldsymbol{A}\|$, $\|\boldsymbol{B}\|$, $\|\boldsymbol{C}\|$.

This is the one and only idea of the Gram-Schmidt process. ***Subtract from every new vector its projections in the directions already set.*** That idea is repeated at every step.

If we had a fourth vector $\boldsymbol{d}$, we would subtract three projections onto $\boldsymbol{A}, \boldsymbol{B}, \boldsymbol{C}$ to get $\boldsymbol{D}$. At the end, *or immediately when each one is found*, divide the orthogonal vectors $\boldsymbol{A}, \boldsymbol{B}, \boldsymbol{C}, \boldsymbol{D}$ by their lengths. The resulting vectors $\boldsymbol{q}_1, \boldsymbol{q}_2, \boldsymbol{q}_3, \boldsymbol{q}_4$ are orthonormal.

Example of **Gram-Schmidt** Suppose the independent non-orthogonal vectors $\boldsymbol{a}, \boldsymbol{b}, \boldsymbol{c}$ are

$$\boldsymbol{a} = \begin{bmatrix} \mathbf{1} \\ \mathbf{-1} \\ \mathbf{0} \end{bmatrix} \quad \text{and} \quad \boldsymbol{b} = \begin{bmatrix} 2 \\ 0 \\ -2 \end{bmatrix} \quad \text{and} \quad \boldsymbol{c} = \begin{bmatrix} 3 \\ -3 \\ 3 \end{bmatrix}.$$

Then $\boldsymbol{A} = \boldsymbol{a}$ has $\boldsymbol{A}^{\mathrm{T}}\boldsymbol{A} = 2$ and $\boldsymbol{A}^{\mathrm{T}}\boldsymbol{b} = 2$. Subtract from $\boldsymbol{b}$ its projection $\boldsymbol{p}$ along $\boldsymbol{A}$:

First step $$\boldsymbol{B} = \boldsymbol{b} - \frac{\boldsymbol{A}^{\mathrm{T}}\boldsymbol{b}}{\boldsymbol{A}^{\mathrm{T}}\boldsymbol{A}}\boldsymbol{A} = \boldsymbol{b} - \frac{2}{2}\boldsymbol{A} = \begin{bmatrix} \mathbf{1} \\ \mathbf{1} \\ \mathbf{-2} \end{bmatrix}.$$

Check: $\boldsymbol{A}^{\mathrm{T}}\boldsymbol{B} = 0$ as required. Now subtract the projections of $\boldsymbol{c}$ on $\boldsymbol{A}$ and $\boldsymbol{B}$ to get $\boldsymbol{C}$:

Next step $$\boldsymbol{C} = \boldsymbol{c} - \frac{\boldsymbol{A}^{\mathrm{T}}\boldsymbol{c}}{\boldsymbol{A}^{\mathrm{T}}\boldsymbol{A}}\boldsymbol{A} - \frac{\boldsymbol{B}^{\mathrm{T}}\boldsymbol{c}}{\boldsymbol{B}^{\mathrm{T}}\boldsymbol{B}}\boldsymbol{B} = \boldsymbol{c} - \frac{6}{2}\boldsymbol{A} + \frac{6}{6}\boldsymbol{B} = \begin{bmatrix} \mathbf{1} \\ \mathbf{1} \\ \mathbf{1} \end{bmatrix}.$$

Check: $\boldsymbol{C} = (1, 1, 1)$ is perpendicular to both $\boldsymbol{A}$ and $\boldsymbol{B}$. Finally convert $\boldsymbol{A}, \boldsymbol{B}, \boldsymbol{C}$ to unit vectors (length 1, orthonormal). The lengths of $\boldsymbol{A}$, $\boldsymbol{B}$, $\boldsymbol{C}$ are $\sqrt{2}$ and $\sqrt{6}$ and $\sqrt{3}$. Divide by those lengths for an orthonormal basis $\boldsymbol{q}_1, \boldsymbol{q}_2, \boldsymbol{q}_3$.

$$\boldsymbol{q}_1 = \frac{1}{\sqrt{2}}\begin{bmatrix} 1 \\ -1 \\ 0 \end{bmatrix} \quad \text{and} \quad \boldsymbol{q}_2 = \frac{1}{\sqrt{6}}\begin{bmatrix} 1 \\ 1 \\ -2 \end{bmatrix} \quad \text{and} \quad \boldsymbol{q}_3 = \frac{1}{\sqrt{3}}\begin{bmatrix} 1 \\ 1 \\ 1 \end{bmatrix}.$$

Usually $\boldsymbol{A}, \boldsymbol{B}, \boldsymbol{C}$ contain fractions. Almost always $\boldsymbol{q}_1, \boldsymbol{q}_2, \boldsymbol{q}_3$ contain square roots.

The Factorization $A = QR$

We started with a matrix A, whose columns were $\boldsymbol{a}, \boldsymbol{b}, \boldsymbol{c}$. We ended with a matrix Q, whose columns are $\boldsymbol{q}_1, \boldsymbol{q}_2, \boldsymbol{q}_3$. How are those matrices related? Since the vectors $\boldsymbol{a}, \boldsymbol{b}, \boldsymbol{c}$ are combinations of the $\boldsymbol{q}$'s (and vice versa), there must be a third matrix connecting A to Q. This third matrix is the triangular R in $A = QR$. (*Not the R in Chapter* 1.)

The first step was $\boldsymbol{q}_1 = \boldsymbol{a}/\|\boldsymbol{a}\|$ (other vectors not involved). The second step was equation (6), where $\boldsymbol{b}$ is a combination of $\boldsymbol{A}$ and $\boldsymbol{B}$. At that stage $\boldsymbol{C}$ and $\boldsymbol{q}_3$ were not involved. This non-involvement of later vectors is the key point of Gram-Schmidt:

- The vectors $\boldsymbol{a}$ and $\boldsymbol{A}$ and $\boldsymbol{q}_1$ are all along a single line.
- The vectors $\boldsymbol{a}, \boldsymbol{b}$ and $\boldsymbol{A}, \boldsymbol{B}$ and $\boldsymbol{q}_1, \boldsymbol{q}_2$ are all in the same plane.
- The vectors $\boldsymbol{a}, \boldsymbol{b}, \boldsymbol{c}$ and $\boldsymbol{A}, \boldsymbol{B}, \boldsymbol{C}$ and $\boldsymbol{q}_1, \boldsymbol{q}_2, \boldsymbol{q}_3$ are in one subspace (dimension 3).

At every step $\boldsymbol{a}_1, \ldots, \boldsymbol{a}_k$ are combinations of $\boldsymbol{q}_1, \ldots, \boldsymbol{q}_k$. Later $\boldsymbol{q}$'s are not involved. The connecting matrix R is ***triangular***, and we have $A = QR$:

$$\begin{bmatrix} & & \\ \boldsymbol{a} & \boldsymbol{b} & \boldsymbol{c} \\ & & \end{bmatrix} = \begin{bmatrix} & & \\ \boldsymbol{q}_1 & \boldsymbol{q}_2 & \boldsymbol{q}_3 \\ & & \end{bmatrix} \begin{bmatrix} \boldsymbol{q}_1^{\mathrm{T}}\boldsymbol{a} & \boldsymbol{q}_1^{\mathrm{T}}\boldsymbol{b} & \boldsymbol{q}_1^{\mathrm{T}}\boldsymbol{c} \\ 0 & \boldsymbol{q}_2^{\mathrm{T}}\boldsymbol{b} & \boldsymbol{q}_2^{\mathrm{T}}\boldsymbol{c} \\ 0 & 0 & \boldsymbol{q}_3^{\mathrm{T}}\boldsymbol{c} \end{bmatrix} \quad \text{or} \quad \boxed{\boldsymbol{A} = \boldsymbol{QR}.} \qquad (8)$$

$A = QR$ is Gram-Schmidt in a nutshell. Multiply by Q^{T} to recognize that $\boldsymbol{R} = \boldsymbol{Q}^{\mathrm{T}}\boldsymbol{A}$.

> **(Gram-Schmidt)** From independent vectors $\boldsymbol{a}_1, \ldots, \boldsymbol{a}_n$, Gram-Schmidt constructs orthonormal vectors $\boldsymbol{q}_1, \ldots, \boldsymbol{q}_n$. The matrices with these columns satisfy $A = QR$. Then $R = Q^{\mathrm{T}}A$ is ***upper triangular*** because later $\boldsymbol{q}$'s are orthogonal to earlier $\boldsymbol{a}$'s.

Here are the original $\boldsymbol{a}$'s and the final $\boldsymbol{q}$'s from the example. The i, j entry of $R = Q^{\mathrm{T}}A$ is row i of Q^{T} times column j of A. The dot products $\boldsymbol{q}_i^{\mathrm{T}}\boldsymbol{a}_j$ go into R. Then $A = QR$:

$$\boldsymbol{A} = \begin{bmatrix} 1 & 2 & 3 \\ -1 & 0 & -3 \\ 0 & -2 & 3 \end{bmatrix} = \begin{bmatrix} 1/\sqrt{2} & 1/\sqrt{6} & 1/\sqrt{3} \\ -1/\sqrt{2} & 1/\sqrt{6} & 1/\sqrt{3} \\ 0 & -2/\sqrt{6} & 1/\sqrt{3} \end{bmatrix} \begin{bmatrix} \boldsymbol{\sqrt{2}} & \sqrt{2} & \sqrt{18} \\ \boldsymbol{0} & \boldsymbol{\sqrt{6}} & -\sqrt{6} \\ \boldsymbol{0} & \boldsymbol{0} & \boldsymbol{\sqrt{3}} \end{bmatrix} = \boldsymbol{QR}.$$

Look closely at Q and R. The lengths of $\boldsymbol{A}, \boldsymbol{B}, \boldsymbol{C}$ are $\sqrt{2}$, $\sqrt{6}$, $\sqrt{3}$ on the diagonal of R. The columns of Q are orthonormal. Because of the square roots, QR might look harder than LU. Both factorizations are absolutely central to calculations in linear algebra.

Any m by n matrix A with independent columns can be factored into $A = QR$. The m by n matrix Q has orthonormal columns, and the square matrix R is upper triangular with positive diagonal. We must not forget why this is useful for least squares: $\boldsymbol{A}^{\mathrm{T}}\boldsymbol{A} = (\boldsymbol{QR})^{\mathrm{T}}\boldsymbol{QR} = \boldsymbol{R}^{\mathrm{T}}\boldsymbol{Q}^{\mathrm{T}}\boldsymbol{QR} = \boldsymbol{R}^{\mathrm{T}}\boldsymbol{R}$. The least squares equation $A^{\mathrm{T}}A\widehat{\boldsymbol{x}} = A^{\mathrm{T}}\boldsymbol{b}$ simplifies to $R^{\mathrm{T}}R\widehat{\boldsymbol{x}} = R^{\mathrm{T}}Q^{\mathrm{T}}\boldsymbol{b}$. Then finally we reach $R\widehat{\boldsymbol{x}} = Q^{\mathrm{T}}\boldsymbol{b}$: success.

$$\boxed{\textbf{Least squares} \qquad R^{\mathrm{T}}R\widehat{\boldsymbol{x}} = R^{\mathrm{T}}Q^{\mathrm{T}}\boldsymbol{b} \;\text{ or }\; R\widehat{\boldsymbol{x}} = Q^{\mathrm{T}}\boldsymbol{b} \;\text{ or }\; \widehat{\boldsymbol{x}} = R^{-1}Q^{\mathrm{T}}\boldsymbol{b}} \qquad (9)$$

Instead of solving $A\boldsymbol{x} = \boldsymbol{b}$, which is impossible, we solve $R\widehat{\boldsymbol{x}} = Q^{\mathrm{T}}\boldsymbol{b}$ by back substitution—which is very fast. The real cost is the mn^2 multiplications needed by Gram-Schmidt.

The next page has an informal code. It projects each new column $\boldsymbol{v} = \boldsymbol{a}_j$ onto the known orthonormal columns $\boldsymbol{q}_1, \ldots, \boldsymbol{q}_{j-1}$. After subtracting those projections from $\boldsymbol{a}_j$, the last line divides the new orthogonalized vector (still called $\boldsymbol{v}$) by its length r_{jj}. This produces the next orthonormal vector $\boldsymbol{q}_j$.

Starting from $\boldsymbol{a}, \boldsymbol{b}, \boldsymbol{c} = \boldsymbol{a}_1, \boldsymbol{a}_2, \boldsymbol{a}_3$ the code will construct $\boldsymbol{q}_1$, $\boldsymbol{B}$, $\boldsymbol{q}_2$, $\boldsymbol{C}^*$, $\boldsymbol{C}$, $\boldsymbol{q}_3$:

$$\boldsymbol{q}_1 = \boldsymbol{a}_1/\|\boldsymbol{a}_1\| \qquad \boldsymbol{B} = \boldsymbol{a}_2 - (\boldsymbol{q}_1^{\mathrm{T}}\boldsymbol{a}_2)\boldsymbol{q}_1 \qquad \boldsymbol{q}_2 = \boldsymbol{B}/\|\boldsymbol{B}\|$$

$$\boldsymbol{C}^* = \boldsymbol{a}_3 - (\boldsymbol{q}_1^{\mathrm{T}}\boldsymbol{a}_3)\boldsymbol{q}_1 \quad \boldsymbol{C} = \boldsymbol{C}^* - (\boldsymbol{q}_2^{\mathrm{T}}\boldsymbol{C}^*)\boldsymbol{q}_2 \quad \boldsymbol{q}_3 = \boldsymbol{C}/\|\boldsymbol{C}\| \tag{10}$$

Notice: We subtract **one projection at a time** in $\boldsymbol{C}^*$ and then $\boldsymbol{C}$. That change is called ***modified Gram-Schmidt***. This code is numerically more stable than equation (7) which subtracts all projections at once.

```
for j = 1:n                       % modified Gram-Schmidt
  v = A(:,j);                     % v begins as column j of the original A
  for i = 1:j-1                   % columns q_1 to q_{j-1} are already settled in Q
    R(i,j) = Q(:,i)'*v;           % compute R_ij = q_i^T a_j which is q_i^T v
    v = v-R(i,j)*Q(:,i);          % subtract the projection (q_i^T v) q_i
  end                             % v is now perpendicular to all of q_1, ..., q_{j-1}
  R(j,j) = norm(v);               % the diagonal entries R_jj are lengths
  Q(:,j) = v/R(j,j);              % divide v by its length to get the next q_j
end                               % the loop "for j = 1 : n" produces all of the q_j
```

To recover column j of A, undo the last step and the middle steps of the code:

$$R(j,j)\boldsymbol{q_j} = (\boldsymbol{v}\ \textbf{\textit{minus its projections}}) = (\text{column } j \text{ of } A) - \sum_{i=1}^{j-1} R(i,j)\boldsymbol{q_i}\,. \tag{11}$$

Moving the sum to the far left, this is column j in the multiplication $QR = A$.

Note Good software like LAPACK, used in good systems like MATLAB and Julia and Python, will offer alternative ways to factor $A = QR$. "Householder reflections" act on A to produce the upper triangular R. This happens one column at a time in the same way that elimination produces the upper triangular U in LU.

Those reflection matrices $I - 2\boldsymbol{u}\boldsymbol{u}^{\mathrm{T}}$ will be described in Section 7.4. If A is tridiagonal we can simplify even more to use 2 by 2 rotations. The result is always $A = QR$ and the MATLAB command to orthogonalize A is $[Q, R] = \mathrm{qr}(A)$.

Here is a further way to reduce roundoff error: Allow reordering of the columns of A. When each new $\boldsymbol{q}_j$ is found, subtract its projections from all remaining columns $j+1$ to n. Then choose the largest of the resulting vectors as $\boldsymbol{a}_{j+1}$ (leading to $\boldsymbol{q}_{j+1}$). We are exchanging columns just as elimination exchanged rows. So a permutation P is allowed on the column side of A, and $\boldsymbol{AP} = \boldsymbol{QR}$.

I believe that Gram-Schmidt is still the good process to understand, even if reflections or rotations or column exchanges lead to a more perfect Q.

■ REVIEW OF THE KEY IDEAS ■

1. If the orthonormal vectors $\boldsymbol{q}_1, \ldots, \boldsymbol{q}_n$ are the columns of Q, then $\boldsymbol{q}_i^{\mathrm{T}}\boldsymbol{q}_j = 0$ and $\boldsymbol{q}_i^{\mathrm{T}}\boldsymbol{q}_i = 1$ translate into the matrix multiplication $Q^{\mathrm{T}}Q = I$.
2. If Q is square (an ***orthogonal matrix***) then $Q^{\mathrm{T}} = Q^{-1}$: ***transpose*** $=$ ***inverse***.
3. The length of $Q\boldsymbol{x}$ equals the length of $\boldsymbol{x}$: $\|Q\boldsymbol{x}\| = \|\boldsymbol{x}\|$.
4. The projection onto the column space of Q spanned by the $\boldsymbol{q}$'s is $P = QQ^{\mathrm{T}}$.
5. If Q is square then $P = QQ^{\mathrm{T}} = I$ and every $\boldsymbol{b} = \boldsymbol{q}_1(\boldsymbol{q}_1^{\mathrm{T}}\boldsymbol{b}) + \cdots + \boldsymbol{q}_n(\boldsymbol{q}_n^{\mathrm{T}}\boldsymbol{b})$.
6. Gram-Schmidt produces orthonormal vectors $\boldsymbol{q}_1, \boldsymbol{q}_2, \boldsymbol{q}_3$ from independent $\boldsymbol{a}, \boldsymbol{b}, \boldsymbol{c}$. In matrix form this is the $\boldsymbol{QR}$ factorization $A =$ (orthogonal Q)(triangular R).

■ WORKED EXAMPLE ■

Add two more rows and columns with all entries 1 or -1, so the columns of this 4 by 4 **Hadamard matrix** are orthogonal. How do you turn H_4 into an *orthogonal matrix* Q?

$$H_2 = \begin{bmatrix} 1 & 1 \\ 1 & -1 \end{bmatrix} \qquad H_4 = \begin{bmatrix} 1 & 1 & x & x \\ 1 & -1 & x & x \\ 1 & 1 & x & x \\ 1 & -1 & x & x \end{bmatrix} \qquad \text{and} \qquad H_8 = \begin{bmatrix} H_4 & H_4 \\ H_4 & -H_4 \end{bmatrix}$$

The projection of $\boldsymbol{b} = (6, 0, 0, 2)$ onto the first column of H_4 is $\boldsymbol{p}_1 = (2, 2, 2, 2)$. The projection onto the second column is $\boldsymbol{p}_2 = (1, -1, 1, -1)$. What is the projection $\boldsymbol{p}_{1,2}$ of $\boldsymbol{b}$ onto the 2-dimensional space spanned by the first two columns?

Solution H_4 is built from H_2 just as H_8 is built from H_4:

$$H_4 = \begin{bmatrix} H_2 & H_2 \\ H_2 & -H_2 \end{bmatrix} = \begin{bmatrix} 1 & 1 & 1 & 1 \\ 1 & -1 & 1 & -1 \\ 1 & 1 & -1 & -1 \\ 1 & -1 & -1 & 1 \end{bmatrix} \text{ has orthogonal columns.}$$

Then $Q = H/2$ has *orthonormal columns*. Dividing by 2 gives unit vectors in Q. A 5 by 5 Hadamard matrix is impossible because the dot product of columns would have five 1's and/or -1's and could not add to zero. H_8 has orthogonal columns of length $\sqrt{8}$.

$$H_8^{\mathrm{T}}H_8 = \begin{bmatrix} H^{\mathrm{T}} & H^{\mathrm{T}} \\ H^{\mathrm{T}} & -H^{\mathrm{T}} \end{bmatrix} \begin{bmatrix} H & H \\ H & -H \end{bmatrix} = \begin{bmatrix} 2H^{\mathrm{T}}H & 0 \\ 0 & 2H^{\mathrm{T}}H \end{bmatrix} = \begin{bmatrix} 8I & 0 \\ 0 & 8I \end{bmatrix}. \quad Q_8 = \frac{H_8}{\sqrt{8}}$$

What is the key point of orthogonal columns? Answer: $A^{\mathrm{T}}A$ is diagonal and easy to invert. **We can project onto lines and just add**. The axes are orthogonal.

Problem Set 4.4

Problems 1–12 are about orthogonal vectors and orthogonal matrices.

1 Are these pairs of vectors orthonormal or only orthogonal or only independent?

$$\text{(a)} \begin{bmatrix}1\\0\end{bmatrix} \text{ and } \begin{bmatrix}-1\\1\end{bmatrix} \qquad \text{(b)} \begin{bmatrix}.6\\.8\end{bmatrix} \text{ and } \begin{bmatrix}.4\\-.3\end{bmatrix} \qquad \text{(c)} \begin{bmatrix}\cos\theta\\\sin\theta\end{bmatrix} \text{ and } \begin{bmatrix}-\sin\theta\\\cos\theta\end{bmatrix}.$$

Change the second vector when necessary to produce orthonormal vectors.

2 The vectors $(2, 2, -1)$ and $(-1, 2, 2)$ are orthogonal. Divide them by their lengths to find orthonormal vectors $\boldsymbol{q}_1$ and $\boldsymbol{q}_2$. Put those into the columns of Q and multiply $Q^{\mathrm{T}}Q$ and QQ^{T}.

3 (a) If A has three orthogonal columns each of length 4, what is $A^{\mathrm{T}}A$?

(b) If A has three orthogonal columns of lengths $1, 2, 3$, what is $A^{\mathrm{T}}A$?

4 Give an example of each of the following:

(a) A matrix Q that has orthonormal columns but $QQ^{\mathrm{T}} \neq I$.

(b) Two orthogonal vectors that are not linearly independent.

(c) An orthonormal basis for $\mathbf{R}^3$, including the vector $\boldsymbol{q}_1 = (1, 1, 1)/\sqrt{3}$.

5 Find two orthogonal vectors in the plane $x + y + 2z = 0$. Make them orthonormal.

6 If Q_1 and Q_2 are orthogonal matrices, show that their product Q_1Q_2 is also an orthogonal matrix. (Use $Q^{\mathrm{T}}Q = I$.)

7 If Q has orthonormal columns, what is the least squares solution $\widehat{\boldsymbol{x}}$ to $Q\boldsymbol{x} = \boldsymbol{b}$?

8 If $\boldsymbol{q}_1$ and $\boldsymbol{q}_2$ are orthonormal vectors in $\mathbf{R}^5$, what combination _____ $\boldsymbol{q}_1 +$ _____ $\boldsymbol{q}_2$ is closest to a given vector $\boldsymbol{b}$?

9 (a) Compute $P = QQ^{\mathrm{T}}$ when $\boldsymbol{q}_1 = (.8, .6, 0)$ and $\boldsymbol{q}_2 = (-.6, .8, 0)$. Verify that $P^2 = P$.

(b) Prove that always $(QQ^{\mathrm{T}})^2 = QQ^{\mathrm{T}}$ by using $Q^{\mathrm{T}}Q = I$. Then $P = QQ^{\mathrm{T}}$ is the projection matrix onto the column space of Q.

10 Orthonormal vectors $\boldsymbol{q}_1, \boldsymbol{q}_2, \boldsymbol{q}_3$ are automatically linearly independent.

(a) Vector proof: When $c_1\boldsymbol{q}_1 + c_2\boldsymbol{q}_2 + c_3\boldsymbol{q}_3 = \boldsymbol{0}$, what dot product leads to $c_1 = 0$? Similarly $c_2 = 0$ and $c_3 = 0$. Thus the $\boldsymbol{q}$'s are independent.

(b) Matrix proof: Show that $Q\boldsymbol{x} = \boldsymbol{0}$ leads to $\boldsymbol{x} = \boldsymbol{0}$. Since Q may be rectangular, you can use Q^{T} but not Q^{-1}.

11 Find orthonormal vectors $\boldsymbol{q}_1$ and $\boldsymbol{q}_2$ in the plane spanned by $\boldsymbol{a} = (1, 3, 4, 5, 7)$ and $\boldsymbol{b} = (-6, 6, 8, 0, 8)$. Which combination is closest to $(1, 0, 0, 0, 0)$?

12 If $\boldsymbol{a}_1, \boldsymbol{a}_2, \boldsymbol{a}_3$ is a basis for $\mathbf{R}^3$, any vector $\boldsymbol{b}$ can be written as $\boldsymbol{b} = x_1\boldsymbol{a}_1 + x_2\boldsymbol{a}_2 + x_3\boldsymbol{a}_3$.

(a) Suppose the $\boldsymbol{a}$'s are orthonormal. Show that $x_1 = \boldsymbol{a}_1^{\mathrm{T}}\boldsymbol{b}$.

(b) Suppose the $\boldsymbol{a}$'s are orthogonal. Show that $x_1 = \boldsymbol{a}_1^{\mathrm{T}}\boldsymbol{b}/\boldsymbol{a}_1^{\mathrm{T}}\boldsymbol{a}_1$.

(c) If the $\boldsymbol{a}$'s are independent, x_1 is the first component of _____ times $\boldsymbol{b}$.

13 What multiple of $\boldsymbol{a} = \begin{bmatrix} 1 \\ 1 \end{bmatrix}$ should be subtracted from $\boldsymbol{b} = \begin{bmatrix} 4 \\ 0 \end{bmatrix}$ to make the result $\boldsymbol{B}$ orthogonal to $\boldsymbol{a}$? Sketch a figure to show $\boldsymbol{a}$, $\boldsymbol{b}$, and $\boldsymbol{B}$.

14 Complete the Gram-Schmidt process in Problem 13 by computing $\boldsymbol{q}_1 = \boldsymbol{a}/\|\boldsymbol{a}\|$ and $\boldsymbol{q}_2 = \boldsymbol{B}/\|\boldsymbol{B}\|$ and factoring into QR:

$$\begin{bmatrix} 1 & 4 \\ 1 & 0 \end{bmatrix} = \begin{bmatrix} \boldsymbol{q}_1 & \boldsymbol{q}_2 \end{bmatrix} \begin{bmatrix} \|\boldsymbol{a}\| & ? \\ 0 & \|\boldsymbol{B}\| \end{bmatrix}.$$

15 Find orthonormal vectors $\boldsymbol{q}_1, \boldsymbol{q}_2, \boldsymbol{q}_3$ such that $\boldsymbol{q}_1, \boldsymbol{q}_2$ span the column space of A. Which "fundamental subspace" contains $\boldsymbol{q}_3$? Solve $A\boldsymbol{x} = (1, 2, 7)$ by least squares.

$$A = \begin{bmatrix} 1 & 1 \\ 2 & -1 \\ -2 & 4 \end{bmatrix}.$$

16 What multiple of $\boldsymbol{a} = (4, 5, 2, 2)$ is closest to $\boldsymbol{b} = (1, 2, 0, 0)$? Find orthonormal vectors $\boldsymbol{q}_1$ and $\boldsymbol{q}_2$ in the plane of $\boldsymbol{a}$ and $\boldsymbol{b}$.

17 Project $\boldsymbol{b} = (1, 3, 5)$ onto the line through $\boldsymbol{a} = (1, 1, 1)$. Then find $\boldsymbol{p}$ and $\boldsymbol{e}$. Compute the orthonormal vectors $\boldsymbol{q}_1 = \boldsymbol{a}/\|\boldsymbol{a}\|$ and $\boldsymbol{q}_2 = \boldsymbol{e}/\|\boldsymbol{e}\|$.

18 (Recommended) Find orthogonal vectors $\boldsymbol{A}, \boldsymbol{B}, \boldsymbol{C}$ by Gram-Schmidt from $\boldsymbol{a}, \boldsymbol{b}, \boldsymbol{c}$:

$$\boldsymbol{a} = (1, -1, 0, 0) \qquad \boldsymbol{b} = (0, 1, -1, 0) \qquad \boldsymbol{c} = (0, 0, 1, -1).$$

$\boldsymbol{A}, \boldsymbol{B}, \boldsymbol{C}$ and $\boldsymbol{a}, \boldsymbol{b}, \boldsymbol{c}$ are bases for the vectors perpendicular to $\boldsymbol{d} = (1, 1, 1, 1)$.

19 If $A = QR$ then $A^{\mathrm{T}}A = R^{\mathrm{T}}R =$ _____ triangular times _____ triangular. *Gram-Schmidt on A corresponds to elimination on $A^{\mathrm{T}}A$.* The pivots for $A^{\mathrm{T}}A$ must be the squares of diagonal entries of R. Find Q and R by Gram-Schmidt for this A:

$$A = \begin{bmatrix} -1 & 1 \\ 2 & 1 \\ 2 & 4 \end{bmatrix} \quad \text{and} \quad A^{\mathrm{T}}A = \begin{bmatrix} 9 & 9 \\ 9 & 18 \end{bmatrix} = \begin{bmatrix} 1 & 0 \\ 1 & 1 \end{bmatrix} \begin{bmatrix} 9 & \\ & 9 \end{bmatrix} \begin{bmatrix} 1 & 1 \\ 0 & 1 \end{bmatrix}.$$

20 Find an orthonormal basis for the column space of A. Then project $\boldsymbol{b}$ on $\mathbf{C}(A)$.

$$A = \begin{bmatrix} 1 & -2 \\ 1 & 0 \\ 1 & 1 \\ 1 & 3 \end{bmatrix} \quad \text{and} \quad \boldsymbol{b} = \begin{bmatrix} -4 \\ -3 \\ 3 \\ 0 \end{bmatrix}.$$

21 Find orthogonal vectors $\boldsymbol{A}, \boldsymbol{B}, \boldsymbol{C}$ by Gram-Schmidt from

$$\boldsymbol{a} = \begin{bmatrix} 1 \\ 1 \\ 2 \end{bmatrix} \quad \text{and} \quad \boldsymbol{b} = \begin{bmatrix} 1 \\ -1 \\ 0 \end{bmatrix} \quad \text{and} \quad \boldsymbol{c} = \begin{bmatrix} 1 \\ 0 \\ 4 \end{bmatrix}.$$

22 Find $\boldsymbol{q}_1, \boldsymbol{q}_2, \boldsymbol{q}_3$ (orthonormal) as combinations of $\boldsymbol{a}, \boldsymbol{b}, \boldsymbol{c}$ (independent columns). Then write A as QR:

$$A = \begin{bmatrix} 1 & 2 & 4 \\ 0 & 0 & 5 \\ 0 & 3 & 6 \end{bmatrix}.$$

Problems 23–26 use the QR code above equation (11). It executes Gram-Schmidt.

23 Show why $\boldsymbol{C}$ (found via $\boldsymbol{C}^*$ in equation (10)) is equal to $\boldsymbol{C}$ in equation (7).

24 Equation (7) subtracts from $\boldsymbol{c}$ its components along $\boldsymbol{A}$ and $\boldsymbol{B}$. Why not subtract the components along $\boldsymbol{a}$ and along $\boldsymbol{b}$?

25 Where are the mn^2 small multiplications in executing Gram-Schmidt ?

26 Apply the MATLAB qr code to $\boldsymbol{a} = (2, 2, -1)$, $\boldsymbol{b} = (0, -3, 3)$, $\boldsymbol{c} = (1, 0, 0)$. What are the $\boldsymbol{q}$'s?

27 If $\boldsymbol{u}$ is a unit vector, then $Q = I - 2\boldsymbol{u}\boldsymbol{u}^{\mathrm{T}}$ is a reflection matrix (Example 3). Find Q_1 from $\boldsymbol{u} = (0, 1)$ and Q_2 from $\boldsymbol{u} = (0, \sqrt{2}/2, \sqrt{2}/2)$. Draw the reflections when Q_1 and Q_2 multiply the vectors $(1, 2)$ and $(1, 1, 1)$.

28 Find all matrices that are both orthogonal and lower triangular.

29 $Q = I - 2\boldsymbol{u}\boldsymbol{u}^{\mathrm{T}}$ is a reflection matrix when $\boldsymbol{u}^{\mathrm{T}}\boldsymbol{u} = 1$. Two reflections give $Q^2 = I$.

(a) Show that $Q\boldsymbol{u} = -\boldsymbol{u}$. The mirror is perpendicular to $\boldsymbol{u}$.

(b) Find $Q\boldsymbol{v}$ when $\boldsymbol{u}^{\mathrm{T}}\boldsymbol{v} = 0$. The mirror contains $\boldsymbol{v}$. It reflects to itself.

Challenge Problems

30 (MATLAB) Factor $[\,Q, R\,] = \mathbf{qr}(A)$ if A has columns $(1, -1, 0, 0)$ and $(0, 1, -1, 0)$ and $(0, 0, 1, -1)$ and $(0, 0, 0, 1)$. Can you scale the orthogonal columns of Q to get nice integer components ?

31 If A is m by n with rank n, then $\mathbf{qr}(A)$ produces a *square* Q and zeros below R:

The factors from MATLAB are $(m \text{ by } m)(m \text{ by } n)$ $\qquad A = \begin{bmatrix} Q_1 & Q_2 \end{bmatrix} \begin{bmatrix} R \\ 0 \end{bmatrix}.$

The n columns of Q_1 are an orthonormal basis for which fundamental subspace? The $m-n$ columns of Q_2 are an orthonormal basis for which fundamental subspace?

32 We know that $P = QQ^{\mathrm{T}}$ is the projection onto the column space of $Q(m \text{ by } n)$. Now add another column $\boldsymbol{a}$ to produce $A = [Q \quad \boldsymbol{a}]$. Gram-Schmidt replaces $\boldsymbol{a}$ by what vector $\boldsymbol{q}$? Start with $\boldsymbol{a}$, subtract ______, divide by ______ to find $\boldsymbol{q}$.

5 Determinants and Linear Transformations

The determinant of a square matrix is an astonishing number. If $\det A = 0$, this signals that the column vectors are dependent and A is not invertible. If $\det A$ is not zero, the formula for A^{-1} will have a division by the determinant. Section 5.1 finds 3 by 3 determinants and the "cofactor formula" for $\det A$ and A^{-1}.

Section 5.2 begins with algebra: Cramer's Rule for $\boldsymbol{x} = A^{-1}\boldsymbol{b}$. Then it moves to geometry: *the volume of a tilted box*. The edges of the box are the column vectors in A. When the box is flat, A is singular. In all cases, the volume of the box is $|\det A|$.

When we multiply matrices AB, we multiply their determinants. Somehow we are multiplying boxes and their volumes. This is easiest to see in the plane, where the boxes are parallelograms and $|\det A| =$ area. Section 5.3 multiplies matrices for $\det AB$.

This link to volumes leads naturally to linear transformations. A linear transformation in $\mathbf{R}^3$ takes a cube into a tilted box. If you know the edges of the box, you know everything. In the words of linear algebra, if you know what the linear transformation T does to a basis, then you know $T(\boldsymbol{v})$ for every vector $\boldsymbol{v}$.

There are three useful formulas for $\det A$. The computational formula *multiplies the pivots* from the triangular matrix U in $PA = LU$. Always $\det P = 1$ or -1 and $\det L = 1$. So the product of pivots in U gives $\pm \det A$. This is usually the fastest way.

A second formula uses determinants of size $n-1$: the "cofactors" of A. They give the best formula for A^{-1}. *The i, j entry of A^{-1} is the j, i cofactor divided by* $\det A$.

The "big formula" for $\det A$ adds up $n!$ terms—one for every path down the matrix. Each path chooses one entry from every row and column, and we multiply the n entries on the path. Reverse the sign if the path has an odd permutation of column numbers. A 3 by 3 matrix has six paths and the big formula has six terms—one for every 3 by 3 permutation.

This chapter could easily become full of formulas! The connection to linear transformations shows how an n by n matrix acts on a shape in n dimensions. It produces another shape. And the ratio of the two volumes is $|\det A|$.

Determinants could tell us everything, if only they were not so hard to compute.

5.1 3 by 3 Determinants

1 The **determinant** of $A = \begin{bmatrix} a & b \\ c & d \end{bmatrix}$ is $\boldsymbol{ad - bc}$. The singular matrix $\begin{bmatrix} a & xa \\ c & xc \end{bmatrix}$ has $\mathbf{det} = \mathbf{0}$.

2 **Row exchange reverses signs** $PA = \begin{bmatrix} 0 & 1 \\ 1 & 0 \end{bmatrix} \begin{bmatrix} a & b \\ c & d \end{bmatrix} = \begin{bmatrix} c & d \\ a & b \end{bmatrix}$ has $\det PA = \boldsymbol{bc - ad} = -\,\mathbf{det}\,\boldsymbol{A}$.

3 The determinant of $\begin{bmatrix} xa + yA & xb + yB \\ c & d \end{bmatrix}$ is $\boldsymbol{x(ad - bc) + y(Ad - Bc)}$. **Det is linear in row 1 by itself.**

4 If A is n by n then $\mathbf{1, 2, 3}$ remain true: $\mathbf{det} = \mathbf{0}$ when A is singular, **det reverses sign** when rows are exchanged, **det is linear in row 1**. **Also**, det $=$ **product of the pivots**. Always $\mathbf{det}\,\boldsymbol{BA} = (\mathbf{det}\,\boldsymbol{B})(\mathbf{det}\,\boldsymbol{A})$ and $\mathbf{det}\,\boldsymbol{A}^{\mathrm{T}} = \mathbf{det}\,\boldsymbol{A}$. This is an amazing number.

Determinants lead to beautiful formulas. But often they are not practical for computing. This section will focus on 3 by 3 matrices, to see how determinants produce the matrix A^{-1}. (This formula for A^{-1} was not seen in Chapters 1 to 4.) First will come 2 by 2 matrices:

$$\det\begin{bmatrix} 1 & 0 \\ 0 & 1 \end{bmatrix} = \mathbf{1} \quad \det\begin{bmatrix} 0 & 1 \\ 1 & 0 \end{bmatrix} = \mathbf{-1} \quad \det\begin{bmatrix} a & b \\ c & d \end{bmatrix} = \boldsymbol{ad - bc} \quad \det\begin{bmatrix} c & d \\ a & b \end{bmatrix} = \boldsymbol{bc - ad}$$

We start with those two permutation matrices. Their determinants change sign when the rows are exchanged. The same sign change appears for any matrix. This rule becomes a key to determinants of all sizes. *The matrix has no inverse when* $\det A = 0$.

For 2 by 2 matrices, $\det A = 0$ means $a/c = b/d$. The rows are parallel. For n by n matrices, $\det A = 0$ means that the columns of A are not independent. A combination of columns gives the zero vector: $A\boldsymbol{x} = \mathbf{0}$ with $\boldsymbol{x} \neq \mathbf{0}$. A is not invertible.

These properties and more will follow after we define the determinant.

3 by 3 Determinants

2 by 2 matrices are easy: $ad - bc$. 4 by 4 matrices are hard. Better to use elimination (or a laptop). For 3 by 3 matrices, the determinant has $3! = 6$ terms, and often you can compute it by hand. We will show how.

Start with the identity matrix ($\det I = 1$) and exchange two rows (then $\det = -1$). Exchanging again brings back $\det = +1$. You quickly have all six permutation matrices. Each row exchange will reverse the sign of the determinant ($+1$ to -1 or else -1 to $+1$):

$$\begin{bmatrix} 1 & & \\ & 1 & \\ & & 1 \end{bmatrix} \begin{bmatrix} & 1 & \\ 1 & & \\ & & 1 \end{bmatrix} \begin{bmatrix} & 1 & \\ & & 1 \\ 1 & & \end{bmatrix} \begin{bmatrix} & & 1 \\ & 1 & \\ 1 & & \end{bmatrix} \begin{bmatrix} & & 1 \\ 1 & & \\ & 1 & \end{bmatrix} \begin{bmatrix} 1 & & \\ & & 1 \\ & 1 & \end{bmatrix}$$
$$\mathbf{det} = \mathbf{+1} \qquad \mathbf{-1} \qquad \mathbf{+1} \qquad \mathbf{-1} \qquad \mathbf{+1} \qquad \mathbf{-1}$$

Notice! If I exchange two rows—say row 1 and row 2—**each determinant changes sign**. Permutations 1 and 2 exchange, 3 and 4 exchange, 5 and 6 exchange. This will carry over to all determinants: Row exchange multiplies $\det A$ by -1.

When you multiply a row by a number, this multiplies the determinant by that number. Suppose the three rows are $\boldsymbol{a\,b\,c}$ and $\boldsymbol{p\,q\,r}$ and $\boldsymbol{x\,y\,z}$. Those nine numbers multiply ± 1.

$$\begin{bmatrix} a & & \\ & q & \\ & & z \end{bmatrix} \begin{bmatrix} & b & \\ p & & \\ & & z \end{bmatrix} \begin{bmatrix} & b & \\ & & r \\ x & & \end{bmatrix} \begin{bmatrix} & & c \\ & q & \\ x & & \end{bmatrix} \begin{bmatrix} & & c \\ p & & \\ & y & \end{bmatrix} \begin{bmatrix} a & & \\ & & r \\ & y & \end{bmatrix}$$

$$\text{det} = +\boldsymbol{aqz} \quad -\boldsymbol{bpz} \quad +\boldsymbol{brx} \quad -\boldsymbol{cqx} \quad +\boldsymbol{cpy} \quad -\boldsymbol{ary}$$

Finally we use the most powerful property we have. **The determinant of A is linear in each row separately**. As equation (3) will show, we can add those six determinants. To remember the plus and minus signs, I follow the arrows in this picture of the matrix.

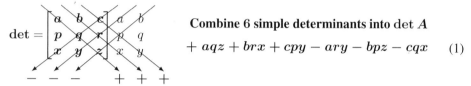

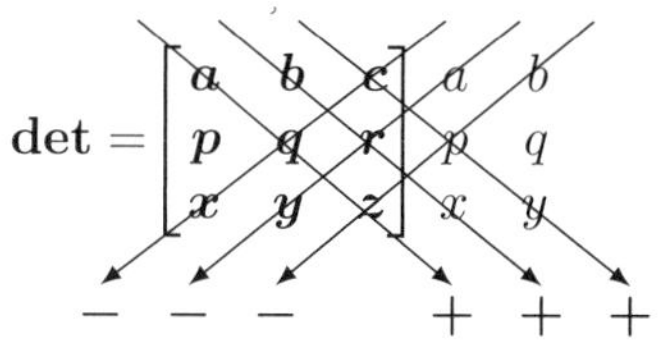

Combine 6 simple determinants into $\det A$

$$+\ \boldsymbol{aqz} + \boldsymbol{brx} + \boldsymbol{cpy} - \boldsymbol{ary} - \boldsymbol{bpz} - \boldsymbol{cqx} \qquad (1)$$

Notice ! Those six terms all have one entry from each row of the matrix. They also have one entry from each column of the matrix. There are $6 = 3\,!$ terms because there are six 3×3 permutation matrices. A 4×4 determinant will have $4\,! = 24$ terms.

This guides us to the "big formula" for the determinant of an n by n matrix. That formula has $n\,!$ terms $a_{1j}\, a_{2k} \ldots a_{nz}$, one for every n by n permutation. Each permutation matrix P picks out n numbers in A (one number from every row and column). *Multiply those n numbers by* $\det \boldsymbol{P} = 1$ *or* -1. *Then add the results like* $a_{11}\, a_{22} - a_{12}\, a_{21} = \boldsymbol{ad - bc}$.

Det P is 1 for even permutations like 231 and -1 for odd permutations like 213. Those are reached from the identity matrix I by an even or an odd number of exchanges. Each permutation P reorders the column numbers $1, 2, \ldots, n$ into some order j, k, z.

The determinant is the sum of n ! simple determinants like $(-1)\boldsymbol{a_{12}a_{21}a_{33}} = -\boldsymbol{bpz}$.

$$\boxed{\begin{aligned} \det \boldsymbol{A} &= \text{sum over all } n\,! \text{ column permutations } P = (j, k, \ldots, z) \\ &= \textstyle\sum (\det \boldsymbol{P})\, \boldsymbol{a_{1j}\, a_{2k} \ldots a_{nz}} = \textbf{BIG FORMULA}. \end{aligned}} \qquad (2)$$

So every term in the big formula picks out one number a_{ij} from each row of A and at the same time one number from each column of A. Multiply those n numbers times 1 or -1. Permutations P and their plus-minus signs are the keys to determinants !

Let me return for a minute to that powerful property: **det A is linear in each row separately**. We can split row 1 into $(a, 0, 0) + (0, b, 0) + (0, 0, c)$. We can split row 2 and row 3 in the same way. This gives us a lot of pieces ($3^3 = 27$ different pieces). But only 6 of those pieces are important and 21 of them are zero automatically (a zero column). $3\,! = 6$ ways to use every row and column once, $3^3 = 27$ ways if columns could repeat.

$$\text{det} \begin{bmatrix} \boldsymbol{a} & 0 & 0 \\ 0 & \boldsymbol{q} & 0 \\ 0 & 0 & \boldsymbol{z} \end{bmatrix} = \begin{array}{l} \boldsymbol{aqz} \text{ is important} \\ \textbf{6 like this in (1)} \end{array} \qquad \text{det} \begin{bmatrix} \boldsymbol{a} & 0 & 0 \\ \boldsymbol{p} & 0 & 0 \\ 0 & 0 & \boldsymbol{z} \end{bmatrix} = \begin{array}{l} \textbf{0} \text{ automatically} \\ \textbf{21 like this} \end{array}$$

Cofactors and a Formula for A^{-1}

For the 3×3 determinant in (1) with 6 terms, I can explain the "cofactor formulas" for the determinant and the inverse of A. *The idea is to reduce from* 3×3 *to* 2×2. Two of the six terms in $\det A$ start with a and with b and with c from row 1.

Cofactor formula $$\boxed{\det A = a\,(\boldsymbol{qz} - \boldsymbol{ry}) + b\,(\boldsymbol{rx} - \boldsymbol{pz}) + c\,(\boldsymbol{py} - \boldsymbol{qx}).} \qquad (3)$$

We have separated the factors a, b, c from their "cofactors". **Each cofactor is a 2 by 2 determinant**. The factor a in the $1,1$ position takes its cofactor from row and column 2 and 3. Every row and column must be used only once! Notice that the cofactor of b in row 1, column 2 is $rx - pz$. There we see the other rows $2, 3$ and columns $1, 3$. But the actual 2 by 2 determinant would be $+\boldsymbol{pz} - \boldsymbol{rx}$. The cofactor reversed those $\pm$ signs.

Here is the definition of cofactors C_{ij} and the $(-1)^{i+j}$ rule for their plus and minus signs.

> For the i, j cofactor C_{ij}, remove row i and column j from the matrix A.
>
> C_{ij} equals $(-1)^{i+j}$ times the determinant of the remaining matrix (size $n-1$).
>
> **The cofactor formula along row i is** $\det A = a_{i1}C_{i1} + \cdots + a_{in}C_{in}$ (4)

Key point: *The cofactor C_{ij} just collects all the terms in* $\det A$ *that are multiplied by* a_{ij}. Thus the cofactors of a 2 by 2 matrix involve 1 by 1 determinants.

$$\text{The cofactors of } A = \begin{bmatrix} a & b \\ c & d \end{bmatrix} \text{ are in the cofactor matrix } C = \begin{bmatrix} d & -c \\ -b & a \end{bmatrix}.$$

Now notice something wonderful. **If A multiplies C^{T} (not C) we get det A times I:**

$$AC^{\mathrm{T}} = \begin{bmatrix} ad - bc & 0 \\ 0 & ad - bc \end{bmatrix} = \begin{bmatrix} \det A & 0 \\ 0 & \det A \end{bmatrix} = (\det A)I. \qquad (5)$$

Dividing by $\det A$, cofactors give our first and best formula for the inverse matrix A^{-1}.

$$\boxed{\textbf{Inverse matrix formula} \qquad A^{-1} = C^{\mathrm{T}} / \det A} \qquad (6)$$

This formula shows why the determinant of an invertible matrix cannot be zero. We need to divide the matrix C^{T} by the number $\det A$. Every entry in the inverse matrix A^{-1} is a ratio of two determinants (size $n-1$ for the cofactor divided by size n for A).

This example has determinant 1, so the inverse is exactly C^{T} (the cofactors). Notice how C_{32} removes row 3 and column 2. That leaves a 2 by 2 matrix with determinant 1. Since $(-1)^{2+3} = \mathbf{-1}$, this becomes -1 in C^{T}.

Example of A^{-1}
Determinant $= 1$
Cofactors in C^{T}

$$A^{-1} = \begin{bmatrix} 1 & 1 & 1 \\ 0 & 1 & 1 \\ 0 & 0 & 1 \end{bmatrix}^{-1} = \begin{bmatrix} 1 & -1 & \mathbf{0} \\ 0 & 1 & \mathbf{-1} \\ 0 & 0 & 1 \end{bmatrix} = C^{\mathrm{T}} \qquad (7)$$

The diagonal entries of AC^{T} are always $\det A$. That is exactly the cofactor formula. Problem 24 will show why the off-diagonal entries of AC^{T} are always zero. Those numbers turn out to be determinants of matrices with two equal rows. Automatically zero.

A typical cofactor C_{31} removes row 3 and column 1. In our 3 by 3 example, that leaves a 2 by 2 matrix of 1's, with determinant $= 0$. This is the bold zero in A^{-1}.

If we change A to $2A$, this determinant is multiplied by $(2)\,(2)\,(2) = 8$. All cofactors in C are multiplied by $(2)\,(2) = 4$. Then $\boldsymbol{A^{-1} = C^{\mathrm{T}}/\det A}$ is divided by 2. Of course.

Section 5.2 will solve $A\boldsymbol{x} = \boldsymbol{b}$ and Section 5.3 will find volumes from $\det A$.

Problem Set 5.1

Questions 1–5 are about the rules for determinants.

1 If a 4 by 4 matrix has $\det A = \frac{1}{2}$, find $\det(2A)$ and $\det(-A)$ and $\det(A^2)$ and $\det(A^{-1})$.

2 If a 3 by 3 matrix has $\det A = -1$, find $\det\left(\frac{1}{2}A\right)$ and $\det(-A)$ and $\det(A^2)$ and $\det(A^{-1})$. What are those answers if $\det A = 0$?

3 True or false, with a reason if true or a counterexample if false:

(a) The determinant of $I + A$ is $1 + \det A$.

(b) The determinant of ABC is $|A|\,|B|\,|C|$.

(c) The determinant of $4A$ is $4|A|$.

(d) The determinant of $AB - BA$ is zero. Try an example with $A = \begin{bmatrix} 0 & 0 \\ 0 & 1 \end{bmatrix}$.

4 Which row exchanges show that these "reverse identity matrices" J_3 and J_4 have $|J_3| = -1$ but $|J_4| = +1$?

$$\det \begin{bmatrix} 0 & 0 & 1 \\ 0 & 1 & 0 \\ 1 & 0 & 0 \end{bmatrix} = \mathbf{-1} \quad \text{but} \quad \det \begin{bmatrix} 0 & 0 & 0 & 1 \\ 0 & 0 & 1 & 0 \\ 0 & 1 & 0 & 0 \\ 1 & 0 & 0 & 0 \end{bmatrix} = \mathbf{+1}.$$

5 For $n = 5, 6, 7$, count the row exchanges to permute the reverse identity J_n to the identity matrix I_n. Propose a rule for every size n and predict whether J_{101} has determinant $+1$ or -1.

6 Find the six terms in equation (1) like $+aqz$ (the main diagonal) and $-cqx$ (the anti-diagonal). Combine those six terms into the determinants of A, B, C.

$$A = \begin{bmatrix} 2 & -1 & 0 \\ -1 & 2 & -1 \\ 0 & -1 & 2 \end{bmatrix} \qquad B = \begin{bmatrix} 2 & 1 & 4 \\ 4 & 2 & 8 \\ 6 & 3 & 12 \end{bmatrix} \qquad C = \begin{bmatrix} 1 & 2 & 3 \\ 4 & 5 & 6 \\ 7 & 8 & 9 \end{bmatrix}$$

7 If you add row 1 $= \begin{bmatrix} a & b & c \end{bmatrix}$ to row 2 $= \begin{bmatrix} p & q & r \end{bmatrix}$ to get $\begin{bmatrix} p+a & q+b & r+c \end{bmatrix}$ in row 2, show that the 3 by 3 determinant in equation (1) *does not change*. Here is another approach:

$$\det \begin{bmatrix} \text{row } 1 \\ \text{row } 1 + \text{row } 2 \\ \text{row } 3 \end{bmatrix} = \det \begin{bmatrix} \text{row } 1 \\ \text{row } 1 \\ \text{row } 3 \end{bmatrix} + \det \begin{bmatrix} \text{row } 1 \\ \text{row } 2 \\ \text{row } 3 \end{bmatrix} = \mathbf{0} + \det \begin{bmatrix} \text{row } 1 \\ \text{row } 2 \\ \text{row } 3 \end{bmatrix}.$$

8 Show that $\det A^{\mathrm{T}} = \det A$ because both of those 3 by 3 determinants come from the same six terms like $\boldsymbol{brx}$. In other words $\det P^{\mathrm{T}} = \det P$ for every permutation P.

9 Do these matrices have determinant $0, 1, 2$, or 3?

$$A = \begin{bmatrix} 0 & 0 & 1 \\ 1 & 0 & 0 \\ 0 & 1 & 0 \end{bmatrix} \quad B = \begin{bmatrix} 0 & 1 & 1 \\ 1 & 0 & 1 \\ 1 & 1 & 0 \end{bmatrix} \quad C = \begin{bmatrix} 1 & 1 & 1 \\ 1 & 1 & 1 \\ 1 & 1 & 1 \end{bmatrix} \quad D = \begin{bmatrix} 1 & 1 & 1 \\ 0 & 1 & 0 \\ 1 & 1 & 1 \end{bmatrix}.$$

10 If the entries in every row of A add to zero, solve $A\boldsymbol{x} = \mathbf{0}$ to prove $\det A = 0$. If those entries add to one, show that $\det(A - I) = 0$. Does this mean $\det A = 1$?

11 Why does $\det(P_1 P_2) = (\det P_1)$ times $(\det P_2)$ for permutations? If P_1 needs 2 row exchanges and P_2 needs 3 row exchanges to reach I, why does $P_1 P_2$ reach I from $2 + 3$ exchanges? Then their determinants will be $(-1)^2(-1)^3 = (-1)^5$.

12 Explain why half of all 5 by 5 permutations are even (with $\det P = 1$).

13 Reduce A to U and find $\det A =$ product of the pivots:

$$A = \begin{bmatrix} 1 & 1 & 1 \\ 1 & 2 & 2 \\ 1 & 2 & 3 \end{bmatrix} \qquad A = \begin{bmatrix} 1 & 2 & 3 \\ 2 & 2 & 3 \\ 3 & 3 & 3 \end{bmatrix}.$$

14 By applying row operations to produce an upper triangular U, compute

$$\det \begin{bmatrix} 1 & 2 & 3 & 0 \\ 2 & 6 & 6 & 1 \\ -1 & 0 & 0 & 3 \\ 0 & 2 & 0 & 7 \end{bmatrix} \quad \text{and} \quad \det \begin{bmatrix} 2 & -1 & 0 & 0 \\ -1 & 2 & -1 & 0 \\ 0 & -1 & 2 & -1 \\ 0 & 0 & -1 & 2 \end{bmatrix}.$$

15 Use row operations to simplify and compute these determinants:

$$\det \begin{bmatrix} 101 & 201 & 301 \\ 102 & 202 & 302 \\ 103 & 203 & 303 \end{bmatrix} \quad \text{and} \quad \det \begin{bmatrix} 1 & t & t^2 \\ t & 1 & t \\ t^2 & t & 1 \end{bmatrix}.$$

16 Find the determinants of a rank one matrix and a skew-symmetric matrix:

$$A = \begin{bmatrix} 1 \\ 2 \\ 3 \end{bmatrix} \begin{bmatrix} 1 & -4 & 5 \end{bmatrix} \quad \text{and} \quad A = \begin{bmatrix} 0 & 1 & 3 \\ -1 & 0 & 4 \\ -3 & -4 & 0 \end{bmatrix}.$$

17 If the i,j entry of A is i times j, show that $\det A = 0$. (Exception when $A = [\,1\,]$.) If the i,j entry of A is $i+j$, show that $\det A = 0$. (Exception when $n = 1$ or 2.)

18 Use row operations to show that the 3 by 3 "Vandermonde determinant" is

$$\det \begin{bmatrix} 1 & a & a^2 \\ 1 & b & b^2 \\ 1 & c & c^2 \end{bmatrix} = (b-a)(c-a)(c-b).$$

19 Place the smallest number of zeros in a 4 by 4 matrix that will guarantee $\det A = 0$. Place as many zeros as possible while still allowing $\det A \neq 0$.

20 (a) If $a_{11} = a_{22} = a_{33} = 0$, how many of the six terms in $\det A$ will be zero?

(b) If $a_{11} = a_{22} = a_{33} = a_{44} = 0$, how many of the 24 products $a_{1j}a_{2k}a_{3l}a_{4m}$ are sure to be zero?

21 If all the cofactors are zero, how do you know that A has no inverse? If none of the cofactors are zero, is A sure to be invertible?

22 The big formula has $n!$ terms. But if an entry of A is zero, $(n-1)!$ terms disappear. If A has only *three nonzero diagonals* (in the center of A), how many terms are left?

For $n = 1,2,3,4$ that tridiagonal determinant has $1,2,3,5$ terms. Those are Fibonacci numbers in Section 6.2! Show why a tridiagonal 5 by 5 determinant has $5+3=8$ nonzero terms (Fibonacci again). Use the cofactors of a_{11} and a_{12}.

23 *Cofactor formula when two rows are equal.* Write out the 6 terms in $\det A$ when a 3 by 3 matrix has row $1 =$ row $2 = a, b, c$. The determinant should be zero.

24 Why is a matrix that has two equal rows always singular? **Then $\det A = 0$.** If we combine the cofactors from one row with the numbers in another row, we will be computing $\det A^*$ when A^* has equal rows. Then $\det A^* = 0$—this is what produces the off-diagonal zeros in $AC^{\mathrm{T}} = (\det A)\, I$.

25 The Big Formula has 24 terms if A is 4 by 4. How many terms

(a) include a_{14}? (b) include a_{13} and a_{22}? (c) are left if $a_{11}=a_{22}=a_{33}=a_{44}=0$?

5.2 Properties and Applications of Determinants

1 Useful properties: $\det A^{\mathrm{T}} = \det A$ and $\det AB = (\det A)\,(\det B)$ and $|\det Q| = \mathbf{1}$.

2 Cramer's Rule finds $\boldsymbol{x} = A^{-1}\boldsymbol{b}$ from ratios of determinants (a slow way).

3 The volume of the box (parallelogram in 2D) with edges $\boldsymbol{e}_1$ to $\boldsymbol{e}_n$ is $|\det \boldsymbol{E}|$.

The determinant of a square matrix is an amazing number. First of all, an invertible matrix has $\det A \neq 0$. A singular matrix has $\det A = 0$. When we come to eigenvalues λ and eigenvectors $\boldsymbol{x}$ with $A\boldsymbol{x} = \lambda\boldsymbol{x}$, we will write that eigenvalue equation as $(A - \lambda I)\boldsymbol{x} = \mathbf{0}$. This tells us that $A - \lambda I$ is singular and $\det(\boldsymbol{A} - \boldsymbol{\lambda I}) = \mathbf{0}$. We have an equation for λ.

Overall, the formulas are useful for small matrices and also for special matrices. And the properties of determinants can make those formulas simpler. If the matrix is triangular or diagonal, we just multiply the diagonal entries to find the determinant:

Triangular matrix
Diagonal matrix

$$\det\begin{bmatrix} \boldsymbol{a} & b & c \\ & \boldsymbol{q} & r \\ & & \boldsymbol{z} \end{bmatrix} = \det\begin{bmatrix} \boldsymbol{a} & & \\ & \boldsymbol{q} & \\ & & \boldsymbol{z} \end{bmatrix} = \boldsymbol{aqz} \qquad (1)$$

If we transpose A, the same formula still takes one number from each row and column:

$$\textbf{Transpose the matrix} \qquad \det(\boldsymbol{A}^{\mathrm{T}}) = \det(\boldsymbol{A}) \qquad (2)$$

If we multiply AB, we just multiply determinants (a wonderful fact):

$$\textbf{Multiply two matrices} \qquad \det(\boldsymbol{AB}) = (\det \boldsymbol{A})\,(\det \boldsymbol{B}) \qquad (3)$$

A proof by algebra can get complicated. We will give a simple proof of (3) by geometry.

When we add matrices, we do not just add determinants! (Try $I + I$) Here are two good consequences of equations (2) and (3):

Orthogonal matrices $\boldsymbol{Q}$ have determinant 1 or -1

We know that $Q^{\mathrm{T}}Q = I$. Then $(\det Q)^2 = (\det Q^{\mathrm{T}})\,(\det Q) = 1$. Therefore $\det Q$ is ± 1.

Invertible matrices have $\det \boldsymbol{A} = \pm$ (product of the pivots)

If $A = LU$ then $\det A = (\det L)\,(\det U) = \det U$. Triangular U: *Multiply the pivots.*

If $PA = LU$ because of row exchanges, then $\det P = 1$ or -1. Permutation matrix!

Multiplying the pivots $U_{11}\ U_{22} \ldots U_{nn}$ on the diagonal reveals the determinant of A. This is how determinants are computed by MATLAB and by all computer systems for linear algebra. The cost to find U in Chapter 2 was only $n^3/3$ multiplications. Notice: The "Big Formula" for $\det A$ will have a much larger cost. It is the sum of $n!$ terms.

We know that exchanging rows will reverse the sign of $\det A$. Equal rows will give $\det A = 0$. Linearity allows us to split up one row as in (4). The key operation in elimination—*subtracting x times row i from row j—does not change the determinant*:

$$\det\begin{bmatrix} a & b \\ c-xa & d-xb \end{bmatrix} = \det\begin{bmatrix} a & b \\ c & d \end{bmatrix} - x\det\begin{bmatrix} a & b \\ a & b \end{bmatrix} = \det\begin{bmatrix} a & b \\ c & d \end{bmatrix} - \text{zero.} \quad (4)$$

This was "linearity in row 2 with row 1 fixed". It means (again) that our elimination steps from the original matrix A to an upper triangular U *do not change the determinant*:

$$\boxed{\det \boldsymbol{A} = \det \boldsymbol{U} = U_{11}\,U_{22}\dots U_{nn} = \textbf{product of the pivots}} \quad (5)$$

Cramer's Rule to Solve $Ax = b$

A neat idea gives the first component x_1 of the solution vector $\boldsymbol{x}$ to $A\boldsymbol{x} = \boldsymbol{b}$. Replace the first column of I by $\boldsymbol{x}$. This new matrix has determinant x_1. When you multiply it by A, *the first column becomes $A\boldsymbol{x}$ which is $\boldsymbol{b}$.* The other columns of B_1 are copied from A:

$$\textbf{Key idea} \qquad \begin{bmatrix} & & \\ & A & \\ & & \end{bmatrix}\begin{bmatrix} \boldsymbol{x_1} & 0 & 0 \\ \boldsymbol{x_2} & 1 & 0 \\ \boldsymbol{x_3} & 0 & 1 \end{bmatrix} = \begin{bmatrix} \boldsymbol{b_1} & a_{12} & a_{13} \\ \boldsymbol{b_2} & a_{22} & a_{23} \\ \boldsymbol{b_3} & a_{32} & a_{33} \end{bmatrix} = \boldsymbol{B_1}. \quad (6)$$

We multiplied a column at a time. *Take determinants of the three matrices to find x_1*:

$$\boxed{\textbf{Product rule} \qquad (\det \boldsymbol{A})\,(\boldsymbol{x_1}) = \det \boldsymbol{B_1} \qquad \text{or} \qquad \boldsymbol{x_1} = \frac{\det \boldsymbol{B_1}}{\det \boldsymbol{A}}.} \quad (7)$$

This is the first component of $\boldsymbol{x}$ in Cramer's Rule. Changing a column of A gave B_1. To find x_2 and B_2, put the vectors $\boldsymbol{x}$ and $\boldsymbol{b}$ into the *second columns* of I and A:

$$\textbf{Same idea} \qquad \begin{bmatrix} & & \\ \boldsymbol{a}_1 & \boldsymbol{a}_2 & \boldsymbol{a}_3 \\ & & \end{bmatrix}\begin{bmatrix} 1 & x_1 & 0 \\ 0 & x_2 & 0 \\ 0 & x_3 & 1 \end{bmatrix} = \begin{bmatrix} & & \\ \boldsymbol{a}_1 & \boldsymbol{b} & \boldsymbol{a}_3 \\ & & \end{bmatrix} = B_2. \quad (8)$$

Take determinants to find $(\det A)(x_2) = \det B_2$. This gives $\boldsymbol{x_2} = \mathbf{(\det \boldsymbol{B_2})/(\det \boldsymbol{A})}$.

Example 1 Solving $\mathbf{3}x_1 + \mathbf{4}x_2 = \mathbf{2}$ and $\mathbf{5}x_1 + \mathbf{6}x_2 = \mathbf{4}$ needs three determinants:

$$\text{Put 2 and 4 into each } B \qquad \det \boldsymbol{A} = \begin{vmatrix} 3 & 4 \\ 5 & 6 \end{vmatrix} \qquad \det \boldsymbol{B_1} = \begin{vmatrix} \mathbf{2} & 4 \\ \mathbf{4} & 6 \end{vmatrix} \qquad \det \boldsymbol{B_2} = \begin{vmatrix} 3 & \mathbf{2} \\ 5 & \mathbf{4} \end{vmatrix}$$

The determinants of B_1 and B_2 are -4 and 2. Those are divided by $\det A = -2$:

$$\textbf{Find } \boldsymbol{x} = \boldsymbol{A}^{-1}\boldsymbol{b} \qquad x_1 = \frac{-4}{-2} = \mathbf{2} \qquad x_2 = \frac{2}{-2} = \mathbf{-1} \qquad \textbf{Check} \begin{bmatrix} 3 & 4 \\ 5 & 6 \end{bmatrix}\begin{bmatrix} \mathbf{2} \\ \mathbf{-1} \end{bmatrix} = \begin{bmatrix} 2 \\ 4 \end{bmatrix}$$

CRAMER's RULE If $\det A$ is not zero, $A\boldsymbol{x} = \boldsymbol{b}$ is solved by determinants:

$$x_1 = \frac{\det B_1}{\det A} \qquad x_2 = \frac{\det B_2}{\det A} \qquad \ldots \qquad x_n = \frac{\det B_n}{\det A} \tag{9}$$

The matrix B_j has the jth column of A replaced by the vector b.

To solve an n by n system, Cramer's Rule evaluates $n+1$ determinants (of A and the n different B's). When each one is the sum of $n!$ terms—applying the "big formula" with all permutations—this makes a total of $(n+1)!$ terms. *It would be crazy to solve equations that way.* But we do finally have an explicit formula for the solution to $A\boldsymbol{x} = \boldsymbol{b}$.

Example 2 Cramer's Rule is inefficient for numbers but it is well suited to letters. For $n = 2$, find the columns $\boldsymbol{x}$ and $\boldsymbol{y}$ of A^{-1} by solving $AA^{-1} = A\,[\boldsymbol{x}\;\;\boldsymbol{y}] = I$.

$$A\boldsymbol{x} = \begin{bmatrix} a & b \\ c & d \end{bmatrix}\begin{bmatrix} x_1 \\ x_2 \end{bmatrix} = \begin{bmatrix} \mathbf{1} \\ \mathbf{0} \end{bmatrix} \qquad A\boldsymbol{y} = \begin{bmatrix} a & b \\ c & d \end{bmatrix}\begin{bmatrix} y_1 \\ y_2 \end{bmatrix} = \begin{bmatrix} \mathbf{0} \\ \mathbf{1} \end{bmatrix}$$

Those share the same matrix A. We need $|A|$ and four determinants for $x_1,\ x_2,\ y_1,\ y_2$:

$$\begin{vmatrix} a & b \\ c & d \end{vmatrix} \quad \text{and} \quad \begin{vmatrix} \mathbf{1} & b \\ \mathbf{0} & \boldsymbol{d} \end{vmatrix} \quad \begin{vmatrix} a & \mathbf{1} \\ \boldsymbol{c} & \mathbf{0} \end{vmatrix} \quad \begin{vmatrix} \mathbf{0} & \boldsymbol{b} \\ \mathbf{1} & d \end{vmatrix} \quad \begin{vmatrix} \boldsymbol{a} & \mathbf{0} \\ c & \mathbf{1} \end{vmatrix}$$

The last four determinants are d, $-c$, $-b$, and a. (They are the cofactors!) Here is A^{-1}:

$$x_1 = \frac{d}{|A|},\; x_2 = \frac{-c}{|A|},\; y_1 = \frac{-b}{|A|},\; y_2 = \frac{a}{|A|} \text{ and then } A^{-1} = \frac{1}{ad-bc}\begin{bmatrix} d & -b \\ -c & a \end{bmatrix}.$$

I chose 2 by 2 so that the main points could come through clearly. The key idea is: $\boldsymbol{A^{-1}}$ **involves the cofactors**. When the right side is a column of the identity matrix I, as in $AA^{-1} = I$, **the determinant of each $\boldsymbol{B_j}$ in Cramer's Rule is a cofactor of $\boldsymbol{A}$.**

You can see those cofactors for $n = 3$. Solve $A\boldsymbol{x} = (1,0,0)$ to find column 1 of A^{-1}:

$$\begin{array}{l} \textbf{Determinants of } \boldsymbol{B}\textbf{'s} \\ \textbf{are Cofactors of } \boldsymbol{A} \end{array} \quad \begin{vmatrix} \mathbf{1} & a_{12} & a_{13} \\ \mathbf{0} & a_{22} & a_{23} \\ \mathbf{0} & a_{32} & a_{33} \end{vmatrix} \quad \begin{vmatrix} a_{11} & \mathbf{1} & a_{13} \\ a_{21} & \mathbf{0} & a_{23} \\ a_{31} & \mathbf{0} & a_{33} \end{vmatrix} \quad \begin{vmatrix} a_{11} & a_{12} & \mathbf{1} \\ a_{21} & a_{22} & \mathbf{0} \\ a_{31} & a_{32} & \mathbf{0} \end{vmatrix}$$

That determinant of B_1 is the cofactor $C_{11} = a_{22}a_{33} - a_{23}a_{32}$. Then $|B_2|$ is the cofactor C_{12}. Notice that the correct minus sign appears in $-(a_{21}a_{33} - a_{23}a_{31})$. This cofactor C_{12} goes into column 1 of A^{-1}. When we divide by $\det A$, we have computed the inverse.

$$\textbf{FORMULA FOR } \boldsymbol{A^{-1}} \qquad (\boldsymbol{A}^{-1})_{ij} = \frac{\boldsymbol{C_{ji}}}{\det \boldsymbol{A}} \qquad \boldsymbol{A}^{-1} = \frac{\boldsymbol{C}^{\mathrm{T}}}{\det \boldsymbol{A}} \tag{10}$$

Areas and Volumes

Determinants lead to formulas for areas and volumes. The regions will have straight sides—but they are not square. A typical region is a parallelogram—or half a parallelogram which is a triangle. The problem is: **Find the area**. For a triangle, the area is $\frac{1}{2}\,\boldsymbol{bh}$: half the base times the height. A parallelogram contains two triangles with equal area, so we omit the $\frac{1}{2}$. Then parallelogram area $= \boldsymbol{bh} =$ *base times height.*

Those formulas are simple to use and remember. But they don't fit our problem, because we are not given the base and height. We only know **the positions of the corners**. For the triangle, suppose the corner points are $(0,0)$ and (a,b) and (c,d). For the parallelogram (twice as large) the fourth corner will be $(a+c, b+d)$. Knowing a, b, c, d, *what is the area*?

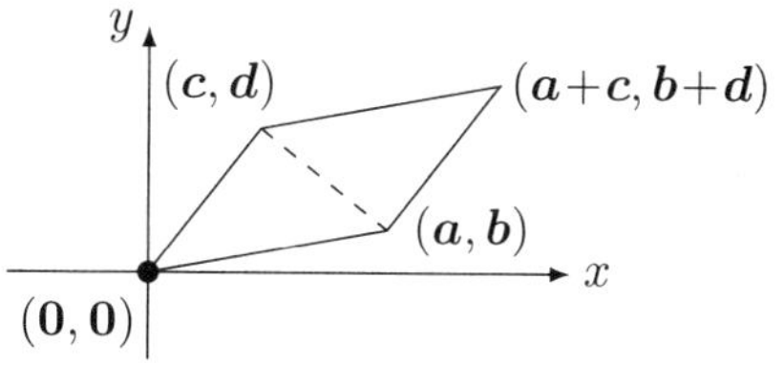

The base could be the lowest side and its length is $\sqrt{a^2+b^2}$. To find the height, we could create a line from (c,d) that is perpendicular to the baseline. The length h of that perpendicular line would involve more square roots. But the area itself does not involve square roots and it has a beautiful formula:

$$\textbf{Area of parallelogram} = \textbf{Determinant of matrix} = \pm \begin{vmatrix} \boldsymbol{a} & \boldsymbol{b} \\ \boldsymbol{c} & \boldsymbol{d} \end{vmatrix} = |\boldsymbol{ad} - \boldsymbol{bc}|. \quad (11)$$

Our goal is to find that formula by linear algebra: no square roots or negative volumes.

We also have a more difficult goal. We need to move into 3 dimensions and eventually into n dimensions. We start with four corners $0,0,0$ and a,b,c and p,q,r and x,y,z of a box. (The box is not rectangular. Every side is a parallelogram. It will look lopsided.) If we use the area formula (11) as a guide, we could guess the correct volume formula:

$$\textbf{Volume of box} = \textbf{Determinant of matrix} = \pm \begin{vmatrix} a & b & c \\ p & q & r \\ x & y & z \end{vmatrix} \quad (12)$$

Our first effort stays in a plane. For this case we use geometry. Figure 5.1 shows how adding pieces to a parallelogram can produce a rectangle. When we subtract the areas of those six pieces, we arrive at the correct parallelogram area $\boldsymbol{ad} - \boldsymbol{bc}$ (no square roots). The picture is not very elegant, but in two dimensions it succeeds.

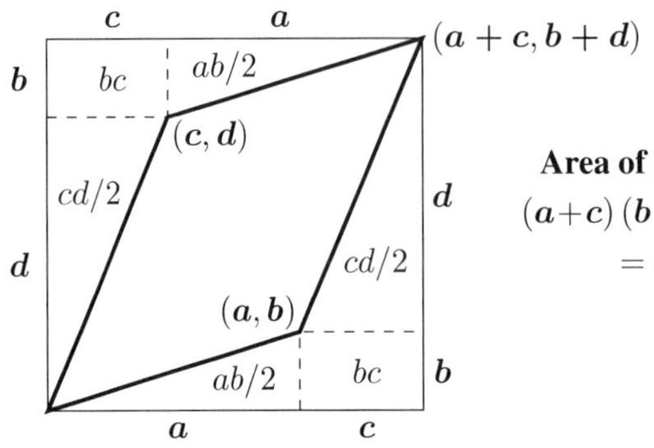

Area of parallelogram

$$(a+c)\,(b+d)-2bc-ab-cd$$
$$= ad - bc$$

Figure 5.1: Adding six simple pieces to a parallelogram produces a rectangle.

Would a similar construction be possible in three dimensions? Following Figure 5.1, I believe we could add simple pieces to make a tilted box into a rectangular box—but it doesn't look easy. And there is a much better way: *Use linear algebra.*

Areas and Volumes by Linear Algebra

In working on this problem, I came to an understanding. *If we do more algebra, then we need less geometry.* Very often, linear algebra comes down to factoring a matrix. We will look there for ideas.

A box in n dimensions has n edges $\boldsymbol{e}_1, \boldsymbol{e}_2, \ldots, \boldsymbol{e}_n$ going out from the origin. The parallelogram in two dimensions had two vectors $\boldsymbol{e}_1 = (a, b)$ and $\boldsymbol{e}_2 = (c, d)$. Those vectors $\boldsymbol{e}$ give two corners or n corners of the "box". In the 2-dimensional picture, the fourth corner was $\boldsymbol{e}_1 + \boldsymbol{e}_2$. In the n-dimensional picture, the other corners of the box would be sums of the $\boldsymbol{e}$'s. *The box is totally determined by the n edges in the matrix E*:

$$\textbf{Edge matrix} \quad \boldsymbol{E_2} = \begin{bmatrix} a & c \\ b & d \end{bmatrix} \quad \text{and} \quad \boldsymbol{E_n} = \begin{bmatrix} | & & | \\ e_1 & \cdots & e_n \\ | & & | \end{bmatrix}$$

Our goal is to prove that the volume of the box is $|\det \boldsymbol{E}|$. We considered three possible factorizations of E, to reach this goal. They are taken from Chapters 2 and 4 and 7. The third factorization is called the Singular Value Decomposition of E: the SVD.

Lower times upper triangular	$\boldsymbol{E = LU}$
Orthogonal times upper triangular	$\boldsymbol{E = QR}$
Orthogonal – Diagonal – Orthogonal	$\boldsymbol{E = U\Sigma V^{\mathrm{T}}}$

Those factors of E are square matrices because E is square (n by n). Remember that the determinant of L is 1 (all ones on its diagonal). The determinant of any orthogonal matrix is 1 or -1. Then $\det L = |\det Q| = |\det U| = |\det V| = 1$. We will certainly depend on the multiplication formula for the determinant of a product, which now tells us that

$$\boxed{|\det \boldsymbol{E}| = |\det \boldsymbol{U}| = |\det \boldsymbol{R}| = |\det \boldsymbol{\Sigma}|.} \tag{13}$$

The problem is to connect the volume of the box to one of those determinants—to connect geometry to linear algebra. Start with the geometry of an orthogonal matrix Q.

Key idea for any region in $\mathbf{R}^n$: Multiplying all points by an orthogonal matrix Q **does not change the volume**. Let me understand this first for $\boldsymbol{E} = \boldsymbol{Q}$ times $\boldsymbol{R}$.

Multiply by any matrix: Straight lines stay straight

Multiply by an orthogonal Q: $\boldsymbol{x}^{\mathrm{T}}\boldsymbol{x}$ and $\boldsymbol{x}^{\mathrm{T}}\boldsymbol{y}$ are the same as $(Q\boldsymbol{x})^{\mathrm{T}}(Q\boldsymbol{x})$ and $(Q\boldsymbol{x})^{\mathrm{T}}(Q\boldsymbol{y})$

Then lengths and angles and box shapes and volumes are not changed by $\boldsymbol{Q}$.

This remains true for curved regions. We divide them into many small cubes plus thin curved pieces. The total volume of those curved pieces can approach zero. The volumes of the cubes are not changed by Q. *The boxes for R and for $E = QR$ have the same volume.*

R is a triangular matrix. Its box has a volume we can compute. For a parallelogram in the xy plane, **the base and height are exactly the diagonal entries of $\boldsymbol{R}$.**

$$\boldsymbol{R} = \begin{bmatrix} u & v \\ 0 & w \end{bmatrix}$$

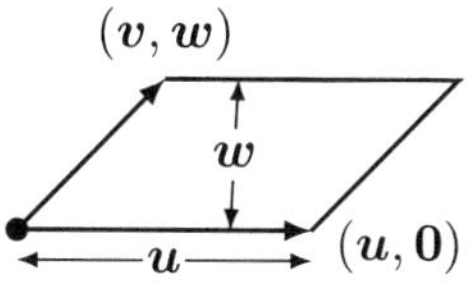

base $= u$, **height** $= w$

$|\textbf{area}| = |\boldsymbol{u} \text{ times } \boldsymbol{w}| = |\det \boldsymbol{R}|$

The key point is: The main diagonal of R shows the height in each new dimension. When we multiply those numbers on the diagonal of R, **we get the volume of the box and also the determinant of the triangular matrix $\boldsymbol{R}$.** The volume formula $|\det E|$ is now proved in all dimensions because $|\det \boldsymbol{Q}| = 1$ and $|\det \boldsymbol{E}| = |\det \boldsymbol{R}|$.

Final comment: The Singular Value Decomposition $E = U\Sigma V^{\mathrm{T}}$ has two orthogonal matrices U and V. The number $|\det E|$ is equal to $|\det \Sigma|$. And this matrix Σ is *diagonal.* It gives a perfectly normal *rectangular box* in $\mathbf{R}^n$. This SVD approach by $U\Sigma V^{\mathrm{T}}$ looks simpler than QR, which had a triangular matrix R producing a tilted box.

But that tilted figure shows a clear geometric meaning for the diagonal entries of R: *base and height.* The geometry of the SVD will be seen in Chapter 7. It is beautifully clear for *ellipses* in n dimensions. But the singular values are not so clear for boxes. Σ gives the lengths of the axes of an *ellipse* but not the sides of a rectangular box. For a box with straight sides, $E = QR$ leads directly to volume $= |\det R|$.

The next section will allow any shape: not just boxes.

Problem Set 5.2

1 If $\det A = 2$, what are $\det A^{-1}$ and $\det A^n$ and $\det A^{\mathrm{T}}$?

2 Compute the determinants of A, B, C, D. Are their columns independent?

$$A = \begin{bmatrix} 1 & 1 & 0 \\ 1 & 0 & 1 \\ 0 & 1 & 1 \end{bmatrix} \quad B = \begin{bmatrix} 1 & 2 & 3 \\ 4 & 5 & 6 \\ 7 & 8 & 9 \end{bmatrix} \quad C = \begin{bmatrix} A & 0 \\ 0 & A \end{bmatrix} \quad D = \begin{bmatrix} A & 0 \\ 0 & B \end{bmatrix}.$$

3 Show that $\det A = 0$, regardless of the five numbers marked by x's:

$$A = \begin{bmatrix} x & x & x \\ 0 & 0 & x \\ 0 & 0 & x \end{bmatrix}.$$

What are the cofactors of row 1?
What is the rank of A?
What are the 6 terms in $\det A$?

4 (a) If $D_n = \det(A^n)$, could $D_n \to \infty$ even if all $|A_{ij}| < 1$?
(b) Could $D_n \to 0$ even if all $|A_{ij}| > 1$?

Problems 5–9 are about Cramer's Rule for $x = A^{-1}b$.

5 Solve these linear equations by Cramer's Rule $x_j = \det B_j / \det A$:

(a) $\begin{aligned} 2x_1 + 5x_2 &= 1 \\ x_1 + 4x_2 &= 2 \end{aligned}$ (b) $\begin{aligned} 2x_1 + x_2 &= 1 \\ x_1 + 2x_2 + x_3 &= 0 \\ x_2 + 2x_3 &= 0. \end{aligned}$

6 Use Cramer's Rule to solve for y (only). Call the 3 by 3 determinant D:

(a) $\begin{aligned} ax + by &= 1 \\ cx + dy &= 0 \end{aligned}$ (b) $\begin{aligned} ax + by + cz &= 1 \\ dx + ey + fz &= 0 \\ gx + hy + iz &= 0. \end{aligned}$

7 Cramer's Rule breaks down when $\det A = 0$. Example (a) has no solution while (b) has infinitely many. What are the ratios $x_j = \det B_j / \det A$ in these two cases?

(a) $\begin{aligned} 2x_1 + 3x_2 &= 1 \\ 4x_1 + 6x_2 &= 1 \end{aligned}$ (parallel lines) (b) $\begin{aligned} 2x_1 + 3x_2 &= 1 \\ 4x_1 + 6x_2 &= 2 \end{aligned}$ (same line)

8 *Quick proof of Cramer's rule.* The determinant is a linear function of column 1. It is zero if two columns are equal. When $\boldsymbol{b} = A\boldsymbol{x} = x_1\boldsymbol{a}_1 + x_2\boldsymbol{a}_2 + x_3\boldsymbol{a}_3$ goes into the first column of A, the determinant of this matrix B_1 is

$$|\boldsymbol{b} \quad \boldsymbol{a}_2 \quad \boldsymbol{a}_3| = |x_1\boldsymbol{a}_1 + x_2\boldsymbol{a}_2 + x_3\boldsymbol{a}_3 \quad \boldsymbol{a}_2 \quad \boldsymbol{a}_3| = x_1|\boldsymbol{a}_1 \quad \boldsymbol{a}_2 \quad \boldsymbol{a}_3| = x_1 \det A.$$

(a) What formula for x_1 comes from left side = right side?
(b) What steps lead to the middle equation?

9 If the right side $\boldsymbol{b}$ is the first column of A, solve the 3 by 3 system $A\boldsymbol{x} = \boldsymbol{b}$. How does each determinant in Cramer's Rule lead to this solution $\boldsymbol{x}$?

10 (a) Find the area of the parallelogram with edges $\boldsymbol{v} = (3, 2)$ and $\boldsymbol{w} = (1, 4)$.

(b) Find the area of the triangle with sides $\boldsymbol{v}$, $\boldsymbol{w}$, and $\boldsymbol{v} + \boldsymbol{w}$. Draw it.

(c) Find the area of the triangle with sides $\boldsymbol{v}$, $\boldsymbol{w}$, and $\boldsymbol{w} - \boldsymbol{v}$. Draw it.

11 (a) The corners of a triangle are $(2, 1)$ and $(3, 4)$ and $(0, 5)$. What is the area?

(b) Add a corner at $(-1, 0)$ to make a lopsided region (four sides). Find the area.

12 The Hadamard matrix H has orthogonal rows. The box is a hypercube!

$$\text{What is} \quad |H| = \begin{vmatrix} 1 & 1 & 1 & 1 \\ 1 & 1 & -1 & -1 \\ 1 & -1 & -1 & 1 \\ 1 & -1 & 1 & -1 \end{vmatrix} = \begin{array}{l} \text{volume of a hypercube in } \mathbf{R}^4? \\ \text{The sides have length 2} \end{array}$$

13 An n-dimensional cube has how many corners? How many edges? How many $(n-1)$-dimensional faces? The cube in $\mathbf{R}^n$ whose edges are the rows of $2I$ has volume _____. A hypercube computer has parallel processors at the corners with connections along the edges.

14 The triangle with corners $(0, 0), (1, 0), (0, 1)$ has area $\frac{1}{2}$. The pyramid in $\mathbf{R}^3$ with four corners $(0, 0, 0), (1, 0, 0), (0, 1, 0), (0, 0, 1)$ has volume _____. What is the volume of a pyramid in $\mathbf{R}^4$ with five corners at $(0, 0, 0, 0)$ and the four columns of I?

15 Suppose E_n is the determinant of the tridiagonal $1, 1, 1$ matrix of order n. By cofactors of row 1 show that $E_n = E_{n-1} - E_{n-2}$. Starting from $E_1 = 1$ and $E_2 = 0$ find $E_3, E_4, \ldots, E_8$. By noticing how the E's repeat, find E_{100}.

$$E_1 = |1| \quad E_2 = \begin{vmatrix} 1 & 1 \\ 1 & 1 \end{vmatrix} \quad E_3 = \begin{vmatrix} 1 & 1 & 0 \\ 1 & 1 & 1 \\ 0 & 1 & 1 \end{vmatrix} \quad E_4 = \begin{vmatrix} 1 & 1 & 0 & 0 \\ 1 & 1 & 1 & 0 \\ 0 & 1 & 1 & 1 \\ 0 & 0 & 1 & 1 \end{vmatrix}.$$

16 From the cofactor formula $AC^{\mathrm{T}} = (\det A)I$ show that $\det C = (\det A)^{n-1}$.

17 Suppose $\det A = 1$ and you know all the cofactors in C. How can you find A?

18 If a 3 by 3 matrix has entries $1, 2, 3, 4, \ldots, 9$, what is the maximum determinant?

19 If the edge matrix E is orthogonal, the box has volume _____.
If the edge matrix E is singular, the box has volume _____.
If the volume in $\mathbf{R}^n$ is V, the box for $2E$ has volume _____.

20 Draw parallelograms for $\begin{bmatrix} 2 & 1 \\ 3 & 4 \end{bmatrix}$ and $\begin{bmatrix} 2 & 3 \\ 1 & 4 \end{bmatrix}$. Can you see any reason for equal areas?

21 Transposing the edge matrix $\begin{bmatrix} u & v \\ 0 & w \end{bmatrix}$ gives a matrix with the same determinant and a new parallelogram with the same area. Can you draw it and recompute its area?

5.3 Linear Transformations

1 Linear transformations T obey the rule $T(c\boldsymbol{v} + d\boldsymbol{w}) = c\,T(\boldsymbol{v}) + d\,T(\boldsymbol{w})$.

2 Derivatives and integrals are linear transformations in function space.

3 Volumes of all shapes are multiplied by $|\det A|$ when every $\boldsymbol{x}$ goes to $A\boldsymbol{x}$.

Transformations T follow the same idea as functions. In goes a number x or a vector $\boldsymbol{v}$, out comes $f(x)$ or $T(\boldsymbol{v})$. For one vector $\boldsymbol{v}$ or one number x, we apply the transformation T or we evaluate $f(x)$. The deeper goal is to see all vectors $\boldsymbol{v}$ at once. We are transforming the whole space $\mathbf{V}$.

Start again with a matrix A. It transforms $\boldsymbol{v}$ to $A\boldsymbol{v}$. It transforms $\boldsymbol{w}$ to $A\boldsymbol{w}$. Then we *know* what happens to $\boldsymbol{u} = \boldsymbol{v} + \boldsymbol{w}$. There is no doubt about $A\boldsymbol{u}$, it has to equal $A\boldsymbol{v} + A\boldsymbol{w}$. Matrix multiplication $T(\boldsymbol{v}) = A\boldsymbol{v}$ is an example of a ***linear transformation*** :

A ***transformation*** T assigns an output $T(\boldsymbol{v})$ to each input vector $\boldsymbol{v}$ in $\mathbf{V}$.

The transformation is ***linear*** if it meets these requirements for all $\boldsymbol{v}$ and $\boldsymbol{w}$:

(a) $T(\boldsymbol{v} + \boldsymbol{w}) = T(\boldsymbol{v}) + T(\boldsymbol{w})$ (b) $T(c\boldsymbol{v}) = cT(\boldsymbol{v})$ for all c.

Those rules tell us : If the input is $\boldsymbol{v} = \mathbf{0}$, then the output must be $T(\mathbf{0}) = \mathbf{0}$. *No shift.*

$T(\mathbf{0} + \boldsymbol{w}) = T(\mathbf{0}) + T(\boldsymbol{w})$ Removing $T(\boldsymbol{w})$ from both sides leaves $\mathbf{0} = T(\mathbf{0})$.

Combining the two rules tells us about linear combinations of $\boldsymbol{v}$ and $\boldsymbol{w}$:

$$\boxed{T(c\boldsymbol{v} + d\boldsymbol{w}) = T(c\boldsymbol{v}) + T(d\boldsymbol{w}) = c\,T(\boldsymbol{v}) + d\,T(\boldsymbol{w})} \qquad (1)$$

Example 1 T_1 rotates the whole xy plane by $90°$ around the center point $(0,0)$.

This is a linear transformation ! Straight lines will rotate into straight lines. A square will rotate into the same size square. The center point $(0,0)$ does not move : $T(\mathbf{0}) = \mathbf{0}$. Requirement (1) for linear combinations $c\boldsymbol{v} + d\boldsymbol{w}$ is satisfied.

The likable part of that example is : No matrix was needed. We can visualize linear geometry without linear algebra. If we have another linear transformation T_2 of the xy plane, then T_2 can follow T_1 to produce T_2T_1 : First find $T_1(\boldsymbol{v})$ and then apply T_2.

Example 2 T_2 reflects each vector (x, y) to its mirror image $(x, -y)$ across the x axis.

This is another linear transformation that doesn't need a matrix. Notice that T_2T_1 *differs from* T_1T_2 : *Reflecting the rotated vector* $\neq$ *rotating the reflected vector.* $(1,0)$ rotates to $(0,1)$ and reflects to $(\mathbf{0}, -\mathbf{1})$. But $(1,0)$ reflects to $(1,0)$ and rotates to $(\mathbf{0}, \mathbf{1})$.

Example 3 The length $T(\boldsymbol{v}) = \|\boldsymbol{v}\|$ is not linear. Requirement (a) for linearity would be $\|\boldsymbol{v}+\boldsymbol{w}\| = \|\boldsymbol{v}\| + \|\boldsymbol{w}\|$. Requirement (b) would be $\|c\boldsymbol{v}\| = c\|\boldsymbol{v}\|$. Both are false!

Not (a): The sides of a triangle satisfy an *inequality* $\|\boldsymbol{v}+\boldsymbol{w}\| \le \|\boldsymbol{v}\| + \|\boldsymbol{w}\|$.

Not (b): The length $\|-\boldsymbol{v}\|$ is $\|\boldsymbol{v}\|$ and not $-\|\boldsymbol{v}\|$. For negative c, linearity fails.

T (every vector) from T (basis vectors)

The rule of linearity extends to combinations of three vectors or n vectors:

$$\textbf{Linearity} \qquad \boldsymbol{u} = c_1\boldsymbol{v}_1 + c_2\boldsymbol{v}_2 + \cdots + c_n\boldsymbol{v}_n \quad \textbf{must transform to} \qquad T(\boldsymbol{u}) = c_1T(\boldsymbol{v}_1) + c_2T(\boldsymbol{v}_2) + \cdots + c_nT(\boldsymbol{v}_n) \tag{2}$$

The 2-vector rule starts the 3-vector proof: $T(c\boldsymbol{u} + d\boldsymbol{v} + e\boldsymbol{w}) = T(c\boldsymbol{u}) + T(d\boldsymbol{v} + e\boldsymbol{w})$. Then linearity applies to both of those parts, to give $c\,T(\boldsymbol{u}) + d\,T(\boldsymbol{v}) + eT(\boldsymbol{w})$.

The n-vector rule (2) leads to the most important fact about linear transformations:

Suppose you know $T(v)$ for all vectors $v_1, \ldots, v_n$ in a basis
Then you know $T(u)$ for every vector u in the space.

You see the reason: Every $\boldsymbol{u}$ in the space is a combination of the basis vectors $\boldsymbol{v}_j$. Then linearity tells us that $T(\boldsymbol{u})$ must be the same combination of the outputs $T(\boldsymbol{v}_j)$.

A key point about linear transformations: **If we choose bases for the input space and the output space, then T can be specified by a matrix A.** The rule for constructing A must use the two bases (which can be the same if output space = input space).

Step 1. Apply the transformation T to each *input basis vector* $\boldsymbol{v}_j$.

Step 2. Write the output $T(\boldsymbol{v}_j)$ as a combination of the *output basis vectors* $\boldsymbol{w}_i$.

Step 3. The coefficients A_{ij} in that combination $T(\boldsymbol{v}_j) = \sum A_{ij}\boldsymbol{w}_i$ go into column j of A.

This matrix A finds the output $T(\boldsymbol{v})$ for any input $\boldsymbol{v}$. If the vector $\boldsymbol{c} = (c_1, \ldots, c_n)$ gives the input coefficients in $\boldsymbol{v} = c_1\boldsymbol{v}_1 + \cdots + c_n\boldsymbol{v}_n$, then $\boldsymbol{b} = A\boldsymbol{c}$ gives the output coefficients in $T(\boldsymbol{v}) = b_1\boldsymbol{w}_1 + \cdots + b_m\boldsymbol{w}_m$. ***T* becomes multiplication by *A*.**

Example 4 T is rotation by θ of the xy plane with basis vectors $\begin{bmatrix} 1 \\ 0 \end{bmatrix}$ and $\begin{bmatrix} 0 \\ 1 \end{bmatrix}$.

$$\boldsymbol{T}\begin{bmatrix} 1 \\ 0 \end{bmatrix} = \begin{bmatrix} \cos\theta \\ \sin\theta \end{bmatrix} \quad \text{and} \quad \boldsymbol{T}\begin{bmatrix} 0 \\ 1 \end{bmatrix} = \begin{bmatrix} -\sin\theta \\ \cos\theta \end{bmatrix} \quad \text{produce} \quad \boldsymbol{A} = \begin{bmatrix} \cos\theta & -\sin\theta \\ \sin\theta & \cos\theta \end{bmatrix}$$

The Derivative is a Linear Transformation

It is linearity that you use, to find the derivative of $u(x) = 6 - 4x + 3x^2$. Start with the derivatives of 1 and x and x^2. Those functions are the basis vectors. Their derivatives are 0 and 1 and $2x$. Then use linearity for the derivative of any combination like $u(x)$:

$$\frac{du}{dx} = \mathbf{6}\,(\text{derivative of } 1) - \mathbf{4}\,(\text{derivative of } x) + \mathbf{3}\,(\text{derivative of } x^2) = \boldsymbol{-4 + 6x}.$$

All of calculus depends on linearity! Precalculus finds a few essential derivatives, for $\boldsymbol{x^n}$ and $\boldsymbol{\sin x}$ and $\boldsymbol{\cos x}$ and $\boldsymbol{e^x}$. Then linearity applies to all their combinations.

I would say that the only rule special to calculus is the *chain rule*. That produces the derivative of a chain of functions $f(g(x))$ or $f(g(h(x)))$. Needed in deep learning!

Nullspace of $\boldsymbol{T(u) = du/dx}$. For the nullspace we solve $T(u) = 0$. The derivative is zero when u *is a constant function*. So the nullspace of d/dx is a line in function space—all multiples of the special solution $u = 1$. "The derivative operator is not invertible."

Column space of $\boldsymbol{T(u) = du/dx}$. In our example the input space contains all quadratics $a + bx + cx^2$. Then the outputs (the column space of T) are all linear functions $b + 2cx$. Notice that the **Counting Theorem** is still true: $\boldsymbol{r + (n - r) = (n - 1) + 1 = n}$.

$$\text{dimension}\,(\textbf{column space}) + \text{dimension}\,(\textbf{nullspace}) = \mathbf{2 + 1} = \text{dimension}\,(\textbf{input space})$$

What is the matrix for d/dx? I can't leave derivatives without asking for a matrix. $T = d/dx$ is a linear transformation. We know what T does to these basis functions:

$$\boldsymbol{v}_1, \boldsymbol{v}_2, \boldsymbol{v}_3 = \mathbf{1}, \boldsymbol{x}, \boldsymbol{x^2} \qquad \frac{d\boldsymbol{v}_1}{dx} = \mathbf{0} \qquad \frac{d\boldsymbol{v}_2}{dx} = 1 = \boldsymbol{v}_1 \qquad \frac{d\boldsymbol{v}_3}{dx} = 2x = \mathbf{2}\boldsymbol{v}_2. \tag{3}$$

The 3-dimensional input space **V** (= quadratics) transforms to the 2-dimensional output space **W** (= linear functions). If $\boldsymbol{v}_1, \boldsymbol{v}_2, \boldsymbol{v}_3$ were vectors, we would know the matrix.

$$\boxed{\boldsymbol{A} = \begin{bmatrix} 0 & \mathbf{1} & 0 \\ 0 & 0 & \mathbf{2} \end{bmatrix} = \textbf{matrix form of the derivative } \boldsymbol{T = \frac{d}{dx}}.} \tag{4}$$

The linear transformation du/dx is perfectly copied by the matrix multiplication $A\boldsymbol{u}$.

$$\textbf{Input } \boldsymbol{u} \quad a + bx + cx^2 \qquad \textbf{Multiplication } A\boldsymbol{u} = \begin{bmatrix} 0 & 1 & 0 \\ 0 & 0 & 2 \end{bmatrix} \begin{bmatrix} a \\ b \\ c \end{bmatrix} = \begin{bmatrix} \boldsymbol{b} \\ \mathbf{2}\boldsymbol{c} \end{bmatrix} \qquad \textbf{Output } \boldsymbol{\frac{du}{dx}} = b + 2cx.$$

The Integral is a Linear Transformation

Next we look at integrals from 0 to x. **They give the pseudoinverse T^+ of the derivative!** I can't write T^{-1} and I can't say "*the inverse of* T" when the derivative of 1 is $T(1) = 0$.

Example 5 **Integration T^+ is also linear**: $\int_0^x (D + Ex)\,dx = Dx + \frac{1}{2}Ex^2$.

The input basis is now $1, x$. The output basis is $1, x, x^2$. The matrix A for T is 2 by 3. The matrix A^+ for T^+ is 3 by 2. A^+ inverts A where possible.

Input $v = D + Ex$ **Multiply** $A^+v = \begin{bmatrix} 0 & 0 \\ 1 & 0 \\ 0 & \frac{1}{2} \end{bmatrix} \begin{bmatrix} D \\ E \end{bmatrix} = \begin{bmatrix} \mathbf{0} \\ \boldsymbol{D} \\ \frac{\mathbf{1}}{\mathbf{2}}\boldsymbol{E} \end{bmatrix}$ **Output = Integral of** $\boldsymbol{v}$: $T^+(\boldsymbol{v}) = Dx + \frac{1}{2}Ex^2$

The Fundamental Theorem of Calculus says that integration is the pseudoinverse of differentiation. For linear algebra, the matrix A^+ is the pseudoinverse of A.

$$A^+A = \begin{bmatrix} 0 & 0 \\ 1 & 0 \\ 0 & \frac{1}{2} \end{bmatrix} \begin{bmatrix} 0 & 1 & 0 \\ 0 & 0 & 2 \end{bmatrix} = \begin{bmatrix} \mathbf{0} & 0 & 0 \\ 0 & \mathbf{1} & 0 \\ 0 & 0 & \mathbf{1} \end{bmatrix} \quad \text{and} \quad AA^+ = \begin{bmatrix} \mathbf{1} & 0 \\ 0 & \mathbf{1} \end{bmatrix}. \tag{5}$$

The derivative of a constant function is zero. That zero is on the diagonal of A^+A. Calculus wouldn't be calculus without that 1-dimensional nullspace of $T = d/dx$.

Example 6 Suppose A is an *invertible matrix*. Certainly $T(\boldsymbol{v} + \boldsymbol{w}) = A\boldsymbol{v} + A\boldsymbol{w} = T(\boldsymbol{v}) + T(\boldsymbol{w})$. Another linear transformation is multiplication by A^{-1}. This produces the ***inverse transformation*** T^{-1}, which brings every vector $T(\boldsymbol{v})$ back to $\boldsymbol{v}$:

$$T^{-1}(T(\boldsymbol{v})) = \boldsymbol{v} \quad \text{matches the matrix multiplication} \quad A^{-1}(A\boldsymbol{v}) = \boldsymbol{v}.$$

If $T(v) = Av$ and $S(u) = Bu$, then $T(S(u))$ matches the product ABu.

We are reaching an unavoidable question. ***Are all linear transformations from* V *to* W *produced by matrices?*** When a linear T is described as a "rotation" or a "projection" or ". . .", is there always a matrix A hiding behind T? Is $T(\boldsymbol{v})$ always $A\boldsymbol{v}$?

The answer is *yes*! This is an approach to linear algebra that doesn't start with matrices. We still end up with matrices—*after we choose an input basis and output basis.*

Example 7 Suppose T shifts the whole plane across by 1. **This is not linear**. Computer graphics has found a way to use a matrix—but it is **3 by 3**. Every point (x, y) is given the "*homogeneous coordinates*" $(x, y, 1)$. Then you can shift a page by using matrices.

A diagonal matrix rescales a page and an orthogonal matrix rotates a page.

The Geometry of Linear Transformations

Suppose T is an invertible linear transformation from the xy plane to itself. Then every straight line is transformed into a straight line. Every triangle is transformed into a triangle.

The vectors $\boldsymbol{v}_1$ and $\boldsymbol{v}_2$ give a basis. If we know $T(\boldsymbol{v}_1)$ and $T(\boldsymbol{v}_2)$, then we know $T(\boldsymbol{u})$ for all other points $\boldsymbol{u}$. The triangle $\mathbf{0}-\boldsymbol{v}_1-\boldsymbol{v}_2$ transforms into a triangle. The area of every triangle (wherever it is) will be connected by the area/determinant formula:

area of transformed triangle = $|\det T|$ times area of original triangle

What is the determinant of a linear transformation T of the xy plane? Use its matrix!

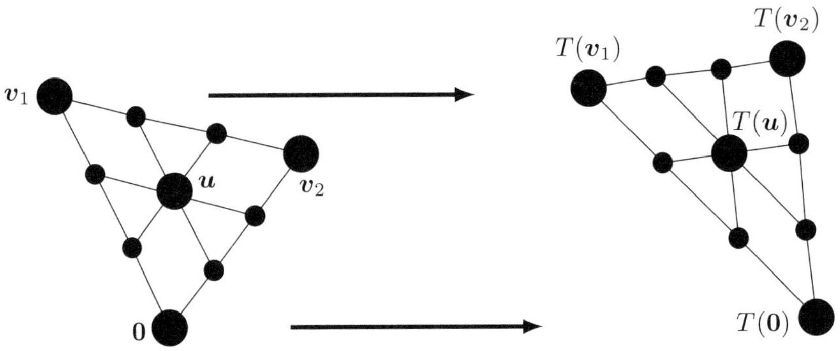

Figure 5.2: Lines to lines, equal spacing to equal spacing, $\boldsymbol{u} = \mathbf{0}$ to $T(\boldsymbol{u}) = \mathbf{0}$.

One key point about areas, when a linear T transforms one part of a plane to another part in Figure 5.2. The area of the big triangle is multiplied by a number. The area of every small triangle is also *multiplied by that number*. More is true: Every circle and square and every shape whatsoever has its area **multiplied by that same number**.

We just fill the shape with small squares, as closely as we want. Their areas are all multiplied the same way. Section 5.2 discovered that the area multiplier is the determinant of the matrix E. Then E takes squares into parallelograms and circles into ellipses.

Why Do Determinants Multiply?

This valuable property $|\det \boldsymbol{AB}| = |\det \boldsymbol{A}|\,|\det \boldsymbol{B}|$ looks messy to prove. It is buried somehow in the big formula for the determinant. Here are two very different proofs, one from geometry using volumes, and one from logic using properties of the determinant.

Geometry Start with a standard cube in n dimensions. Each edge has length 1. The n edges going out from $(0, \ldots, 0)$ are the rows of the n by n identity matrix I. The volume of the cube is 1, which is the determinant of I.

Now multiply every point in the cube by the matrix B. This gives a box with sides from B. We know that *volume of box = determinant of* B. No problem so far, except for some risk that our logic might become circular.

Now multiply every point in that B-box by the matrix A. This gives a new AB-box. Altogether we have multiplied the columns of the identity matrix by AB, so the AB-box has volume $= |\mathbf{det}\, \boldsymbol{AB}\,|$.

But also we have multiplied the B-box by A. When any shape is multiplied by A, its volume is multiplied by $|\det A\,|$. Since the B-box had volume $|\det B\,|$, the AB-box has volume $= |\mathbf{det}\, \boldsymbol{A}\,|\,|\mathbf{det}\, \boldsymbol{B}\,|$.

The two calculations of the same volume tell us that $|\,\mathbf{det}\, \boldsymbol{AB}\,| = |\,\mathbf{det}\, \boldsymbol{A}\,|\,|\,\mathbf{det}\, \boldsymbol{B}\,|$.

Logic This is a different approach to the determinant. Instead of a formula to compute it (the Big Formula was hard to work with), we give three properties of all determinants.

1. The determinant of $A = I$ is 1.
2. The determinant reverses sign if two rows of A are exchanged.
3. The determinant is a linear function of each row separately.

From those properties $\mathbf{1}, \mathbf{2}, \mathbf{3}$ we can make any matrix triangular and find its determinant. Rule 3 will allow row operations like elimination. Rule 2 allows us to exchange rows. Rule 1 sets a multiplying factor to 1, for the volume of a unit cube.

The product rule comes from checking that the ratio $\det AB/\det B$ *has those same three properties.* Then that ratio must equal $\det A$. Thus $\mathbf{det}\, \boldsymbol{AB} = (\mathbf{det}\, \boldsymbol{A})\,(\mathbf{det}\, \boldsymbol{B})$.

Change of Bases

Suppose the input space and output space are both $\mathbf{R}^2$. Suppose $\boldsymbol{v}_1, \boldsymbol{v}_2$ is the input basis and $\boldsymbol{w}_1, \boldsymbol{w}_2$ is the output basis. What is the matrix B (using these bases) for the identity transformation $T(\boldsymbol{v}) = \boldsymbol{v}$? *Not always* I! B is the "change of basis matrix".

> **When $\boldsymbol{V} = \begin{bmatrix} \boldsymbol{v}_1 & \boldsymbol{v}_2 \end{bmatrix}$ and $\boldsymbol{W} = \begin{bmatrix} \boldsymbol{w}_1 & \boldsymbol{w}_2 \end{bmatrix}$, the change of basis matrix is $\boldsymbol{B} = \boldsymbol{W}^{-1}\boldsymbol{V}$.**
>
> **For any linear transformation, its matrix A (in the old bases) changes to $\boldsymbol{W}^{-1}\boldsymbol{A}\boldsymbol{V}$.**

I see a clear way to understand those rules. Suppose the same vector $\boldsymbol{u}$ is written in the input basis of $\boldsymbol{v}$'s and the output basis of $\boldsymbol{w}$'s. Then $T = I$. Its matrix gives $\boldsymbol{d} = W^{-1}V\boldsymbol{c}$.

$$\begin{array}{l} \boldsymbol{u} = c_1\boldsymbol{v}_1 + \cdots + c_n\boldsymbol{v}_n \\ \boldsymbol{u} = d_1\boldsymbol{w}_1 + \cdots + d_n\boldsymbol{w}_n \end{array} \quad \text{is} \quad \begin{bmatrix} \\ \boldsymbol{v}_1 & \cdots & \boldsymbol{v}_n \\ \\ \end{bmatrix} \begin{bmatrix} c_1 \\ \vdots \\ c_n \end{bmatrix} = \begin{bmatrix} \\ \boldsymbol{w}_1 & \cdots & \boldsymbol{w}_n \\ \\ \end{bmatrix} \begin{bmatrix} d_1 \\ \vdots \\ d_n \end{bmatrix} \quad \text{and } \boldsymbol{Vc} = \boldsymbol{Wd}.$$

The coefficients $\boldsymbol{d}$ in the new basis of $\boldsymbol{w}$'s are $\boldsymbol{d} = \boldsymbol{W}^{-1}\boldsymbol{V}\boldsymbol{c}$. Then $\boldsymbol{B}$ is $\boldsymbol{W}^{-1}\boldsymbol{V}$. For a transformation T between spaces $\mathbf{V}$ and $\mathbf{W}$, we insert the matrix for T to get $\boldsymbol{W}^{-1}\boldsymbol{A}\boldsymbol{V}$:

$$T(\boldsymbol{v}) = \boldsymbol{w} \text{ is } \begin{bmatrix} \\ T(\boldsymbol{v}_1) & \cdots & T(\boldsymbol{v}_m) \\ \\ \end{bmatrix} \begin{bmatrix} c_1 \\ \vdots \\ c_n \end{bmatrix} = \begin{bmatrix} \\ \boldsymbol{w}_1 & \cdots & \boldsymbol{w}_n \\ \\ \end{bmatrix} \begin{bmatrix} d_1 \\ \vdots \\ d_n \end{bmatrix} \quad \text{and } \boldsymbol{AVc} = \boldsymbol{Wd}.$$

Change of basis leads to $W^{-1}AV$ not WAV. Larger vectors $\boldsymbol{w}$ have smaller coefficients d!

The best bases are eigenvectors of A (they are orthogonal when A is symmetric) and singular vectors (they are orthogonal for every A). The matrix becomes **diagonal**.

Eigenvector basis
Eigenvalue matrix
Chapter 6 $\begin{bmatrix} \lambda_1 & & \\ & \ddots & \\ & & \lambda_n \end{bmatrix}$

Singular vectors $\boldsymbol{u}, \boldsymbol{v}$
Orthogonal for all A
Chapter 7 $\begin{bmatrix} \sigma_1 & & & \\ & \ddots & & \\ & & \sigma_r & \\ & & & \end{bmatrix}$

Linear Transformations of a House

It is more interesting to *see* a transformation than to define it. When a 2 by 2 matrix A multiplies all vectors in $\mathbf{R}^2$, we can watch how it acts. Start with a "house" that has eleven endpoints. Those eleven vectors $\boldsymbol{v}$ are transformed into eleven vectors $A\boldsymbol{v}$. Straight lines between $\boldsymbol{v}$'s become straight lines between the transformed vectors $A\boldsymbol{v}$. (The transformation from house to house is linear!) Applying A to a standard house produces a new house—possibly stretched or rotated or otherwise unlivable.

This part of the book is visual, not theoretical. We will show four houses and the matrices that produce them. The columns of H are the eleven corners of the first house. (H is 2 by 12, so **plot2d** on the website will connect the 11th corner to the first.) A multiplies the 11 points in the house matrix H to produce the corners AH of the other houses.

House matrix $$\boldsymbol{H} = \begin{bmatrix} -6 & -6 & -7 & 0 & 7 & 6 & 6 & -3 & -3 & 0 & 0 & -6 \\ -7 & 2 & 1 & 8 & 1 & 2 & -7 & -7 & -2 & -2 & -7 & -7 \end{bmatrix}.$$

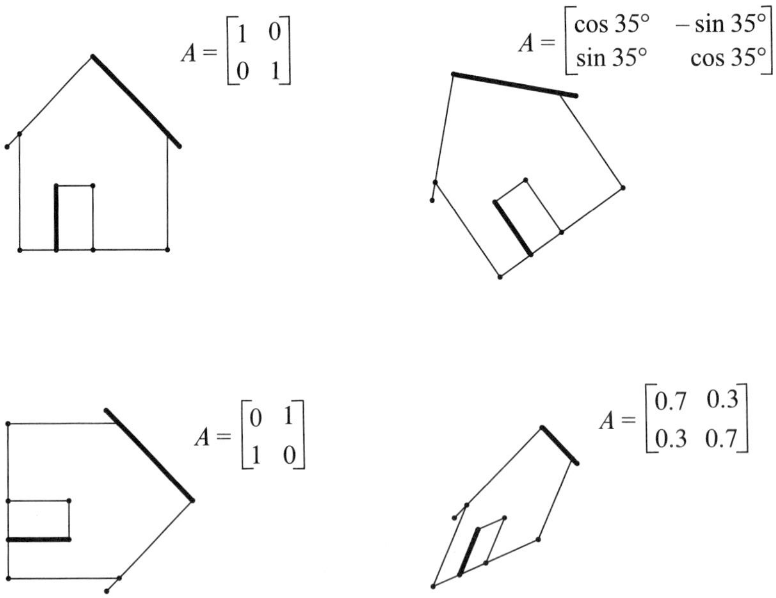

Figure 5.3: Linear transformations of a house drawn by **plot2d**$(A * H)$.

Problem Set 5.3

1 A linear transformation must leave the zero vector fixed: $T(\mathbf{0}) = \mathbf{0}$. Prove this from $T(\boldsymbol{v}+\boldsymbol{w}) = T(\boldsymbol{v}) + T(\boldsymbol{w})$ by choosing $\boldsymbol{w} =$ _____ (and finish the proof). Prove it also from $T(c\boldsymbol{v}) = cT(\boldsymbol{v})$ by choosing $c =$ _____.

2 Suppose a linear T transforms $(1,1)$ to $(2,2)$ and $(2,0)$ to $(0,0)$. Find $T(\boldsymbol{v})$:

(a) $\boldsymbol{v} = (2,2)$ (b) $\boldsymbol{v} = (3,1)$ (c) $\boldsymbol{v} = (-1,1)$ (d) $\boldsymbol{v} = (a,b)$.

3 Which of these transformations are not linear? The input is $\boldsymbol{v} = (v_1, v_2)$:

(a) $T(\boldsymbol{v}) = (v_2, v_1)$ (b) $T(\boldsymbol{v}) = (v_1, v_1)$ (c) $T(\boldsymbol{v}) = (0, v_1)$

(d) $T(\boldsymbol{v}) = (0, 1)$ (e) $T(\boldsymbol{v}) = v_1 - v_2$ (f) $T(\boldsymbol{v}) = v_1 v_2$.

4 If S and T are linear transformations, is $T(S(\boldsymbol{v}))$ linear or quadratic?

(a) (Special case) If $S(\boldsymbol{v}) = \boldsymbol{v}$ and $T(\boldsymbol{v}) = \boldsymbol{v}$, then $T(S(\boldsymbol{v})) = \boldsymbol{v}$ or $\boldsymbol{v}^2$?

(b) (General case) $S(\boldsymbol{v}_1+\boldsymbol{v}_2) = S(\boldsymbol{v}_1)+S(\boldsymbol{v}_2)$ and $T(\boldsymbol{v}_1+\boldsymbol{v}_2) = T(\boldsymbol{v}_1)+T(\boldsymbol{v}_2)$ combine into $T(S(\boldsymbol{v}_1+\boldsymbol{v}_2)) = T($_____$) =$ _____ $+$ _____.

5 Suppose $T(\boldsymbol{v}) = \boldsymbol{v}$ except that $T(0, v_2) = (0,0)$. Show that this transformation satisfies $T(c\boldsymbol{v}) = cT(\boldsymbol{v})$ but does not satisfy $T(\boldsymbol{v}+\boldsymbol{w}) = T(\boldsymbol{v}) + T(\boldsymbol{w})$.

6 True or False: If we know $T(\boldsymbol{v})$ for n different nonzero vectors in $\mathbf{R}^n$, then we know $T(\boldsymbol{v})$ for every vector $\boldsymbol{v}$ in $\mathbf{R}^n$.

7 Which of these transformations satisfy $T(\boldsymbol{v}+\boldsymbol{w}) = T(\boldsymbol{v})+T(\boldsymbol{w})$ and which satisfy $T(c\boldsymbol{v}) = cT(\boldsymbol{v})$?

(a) $T(\boldsymbol{v}) = \boldsymbol{v}/\|\boldsymbol{v}\|$ (b) $T(\boldsymbol{v}) = v_1+v_2+v_3$ (c) $T(\boldsymbol{v}) = (v_1, 2v_2, 3v_3)$

(d) $T(\boldsymbol{v}) =$ largest component of $\boldsymbol{v}$.

8 How can you tell from the picture of T (house) that A is

(a) a diagonal matrix? The house expands or contracts along each axis.

(b) a rank-one matrix?

(c) a lower triangular matrix?

9 Draw a picture of T (house) for these matrices:

$$D = \begin{bmatrix} 2 & 0 \\ 0 & 1 \end{bmatrix} \quad \text{and} \quad A = \begin{bmatrix} .7 & .7 \\ .3 & .3 \end{bmatrix} \quad \text{and} \quad U = \begin{bmatrix} 1 & 1 \\ 0 & 1 \end{bmatrix}.$$

10 What are the conditions on $A = \left[\begin{smallmatrix} a & b \\ c & d \end{smallmatrix}\right]$ to ensure that T (house) will

(a) sit straight up?

(b) expand the house by 3 in all directions?

(c) rotate the house with no change in its shape?

11 Without a computer sketch the houses $A * H$ for these matrices A:

$$\begin{bmatrix} 1 & 0 \\ 0 & .1 \end{bmatrix} \quad \text{and} \quad \begin{bmatrix} .5 & .5 \\ .5 & .5 \end{bmatrix} \quad \text{and} \quad \begin{bmatrix} .5 & .5 \\ -.5 & .5 \end{bmatrix} \quad \text{and} \quad \begin{bmatrix} 1 & 1 \\ 1 & 0 \end{bmatrix}.$$

12 What conditions on $\det A = ad - bc$ ensure that the output house AH will

(a) be squashed onto a line?

(b) keep its endpoints in clockwise order (not reflected)?

(c) have the same area as the original house?

13 This code creates a vector theta of 50 angles. It draws the unit circle and then it draws T (circle) = ellipse. **The multiplication Av takes circles to ellipses.**

```
A = [2 1;1 2]    % You can change A
theta = [0:2 * pi/50:2 * pi]; % 50 angles
circle = [cos(theta); sin(theta)]; % 50 points
ellipse = A * circle; % circle to ellipse
axis([–4 4 –4 4]); axis('square')
plot(circle(1,:), circle(2,:), ellipse(1,:), ellipse(2,:))
```

14 Suppose the spaces $\mathbf{V}$ and $\mathbf{W}$ have the same basis $\boldsymbol{v}_1, \boldsymbol{v}_2$.

(a) Describe a transformation T (not I) that is its own inverse.

(b) Describe a transformation T (not I) that equals T^2.

(c) Why can't the same T be used for both (a) and (b)?

Questions 15–18 are about changing the basis.

15 (a) What matrix B transforms $(1,0)$ into $(2,5)$ and transforms $(0,1)$ to $(1,3)$?

(b) What matrix C transforms $(2,5)$ to $(1,0)$ and $(1,3)$ to $(0,1)$?

(c) Why does no matrix transform $(2,6)$ to $(1,0)$ and $(1,3)$ to $(0,1)$?

16 (a) What matrix M transforms $(1,0)$ and $(0,1)$ to (r,t) and (s,u)?

(b) What matrix N transforms (a,c) and (b,d) to $(1,0)$ and $(0,1)$?

(c) What condition on a, b, c, d will make part (b) impossible?

17 (a) How do M and N in Problem 16 yield the matrix that transforms (a,c) to (r,t) and (b,d) to (s,u)?

(b) What matrix transforms $(2,5)$ to $(1,1)$ and $(1,3)$ to $(0,2)$?

18 If you keep the same basis vectors but put them in a different order, the change of basis matrix B is a _____ matrix. If you keep the basis vectors in order but change their lengths, B is a _____ matrix.

19 Why is integration not the inverse of differentiation?

6 Eigenvalues and Eigenvectors

Eigenvalues λ and eigenvectors $\boldsymbol{x}$ obey the equation $A\boldsymbol{x} = \lambda\boldsymbol{x}$. Here A is a square matrix and λ is a number. The vector $A\boldsymbol{x}$ is in the same direction as $\boldsymbol{x}$: unusual. If we find n of these pairs $A\boldsymbol{x}_1 = \lambda_1\boldsymbol{x}_1$ up to $A\boldsymbol{x}_n = \lambda_n\boldsymbol{x}_n$, then an n-dimensional problem turns into n simple one-dimensional problems. We write the input as a combination of eigenvectors, and we solve for the output as a combination of eigenvectors.

Here is an example: **Solve** $\boldsymbol{u}_{k+1} = \boldsymbol{A}\boldsymbol{u}_k$. Each step multiplies by A. So each eigenvector $\boldsymbol{x}_k$ is multiplied by its eigenvalue λ_k : n simple pieces.

Input $\boldsymbol{u}_0 = c_1\boldsymbol{x}_1 + \cdots + c_n\boldsymbol{x}_n$ **Output** $\boldsymbol{u}_N = A^N\boldsymbol{u} = c_1\lambda_1^N\boldsymbol{x}_1 + \cdots + c_n\lambda_n^N\boldsymbol{x}_n$

The input combination $\boldsymbol{u}_0$ leads to the output combination $\boldsymbol{u}_N$. The differential equation $d\boldsymbol{u}/dt = A\boldsymbol{u}$ is solved at time t in exactly the same way. The numbers λ^N change to $e^{\lambda t}$.

In matrix language, the matrix X of eigenvectors turns A into $X^{-1}AX = \Lambda$. This is the diagonal matrix of eigenvalues. A diagonal matrix means that the system is uncoupled into n easy equations like $d\boldsymbol{u}/dt = \lambda\boldsymbol{u}$, and solved by $\boldsymbol{u} = e^{\lambda t}$.

Sections 6.1 and 6.2 present the key ideas of eigenvalues. Then Section 6.3 goes from general matrices A to *symmetric matrices* S. The eigenvectors of S are *orthogonal.* Their matrix X becomes Q with $Q^{\mathrm{T}}Q = I$. And with positive eigenvalues $\lambda > 0$, we have the best matrices in pure and applied mathematics: **symmetric and positive definite.**

Positive definiteness can be tested *five ways*—by positive pivots and determinants and eigenvalues and energy (and by $S = A^{\mathrm{T}}A$). Those connect to the first five chapters of the book. The central ideas are coming together for the best matrices.

6.1 Introduction to Eigenvalues

1 An **eigenvector** x lies along the same line as A times x : $\boxed{Ax = \lambda x.}$ **The eigenvalue is λ.**

2 If $Ax = \lambda x$ then $A^2 x = \lambda^2 x$ and $A^{-1}x = \lambda^{-1}x$ and $(A + cI)x = (\lambda + c)x$: the same x.

3 If $Ax = \lambda x$ then $(A - \lambda I)x = \mathbf{0}$ and $A - \lambda I$ is singular: $\boxed{\det(A - \lambda I) = 0 \text{ gives } n \text{ eigenvalues.}}$

4 Check λ's by $\det A = (\lambda_1)(\lambda_2)\cdots(\lambda_n)$ and diagonal sum $a_{11} + a_{22} + \cdots + a_{nn} =$ sum of λ's.

5 Projections have $\lambda = 1$ and 0. Reflections have 1 and -1. Rotations have $e^{i\theta}$ and $e^{-i\theta}$: *complex*!

This chapter enters a new part of linear algebra. The first part was about $Ax = b$: linear equations to find a steady state. Now the second part is about **change**. Time enters the picture—continuous time in a differential equation $du/dt = Au$ or time steps in a difference equation $u_{k+1} = Au_k$. Those equations are NOT solved by elimination. The key idea is to find solutions $u(t)$ that stay in the direction of a fixed vector x.

We want "eigenvectors" that don't change direction when you multiply by A. The eigenvector-eigenvalue equation is $Ax = \lambda x$. We look for n eigenvectors x and their eigenvalues λ. Then A^2 also has those eigenvectors: $A^2 x = A(\lambda x) = \lambda^2 x$.

A good model comes from the powers $A, A^2, A^3, \ldots$ of a matrix. Suppose you need the hundredth power A^{100}. Its columns are very close to the eigenvector $x = (.6,\ .4)$:

$$\boldsymbol{A}, A^2, A^3 = \begin{bmatrix} .8 & .3 \\ .2 & .7 \end{bmatrix} \begin{bmatrix} .70 & .45 \\ .30 & .55 \end{bmatrix} \begin{bmatrix} .650 & .525 \\ .350 & .475 \end{bmatrix} \qquad \boldsymbol{A}^{100} \approx \begin{bmatrix} .6000 & .6000 \\ .4000 & .4000 \end{bmatrix}$$

A^{100} was found by using the **eigenvalues of A**, not by multiplying 100 matrices. Again: Each eigenvector x has an eigenvalue λ with $Ax = \lambda x$. Then $A^{100}x = \lambda^{100}x$.

To explain eigenvalues, we first explain eigenvectors. Almost all vectors will change direction, when they are multiplied by A. ***Certain exceptional vectors x are in the same direction as Ax. Those are the eigenvectors***. Multiply an eigenvector by A, and the vector Ax is a number λ times the original x.

$$\boxed{\textbf{The basic equation is } Ax = \lambda x. \textbf{ This leads to } (A - \lambda I)\, x = 0.}$$

The eigenvalue λ tells whether the eigenvector x is stretched or shrunk or reversed or left unchanged—when it is multiplied by A. We may find $\lambda = 2$ or $\frac{1}{2}$ or -1 or 1. If $\lambda = 0$ then $Ax = 0x$ means that this eigenvector x is in the nullspace of A.

If A is the identity matrix, every vector has $Ax = x$. All vectors are eigenvectors of I. All eigenvalues "lambda" are $\lambda = 1$. This is unusual to say the least. Most 2 by 2 matrices have *two* eigenvector directions and *two* eigenvalues: $Ax_1 = \lambda_1 x_1$ and $Ax_2 = \lambda_2 x_2$.

This section will explain how to compute the x's and λ's. It can come early in the course because we only need the determinant of a 2 by 2 matrix $A - \lambda I$.

Example 1 A has two eigenvalues $\lambda = 1$ and $\lambda = 1/2$. **Solve** $\det(\boldsymbol{A} - \boldsymbol{\lambda I}) = \mathbf{0}$:

$$A = \begin{bmatrix} .8 & .3 \\ .2 & .7 \end{bmatrix} \quad \det \begin{bmatrix} .8-\lambda & .3 \\ .2 & .7-\lambda \end{bmatrix} = \boldsymbol{\lambda^2 - \frac{3}{2}\lambda + \frac{1}{2}} = (\lambda - 1)\left(\lambda - \frac{1}{2}\right).$$

I factored the quadratic into $\lambda - 1$ times $\lambda - \frac{1}{2}$, to see the eigenvalues $\boldsymbol{\lambda} = \mathbf{1}$ and $\frac{\mathbf{1}}{\mathbf{2}}$. For those numbers, the matrix $A - \lambda I$ becomes *singular* (zero determinant). Then the *eigenvectors* $\boldsymbol{x}_1$ *and* $\boldsymbol{x}_2$ *solve* $(A - \lambda_1 I)\boldsymbol{x}_1 = \mathbf{0}$ *and* $(A - \lambda_2 I)\boldsymbol{x}_2 = \mathbf{0}$.

$$(A - I)\boldsymbol{x}_1 = 0 \text{ is } \begin{bmatrix} -.2 & .3 \\ .2 & -.3 \end{bmatrix} \boldsymbol{x}_1 = \begin{bmatrix} 0 \\ 0 \end{bmatrix} \text{ and the first eigenvector is } \boldsymbol{x}_1 = \begin{bmatrix} .6 \\ .4 \end{bmatrix}.$$

$$(A - \tfrac{1}{2}I)\boldsymbol{x}_2 = 0 \text{ is } \begin{bmatrix} .3 & .3 \\ .2 & .2 \end{bmatrix} \boldsymbol{x}_1 = \begin{bmatrix} 0 \\ 0 \end{bmatrix} \text{ and the second eigenvector is } \boldsymbol{x}_2 = \begin{bmatrix} 1 \\ -1 \end{bmatrix}.$$

If $\boldsymbol{x}_1$ is multiplied again by A, we still get $\boldsymbol{x}_1$. Every power of A will give $A^n\boldsymbol{x}_1 = \boldsymbol{x}_1$. Multiplying $\boldsymbol{x}_2$ by A gave $\frac{1}{2}\boldsymbol{x}_2$, and if we multiply again we get $(\frac{1}{2})^2 = \frac{1}{4}$ times $\boldsymbol{x}_2$.

When A is squared, the eigenvectors stay the same. The eigenvalues are squared.

This pattern keeps going, because the eigenvectors stay in their own directions (Figure 6.1). They never get mixed. The eigenvectors of A^{100} are the same $\boldsymbol{x}_1$ and $\boldsymbol{x}_2$. The eigenvalues of A^{100} are $1^{100} = 1$ and $(\frac{1}{2})^{100} =$ very small number.

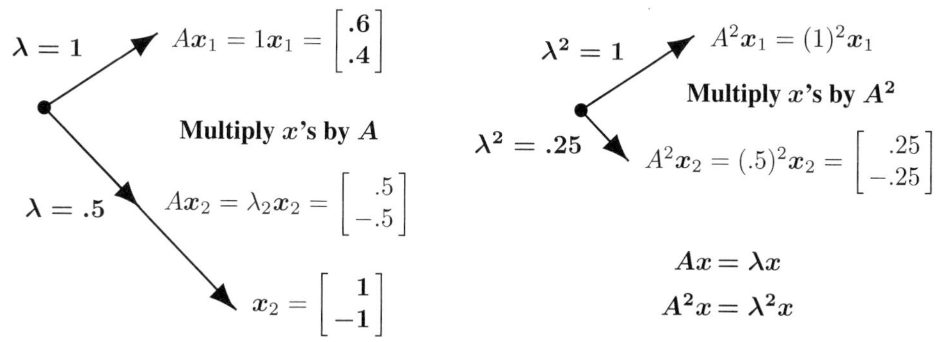

Figure 6.1: The eigenvectors of A are also eigenvectors of A^2 : eigenvalue $= \lambda^2$.

Other vectors like $(.8, .2)$ do change direction. But all other vectors are combinations of the two eigenvectors $\boldsymbol{x}_1 = (.6, .4)$ and $\boldsymbol{x}_2 = (1, -1)$: The first column of A is $\boldsymbol{x}_1 + (.2)\boldsymbol{x}_2$:

Separate other vectors into eigenvectors $$\begin{bmatrix} .8 \\ .2 \end{bmatrix} = \boldsymbol{x}_1 + (.2)\boldsymbol{x}_2 = \begin{bmatrix} .6 \\ .4 \end{bmatrix} + \begin{bmatrix} .2 \\ -.2 \end{bmatrix}. \qquad (1)$$

When we multiply $\boldsymbol{x}_1 + (.2)\boldsymbol{x}_2$ by A, ***each eigenvector is multiplied by its eigenvalue***:

$$\textbf{Multiply } \boldsymbol{x}_1 \textbf{ and } \boldsymbol{x}_2 \textbf{ by } \boldsymbol{\lambda} = \mathbf{1} \textbf{ and } \boldsymbol{\lambda} = \tfrac{\mathbf{1}}{\mathbf{2}} \qquad A\begin{bmatrix} .8 \\ .2 \end{bmatrix} = A\begin{bmatrix} .6 \\ .4 \end{bmatrix} + A\begin{bmatrix} .2 \\ -.2 \end{bmatrix} = \begin{bmatrix} .6 \\ .4 \end{bmatrix} + \begin{bmatrix} .1 \\ -.1 \end{bmatrix} = \begin{bmatrix} .7 \\ .3 \end{bmatrix}.$$

For every multiplication by A, $\boldsymbol{x}_2$ is multiplied by $\left(\frac{1}{2}\right)$. The first eigenvector is multiplied by 1. Then 99 steps give $(1)^{99} = 1$. The very small number $\left(\frac{1}{2}\right)^{99}$ appears in A^{100}:

$$\boxed{A^{99}\begin{bmatrix} .8 \\ .2 \end{bmatrix} \quad \text{is really} \quad (1)^{99}\,\boldsymbol{x}_1 + (.2)\left(\frac{1}{2}\right)^{99}\boldsymbol{x}_2 = \begin{bmatrix} \mathbf{.6} \\ \mathbf{.4} \end{bmatrix} + \begin{bmatrix} \text{very} \\ \text{small} \\ \text{vector} \end{bmatrix}.}$$

This is the first column of A^{100}. The number we originally wrote as .6000 was not exact. We left out $(.2)\left(\frac{1}{2}\right)^{99}$ which wouldn't show up for 30 decimal places.

The eigenvector $\boldsymbol{x}_1$ is a "steady state" that doesn't change (because $\lambda_1 = 1$).
The eigenvector $\boldsymbol{x}_2$ is "decaying" and virtually disappears (because $\lambda_2 = \frac{1}{2}$).

The higher the power of A, the more closely its columns approach the steady state. This particular A is a ***Markov matrix***: Columns add to 1. Its largest eigenvalue is $\lambda = 1$. Its eigenvector $\boldsymbol{x}_1 = (.6, .4)$ is the *steady state*—which all columns of A^k will approach. A giant Markov matrix is the key to Google's search algorithm—which is truly fast.

Other matrices have other eigenvalues. Projection matrices have $\boldsymbol{\lambda} = \mathbf{1}$ for vectors in the column space and $\boldsymbol{\lambda} = \mathbf{0}$ for vectors in the nullspace (projected to the zero vector). Then $P\boldsymbol{x}_1 = \boldsymbol{x}_1$ and $P\boldsymbol{x}_2 = \mathbf{0}$. We have $P^2 = P$ because $1^2 = 1$ and $0^2 = 0$.

Example 2 **The projection matrix** $P = \begin{bmatrix} .5 & .5 \\ .5 & .5 \end{bmatrix}$ has eigenvalues $\lambda = \mathbf{1}$ and $\lambda = \mathbf{0}$.

Its eigenvectors are $\boldsymbol{x}_1 = (1, 1)$ and $\boldsymbol{x}_2 = (1, -1)$. For those vectors, $P\boldsymbol{x}_1$ equals $\boldsymbol{x}_1$ (steady state) and $P\boldsymbol{x}_2 = \mathbf{0}$ (nullspace). Then $P^2\boldsymbol{x}_1 = \boldsymbol{x}_1$ and $P^2\boldsymbol{x}_2 = \mathbf{0}$ and $P^2 = P$. Our examples illustrate Markov matrices and singular matrices and symmetric matrices: Those matrices have special λ's and special eigenvectors:

1. **Markov matrix**: Each column adds to 1. This makes $\lambda = 1$ an eigenvalue.

2. **Singular matrix**: Some vector has $A\boldsymbol{x} = \mathbf{0}$. Then $\lambda = 0$ is an eigenvalue.

3. **Symmetric matrix**: The eigenvectors $(1, 1)$ and $(1, -1)$ are perpendicular.

The only eigenvalues of a projection matrix are 0 and 1. The nullspace is projected to zero. The column space projects onto itself. The projection keeps the column space and destroys the nullspace, so $\lambda = 1$ and $\lambda = 0$:

$$\textbf{Project each part} \quad \boldsymbol{v} = \begin{bmatrix} 1 \\ -1 \end{bmatrix} + \begin{bmatrix} 2 \\ 2 \end{bmatrix} \quad \textbf{projects onto} \quad P\boldsymbol{v} = \begin{bmatrix} \mathbf{0} \\ \mathbf{0} \end{bmatrix} + \begin{bmatrix} \mathbf{2} \\ \mathbf{2} \end{bmatrix}.$$

The next matrix R is a reflection matrix and also a permutation matrix.

Example 3 **The reflection matrix $R = \left[\begin{smallmatrix} 0 & 1 \\ 1 & 0 \end{smallmatrix}\right]$ has eigenvalues 1 and -1.**

The eigenvector $(1, 1)$ is unchanged by R. The second eigenvector is $(1, -1)$. Its signs are reversed by R. A matrix with no negative entries can still have a negative eigenvalue! The eigenvectors for R are the same as for P, because *reflection* $= 2$(*projection*) $- I$:

$$\textbf{Eigenvalues of } \boldsymbol{R = 2P - I} \qquad \lambda = 2(1) - 1 = \mathbf{1} \text{ and } \lambda = 2(0) - 1 = \mathbf{-1} \tag{2}$$

When a matrix is shifted by I, each λ is shifted by 1. No change in its eigenvectors.

The Equation for the Eigenvalues

For projection matrices we found λ's and $\boldsymbol{x}$'s by geometry: $P\boldsymbol{x} = \boldsymbol{x}$ and $P\boldsymbol{x} = \mathbf{0}$. For other matrices we use determinants and linear algebra. *This is the key calculation in the chapter*—almost every application starts by solving $A\boldsymbol{x} = \lambda\boldsymbol{x}$.

First move $\lambda\boldsymbol{x}$ to the left side. Write the equation $A\boldsymbol{x} = \lambda\boldsymbol{x}$ as $(A - \lambda I)\boldsymbol{x} = \mathbf{0}$. ***The eigenvectors are in the nullspace of $A - \lambda I$.*** When we know an eigenvalue λ, we find $\boldsymbol{x}$ by solving $(A - \lambda I)\boldsymbol{x} = \mathbf{0}$.

Eigenvalues first. If $(A - \lambda I)\boldsymbol{x} = \mathbf{0}$ has a nonzero solution, $A - \lambda I$ is not invertible. ***The determinant of $A - \lambda I$ must be zero.*** This is how to recognize an eigenvalue λ:

Eigenvalues	**λ is an eigenvalue of $A \Leftrightarrow A - \lambda I$ is singular**
Equation for λ	$\det(A - \lambda I) = 0$

(3)

This "*characteristic polynomial*" $\det(A - \lambda I)$ involves only λ, not $\boldsymbol{x}$. When A is n by n, equation (3) contains λ^n. So A has n eigenvalues (repeats possible!) Each λ leads to $\boldsymbol{x}$:

For each eigenvalue λ solve $(A - \lambda I)\boldsymbol{x} = \mathbf{0}$ or $A\boldsymbol{x} = \lambda\boldsymbol{x}$ **to find an eigenvector $\boldsymbol{x}$.**

Example 4 $A = \begin{bmatrix} \mathbf{1} & \mathbf{2} \\ \mathbf{2} & \mathbf{4} \end{bmatrix}$ is already singular (zero determinant). Find its λ's and $\boldsymbol{x}$'s.

When A is singular, $\lambda = 0$ is one of the eigenvalues. The equation $A\boldsymbol{x} = 0\boldsymbol{x}$ has solutions. They are the eigenvectors for $\lambda = 0$. Solving $\det(A - \lambda I) = 0$ is the way to find *all* λ's and $\boldsymbol{x}$'s. Always subtract λI from A:

$$\textit{\textbf{Subtract λ from the diagonal of A to find}} \quad A - \lambda I = \begin{bmatrix} 1 - \lambda & 2 \\ 2 & 4 - \lambda \end{bmatrix}. \tag{4}$$

Take the determinant "$ad - bc$" of this 2 by 2 matrix. From $1 - \lambda$ times $4 - \lambda$, the "ad" part is $\lambda^2 - 5\lambda + 4$. The "bc" part without λ is 2 times 2. Subtract:

$$\det \begin{bmatrix} 1 - \lambda & 2 \\ 2 & 4 - \lambda \end{bmatrix} = (1 - \lambda)(4 - \lambda) - (2)(2) = \boldsymbol{\lambda^2 - 5\lambda}. \tag{5}$$

Set this determinant $\boldsymbol{\lambda^2 - 5\lambda}$ to zero. One solution is $\lambda_1 = 0$. This was expected, since A is singular. Factoring $\lambda^2 - 5\lambda$ into λ times $\lambda - 5$, the other eigenvalue is $\lambda_2 = 5$:

$$\det(\boldsymbol{A} - \boldsymbol{\lambda I}) = \boldsymbol{\lambda^2 - 5\lambda = 0} \quad \text{yields the eigenvalues} \quad \boldsymbol{\lambda_1 = 0} \quad \textbf{and} \quad \boldsymbol{\lambda_2 = 5.}$$

Now find the eigenvectors. Solve $(A - \lambda I)\boldsymbol{x} = \mathbf{0}$ separately for $\lambda_1 = 0$ and $\lambda_2 = 5$.

$$(A - 0I)\boldsymbol{x} = \begin{bmatrix} 1 & 2 \\ 2 & 4 \end{bmatrix}\begin{bmatrix} y \\ z \end{bmatrix} = \begin{bmatrix} 0 \\ 0 \end{bmatrix} \text{ yields an eigenvector } \begin{bmatrix} y \\ z \end{bmatrix} = \begin{bmatrix} \mathbf{2} \\ \mathbf{-1} \end{bmatrix} \text{ for } \lambda_1 = 0$$

$$(A - 5I)\boldsymbol{x} = \begin{bmatrix} -4 & 2 \\ 2 & -1 \end{bmatrix}\begin{bmatrix} y \\ z \end{bmatrix} = \begin{bmatrix} 0 \\ 0 \end{bmatrix} \text{ yields an eigenvector } \begin{bmatrix} y \\ z \end{bmatrix} = \begin{bmatrix} \mathbf{1} \\ \mathbf{2} \end{bmatrix} \text{ for } \lambda_2 = 5.$$

The matrices $A - 0I$ and $A - 5I$ are singular (because $\lambda = 0$ and 5 are eigenvalues). The eigenvectors $(2, -1)$ and $(1, 2)$ are in the nullspaces: $(A - \lambda I)\boldsymbol{x} = \mathbf{0}$ is $A\boldsymbol{x} = \lambda\boldsymbol{x}$.

We need to emphasize: *There is nothing exceptional about* $\lambda = 0$. Like every other number, zero might be an eigenvalue and it might not. If A is singular, the eigenvectors for $\lambda = 0$ fill the nullspace: $A\boldsymbol{x} = 0\boldsymbol{x} = \mathbf{0}$. If A is invertible, zero is not an eigenvalue. We shift A by a multiple of I to *make it singular.*

In Example 4 the shifted matrix $A - 5I$ is singular. Then 5 is the other eigenvalue of A.

Summary To find the eigenvalues of an n by n matrix, follow these steps: best if $n = 2$.

1. ***Compute the determinant of*** $A - \lambda I$. With λ subtracted along the diagonal, this determinant starts with λ^n or $-\lambda^n$. It is a polynomial in λ of degree n.
2. ***Find the roots of this polynomial***, by solving $\det(A - \lambda I) = 0$. The n roots are the n eigenvalues of A. Those n numbers make $A - \lambda I$ singular.
3. For each eigenvalue λ: ***Solve*** $(A - \lambda I)\boldsymbol{x} = \mathbf{0}$ ***to find an eigenvector***: $A\boldsymbol{x} = \lambda\boldsymbol{x}$.
4. The eigenvalues of a **triangular matrix** are the numbers on its diagonal.

A note on the eigenvectors of 2 by 2 matrices. When $A - \lambda I$ is singular, both rows are multiples of a vector (a, b). *The eigenvector direction is* $(b, -a)$. The example had

$\lambda = 0$: rows of $A - 0I$ in the direction $(1, 2)$; eigenvector in the direction $(2, -1)$

$\lambda = 5$: rows of $A - 5I$ in the direction $(-4, 2)$; eigenvector in the direction $(2, 4)$

Previously we wrote that last eigenvector as $(1, 2)$. Both $(1, 2)$ and $(2, 4)$ are correct. There is a whole *line of eigenvectors*—any nonzero multiple of $\boldsymbol{x}$ is as good as $\boldsymbol{x}$. MATLAB's **eig**(A) divides by the length of $\boldsymbol{x}$, to make the eigenvector into a unit vector.

We must add a warning. Some 2 by 2 matrices have only *one line of eigenvectors*. This can happen only when two eigenvalues are equal. (On the other hand $A = I$ has equal eigenvalues and plenty of eigenvectors.) Without a full set of eigenvectors, we don't have a basis. We can't write every $\boldsymbol{v}$ as a combination of eigenvectors. In the language of the next section, *we can't diagonalize a matrix without n independent eigenvectors.*

Determinant and Trace

Bad news first: If you add a row of A to another row, or exchange rows, then the eigenvalues usually change. *Elimination does not preserve the* λ's. The upper triangular U has its eigenvalues sitting along the diagonal—they are the pivots. Those pivots are eigenvalues of U but not of A! Eigenvalues are changed when row 1 is added to row 2:

$$U = \begin{bmatrix} 1 & 3 \\ 0 & 0 \end{bmatrix} \quad \text{has } \lambda = 0 \text{ and } \lambda = 1; \quad A = \begin{bmatrix} 1 & 3 \\ 1 & 3 \end{bmatrix} \quad \text{has } \lambda = 0 \text{ and } \lambda = 4.$$

Good news second: The *product* λ_1 *times* λ_2 *and the sum* $\lambda_1 + \lambda_2$ *can be found quickly from the matrix.* For this A, the product is 0 times 4. That agrees with the determinant (which is 0). The sum of eigenvalues is $0 + 4$. That agrees with the sum down the main diagonal of A (the **trace** is $1 + 3$). These quick checks always work:

> ***The product of the n eigenvalues*** *equals* ***the determinant.***
>
> ***The sum of the n eigenvalues*** *equals* ***the sum of the n diagonal entries.***

The sum of the entries of A along the main diagonal is called the ***trace*** of A:

$$\boxed{\lambda_1 + \lambda_2 + \cdots + \lambda_n = \textbf{\textit{trace}} = a_{11} + a_{22} + \cdots + a_{nn}.} \qquad (6)$$

Those checks are very useful. They are proved in PSet 6.1 and again in Section 6.2. They don't remove the pain of computing λ's. But when the computation is wrong, they generally tell us so. To compute the correct λ's, go back to $\det(A - \lambda I) = 0$.

The trace and determinant *do* tell everything when the matrix is 2 by 2. We never want to get those wrong! Here trace $= \mathbf{3}$ and det $= \mathbf{2}$, so these matrices have $\lambda = \mathbf{1}$ and $\lambda = \mathbf{2}$:

$$A = \begin{bmatrix} 1 & 9 \\ 0 & 2 \end{bmatrix} \text{ or } \begin{bmatrix} 3 & 1 \\ -2 & 0 \end{bmatrix} \text{ or } \begin{bmatrix} 7 & -3 \\ 10 & -4 \end{bmatrix}. \qquad (7)$$

And here is a question about the easiest matrices for finding eigenvalues: *triangular A.*

Why do the eigenvalues of a triangular matrix A lie along its diagonal?

Imaginary Eigenvalues

One more possibility, not too terrible. The eigenvalues might not be real numbers.

Example 5 ***The*** $90°$ ***rotation*** $Q = \begin{bmatrix} 0 & -1 \\ 1 & 0 \end{bmatrix}$ ***has no real eigenvectors. Its eigenvalues are*** $\lambda_1 = i$ ***and*** $\lambda_2 = -i$. ***Then*** $\lambda_1 + \lambda_2 =$ **trace** $= 0$ ***and*** $\lambda_1\lambda_2 =$ ***determinant*** $= 1$.

After a rotation, *no real vector $Q\boldsymbol{x}$ stays in the same direction as $\boldsymbol{x}$* ($\boldsymbol{x} = \mathbf{0}$ is useless). There cannot be an eigenvector, unless we go to imaginary numbers. Which we do.

To see how $i = \sqrt{-1}$ can help, look at a rotation Q through $90°$. Then Q^2 is rotation through $180°$. Its eigenvalues are -1 and -1 because $-I\boldsymbol{x} = -1\boldsymbol{x}$. Squaring Q will square each λ, so we must have $\lambda^2 = -1$. *The eigenvalues of the* $90°$ *rotation matrix Q are $+i$ and $-i$,* because $i^2 = -1$.

Those λ's come as usual from $\det(Q - \lambda I) = 0$. This equation is $\lambda^2 + 1 = 0$. Its roots are i and $-i$. We meet the imaginary number i also in the eigenvectors:

$$\textbf{Complex eigenvectors} \qquad \begin{bmatrix} 0 & -1 \\ 1 & 0 \end{bmatrix} \begin{bmatrix} 1 \\ i \end{bmatrix} = -\boldsymbol{i} \begin{bmatrix} 1 \\ i \end{bmatrix} \quad \text{and} \quad \begin{bmatrix} 0 & -1 \\ 1 & 0 \end{bmatrix} \begin{bmatrix} i \\ 1 \end{bmatrix} = \boldsymbol{i} \begin{bmatrix} i \\ 1 \end{bmatrix}.$$

Somehow the complex vectors $\boldsymbol{x}_1 = (1, i)$ and $\boldsymbol{x}_2 = (i, 1)$ keep their direction as they are rotated. Don't ask me how. This example makes the important point that real matrices can have complex eigenvalues and eigenvectors. The particular eigenvalues i and $-i$ also illustrate two properties of this matrix Q:

1. The absolute value of each λ is $|\lambda| = 1$. Orthogonal matrices Q have $||Q\boldsymbol{x}|| = ||\boldsymbol{x}||$.
2. This Q is a skew-symmetric matrix ($Q^{\mathrm{T}} = -Q$) so each λ is pure imaginary.

A symmetric matrix ($S^{\mathrm{T}} = S$) can be compared to a real number: λ is real.

A skew-symmetric matrix ($A^{\mathrm{T}} = -A$) is like an imaginary number: λ is imaginary.

An orthogonal matrix ($Q^{\mathrm{T}}Q = I$) corresponds to a complex number with $|\lambda| = 1$.

The eigenvectors for these special matrices S and A and Q are perpendicular. Somehow $(i, 1)$ and $(1, i)$ are perpendicular when we take the dot product of complex vectors.

Eigenvalues of AB and $A+B$

The first guess about the eigenvalues of AB is **not true**. An eigenvalue λ of A times an eigenvalue β of B usually does *not* give an eigenvalue of AB:

$$\textbf{False proof} \qquad AB\boldsymbol{x} = A\beta\boldsymbol{x} = \beta A\boldsymbol{x} = \beta\lambda\boldsymbol{x}. \tag{8}$$

When $\boldsymbol{x}$ is an eigenvector for A and B, this equation is correct. ***The mistake is to expect that A and B automatically share the same eigenvector $\boldsymbol{x}$.*** Usually they don't. Eigenvectors of A are not generally eigenvectors of B. These singular matrices A and B have all zero eigenvalues while 1 is an eigenvalue of AB and $A + B$:

$$A = \begin{bmatrix} 0 & 1 \\ 0 & 0 \end{bmatrix} \quad \text{and} \quad B = \begin{bmatrix} 0 & 0 \\ 1 & 0 \end{bmatrix}; \quad \text{then} \quad AB = \begin{bmatrix} 1 & 0 \\ 0 & 0 \end{bmatrix} \quad \text{and} \quad A+B = \begin{bmatrix} 0 & 1 \\ 1 & 0 \end{bmatrix}.$$

The eigenvalues of $A + B$ are generally not $\lambda + \beta$. Here $\lambda + \beta = 0$ while $A + B$ has eigenvalues 1 and -1: trace $= 0$ and determinant $= -1$.

The false proof suggests what is true. Suppose $\boldsymbol{x}$ really is an eigenvector for A and B. Then we do have $AB\boldsymbol{x} = \lambda\beta\boldsymbol{x}$ and $BA\boldsymbol{x} = \lambda\beta\boldsymbol{x}$. When all n eigenvectors are shared by A and B, we *can* multiply eigenvalues. The test $AB = BA$ for shared eigenvectors is important in quantum mechanics—time out to mention this application of linear algebra.

A and B share the same $\boldsymbol{n}$ *independent* eigenvectors if and only if $\boldsymbol{AB} = \boldsymbol{BA}$.

Heisenberg's uncertainty principle In quantum mechanics, the position matrix P and the momentum matrix Q do not commute. In fact $QP - PQ = I$ (these are infinite matrices). To have $P\boldsymbol{x} = \boldsymbol{0}$ at the same time as $Q\boldsymbol{x} = \boldsymbol{0}$ would require $\boldsymbol{x} = I\boldsymbol{x} = \boldsymbol{0}$. But if we knew the position exactly, we could not also know the momentum exactly. Heisenberg's uncertainty principle $\|P\boldsymbol{x}\|\,\|Q\boldsymbol{x}\| \geq \frac{1}{2}\|\boldsymbol{x}\|^2$ is in the Problem Set.

■ WORKED EXAMPLES ■

6.1 A Find the eigenvalues and eigenvectors of A and A^2 and A^{-1} and $A + 4I$:

$$A = \begin{bmatrix} 2 & -1 \\ -1 & 2 \end{bmatrix} \quad \text{and} \quad A^2 = \begin{bmatrix} 5 & -4 \\ -4 & 5 \end{bmatrix}.$$

Check that the trace $\lambda_1 + \lambda_2 = 2 + 2 = 4$ and the determinant is $\lambda_1\lambda_2 = 4 - 1 = 3$. We don't need to compute A^2 to find its eigenvalues λ^2.

Solution The eigenvalues of A come from $\det(A - \lambda I) = 0$:

$$A = \begin{bmatrix} 2 & -1 \\ -1 & 2 \end{bmatrix} \qquad \det(A - \lambda I) = \begin{vmatrix} \mathbf{2-\boldsymbol{\lambda}} & \mathbf{-1} \\ \mathbf{-1} & \mathbf{2-\boldsymbol{\lambda}} \end{vmatrix} = \boldsymbol{\lambda}^{\mathbf{2}} - \mathbf{4}\boldsymbol{\lambda} + \mathbf{3} = 0.$$

This factors into $(\lambda - 1)(\lambda - 3) = 0$ so the eigenvalues of A are $\lambda_1 = \mathbf{1}$ and $\lambda_2 = \mathbf{3}$. For the trace, the sum $2+2$ agrees with $1+3$. The determinant 3 agrees with the product $\lambda_1\lambda_2$.

The eigenvectors come separately by solving $(A - \lambda I)\boldsymbol{x} = \boldsymbol{0}$ which is $A\boldsymbol{x} = \lambda\boldsymbol{x}$:

$$\boldsymbol{\lambda} = \mathbf{1}\text{:} \quad (A - I)\boldsymbol{x} = \begin{bmatrix} 1 & -1 \\ -1 & 1 \end{bmatrix}\begin{bmatrix} x \\ y \end{bmatrix} = \begin{bmatrix} 0 \\ 0 \end{bmatrix} \text{ gives the eigenvector } \boldsymbol{x}_1 = \begin{bmatrix} 1 \\ 1 \end{bmatrix}$$

$$\boldsymbol{\lambda} = \mathbf{3}\text{:} \quad (A - 3I)\boldsymbol{x} = \begin{bmatrix} -1 & -1 \\ -1 & -1 \end{bmatrix}\begin{bmatrix} x \\ y \end{bmatrix} = \begin{bmatrix} 0 \\ 0 \end{bmatrix} \text{ gives the eigenvector } \boldsymbol{x}_2 = \begin{bmatrix} 1 \\ -1 \end{bmatrix}$$

A^2 and A^{-1} keep the *same eigenvectors as* A. Their eigenvalues are λ^2 and λ^{-1}.

A^2 has eigenvalues $1^2 = \mathbf{1}$ and $3^2 = \mathbf{9}$ $\quad A^{-1}$ has $\dfrac{\mathbf{1}}{\mathbf{1}}$ and $\dfrac{\mathbf{1}}{\mathbf{3}}$ $\quad A + 4I$ has $\begin{matrix} 1+4 = \mathbf{5} \\ 3+4 = \mathbf{7} \end{matrix}$

Notes for later sections: A has *orthogonal eigenvectors* (Section 6.3 on symmetric matrices). A can be *diagonalized* since $\lambda_1 \neq \lambda_2$ (Section 6.2). A is *similar* to any 2 by 2 matrix with eigenvalues 1 and 3 (Section 6.2). A is a *positive definite matrix* (Section 6.3).

6.1 B **How can you estimate the eigenvalues of any A?** Gershgorin gave this answer.

Every eigenvalue of A must be "near" at least one of the entries a_{ii} on the main diagonal. For λ to be "near a_{ii}" means that $|a_{ii} - \lambda|$ is **no more than the sum R_i of all other $|a_{ij}|$ in that row i of the matrix**. Then $R_i = \Sigma_{j \neq i} |a_{ij}|$ is the radius of a circle centered at a_{ii}.

Every λ is in the circle around one or more diagonal entries a_{ii} : $|a_{ii} - \lambda| \leq R_i$.

Proof. Suppose $(A - \lambda I)\boldsymbol{x} = \mathbf{0}$ and the largest component of $\boldsymbol{x}$ is x_2. Then

$$a_{21}x_1 + (a_{22} - \lambda)\,x_2 + a_{23}\,x_3 = 0 \quad \text{gives} \quad |a_{22} - \lambda|\,|x_2| \leq |a_{21}|\,|x_2| + |a_{23}|\,|x_2|.$$

Dividing by $|x_2|$ leaves $|a_{22} - \lambda| \leq R_2$ and λ is inside the second Gershgorin circle.

Example 1. Every eigenvalue λ of this A falls into one or both of the **Gershgorin circles**: The centers are a and d, the radii are $R_1 = |b|$ and $R_2 = |c|$.

$$A = \begin{bmatrix} a & b \\ c & d \end{bmatrix} \qquad \begin{array}{ll} \text{First circle:} & |\lambda - a| \leq |b| \\ \text{Second circle:} & |\lambda - d| \leq |c| \end{array}$$

Those are circles in the complex plane, since λ could certainly be complex.

Example 2. All eigenvalues of this A lie in a circle of radius $R = 3$ around *one or more* of the diagonal entries d_1, d_2, d_3:

$$A = \begin{bmatrix} d_1 & 1 & 2 \\ 2 & d_2 & 1 \\ -1 & 2 & d_3 \end{bmatrix} \qquad \begin{array}{l} |\lambda - d_1| \leq 1 + 2 = R_1 \\ |\lambda - d_2| \leq 2 + 1 = R_2 \\ |\lambda - d_3| \leq 1 + 2 = R_3 \end{array}$$

You see that "near" means not more than 3 away from d_1 or d_2 or d_3, for this example.

6.1 C Find the eigenvalues $0, 1, 3$ and eigenvectors of this symmetric 3 by 3 matrix S:

Symmetric matrix
Singular matrix
Trace $1 + 2 + 1 = 4$

$$S = \begin{bmatrix} 1 & -1 & 0 \\ -1 & 2 & -1 \\ 0 & -1 & 1 \end{bmatrix}$$

All eigenvalues are in the Gershgorin circle $|\lambda - 2| \leq 1 + 1$.

Solution Since all rows of S add to zero, the vector $\boldsymbol{x} = (1, 1, 1)$ gives $S\boldsymbol{x} = \mathbf{0}$. This is an eigenvector for $\lambda = 0$. To find λ_2 and λ_3 I will compute the 3 by 3 determinant:

$$\det(S - \lambda I) = \begin{vmatrix} 1-\lambda & -1 & 0 \\ -1 & 2-\lambda & -1 \\ 0 & -1 & 1-\lambda \end{vmatrix} \begin{array}{l} = (1-\lambda)(2-\lambda)(1-\lambda) - 2(1-\lambda) \\ = (1-\lambda)[(2-\lambda)(1-\lambda) - 2] \\ = \boldsymbol{(1-\lambda)(-\lambda)(3-\lambda)}. \end{array}$$

Those three factors give $\boldsymbol{\lambda = 0, 1, 3}$. Each eigenvalue corresponds to an eigenvector:

$$\boldsymbol{x}_1 = \begin{bmatrix} 1 \\ 1 \\ 1 \end{bmatrix} \quad S\boldsymbol{x}_1 = \mathbf{0}\boldsymbol{x}_1 \qquad \boldsymbol{x}_2 = \begin{bmatrix} 1 \\ 0 \\ -1 \end{bmatrix} \quad S\boldsymbol{x}_2 = \mathbf{1}\boldsymbol{x}_2 \qquad \boldsymbol{x}_3 = \begin{bmatrix} 1 \\ -2 \\ 1 \end{bmatrix} \quad S\boldsymbol{x}_3 = \mathbf{3}\boldsymbol{x}_3 \,.$$

I notice again that eigenvectors are perpendicular when S is symmetric. We were lucky to find $\lambda = 0, 1, 3$. For a larger matrix I would use **eig**(A), and never touch determinants.

The full command $[\boldsymbol{X}, \boldsymbol{E}] = \textbf{eig}(A)$ will produce unit eigenvectors in the columns of X.

Problem Set 6.1

1 The example at the start of the chapter has powers of this matrix A:

$$A = \begin{bmatrix} .8 & .3 \\ .2 & .7 \end{bmatrix} \quad \text{and} \quad A^2 = \begin{bmatrix} .70 & .45 \\ .30 & .55 \end{bmatrix} \quad \text{and} \quad A^\infty = \begin{bmatrix} .6 & .6 \\ .4 & .4 \end{bmatrix}.$$

Find the eigenvalues of these matrices. All powers have the same eigenvectors.

(a) Show from A how a row exchange can produce different eigenvalues.

(b) Why is a zero eigenvalue *not* changed by the steps of elimination?

2 Find the eigenvalues and the eigenvectors of these two matrices:

$$A = \begin{bmatrix} 1 & 4 \\ 2 & 3 \end{bmatrix} \quad \text{and} \quad A + I = \begin{bmatrix} 2 & 4 \\ 2 & 4 \end{bmatrix}.$$

$A + I$ has the ______ eigenvectors as A. Its eigenvalues are ______ by 1.

3 Compute the eigenvalues and eigenvectors of A and A^{-1}. Check the trace!

$$A = \begin{bmatrix} 0 & 2 \\ 1 & 1 \end{bmatrix} \quad \text{and} \quad A^{-1} = \begin{bmatrix} -1/2 & 1 \\ 1/2 & 0 \end{bmatrix}.$$

A^{-1} has the ______ eigenvectors as A. When A has eigenvalues λ_1 and λ_2, its inverse has eigenvalues ______.

4 Compute the eigenvalues and eigenvectors of A and A^2:

$$A = \begin{bmatrix} -1 & 3 \\ 2 & 0 \end{bmatrix} \quad \text{and} \quad A^2 = \begin{bmatrix} 7 & -3 \\ -2 & 6 \end{bmatrix}.$$

A^2 has the same ______ as A. When A has eigenvalues λ_1 and λ_2, A^2 has eigenvalues ______. In this example, why is $\lambda_1^2 + \lambda_2^2 = 13$?

5 Find the eigenvalues of A and B (easy for triangular matrices) and $A + B$:

$$A = \begin{bmatrix} 3 & 0 \\ 1 & 1 \end{bmatrix} \quad \text{and} \quad B = \begin{bmatrix} 1 & 1 \\ 0 & 3 \end{bmatrix} \quad \text{and} \quad A + B = \begin{bmatrix} 4 & 1 \\ 1 & 4 \end{bmatrix}.$$

Eigenvalues of $A + B$ *are / are not* equal to eigenvalues of A + eigenvalues of B.

6 Find the eigenvalues of A and B and AB and BA:

$$A = \begin{bmatrix} 1 & 0 \\ 1 & 1 \end{bmatrix} \quad \text{and} \quad B = \begin{bmatrix} 1 & 2 \\ 0 & 1 \end{bmatrix} \quad \text{and} \quad AB = \begin{bmatrix} 1 & 2 \\ 1 & 3 \end{bmatrix} \quad \text{and} \quad BA = \begin{bmatrix} 3 & 2 \\ 1 & 1 \end{bmatrix}.$$

(a) Are the eigenvalues of AB equal to eigenvalues of A times eigenvalues of B?

(b) Are the eigenvalues of AB equal to the eigenvalues of BA?

7 Elimination produces $A = LU$. The eigenvalues of U are on its diagonal; they are the ______. The eigenvalues of L are on its diagonal; they are all ______. The eigenvalues of A are not the same as ______.

8 (a) If you know that $\boldsymbol{x}$ is an eigenvector, the way to find λ is to ______.

(b) If you know that λ is an eigenvalue, the way to find $\boldsymbol{x}$ is to ______.

9 What do you do to the equation $A\boldsymbol{x} = \lambda\boldsymbol{x}$, in order to prove (a), (b), and (c)?

(a) λ^2 is an eigenvalue of A^2, as in Problem 4.

(b) λ^{-1} is an eigenvalue of A^{-1}, as in Problem 3.

(c) $\lambda + 1$ is an eigenvalue of $A + I$, as in Problem 2.

10 Find the eigenvalues and eigenvectors for both of these Markov matrices A and A^∞. Explain from those answers why A^{100} is close to A^∞:

$$A = \begin{bmatrix} .6 & .2 \\ .4 & .8 \end{bmatrix} \quad \text{and} \quad A^\infty = \begin{bmatrix} 1/3 & 1/3 \\ 2/3 & 2/3 \end{bmatrix}.$$

11 Here is a strange fact about 2 by 2 matrices with eigenvalues $\lambda_1 \neq \lambda_2$: The columns of $A - \lambda_1 I$ are multiples of the eigenvector $\boldsymbol{x}_2$. Any idea why this should be?

12 Find three eigenvectors for this matrix P (projection matrices have $\lambda = 1$ and 0):

$$\textbf{Projection matrix} \qquad P = \begin{bmatrix} .2 & .4 & 0 \\ .4 & .8 & 0 \\ 0 & 0 & 1 \end{bmatrix}.$$

If two eigenvectors share the same λ, so do all their linear combinations. Find an eigenvector of P with no zero components.

13 From the unit vector $\boldsymbol{u} = (1, 1, 3, 5)/6$ construct the rank one projection matrix $P = \boldsymbol{u}\boldsymbol{u}^{\mathrm{T}}$. This matrix has $P^2 = P$ because $\boldsymbol{u}^{\mathrm{T}}\boldsymbol{u} = 1$.

(a) $P\boldsymbol{u} = \boldsymbol{u}$ comes from $(\boldsymbol{u}\boldsymbol{u}^{\mathrm{T}})\boldsymbol{u} = \boldsymbol{u}(\boldsymbol{u}^{\mathrm{T}}\boldsymbol{u}) = \boldsymbol{u}$. Then $\boldsymbol{u}$ is an eigenvector with eigenvalue $\lambda = 1$. In that case find $P^{100}\boldsymbol{u}$.

(b) If $\boldsymbol{v}$ is perpendicular to $\boldsymbol{u}$ show that $P\boldsymbol{v} = \boldsymbol{0}$. Then $\lambda = 0$.

(c) Find three independent eigenvectors of P all with eigenvalue $\lambda = 0$.

14 Solve $\det(Q - \lambda I) = 0$ by the quadratic formula to reach $\lambda = \cos\theta \pm i \sin\theta$:

$$Q = \begin{bmatrix} \cos\theta & -\sin\theta \\ \sin\theta & \cos\theta \end{bmatrix} \quad \text{rotates the } xy \text{ plane by the angle } \theta. \text{ No real } \lambda\text{'s.}$$

Find the eigenvectors of Q by solving $(Q - \lambda I)\boldsymbol{x} = \boldsymbol{0}$. Use $i^2 = -1$.

15 Every permutation matrix leaves $\boldsymbol{x} = (1, 1, \ldots, 1)$ unchanged. Then $\lambda = 1$. Find two more λ's (possibly complex) for these permutations, from $\det(P - \lambda I) = 0$:

$$P = \begin{bmatrix} 0 & 1 & 0 \\ 0 & 0 & 1 \\ 1 & 0 & 0 \end{bmatrix} \quad \text{and} \quad P = \begin{bmatrix} 0 & 0 & 1 \\ 0 & 1 & 0 \\ 1 & 0 & 0 \end{bmatrix}.$$

16 **The determinant of A equals the product $\boldsymbol{\lambda_1 \lambda_2 \cdots \lambda_n}$.** Start with the polynomial $\det(A - \lambda I)$ separated into its n factors (always possible). Then set $\lambda = 0$:

$$\det(A - \lambda I) = (\lambda_1 - \lambda)(\lambda_2 - \lambda) \cdots (\lambda_n - \lambda) \quad \text{so} \quad \det A = \underline{\qquad}.$$

Check this rule in Example 1 where the Markov matrix has $\lambda = 1$ and $\frac{1}{2}$.

17 If A has $\lambda_1 = 4$ and $\lambda_2 = 5$ then $\det(A - \lambda I) = (\lambda - 4)(\lambda - 5) = \lambda^2 - 9\lambda + 20$. Find three matrices that have trace $a + d = 9$ and determinant 20 and $\lambda = 4, 5$.

18 A 3 by 3 matrix B is known to have eigenvalues $0, 1, 2$. This information is enough to find three of these (give the answers where possible):

(a) the rank of B (b) the determinant of $B^{\mathrm{T}}B$

(c) the eigenvalues of $B^{\mathrm{T}}B$ (d) the eigenvalues of $(B^2 + I)^{-1}$.

19 Choose the last rows of A and C to give eigenvalues $4, 7$ and $1, 2, 3$:

Companion matrices $$A = \begin{bmatrix} 0 & 1 \\ * & * \end{bmatrix} \quad C = \begin{bmatrix} 0 & 1 & 0 \\ 0 & 0 & 1 \\ * & * & * \end{bmatrix}.$$

20 ***The eigenvalues of A equal the eigenvalues of $\boldsymbol{A^{\mathrm{T}}}$***. This is because $\det(A - \lambda I)$ equals $\det(A^{\mathrm{T}} - \lambda I)$. That is true because _____. Show by an example that the eigenvectors of A and A^{T} are *not* the same.

21 Find three 2 by 2 matrices that have $\lambda_1 = \lambda_2 = 0$. The trace is zero and the determinant is zero. A might not be the zero matrix but check that $A^2 = 0$.

22 This matrix is singular with rank one. Find three λ's and three eigenvectors:

$$A = \begin{bmatrix} 1 \\ 2 \\ 1 \end{bmatrix} \begin{bmatrix} 2 & 1 & 2 \end{bmatrix} = \begin{bmatrix} 2 & 1 & 2 \\ 4 & 2 & 4 \\ 2 & 1 & 2 \end{bmatrix}$$

All eigenvalues are in the Gershgorin circle $|\lambda - 2| \leq 8$.

23 Suppose A and B have the same eigenvalues $\lambda_1, \ldots, \lambda_n$ with the same independent eigenvectors $\boldsymbol{x}_1, \ldots, \boldsymbol{x}_n$. Then $A = B$. *Reason*: Any vector $\boldsymbol{x}$ is a combination $c_1\boldsymbol{x}_1 + \cdots + c_n\boldsymbol{x}_n$. What is $A\boldsymbol{x}$? What is $B\boldsymbol{x}$?

24 Find the rank and the four eigenvalues of A and C:

$$A = \begin{bmatrix} 1 & 1 & 1 & 1 \\ 1 & 1 & 1 & 1 \\ 1 & 1 & 1 & 1 \\ 1 & 1 & 1 & 1 \end{bmatrix} \quad \text{and} \quad C = \begin{bmatrix} 1 & 0 & 1 & 0 \\ 0 & 1 & 0 & 1 \\ 1 & 0 & 1 & 0 \\ 0 & 1 & 0 & 1 \end{bmatrix}.$$

25 (Review) Find the eigenvalues of A, B, and C:

$$A = \begin{bmatrix} 1 & 2 & 3 \\ 0 & 4 & 5 \\ 0 & 0 & 6 \end{bmatrix} \quad \text{and} \quad B = \begin{bmatrix} 0 & 0 & 1 \\ 0 & 2 & 0 \\ 3 & 0 & 0 \end{bmatrix} \quad \text{and} \quad C = \begin{bmatrix} 2 & 2 & 2 \\ 2 & 2 & 2 \\ 2 & 2 & 2 \end{bmatrix}.$$

26 Suppose A has eigenvalues $0, 3, 5$ with independent eigenvectors $\boldsymbol{u}, \boldsymbol{v}, \boldsymbol{w}$.

(a) Give a basis for the nullspace and a basis for the column space.

(b) Find a particular solution $\boldsymbol{x}$ to $A\boldsymbol{x} = \boldsymbol{v} + \boldsymbol{w}$. Find all solutions.

(c) $A\boldsymbol{x} = \boldsymbol{u}$ has no solution. If it did then _____ would be in the column space.

Challenge Problems

27 Show that $\boldsymbol{u}$ is an eigenvector of the rank one 2 × 2 matrix $A = \boldsymbol{u}\boldsymbol{v}^{\mathrm{T}}$. Find both eigenvalues of A. Check that $\lambda_1 + \lambda_2$ agrees with the trace $u_1v_1 + u_2v_2 = \boldsymbol{u}^{\mathrm{T}}\boldsymbol{v}$.

28 There are six 3 by 3 permutation matrices P. What numbers can be the *determinants* of P? What numbers can be *pivots*? What numbers can be the *trace* of P? What *four numbers* can be eigenvalues of P, as in Problem 15?

29 (**Heisenberg's Uncertainty Principle**) $AB - BA = I$ can happen for infinite matrices with $A = A^{\mathrm{T}}$ and $B = -B^{\mathrm{T}}$. Then

$$\boldsymbol{x}^{\mathrm{T}}\boldsymbol{x} = \boldsymbol{x}^{\mathrm{T}}AB\boldsymbol{x} - \boldsymbol{x}^{\mathrm{T}}BA\boldsymbol{x} \le 2\,\|A\boldsymbol{x}\|\,\|B\boldsymbol{x}\|.$$

Explain that last step by using the Schwarz inequality $|\boldsymbol{u}^{\mathrm{T}}\boldsymbol{v}| \le \|\boldsymbol{u}\|\,\|\boldsymbol{v}\|$. Then Heisenberg's inequality says that $\|A\boldsymbol{x}\|/\|\boldsymbol{x}\|$ times $\|B\boldsymbol{x}\|/\|\boldsymbol{x}\|$ is at least $\frac{1}{2}$. It is impossible to get the position error and momentum error both very small.

30 *Find a 2 by 2 rotation matrix* (other than I) *with* $A^3 = I$. Its eigenvalues must satisfy $\lambda^3 = 1$. They can be $e^{2\pi i/3}$ and $e^{-2\pi i/3}$. What are the trace and determinant?

6.2 Diagonalizing a Matrix

1 The columns of $AX = X\Lambda$ are $A\boldsymbol{x}_k = \lambda_k \boldsymbol{x}_k$. The eigenvalue matrix Λ is diagonal.

2 $\boldsymbol{n}$ independent eigenvectors in X diagonalize A $\boxed{\boldsymbol{A = X\Lambda X^{-1}} \text{ and } \boldsymbol{\Lambda = X^{-1}AX}}$

3 Solve $\boldsymbol{u}_{k+1} = A\boldsymbol{u}_k$ by $\boldsymbol{u}_k = A^k\boldsymbol{u}_0 = X\Lambda^k X^{-1}\boldsymbol{u}_0 = \boxed{\boldsymbol{c_1(\lambda_1)^k x_1 + \cdots + c_n(\lambda_n)^k x_n}}$

4 No equal eigenvalues $\Rightarrow$ n independent eigenvectors in X. Then A can be diagonalized. Equal eigenvalues $\Rightarrow$ A *might* have too few independent eigenvectors. Then X^{-1} fails.

5 Every matrix $C = B^{-1}AB$ has the same eigenvalues as A. These C's are "**similar**" to A.

When $\boldsymbol{x}$ is an eigenvector, multiplication by A is just multiplication by a number λ: $A\boldsymbol{x} = \lambda\boldsymbol{x}$. All the difficulties of matrices are swept away. Instead of an interconnected system, we can follow the eigenvectors separately. It is like having a *diagonal matrix*, with no off-diagonal interconnections. The 100th power of a diagonal matrix is easy.

The point of this section is very direct. ***A turns into a diagonal matrix* $\boldsymbol{\Lambda}$ *when we use the eigenvectors properly***. Diagonalizing A is the matrix form of our key idea.

Suppose the n by n matrix A has n linearly independent eigenvectors $\boldsymbol{x}_1, \ldots, \boldsymbol{x}_n$. Put them into the columns of an ***eigenvector matrix*** $\boldsymbol{X}$. We will prove $\boldsymbol{AX = X\Lambda}$. Therefore $\boldsymbol{X^{-1}AX}$ ***is the eigenvalue matrix*** $\boldsymbol{\Lambda}$:

Eigenvector matrix $\boldsymbol{X}$

Eigenvalue matrix $\boldsymbol{\Lambda}$

$$\boldsymbol{X^{-1}AX = \Lambda} = \begin{bmatrix} \lambda_1 & & \\ & \ddots & \\ & & \lambda_n \end{bmatrix}. \qquad (1)$$

The matrix A is "diagonalized." We use capital lambda for the eigenvalue matrix, because the small λ's (the eigenvalues of A) are on its diagonal.

Example 1 This A is triangular so its eigenvalues 1 and 6 are on the main diagonal:

Eigenvectors go into $\boldsymbol{X}$ $\begin{bmatrix} \mathbf{1} \\ \mathbf{0} \end{bmatrix} \begin{bmatrix} \mathbf{1} \\ \mathbf{1} \end{bmatrix}$

$$\underset{\boldsymbol{X^{-1}}}{\begin{bmatrix} 1 & -1 \\ 0 & 1 \end{bmatrix}} \underset{\boldsymbol{A}}{\begin{bmatrix} \mathbf{1} & \mathbf{5} \\ \mathbf{0} & \mathbf{6} \end{bmatrix}} \underset{\boldsymbol{X}}{\begin{bmatrix} 1 & 1 \\ 0 & 1 \end{bmatrix}} = \underset{\boldsymbol{\Lambda}}{\begin{bmatrix} \mathbf{1} & 0 \\ 0 & \mathbf{6} \end{bmatrix}}$$

In other words $A = X\Lambda X^{-1}$. Then $A^2 = X\Lambda X^{-1}X\Lambda X^{-1}$. So $\boldsymbol{A^2}$ **is** $\boldsymbol{X\Lambda^2X^{-1}}$.

$\boldsymbol{A^2}$ ***has the same eigenvectors in*** $\boldsymbol{X}$ ***and its squared eigenvalues are in*** $\boldsymbol{\Lambda^2}$.

Why is $AX = X\Lambda$? A multiplies its eigenvectors, which are the columns of X. The first column of AX is $A\boldsymbol{x}_1$. That is $\lambda_1\boldsymbol{x}_1$. Each column of X is multiplied by its eigenvalue:

$$\textbf{A times X} \qquad AX = A\begin{bmatrix} \boldsymbol{x}_1 & \cdots & \boldsymbol{x}_n \end{bmatrix} = \begin{bmatrix} \lambda_1\boldsymbol{x}_1 & \cdots & \lambda_n\boldsymbol{x}_n \end{bmatrix}.$$

The key idea is to split this matrix AX into X times Λ:

$$\textbf{X times }\boldsymbol{\Lambda} \qquad \begin{bmatrix} \lambda_1\boldsymbol{x}_1 & \cdots & \lambda_n\boldsymbol{x}_n \end{bmatrix} = \begin{bmatrix} \boldsymbol{x}_1 & \cdots & \boldsymbol{x}_n \end{bmatrix}\begin{bmatrix} \lambda_1 & & \\ & \ddots & \\ & & \lambda_n \end{bmatrix} = X\Lambda.$$

Keep those matrices in the right order! Then λ_1 multiplies the first column $\boldsymbol{x}_1$, as shown. The diagonalization is complete, and we can write $AX = X\Lambda$ in two good ways:

$$\boxed{AX = X\Lambda \quad \text{is} \quad \boldsymbol{X^{-1}AX = \Lambda} \quad \text{or} \quad \boldsymbol{A = X\Lambda X^{-1}}} \qquad (2)$$

The matrix X has an inverse, because its columns (the eigenvectors of A) were assumed to be linearly independent. *Without n independent eigenvectors, we can't diagonalize A.*

A and Λ have the same eigenvalues $\lambda_1, \ldots, \lambda_n$. Their eigenvectors are different. The job of the original eigenvectors $\boldsymbol{x}_1, \ldots, \boldsymbol{x}_n$ was to diagonalize A. Those eigenvectors in X produce $A = X\Lambda X^{-1}$. You will soon see their simplicity and importance and meaning. The kth power will be $A^k = X\Lambda^k X^{-1}$ which is easy to compute:

$$\boxed{\boldsymbol{A^k = (X\Lambda X^{-1})(X\Lambda X^{-1})\ldots(X\Lambda X^{-1}) = X\Lambda^k X^{-1}}}$$

Example 1
$\boldsymbol{\lambda}$ = 1 and 6

$$\begin{bmatrix} 1 & 5 \\ 0 & 6 \end{bmatrix}^k = \begin{bmatrix} 1 & 1 \\ 0 & 1 \end{bmatrix}\begin{bmatrix} 1 & \\ & 6^k \end{bmatrix}\begin{bmatrix} 1 & -1 \\ 0 & 1 \end{bmatrix} = \begin{bmatrix} \mathbf{1} & \mathbf{6^k - 1} \\ 0 & \mathbf{6^k} \end{bmatrix} = A^k.$$

With $k = 1$ we get A. With $k = 0$ we get $A^0 = I$ (and $\lambda^0 = 1$). *With $k = -1$ we get A^{-1}* with eigenvalues 1 and $\frac{1}{6}$. You see how $A^2 = [1\ \ 35;\ \ 0\ \ 36]$ fits that formula when $k = 2$.

Here are four small remarks before we use Λ again in Example 2.

Remark 1 Suppose the eigenvalues $\lambda_1, \ldots, \lambda_n$ are all different. Then it is automatic that the eigenvectors $\boldsymbol{x}_1, \ldots, \boldsymbol{x}_n$ are independent. The eigenvector matrix X will be *invertible*. ***Every matrix that has no repeated eigenvalues can be diagonalized.***

Remark 2 ***We can multiply eigenvectors by any nonzero constants.*** $A(c\boldsymbol{x}) = \lambda(c\boldsymbol{x})$ is still true. In Example 1, we can divide $\boldsymbol{x} = (1, 1)$ by $\sqrt{2}$ to produce a unit eigenvector.

MATLAB and virtually all other codes produce eigenvectors of length $||\boldsymbol{x}|| = 1$.

Remark 3 The eigenvectors in X come in the same order as the eigenvalues in Λ. To reverse the order in Λ, put the eigenvector $(1,1)$ before the eigenvector $(1,0)$ in X :

$$\textbf{New order } 6,1 \qquad \begin{bmatrix} 0 & 1 \\ 1 & -1 \end{bmatrix}\begin{bmatrix} 1 & 5 \\ 0 & 6 \end{bmatrix}\begin{bmatrix} 1 & 1 \\ 1 & 0 \end{bmatrix} = \begin{bmatrix} \mathbf{6} & 0 \\ 0 & \mathbf{1} \end{bmatrix} = \Lambda_{\textbf{new}}$$

To diagonalize A we *must* use an eigenvector matrix. From $X^{-1}AX = \Lambda$ we know that $AX = X\Lambda$. Suppose the first column of X is $\boldsymbol{x}$. Then the first columns of AX and $X\Lambda$ are $A\boldsymbol{x}$ and $\lambda_1\boldsymbol{x}$. For those to be equal, $\boldsymbol{x}$ must be an eigenvector.

Remark 4 (repeated warning for repeated eigenvalues) Some matrices have too few eigenvectors. *Those matrices cannot be diagonalized.* Here are two examples:

$$\begin{array}{l}\textbf{\textit{A} and \textit{B} are not diagonalizable} \\ \textbf{They have only one eigenvector}\end{array} \qquad A = \begin{bmatrix} 1 & -1 \\ 1 & -1 \end{bmatrix} \quad \text{and} \quad B = \begin{bmatrix} 0 & 1 \\ 0 & 0 \end{bmatrix}.$$

Their eigenvalues happen to be 0 and 0. Nothing is special about $\lambda = 0$, the problem is the repetition of λ. All eigenvectors of the first matrix are multiples of $(1,1)$:

$$\begin{array}{l}\textbf{Only one line} \\ \textbf{of eigenvectors}\end{array} \qquad A\boldsymbol{x} = 0\boldsymbol{x} \quad \text{means} \quad \begin{bmatrix} 1 & -1 \\ 1 & -1 \end{bmatrix}\begin{bmatrix} \boldsymbol{x} \end{bmatrix} = \begin{bmatrix} 0 \\ 0 \end{bmatrix} \quad \text{and} \quad \boldsymbol{x} = c\begin{bmatrix} 1 \\ 1 \end{bmatrix}.$$

There is no second eigenvector, so this unusual matrix A cannot be diagonalized.

Those matrices are the best examples to test any statement about eigenvectors. In many true-false questions, non-diagonalizable matrices lead to *false*.

Remember that there is no connection between invertibility and diagonalizability:

- ***Invertibility*** is concerned with the ***eigenvalues*** ($\lambda = 0$ or $\lambda \neq 0$).
- ***Diagonalizability*** is concerned with the ***eigenvectors*** (too few or enough for X).

Each eigenvalue has at least one eigenvector! $A - \lambda I$ is singular. If $(A - \lambda I)\boldsymbol{x} = \mathbf{0}$ leads you to $\boldsymbol{x} = \mathbf{0}$, λ is *not* an eigenvalue. Look for a mistake in solving $\det(A - \lambda I) = 0$.

Eigenvectors for n different λ's are independent. Then we can diagonalize A.

> **Independent $\boldsymbol{x}$ from different λ** Eigenvectors $\boldsymbol{x}_1, \ldots, \boldsymbol{x}_j$ that correspond to distinct (all different) eigenvalues are linearly independent. An n by n matrix that has n different eigenvalues (no repeated λ's) must be diagonalizable.

Proof Suppose $c_1\boldsymbol{x}_1 + c_2\boldsymbol{x}_2 = \mathbf{0}$. Multiply by A to find $c_1\lambda_1\boldsymbol{x}_1 + c_2\lambda_2\boldsymbol{x}_2 = \mathbf{0}$. Multiply by λ_2 to find $c_1\lambda_2\boldsymbol{x}_1 + c_2\lambda_2\boldsymbol{x}_2 = \mathbf{0}$. Now subtract one from the other to show $c_1 = 0$.

$$\text{Subtraction leaves} \quad (\lambda_1 - \lambda_2)c_1\boldsymbol{x}_1 = \mathbf{0}. \quad \text{Then } \boldsymbol{c_1} = \mathbf{0} \text{ because } \lambda_1 \neq \lambda_2.$$

Similarly $c_2 = 0$. Only the combination of $\boldsymbol{x}$'s with $c_1 = c_2 = 0$ gives $c_1\boldsymbol{x}_1 + c_2\boldsymbol{x}_2 = \mathbf{0}$. So the eigenvectors $\boldsymbol{x}_1$ and $\boldsymbol{x}_2$ must be independent.

This proof extends directly to 3 eigenvectors. Suppose that $c_1\boldsymbol{x}_1 + c_2\boldsymbol{x}_2 + c_3\boldsymbol{x}_3 = \mathbf{0}$. Multiply by $A - \lambda_3 I$ and $\boldsymbol{x}_3$ is gone. Multiply by $A - \lambda_2 I$ and $\boldsymbol{x}_2$ is gone:

$$\textbf{Only } \boldsymbol{x}_1 \textbf{ is left} \quad (\lambda_1 - \lambda_2)(\lambda_1 - \lambda_3)c_1\boldsymbol{x}_1 = \mathbf{0} \quad \text{which forces} \quad c_1 = 0. \tag{3}$$

Similarly every $c_i = 0$. When the λ's are all different, the eigenvectors are independent. A full set of n eigenvectors can go into the n columns of the eigenvector matrix X.

Example 2 **Powers of A** The Markov matrix $A = \begin{bmatrix} .8 & .3 \\ .2 & .7 \end{bmatrix}$ has $\lambda_1 = 1$ and $\lambda_2 = .5$. Here is $A = X\Lambda X^{-1}$ with those eigenvalues in the diagonal Λ:

$$\textbf{Markov example} \quad \begin{bmatrix} .8 & .3 \\ .2 & .7 \end{bmatrix} = \begin{bmatrix} .6 & 1 \\ .4 & -1 \end{bmatrix} \begin{bmatrix} 1 & 0 \\ 0 & .5 \end{bmatrix} \begin{bmatrix} 1 & 1 \\ .4 & -.6 \end{bmatrix} = X\Lambda X^{-1}.$$

The eigenvectors $(.6, .4)$ and $(1, -1)$ are in the columns of X. They are also the eigenvectors of A^2. Watch how A^2 has the same X, and ***the eigenvalue matrix of A^2 is Λ^2***:

$$\textbf{Same } X \textbf{ for } A^2 \qquad \boxed{\boldsymbol{A^2} = X\Lambda X^{-1}X\Lambda X^{-1} = \boldsymbol{X\Lambda^2 X^{-1}}.} \tag{4}$$

Just keep going, and you see why the high powers A^k approach a "steady state":

$$\textbf{Powers of } A \qquad A^k = X\Lambda^k X^{-1} = \begin{bmatrix} .6 & 1 \\ .4 & -1 \end{bmatrix} \begin{bmatrix} 1^k & 0 \\ 0 & (.5)^k \end{bmatrix} \begin{bmatrix} 1 & 1 \\ .4 & -.6 \end{bmatrix}.$$

As k gets larger, $(.5)^k$ gets smaller. In the limit it disappears completely. That limit is A^∞:

$$\textbf{Limit } k \to \infty \qquad A^\infty = \begin{bmatrix} .6 & 1 \\ .4 & -1 \end{bmatrix} \begin{bmatrix} 1 & 0 \\ 0 & 0 \end{bmatrix} \begin{bmatrix} 1 & 1 \\ .4 & -.6 \end{bmatrix} = \begin{bmatrix} \mathbf{.6} & \mathbf{.6} \\ \mathbf{.4} & \mathbf{.4} \end{bmatrix}.$$

The limit has the eigenvector $\boldsymbol{x}_1$ in both columns. We saw this A^∞ on the very first page of Chapter 6. Now we see it coming from powers like $A^{100} = X\Lambda^{100}X^{-1}$.

Question ***When does $A^k \to$ zero matrix?*** **Answer** ***All $|\lambda| < 1$.***

Similar Matrices: Same Eigenvalues

Suppose the eigenvalue matrix Λ is *fixed*. As we change the eigenvector matrix X, we get a whole family of different matrices $A = X\Lambda X^{-1}$—*all with the same eigenvalues in* Λ. All those matrices A (with the same eigenvalues in Λ) are called **similar**.

This idea extends to matrices C that can't be diagonalized. Again we look at the whole family of matrices $\boldsymbol{A = BCB^{-1}}$, allowing all invertible matrices B. Again all those matrices A and C are **similar**.

We are using C instead of Λ because C might not be diagonal. We are using B instead of X because the columns of B might not be eigenvectors. We only require that B is invertible. **Similar matrices C and BCB^{-1} have the same eigenvalues.**

All the matrices $A = BCB^{-1}$ are "similar." They all share the eigenvalues of C.

Suppose $Cx = \lambda x$. *Then* BCB^{-1} *also has the eigenvalue* λ. The new eigenvector is Bx :

$$\textbf{\textit{Same}}\ \boldsymbol{\lambda} \qquad (BCB^{-1})(\boldsymbol{Bx}) = BC\boldsymbol{x} = B\lambda\boldsymbol{x} = \lambda(\boldsymbol{Bx}). \tag{5}$$

A fixed matrix C produces a family of similar matrices BCB^{-1}, allowing all B. When C is the identity matrix, the "family" is very small. The only member is $BIB^{-1} = I$. The identity matrix is the only diagonalizable matrix with all eigenvalues $\lambda = 1$.

The family is larger when $\lambda = 1$ and 1 *with only one eigenvector* (not diagonalizable). The simplest C in this family is called the *Jordan form*. Every matrix A in the family has determinant = 1 and trace = 2 and this special form with $A = I$ excluded:

$$C = \begin{bmatrix} \mathbf{1} & \mathbf{1} \\ 0 & \mathbf{1} \end{bmatrix} = \text{Jordan form for every } A = BCB^{-1} = \begin{bmatrix} 1 - rs & r^2 \\ -s^2 & 1 + rs \end{bmatrix}. \tag{6}$$

For an important example I will take eigenvalues $\lambda = 1$ and 0 (not repeated!). Now the whole family is diagonalizable with the same eigenvalue matrix Λ. We get every 2 by 2 matrix A that has eigenvalues 1 and 0. The trace of A is 1 and the determinant is zero:

$$\textbf{All similar} \quad \Lambda = \begin{bmatrix} \mathbf{1} & 0 \\ 0 & \mathbf{0} \end{bmatrix} \quad A = \begin{bmatrix} 1 & 1 \\ 0 & 0 \end{bmatrix} \ \text{or}\ A = \begin{bmatrix} .5 & .5 \\ .5 & .5 \end{bmatrix} \ \text{or any}\ A = \frac{\boldsymbol{x}\boldsymbol{y}^{\mathrm{T}}}{\boldsymbol{x}^{\mathrm{T}}\boldsymbol{y}}.$$

The family contains all matrices with $A^2 = A$, including $A = \Lambda$. When A is symmetric these are projection matrices $P^2 = P$. Eigenvalues 1 and 0 make life easy.

Fibonacci Numbers

We present a famous example, where ***every new Fibonacci number is the sum of the two previous F's***. Then eigenvalues of A tell how fast the Fibonacci numbers grow.

The sequence $0, 1, 1, 2, 3, 5, 8, 13, \ldots$ ***comes from*** $F_{k+2} = F_{k+1} + F_k$.

Problem: Find the Fibonacci number F_{100} The slow way is to apply the rule $F_{k+2} = F_{k+1} + F_k$ one step at a time. By adding $F_6 = 8$ to $F_7 = 13$ we reach $F_8 = 21$. Eventually we come to F_{100}. Linear algebra gives a better way.

The key is to begin with a matrix equation $\mathbf{u}_{k+1} = A\boldsymbol{u}_k$. That is a *one-step* rule for vectors, while Fibonacci gave a two-step rule for scalars. We match those rules by putting two Fibonacci numbers into a vector $\boldsymbol{u}$. Then you will see the matrix A.

$$\text{Let } \boldsymbol{u}_k = \begin{bmatrix} F_{k+1} \\ F_k \end{bmatrix}. \text{ The rule } \begin{matrix} F_{k+2} = F_{k+1} + F_k \\ F_{k+1} = F_{k+1} \end{matrix} \text{ is } \boldsymbol{u}_{k+1} = \begin{bmatrix} \mathbf{1} & \mathbf{1} \\ \mathbf{1} & \mathbf{0} \end{bmatrix} \boldsymbol{u}_k. \tag{7}$$

Every step multiplies by $A = \begin{bmatrix} 1 & 1 \\ 1 & 0 \end{bmatrix}$. After 100 steps we reach $\boldsymbol{u}_{100} = A^{100}\boldsymbol{u}_0$:

$$\boldsymbol{u}_0 = \begin{bmatrix} 1 \\ 0 \end{bmatrix}, \quad \boldsymbol{u}_1 = \begin{bmatrix} 1 \\ 1 \end{bmatrix}, \quad \boldsymbol{u}_2 = \begin{bmatrix} 2 \\ 1 \end{bmatrix}, \quad \boldsymbol{u}_3 = \begin{bmatrix} 3 \\ 2 \end{bmatrix}, \quad \ldots, \quad \boldsymbol{u}_{100} = \begin{bmatrix} F_{101} \\ F_{100} \end{bmatrix}.$$

This problem is perfect for eigenvalues. Take the determinant of $A - \lambda I$:

$$A - \lambda I = \begin{bmatrix} 1-\lambda & 1 \\ 1 & -\lambda \end{bmatrix} \quad \text{leads to} \quad \det(A - \lambda I) = \lambda^2 - \lambda - 1.$$

The equation $\lambda^2 - \lambda - 1 = 0$ is solved by the quadratic formula $\left(-b \pm \sqrt{b^2 - 4ac}\right)/2a$:

$$\textbf{Eigenvalues} \quad \lambda_1 = \frac{1+\sqrt{5}}{2} \approx \mathbf{1.618} \quad \text{and} \quad \lambda_2 = \frac{1-\sqrt{5}}{2} \approx \mathbf{-0.618}.$$

These eigenvalues lead to eigenvectors $\boldsymbol{x}_1 = (\lambda_1, 1)$ and $\boldsymbol{x}_2 = (\lambda_2, 1)$. Step 2 finds the combination of those eigenvectors that gives the starting vector $\boldsymbol{u}_0 = (1, 0)$:

$$\begin{bmatrix} 1 \\ 0 \end{bmatrix} = \frac{1}{\lambda_1 - \lambda_2}\left(\begin{bmatrix} \lambda_1 \\ 1 \end{bmatrix} - \begin{bmatrix} \lambda_2 \\ 1 \end{bmatrix}\right) \quad \text{or} \quad \boldsymbol{u}_0 = \frac{\boldsymbol{x}_1 - \boldsymbol{x}_2}{\lambda_1 - \lambda_2}. \tag{8}$$

Step 3 multiplies $\boldsymbol{u}_0$ by A^{100} to find $\boldsymbol{u}_{100}$. The eigenvectors $\boldsymbol{x}_1$ and $\boldsymbol{x}_2$ stay separate! They are multiplied by $(\lambda_1)^{100}$ and $(\lambda_2)^{100}$:

$$\boxed{\textbf{100 steps from } \boldsymbol{u}_0 \qquad \boldsymbol{u}_{100} = \frac{(\lambda_1)^{100}\boldsymbol{x}_1 - (\lambda_2)^{100}\boldsymbol{x}_2}{\lambda_1 - \lambda_2}} \tag{9}$$

We want F_{100} = second component of $\boldsymbol{u}_{100}$. The second components of $\boldsymbol{x}_1$ and $\boldsymbol{x}_2$ are 1. The difference between $\lambda_1 = (1+\sqrt{5})/2$ and $\lambda_2 = (1-\sqrt{5})/2$ is $\sqrt{5}$. And $\lambda_2^{100} \approx 0$.

$$\boxed{\text{100th Fibonacci number} = \frac{\boldsymbol{\lambda_1^{100} - \lambda_2^{100}}}{\boldsymbol{\lambda_1 - \lambda_2}} = \text{nearest integer to } \frac{1}{\sqrt{5}}\left(\frac{1+\sqrt{5}}{2}\right)^{100}} \tag{10}$$

Every F_k is a whole number. The ratio F_{101}/F_{100} must be very close to the limiting ratio $\left(1+\sqrt{5}\right)/2$. The Greeks called this number the *"golden mean"*. For some reason a rectangle with sides 1.618 and 1 looks especially graceful.

Matrix Powers A^k

Fibonacci's example is a typical difference equation $\boldsymbol{u}_{k+1} = A\boldsymbol{u}_k$. ***Each step multiplies by A.*** The solution is $\boldsymbol{u}_k = A^k\boldsymbol{u}_0$. We want to make clear how diagonalizing the matrix gives a quick way to compute A^k and find $\boldsymbol{u}_k$ in three steps.

The eigenvector matrix X produces $A = X\Lambda X^{-1}$. This is a factorization of the matrix, like $A = LU$ or $A = QR$. The new factorization is perfectly suited to computing powers, because *every time* X^{-1} *multiplies* X *we get* I:

$$\textbf{Powers of } \boldsymbol{A} \qquad A^k\boldsymbol{u}_0 = (X\Lambda X^{-1})\cdots(X\Lambda X^{-1})\boldsymbol{u}_0 = \boldsymbol{X\Lambda^k X^{-1}u_0}$$

I will split $X\Lambda^k X^{-1}\boldsymbol{u}_0$ into three steps. Equation (11) will show how eigenvalues work.

1. Write $\boldsymbol{u}_0$ as a combination $c_1\boldsymbol{x}_1 + \cdots + c_n\boldsymbol{x}_n$ of the eigenvectors. Then $\boldsymbol{c} = X^{-1}\boldsymbol{u}_0$.
2. Multiply each eigenvector $\boldsymbol{x}_i$ by $(\lambda_i)^k$. Now we have $\Lambda^k X^{-1}\boldsymbol{u}_0$.
3. Add up the pieces $c_i(\lambda_i)^k\boldsymbol{x}_i$ to find the solution $\boldsymbol{u}_k = A^k\boldsymbol{u}_0$. This is $X\Lambda^k X^{-1}\boldsymbol{u}_0$.

$$\textbf{Solution for } \boldsymbol{u}_{k+1} = A\boldsymbol{u}_k \qquad \boldsymbol{u}_k = A^k\boldsymbol{u}_0 = c_1(\lambda_1)^k\boldsymbol{x}_1 + \cdots + c_n(\lambda_n)^k\boldsymbol{x}_n. \tag{11}$$

In matrix language A^k equals $(X\Lambda X^{-1})^k$ which is X times Λ^k times X^{-1}. In Step 1, the eigenvectors in X lead to the c's in the combination $\boldsymbol{u}_0 = c_1\boldsymbol{x}_1 + \cdots + c_n\boldsymbol{x}_n$:

$$\text{Step 1} \qquad \boldsymbol{u}_0 = \begin{bmatrix} & & \\ \boldsymbol{x}_1 & \cdots & \boldsymbol{x}_n \\ & & \end{bmatrix} \begin{bmatrix} c_1 \\ \vdots \\ c_n \end{bmatrix}. \quad \text{This says that } \boldsymbol{u}_0 = X\boldsymbol{c}. \tag{12}$$

The coefficients in Step 1 are $\boldsymbol{c} = X^{-1}\boldsymbol{u}_0$. Then Step 2 multiplies each c_k by Λ^k. The final result $\boldsymbol{u}_k = \sum c_i(\lambda_i)^k\boldsymbol{x}_i$ in Step 3 is the product of X and Λ^k and $\boldsymbol{c} = X^{-1}\boldsymbol{u}_0$:

$$\boldsymbol{A^k u_0} = X\Lambda^k X^{-1}\boldsymbol{u}_0 = X\Lambda^k\boldsymbol{c} = \begin{bmatrix} & & \\ \boldsymbol{x}_1 & \ldots & \boldsymbol{x}_n \\ & & \end{bmatrix} \begin{bmatrix} (\boldsymbol{\lambda}_1)^k & & \\ & \ddots & \\ & & (\boldsymbol{\lambda}_n)^k \end{bmatrix} \begin{bmatrix} \boldsymbol{c}_1 \\ \vdots \\ \boldsymbol{c}_n \end{bmatrix}. \tag{13}$$

This result is exactly $\boldsymbol{u}_k = \boldsymbol{c_1(\lambda_1)^k x_1} + \cdots + \boldsymbol{c_n(\lambda_n)^k x_n}$. It solves $\boldsymbol{u}_{k+1} = A\boldsymbol{u}_k$.

Nondiagonalizable Matrices (Optional)

Suppose λ is an eigenvalue of A. We discover that fact in two ways:

1. **Eigenvectors** (geometric) There are nonzero solutions to $A\boldsymbol{x} = \lambda\boldsymbol{x}$.
2. **Eigenvalues** (algebraic) The determinant of $A - \lambda I$ is zero.

The number λ may be a simple eigenvalue or a multiple eigenvalue, and we want to know its ***multiplicity***. Most eigenvalues have multiplicity $M = 1$ (simple eigenvalues). Then there is a single line of eigenvectors, and $\det(A - \lambda I)$ does not have a double factor.

For exceptional matrices, an eigenvalue can be ***repeated***. Then there are two different ways to count its multiplicity. Always **GM** $\leq$ **AM** for each λ:

1. (Geometric Multiplicity = **GM**) Count the **independent eigenvectors for $\boldsymbol{\lambda}$**. Then GM is the dimension of the nullspace of $A - \lambda I$.
2. (Algebraic Multiplicity = **AM**) AM counts the **repetitions of $\boldsymbol{\lambda}$** among the eigenvalues of A. Look at the n roots of $\det(A - \lambda I) = 0$.

If not enough eigenvectors → NOT DIAGONIZIBLE
If eigen value is different, → DIAGONIZABLE

If A has $\lambda = 4, 4, 4$, then that eigenvalue has AM $= 3$ and GM $= 1$ or 2 or 3.

The following matrix A is an example of trouble. Its eigenvalue $\lambda = 0$ is repeated. It is a double eigenvalue (AM $= 2$) with only one independent eigenvector $\boldsymbol{x} = (1, 0)$.

$$\begin{array}{l}\textbf{AM} = \textbf{2}\\ \textbf{GM} = \textbf{1}\end{array} \qquad A = \begin{bmatrix} 0 & 1 \\ 0 & 0 \end{bmatrix} \text{ has } \det(A - \lambda I) = \begin{vmatrix} -\lambda & 1 \\ 0 & -\lambda \end{vmatrix} = \lambda^2. \qquad \begin{array}{l}\boldsymbol{\lambda = 0, 0} \textbf{ but}\\ \textbf{1 eigenvector}\end{array}$$

***This shortage of eigenvectors when* GM *is below* AM *means that A is not diagonalizable*.**

■ WORKED EXAMPLES ■

6.2 A Find the inverse and the eigenvalues and the determinant of this matrix A:

$$\boldsymbol{A} = 5 * \textbf{eye}(4) - \textbf{ones}(4) = \begin{bmatrix} 4 & -1 & -1 & -1 \\ -1 & 4 & -1 & -1 \\ -1 & -1 & 4 & -1 \\ -1 & -1 & -1 & 4 \end{bmatrix}.$$

Describe an eigenvector matrix X that gives $X^{-1}AX = \Lambda$.

Solution What are the eigenvalues of the all-ones matrix? Its rank is certainly 1, so three eigenvalues are $\lambda = 0, 0, 0$. Its trace is 4, so the other eigenvalue is $\lambda = 4$. Subtract this all-ones matrix from $5I$ to get our matrix A:

Subtract the eigenvalues $4, 0, 0, 0$ from $5, 5, 5, 5$. The eigenvalues of A are $1, 5, 5, 5$.

The determinant of A is 125, the product of those four eigenvalues. The eigenvector for $\lambda = 1$ is $\boldsymbol{x} = (1, 1, 1, 1)$ or (c, c, c, c). The other eigenvectors are perpendicular to $\boldsymbol{x}$ (since A is symmetric). The nicest eigenvector matrix X is the symmetric orthogonal **Hadamard matrix $\boldsymbol{H}$**. The factor $\frac{1}{2}$ produces unit column vectors = eigenvectors of A.

$$\textbf{Orthonormal eigenvectors} \qquad X = H = \frac{1}{2}\begin{bmatrix} 1 & 1 & 1 & 1 \\ 1 & -1 & 1 & -1 \\ 1 & 1 & -1 & -1 \\ 1 & -1 & -1 & 1 \end{bmatrix} = H^{\mathrm{T}} = H^{-1}.$$

The eigenvalues of A^{-1} are $1, \frac{1}{5}, \frac{1}{5}, \frac{1}{5}$. The eigenvectors are not changed so $A^{-1} = H\Lambda^{-1}H^{-1}$. The inverse matrix is surprisingly neat: $A^{-1} = (I + \text{ all ones})/5$.

A is a rank-one change from $5I$. So A^{-1} is a rank-one change from $I/5$.

In a graph with 5 nodes, the determinant 125 counts the "spanning trees". Those trees have **no loops** and they touch all 5 nodes.In a graph with 5 nodes, the determinant 125 counts the "spanning trees". Those trees have **no loops** and they touch all 5 nodes.

With 6 nodes, the matrix $6 * \textbf{eye}(5) - \textbf{ones}(5)$ has the five eigenvalues $1, 6, 6, 6, 6$.

Problem Set 6.2

Questions 1–7 are about the eigenvalue and eigenvector matrices Λ and X.

1 (a) Factor these two matrices into $A = X\Lambda X^{-1}$:

$$A = \begin{bmatrix} 1 & 2 \\ 0 & 3 \end{bmatrix} \quad \text{and} \quad A = \begin{bmatrix} 1 & 1 \\ 3 & 3 \end{bmatrix}.$$

(b) If $A = X\Lambda X^{-1}$ then $A^3 = (\quad)(\quad)(\quad)$ and $A^{-1} = (\quad)(\quad)(\quad)$.

2 If A has $\lambda_1 = 2$ with eigenvector $\boldsymbol{x}_1 = \begin{bmatrix} 1 \\ 0 \end{bmatrix}$ and $\lambda_2 = 5$ with $\boldsymbol{x}_2 = \begin{bmatrix} 1 \\ 1 \end{bmatrix}$, use $X\Lambda X^{-1}$ to find A. No other matrix has the same λ's and $\boldsymbol{x}$'s.

3 Suppose $A = X\Lambda X^{-1}$. What is the eigenvalue matrix for $A + 2I$? What is the eigenvector matrix? Check that $A + 2I = (\quad)(\quad)(\quad)^{-1}$.

4 True or false : If the columns of X (eigenvectors of A) are linearly independent, then

(a) A is invertible (b) A is diagonalizable

(c) X is invertible (d) X is diagonalizable.

5 If the eigenvectors of A are the columns of I, then A is a _____ matrix. If the eigenvector matrix X is triangular, then X^{-1} is triangular. Prove that A is also triangular.

6 Describe all matrices X that diagonalize this matrix A (find all eigenvectors):

$$A = \begin{bmatrix} 4 & 0 \\ 1 & 2 \end{bmatrix}. \text{ Why does this } X \text{ also diagonize } A^{-1} = \frac{1}{8}\begin{bmatrix} 2 & 0 \\ -1 & 4 \end{bmatrix}$$

7 Diagonalize the Fibonacci matrix by completing X^{-1}:

$$\begin{bmatrix} 1 & 1 \\ 1 & 0 \end{bmatrix} = \begin{bmatrix} \lambda_1 & \lambda_2 \\ 1 & 1 \end{bmatrix} \begin{bmatrix} \lambda_1 & 0 \\ 0 & \lambda_2 \end{bmatrix} \begin{bmatrix} & \\ & \end{bmatrix}.$$

Do the multiplication $\boldsymbol{X\Lambda^k X^{-1}}\begin{bmatrix} 1 \\ 0 \end{bmatrix}$ to find its second component. This is the kth Fibonacci number $F_k = (\lambda_1^k - \lambda_2^k)/(\lambda_1 - \lambda_2)$.

8 Suppose G_{k+2} is the *average* of the two previous numbers G_{k+1} and G_k:

$$\begin{aligned} G_{k+2} &= \tfrac{1}{2}G_{k+1} + \tfrac{1}{2}G_k \\ G_{k+1} &= G_{k+1} \end{aligned} \quad \text{is} \quad \begin{bmatrix} G_{k+2} \\ G_{k+1} \end{bmatrix} = \begin{bmatrix} & A & \end{bmatrix} \begin{bmatrix} G_{k+1} \\ G_k \end{bmatrix}.$$

(a) Find the eigenvalues and eigenvectors of A.

(b) Find the limit as $n \to \infty$ of the matrices $A^n = X\Lambda^n X^{-1}$.

(c) If $G_0 = 0$ and $G_1 = 1$ show that the Gibonacci numbers approach $\frac{2}{3}$.

9 Prove that every third Fibonacci number in $0, 1, 1, 2, 3, \ldots$ is even.

10 Write down the most general matrix that has eigenvectors $\begin{bmatrix} 1 \\ 1 \end{bmatrix}$ and $\begin{bmatrix} 1 \\ -1 \end{bmatrix}$.

Questions 11–14 are about diagonalizability.

11 True or false: If the eigenvalues of A are $2, 2, 5$ then the matrix is certainly

(a) invertible (b) diagonalizable (c) not diagonalizable.

12 True or false: If the only eigenvectors of A are multiples of $(1, 4)$ then A has

(a) no inverse (b) a repeated eigenvalue (c) no diagonalization $X\Lambda X^{-1}$.

13 Complete these matrices so that $\det A = 25$. Then check that $\lambda = 5$ is repeated—the trace is 10 so the determinant of $A - \lambda I$ is $(\lambda - 5)^2$. Find an eigenvector with $A\boldsymbol{x} = 5\boldsymbol{x}$. These matrices will not be diagonalizable because there is no second line of eigenvectors.

$$A = \begin{bmatrix} 8 & \\ & 2 \end{bmatrix} \quad \text{and} \quad A = \begin{bmatrix} 9 & 4 \\ & 1 \end{bmatrix} \quad \text{and} \quad A = \begin{bmatrix} 10 & 5 \\ -5 & \end{bmatrix}$$

14 The matrix $A = \begin{bmatrix} 3 & 1 \\ 0 & 3 \end{bmatrix}$ is not diagonalizable because the rank of $A - 3I$ is ____. Change one entry to make A diagonalizable. Which entries could you change?

Questions 15–19 are about powers of matrices.

15 $A^k = X\Lambda^k X^{-1}$ approaches the zero matrix as $k \to \infty$ if and only if every λ has absolute value less than ____. Which of these matrices has $A^k \to 0$?

$$A_1 = \begin{bmatrix} .6 & .9 \\ .4 & .1 \end{bmatrix} \quad \text{and} \quad A_2 = \begin{bmatrix} .6 & .9 \\ .1 & .6 \end{bmatrix}.$$

16 (Recommended) Find Λ and X to diagonalize A_1 in Problem 15. What is the limit of Λ^k as $k \to \infty$? What is the limit of $X\Lambda^k X^{-1}$? In the columns of this limiting matrix you see the ____.

17 Find Λ and X to diagonalize A_2 in Problem 15. What is $(A_2)^{10}\boldsymbol{u}_0$ for these $\boldsymbol{u}_0$?

$$\boldsymbol{u}_0 = \begin{bmatrix} 3 \\ 1 \end{bmatrix} \quad \text{and} \quad \boldsymbol{u}_0 = \begin{bmatrix} 3 \\ -1 \end{bmatrix} \quad \text{and} \quad \boldsymbol{u}_0 = \begin{bmatrix} 6 \\ 0 \end{bmatrix}.$$

18 Diagonalize A and compute $X\Lambda^k X^{-1}$ to prove this formula for A^k:

$$A = \begin{bmatrix} 2 & -1 \\ -1 & 2 \end{bmatrix} \quad \text{has} \quad A^k = \frac{1}{2}\begin{bmatrix} 1+3^k & 1-3^k \\ 1-3^k & 1+3^k \end{bmatrix}.$$

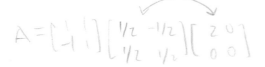

19 Diagonalize B and compute $X\Lambda^k X^{-1}$ to prove this formula for B^k:

$$B = \begin{bmatrix} 5 & 1 \\ 0 & 4 \end{bmatrix} \quad \text{has} \quad B^k = \begin{bmatrix} 5^k & 5^k - 4^k \\ 0 & 4^k \end{bmatrix}.$$

20 Suppose $A = X\Lambda X^{-1}$. Take determinants to prove $\det A = \det \Lambda = \lambda_1 \lambda_2 \cdots \lambda_n$. This quick proof only works when A can be _____.

21 Show that trace XY = trace YX, by adding the diagonal entries of XY and YX:

$$X = \begin{bmatrix} a & b \\ c & d \end{bmatrix} \quad \text{and} \quad Y = \begin{bmatrix} q & r \\ s & t \end{bmatrix}.$$

Now choose Y to be ΛX^{-1}. Then $X\Lambda X^{-1}$ has the same trace as $\Lambda X^{-1} X = \Lambda$. This proves that *the trace of A equals the trace of Λ = sum of the eigenvalues.*

22 **When is a matrix $\boldsymbol{A}$ similar to its eigenvalue matrix $\boldsymbol{\Lambda}$?**

A and Λ always have the same eigenvalues. But similarity requires a matrix B with $A = B\Lambda B^{-1}$. Then B is the _____ matrix and A must have n independent _____.

23 If $A = X\Lambda X^{-1}$, diagonalize the block matrix $B = \begin{bmatrix} A & 0 \\ 0 & 2A \end{bmatrix}$. Find its eigenvalue and eigenvector (block) matrices.

24 Consider all 4 by 4 matrices A that are diagonalized by the same fixed eigenvector matrix X. Show that the A's form a subspace (cA and $A_1 + A_2$ have this same X). What is this subspace when $X = I$? What is its dimension?

25 Suppose $A^2 = A$. On the left side A multiplies each column of A. Which of our four subspaces contains eigenvectors with $\lambda = 1$? Which subspace contains eigenvectors with $\lambda = 0$? From the dimensions of those subspaces, A has a full set of independent eigenvectors. So a matrix with $A^2 = A$ can be diagonalized.

26 (Recommended) Suppose $A\boldsymbol{x} = \lambda\boldsymbol{x}$. If $\lambda = 0$ then $\boldsymbol{x}$ is in the nullspace. If $\lambda \neq 0$ then $\boldsymbol{x}$ is in the column space. Those spaces have dimensions $(n - r) + r = n$. So why doesn't every square matrix have n linearly independent eigenvectors?

27 The eigenvalues of A are 1 and 9, and the eigenvalues of B are -1 and 9:

$$A = \begin{bmatrix} 5 & 4 \\ 4 & 5 \end{bmatrix} \quad \text{and} \quad B = \begin{bmatrix} 4 & 5 \\ 5 & 4 \end{bmatrix}.$$

Find a matrix square root of A from $R = X\sqrt{\Lambda}\, X^{-1}$. Why is there no real matrix square root of B?

28 If A and B have the same λ's with the same independent eigenvectors, their factorizations into _____ are the same. So $A = B$.

29 Suppose the same X diagonalizes both A and B. They have the *same eigenvectors* in $A = X\Lambda_1 X^{-1}$ and $B = X\Lambda_2 X^{-1}$. Prove that $AB = BA$.

30 (a) If $A = \left[\begin{smallmatrix} a & b \\ 0 & d \end{smallmatrix}\right]$ then the determinant of $A - \lambda I$ is $(\lambda - a)(\lambda - d)$. Check the "Cayley-Hamilton Theorem" that $(A - aI)(A - dI) = \textit{zero matrix}$.

(b) Test the Cayley-Hamilton Theorem on Fibonacci's $A = \left[\begin{smallmatrix} 1 & 1 \\ 1 & 0 \end{smallmatrix}\right]$. The theorem predicts that $A^2 - A - I = 0$, since the polynomial $\det(A - \lambda I)$ is $\lambda^2 - \lambda - 1$.

31 Substitute $A = X\Lambda X^{-1}$ into the product $(A - \lambda_1 I)(A - \lambda_2 I)\cdots(A - \lambda_n I)$ and explain why this produces the zero matrix. We are substituting the matrix A for the number λ in the polynomial $p(\lambda) = \det(A - \lambda I)$. The ***Cayley-Hamilton Theorem*** says that this product is always $p(A) = \textit{zero matrix}$, even if A is not diagonalizable.

32 If $A = \left[\begin{smallmatrix} 1 & 0 \\ 0 & 2 \end{smallmatrix}\right]$ and $AB = BA$, show that $B = \left[\begin{smallmatrix} a & b \\ c & d \end{smallmatrix}\right]$ is also a diagonal matrix. B has the same eigen_____ as A but different eigen_____ . These diagonal matrices B form a two-dimensional subspace of matrix space. $AB - BA = 0$ gives four equations for the unknowns $\mathbf{a}, \mathbf{b}, \mathbf{c}, \mathbf{d}$—find the rank of the 4 by 4 matrix.

33 The powers A^k approach zero if all $|\lambda_i| < 1$ and they blow up if any $|\lambda_i| > 1$. Peter Lax gives these striking examples in his book *Linear Algebra*:

$$A = \begin{bmatrix} 3 & 2 \\ 1 & 4 \end{bmatrix} \qquad B = \begin{bmatrix} 3 & 2 \\ -5 & -3 \end{bmatrix} \quad C = \begin{bmatrix} 5 & 7 \\ -3 & -4 \end{bmatrix} \quad D = \begin{bmatrix} 5 & 6.9 \\ -3 & -4 \end{bmatrix}$$

$$\|\boldsymbol{A^{1024}}\| > \mathbf{10^{700}} \quad \boldsymbol{B^{1024} = I} \qquad \boldsymbol{C^{1024} = -C} \qquad \|\boldsymbol{D^{1024}}\| < \mathbf{10^{-78}}$$

Find the eigenvalues $\lambda = e^{i\theta}$ of B and C to show $B^4 = I$ and $C^3 = -I$.

34 The nth power of rotation through θ is rotation through $n\theta$:

$$A^n = \begin{bmatrix} \cos\theta & -\sin\theta \\ \sin\theta & \cos\theta \end{bmatrix}^n = \begin{bmatrix} \cos n\theta & -\sin n\theta \\ \sin n\theta & \cos n\theta \end{bmatrix}.$$

Prove that neat formula by diagonalizing $A = X\Lambda X^{-1}$. The eigenvectors (columns of X) are $(1, i)$ and $(i, 1)$. You need to know Euler's formula $e^{i\theta} = \cos\theta + i\sin\theta$.

35 The transpose of $A = X\Lambda X^{-1}$ is $A^{\mathrm{T}} = (X^{-1})^{\mathrm{T}}\Lambda X^{\mathrm{T}}$. The eigenvectors in $A^{\mathrm{T}}\boldsymbol{y} = \lambda\boldsymbol{y}$ are the columns of that matrix $(X^{-1})^{\mathrm{T}}$. They are often called ***left eigenvectors of A***, because $\boldsymbol{y}^{\mathrm{T}}A = \lambda\boldsymbol{y}^{\mathrm{T}}$. How do you multiply matrices to find this formula for A?

Sum of rank-1 matrices $A = X\Lambda X^{-1} = \lambda_1\boldsymbol{x}_1\boldsymbol{y}_1^{\mathrm{T}} + \cdots + \lambda_n\boldsymbol{x}_n\boldsymbol{y}_n^{\mathrm{T}}$.

36 The inverse of $A = \mathbf{eye}(n) + \mathbf{ones}(n)$ is $A^{-1} = \mathbf{eye}(n) + C * \mathbf{ones}(n)$. Multiply AA^{-1} to find that number C (depending on n).

37 Suppose A_1 and A_2 are n by n invertible matrices. What matrix B shows that $A_2A_1 = B(A_1A_2)B^{-1}$? Then A_2A_1 is similar to A_1A_2: *same eigenvalues*.

38 (Pavel Grinfeld) Without writing down any calculations, can you find the eigenvalues of this matrix ? Can you find the 2020th power A^{2020} ?

$$A = \begin{bmatrix} 110 & 55 & -164 \\ 42 & 21 & -62 \\ 88 & 44 & -131 \end{bmatrix}.$$

6.3 Symmetric Positive Definite Matrices

1 A symmetric matrix S has n real eigenvalues λ_i and n orthonormal eigenvectors $\boldsymbol{q}_i$.

2 Then S is diagonalized by an orthogonal matrix Q $\boxed{\boldsymbol{S = Q\Lambda Q^{-1} = Q\Lambda Q^{\mathrm{T}}}}$

3A **Positive definite**: all $\lambda > 0 \Leftrightarrow$ every pivot $> 0 \Leftrightarrow$ all upper left determinants > 0.

3B The **energy test** is $\boldsymbol{x}^{\mathrm{T}} S\boldsymbol{x} > 0$ for all $\boldsymbol{x} \neq \mathbf{0}$. Then $S = A^{\mathrm{T}}A$ with independent columns in A.

4 **Positive semidefinite** allows $\lambda = 0$, pivot $= 0$, determinant $= 0$, energy $\boldsymbol{x}^{\mathrm{T}} S\boldsymbol{x} = 0$, any A.

Symmetric matrices $S = S^{\mathrm{T}}$ deserve all the attention they get. Looking at their eigenvalues and eigenvectors, you see why they are special:

1 All n eigenvalues λ of a symmetric matrix S are real numbers.

2 The n eigenvectors q can be chosen orthogonal (perpendicular to each other).

The identity matrix $S = I$ is an extreme case. All its eigenvalues are $\lambda = 1$. Every nonzero vector $\boldsymbol{x}$ is an eigenvector: $I\boldsymbol{x} = 1\boldsymbol{x}$. This shows why we wrote "can be chosen" in Property 2 above. With repeated eigenvalues like $\lambda_1 = \lambda_2 = 1$, we have a choice of eigenvectors. We can choose them to be orthogonal. And we can rescale them to be unit vectors (length 1). Then those eigenvectors $\boldsymbol{q}_1, \ldots, \boldsymbol{q}_n$ are not just orthogonal, they are **orthonormal. The eigenvector matrix for S has $\boldsymbol{Q^{\mathrm{T}}Q = I}$**: orthonormal columns in Q.

$$\boldsymbol{q_i}^{\mathrm{T}}\boldsymbol{q_j} = \begin{cases} 0 & i \neq j \\ 1 & i = j \end{cases} \quad \text{leads to} \quad \begin{bmatrix} - \boldsymbol{q}_1^{\mathrm{T}} - \\ \vdots \\ - \boldsymbol{q}_n^{\mathrm{T}} - \end{bmatrix} \begin{bmatrix} | & & | \\ \boldsymbol{q}_1 & \cdots & \boldsymbol{q}_n \\ | & & | \end{bmatrix} = \begin{bmatrix} \mathbf{1} & 0 & \cdot & 0 \\ 0 & \mathbf{1} & 0 & \cdot \\ \cdot & 0 & \mathbf{1} & 0 \\ 0 & \cdot & 0 & \mathbf{1} \end{bmatrix}$$

We write Q instead of X for the eigenvector matrix of S, to emphasize that these eigenvectors are orthonormal: $Q^{\mathrm{T}}Q = I$ and $Q^{\mathrm{T}} = Q^{-1}$. This eigenvector matrix is an **orthogonal matrix**. The usual $A = X\Lambda X^{-1}$ becomes $S = Q\Lambda Q^{\mathrm{T}}$:

Spectral Theorem	**Every real symmetric matrix has the form $\boldsymbol{S = Q\Lambda Q^{\mathrm{T}}}$.**

Every matrix of that form is symmetric: Transpose $Q\Lambda Q^{\mathrm{T}}$ to get $Q^{\mathrm{TT}}\Lambda^{\mathrm{T}}Q^{\mathrm{T}} = Q\Lambda Q^{\mathrm{T}}$.

Quick Proofs: Orthogonal Eigenvectors and Real Eigenvalues

Suppose first that $S\boldsymbol{x} = \lambda\boldsymbol{x}$ and $S\boldsymbol{y} = 0\boldsymbol{y}$. The symmetric matrix S has a nonzero eigenvalue λ and a zero eigenvalue. Then $\boldsymbol{y}$ is in the nullspace of S and $\boldsymbol{x}$ is in the column space of S ($\boldsymbol{x} = S\boldsymbol{x}/\lambda$ is a combination of the columns of S). *But S is symmetric: column space = row space!* Since the row space and nullspace are always orthogonal, we have proved that $\boldsymbol{x}$ **is orthogonal to** $\boldsymbol{y}$.

When that second eigenvalue is not zero, we have $S\boldsymbol{y} = \alpha\boldsymbol{y}$. In this case we look at the matrix $S - \alpha I$. Then $(S - \alpha I)\boldsymbol{y} = 0\boldsymbol{y}$ and $(S - \alpha I)\boldsymbol{x} = (\lambda - \alpha)\boldsymbol{x}$ with $\lambda - \alpha \neq 0$. Now $\boldsymbol{y}$ is in the nullspace and $\boldsymbol{x}$ is in the column space (= row space!) of $S - \alpha I$. So $\boldsymbol{y}^{\mathrm{T}}\boldsymbol{x} = 0$: *Orthogonal eigenvectors whenever the eigenvalues* $\lambda \neq \alpha$ *are different.*

Note on complex numbers That proof of orthogonality assumed real eigenvalues and eigenvectors of S. To prove this, suppose they could involve complex numbers. Multiply $S\boldsymbol{x} = \lambda\boldsymbol{x}$ by the complex conjugate vector $\overline{\boldsymbol{x}}^{\mathrm{T}}$ (every i changes to $-i$). That gives $\overline{\boldsymbol{x}}^{\mathrm{T}}S\boldsymbol{x} = \lambda\overline{\boldsymbol{x}}^{\mathrm{T}}\boldsymbol{x}$. When we show that $\overline{\boldsymbol{x}}^{\mathrm{T}}\boldsymbol{x}$ and $\overline{\boldsymbol{x}}^{\mathrm{T}}S\boldsymbol{x}$ are real, we see that $\boldsymbol{\lambda}$ **is real.**

I would like to leave complex numbers and complex matrices for the end of this section. The rules and the matrices are important. But positive definite matrices are so beautiful, and they connect to so many ideas in linear algebra, that they deserve to come first.

Positive Definite Matrices

We are working with real symmetric matrices $S = S^{\mathrm{T}}$. All their eigenvalues are real. Some of those symmetric matrices (*not all*) have an additional powerful property. Here is that important property, which puts S at the center of applied mathematics.

Test 1 **A positive definite matrix has all positive eigenvalues**

We would like to check for positive eigenvalues without computing those numbers λ. You will see four more tests for positive definite matrices, after these five examples.

1 $S = \begin{bmatrix} 2 & 0 \\ 0 & 6 \end{bmatrix}$ is positive definite. Its eigenvalues 2 and 6 are both positive

2 $S = Q\begin{bmatrix} 2 & 0 \\ 0 & 6 \end{bmatrix}Q^{\mathrm{T}}$ is positive definite if $Q^{\mathrm{T}} = Q^{-1}$: same $\lambda = 2$ and 6

3 $S = C\begin{bmatrix} 2 & 0 \\ 0 & 6 \end{bmatrix}C^{\mathrm{T}}$ is positive definite if C is invertible (not obvious)

4 $S = \begin{bmatrix} a & b \\ b & c \end{bmatrix}$ is positive definite exactly when $a > 0$ and $ac > b^2$

5 $S = \begin{bmatrix} 2 & 0 \\ 0 & 0 \end{bmatrix}$ is only **positive semidefinite**: it has all $\boldsymbol{\lambda \geq 0}$ but not $\lambda > 0$

Try Test 1 on these examples. The other tests may give faster answers. **No**, **No**, **Yes**.

$$S = \begin{bmatrix} 1 & 2 \\ 2 & 1 \end{bmatrix} \qquad S = \boldsymbol{v}\boldsymbol{v}^{\mathrm{T}}\,(\text{rank } 1) \qquad S = \begin{bmatrix} 2 & 1 & 0 \\ 1 & 2 & 1 \\ 0 & 1 & 2 \end{bmatrix}$$

The Energy-based Definition

May I bring forward the most important idea about positive definite matrices ? This new approach doesn't directly involve eigenvalues, but it turns out to be a perfect test for $\lambda > 0$. This is a good definition of positive definite matrices : **Test 2 is the energy test.**

$$\boxed{\textbf{S is positive definite if the energy } \boldsymbol{x}^{\mathrm{T}}\boldsymbol{S}\boldsymbol{x} \textbf{ is positive for all vectors } \boldsymbol{x} \neq \mathbf{0}} \quad (1)$$

Of course $S = I$ is positive definite : All $\lambda_i = 1$. The energy $\boldsymbol{x}^{\mathrm{T}} I \boldsymbol{x} = \boldsymbol{x}^{\mathrm{T}} \boldsymbol{x}$ is positive if $\boldsymbol{x} \neq 0$. Let me show you the energy in a 2 by 2 matrix. It depends on $\boldsymbol{x} = (x_1, x_2)$.

$$\textbf{Energy} \quad \boxed{\boldsymbol{x}^{\mathrm{T}} \boldsymbol{S} \boldsymbol{x} = \begin{bmatrix} x_1 & x_2 \end{bmatrix} \begin{bmatrix} 2 & 4 \\ 4 & 9 \end{bmatrix} \begin{bmatrix} x_1 \\ x_2 \end{bmatrix} = \mathbf{2}\, \boldsymbol{x_1^2} + \mathbf{8}\, \boldsymbol{x_1 x_2} + \mathbf{9}\, \boldsymbol{x_2^2}}$$

Is this positive for every x_1 and x_2 except $(x_1, x_2) = (0, 0)$? *Yes, it is a sum of squares* :

$$\boldsymbol{x}^{\mathrm{T}} S \boldsymbol{x} = 2x_1^2 + 8x_1x_2 + 9x_2^2 = 2\,(x_1 + 2x_2)^2 + x_2^2 = \text{ positive energy.}$$

We must connect positive energy $\boldsymbol{x}^{\mathrm{T}} S \boldsymbol{x} > 0$ to positive eigenvalues $\lambda > 0$:

$$\textbf{If } \boldsymbol{Sx} = \boldsymbol{\lambda x} \textbf{ then } \boldsymbol{x}^{\mathrm{T}} \boldsymbol{Sx} = \boldsymbol{\lambda x}^{\mathrm{T}} \boldsymbol{x}. \textbf{ So } \boldsymbol{\lambda} > \mathbf{0} \textbf{ leads to energy } \boldsymbol{x}^{\mathrm{T}} \boldsymbol{Sx} > \mathbf{0}.$$

That line tested $\boldsymbol{x}^{\mathrm{T}} S \boldsymbol{x}$ for each separate eigenvector $\boldsymbol{x}$. But more is true. If every eigenvector has positive energy, **then all nonzero vectors $\boldsymbol{x}$ have positive energy** :

If $\boldsymbol{x}^{\mathrm{T}} S \boldsymbol{x} > 0$ for the eigenvectors of S, then $\boldsymbol{x}^{\mathrm{T}} S \boldsymbol{x} > 0$ for every nonzero vector $\boldsymbol{x}$.

Here is the reason. Every $\boldsymbol{x}$ is a combination $c_1 \boldsymbol{x}_1 + \cdots + c_n \boldsymbol{x}_n$ of the eigenvectors. Those eigenvectors can be chosen orthogonal because S is symmetric. We will now show : $\boldsymbol{x}^{\mathrm{T}} S \boldsymbol{x}$ is a positive combination of the energies $\lambda_k \boldsymbol{x}_k^{\mathrm{T}} \boldsymbol{x}_k > 0$ in the separate eigenvectors.

$$\begin{aligned} \boldsymbol{x}^{\mathrm{T}} S \boldsymbol{x} &= (c_1 \boldsymbol{x}_1^{\mathrm{T}} + \cdots + c_n \boldsymbol{x}_n^{\mathrm{T}})\, S\, (c_1 \boldsymbol{x}_1 + \cdots + c_n \boldsymbol{x}_n) \\ &= (c_1 \boldsymbol{x}_1^{\mathrm{T}} + \cdots + c_n \boldsymbol{x}_n^{\mathrm{T}})\, (c_1 \lambda_1 \boldsymbol{x}_1 + \cdots + c_n \lambda_n \boldsymbol{x}_n) \\ &= c_1^2 \lambda_1 \boldsymbol{x}_1^{\mathrm{T}} \boldsymbol{x}_1 + \cdots + c_n^2 \lambda_n \boldsymbol{x}_n^{\mathrm{T}} \boldsymbol{x}_n > \mathbf{0} \textbf{ if every } \boldsymbol{\lambda_i} > \mathbf{0}. \end{aligned}$$

From line 2 to line 3 we used the orthogonality of the eigenvectors of S : $\boldsymbol{x}_i^{\mathrm{T}} \boldsymbol{x}_j = 0$. Here is a typical use for the energy test, without knowing any eigenvalues or eigenvectors.

$$\boxed{\textbf{If } \boldsymbol{S_1} \textbf{ and } \boldsymbol{S_2} \textbf{ are symmetric positive definite, so is } \boldsymbol{S_1} + \boldsymbol{S_2}}$$

$$\textit{Proof by adding energies}: \quad \boldsymbol{x}^{\mathrm{T}} (S_1 + S_2)\, \boldsymbol{x} = \boldsymbol{x}^{\mathrm{T}} S_1\, \boldsymbol{x} + \boldsymbol{x}^{\mathrm{T}} S_2\, \boldsymbol{x} > \mathbf{0} + \mathbf{0}$$

The eigenvalues and eigenvectors of $S_1 + S_2$ are not easy to find. Energies just add.

Three More Equivalent Tests

So far we have tests **1** and **2**: positive eigenvalues and positive energy. That energy test quickly produces three more useful tests (and probably others, but we stop with three):

Test 3	**$S = A^{\mathrm{T}}A$ for some matrix A with independent columns**
Test 4	**All the leading determinants $D_1, D_2, \ldots, D_n$ of S are positive**
Test 5	**All the pivots of S are positive**

Test 3 applies to $S = A^{\mathrm{T}}A$. Why must columns of A be independent in this test? Watch these parentheses:

$$\boldsymbol{S = A^{\mathrm{T}}A} \qquad \textbf{Energy} = \boldsymbol{x}^{\mathrm{T}}S\boldsymbol{x} = \boldsymbol{x}^{\mathrm{T}}A^{\mathrm{T}}A\boldsymbol{x} = (\boldsymbol{Ax})^{\mathrm{T}}(\boldsymbol{Ax}) = ||\boldsymbol{Ax}||^2. \quad (2)$$

The energy is the **length squared** of the vector $A\boldsymbol{x}$. This energy is positive provided $A\boldsymbol{x}$ is not the zero vector. To assure $A\boldsymbol{x} \neq \mathbf{0}$ when $\boldsymbol{x} \neq \mathbf{0}$, the columns of A must be independent. In this 2 by 3 example, A has *dependent columns*. There $A^{\mathrm{T}}A$ has rank 2, not 3.

$$S = A^{\mathrm{T}}A = \begin{bmatrix} 1 & 1 \\ 1 & 2 \\ 1 & 3 \end{bmatrix} \begin{bmatrix} 1 & 1 & 1 \\ 1 & 2 & 3 \end{bmatrix} = \begin{bmatrix} 2 & 3 & 4 \\ 3 & 5 & 7 \\ 4 & 7 & 10 \end{bmatrix}$$

is **not** positive definite.
It is positive semidefinite.
$||A\boldsymbol{x}||^2 \geq 0$

This A has column $1 +$ column $3 = 2$ (column 2). Then $\boldsymbol{x} = (1, -2, 1)$ has zero energy. It is an eigenvector of $A^{\mathrm{T}}A$ with $\lambda = 0$. Then $S = A^{\mathrm{T}}A$ is only positive semidefinite, because the energy and the eigenvalues touch zero.

Equation (2) says that $A^{\mathrm{T}}A$ is at least *semidefinite*, because $\boldsymbol{x}^{\mathrm{T}}S\boldsymbol{x} = ||A\boldsymbol{x}||^2$ is never negative. ***Semidefinite allows energy / eigenvalues / determinants / pivots of S to be zero.***

Determinant Test and Pivot Test

The determinant test is recommended for a small matrix. I will mark the four "leading determinants" D_1, D_2, D_3, D_4 in this 4 by 4 symmetric second difference matrix.

Test 4 $\quad S = \begin{bmatrix} 2 & -1 & & \\ -1 & 2 & -1 & \\ & -1 & 2 & -1 \\ & & -1 & 2 \end{bmatrix}$ has

1st determinant $D_1 = \mathbf{2}$
2nd determinant $D_2 = \mathbf{3}$
3rd determinant $D_3 = \mathbf{4}$
4th determinant $D_4 = \mathbf{5}$

The determinant test is here passed! The energy $\boldsymbol{x}^{\mathrm{T}}S\boldsymbol{x}$ must be positive too.

Leading determinants are closely related to pivots (the numbers on the diagonal after elimination). Here the first pivot is **2**. The second pivot $\mathbf{\frac{3}{2}}$ appears when $\frac{1}{2}$(row 1) is added to row 2. The third pivot $\mathbf{\frac{4}{3}}$ appears when $\frac{2}{3}$(new row 2) is added to row 3. Those fractions $\frac{2}{1}, \frac{3}{2}, \frac{4}{3}$ are ratios of determinants! The last pivot is $\mathbf{\frac{5}{4}}$.

The kth pivot equals the ratio $\dfrac{D_k}{D_{k-1}}$ of the leading determinants (sizes k and $k-1$)

Test 4 **The pivots are positive** Test 5 **The leading determinants are all positive**

I can quickly connect these tests **4** and **5** to the third test $S = A^{\mathrm{T}}A$. In fact elimination on S produces an important choice of A. Remember that *elimination = triangular factorization* ($\boldsymbol{S = LU}$). Up to now L has had 1's on its diagonal and U contained the pivots. But with symmetric matrices we can balance S as $\boldsymbol{LDL^{\mathrm{T}}}$:

$$\begin{bmatrix} 2 & -1 & 0 \\ -1 & 2 & -1 \\ 0 & -1 & 2 \end{bmatrix} = \begin{bmatrix} 1 & & \\ -\frac{1}{2} & 1 & \\ 0 & -\frac{2}{3} & 1 \end{bmatrix} \begin{bmatrix} 2 & -1 & 0 \\ & \frac{3}{2} & -1 \\ & & \frac{4}{3} \end{bmatrix} \qquad \boldsymbol{S = LU} \quad (3)$$

Put pivots into D Test 5

$$= \begin{bmatrix} 1 & & \\ -\frac{1}{2} & 1 & \\ 0 & -\frac{2}{3} & 1 \end{bmatrix} \begin{bmatrix} \mathbf{2} & & \\ & \mathbf{\frac{3}{2}} & \\ & & \mathbf{\frac{4}{3}} \end{bmatrix} \begin{bmatrix} 1 & -\frac{1}{2} & 0 \\ & 1 & -\frac{2}{3} \\ & & 1 \end{bmatrix} \qquad \boldsymbol{S = LDL^{\mathrm{T}}} \quad (4)$$

Share those pivots between A^{T} and A Test 3

$$= \begin{bmatrix} \sqrt{2} & & \\ -\sqrt{\frac{1}{2}} & \sqrt{\frac{3}{2}} & \\ 0 & -\sqrt{\frac{2}{3}} & \sqrt{\frac{4}{3}} \end{bmatrix} \begin{bmatrix} \sqrt{2} & -\sqrt{\frac{1}{2}} & 0 \\ & \sqrt{\frac{3}{2}} & -\sqrt{\frac{2}{3}} \\ & & \sqrt{\frac{4}{3}} \end{bmatrix} \qquad S = \boldsymbol{A^{\mathrm{T}}A} \, (5)$$

I am sorry about those square roots—but the pattern $S = A^{\mathrm{T}}A$ is beautiful: $\boldsymbol{A = \sqrt{D}L^{\mathrm{T}}}$.

Elimination factors every positive definite S into $A^{\mathrm{T}}A$ (A is upper triangular)

This is the Cholesky factorization $S = A^{\mathrm{T}}A$ with $\sqrt{\text{pivots}}$ on the main diagonal of A.

To apply the $S = A^{\mathrm{T}}A$ test when S is positive definite, we must find at least one possible A. There are many choices for A, including (1) **symmetric** and (2) **triangular**.

1 If $S = Q\Lambda Q^{\mathrm{T}}$, take square roots of those eigenvalues. Then $A = Q\sqrt{\Lambda}Q^{\mathrm{T}} = A^{\mathrm{T}}$.

2 If $S = LU = LDL^{\mathrm{T}}$ with positive pivots in D, then $S = (L\sqrt{D})(\sqrt{D}L^{\mathrm{T}})$.

> *Summary* The 5 tests for positive definiteness of $S = S^{\mathrm{T}}$ involve 5 different parts of linear algebra—pivots, determinants, eigenvalues, $S = A^{\mathrm{T}}A$, **and energy**. Each test gives a complete answer by itself: positive definite or semidefinite or neither.
>
> **Positive energy $\boldsymbol{x^{\mathrm{T}}Sx > 0}$ is the best definition. It connects them all.**

Positive Definite Matrices and Minimum Problems

Suppose S is a symmetric positive definite 2 by 2 matrix. Apply four of the tests:

$$S = \begin{bmatrix} a & b \\ b & c \end{bmatrix} \quad \begin{matrix} \textbf{determinants} & a > 0, ac - b^2 > 0 & \textbf{pivots} & a > 0, (ac - b^2)/a > 0 \\ \textbf{eigenvalues} & \lambda_1 > 0,\ \lambda_2 > 0 & \textbf{energy} & ax^2 + 2bxy + cy^2 > 0 \end{matrix}$$

I will choose an example with $a = c = 5$ and $b = 4$. This matrix S has $\lambda = \mathbf{9}$ and $\lambda = \mathbf{1}$.

$$\textbf{Energy } \boldsymbol{E = x^{\mathrm{T}} S x} \qquad \begin{bmatrix} x & y \end{bmatrix} \begin{bmatrix} \mathbf{5} & \mathbf{4} \\ \mathbf{4} & \mathbf{5} \end{bmatrix} \begin{bmatrix} x \\ y \end{bmatrix} = \mathbf{5\,x^2 + 8\,xy + 5\,y^2 > 0}$$

The graph of that energy function $E(x, y)$ is a **bowl opening upwards**. The bottom point of the bowl has energy $E = 0$ when $x = y = 0$. This connects minimum problems in calculus with positive definite matrices in linear algebra.

Chapter 8 of this book describes numerical minimization. For the best problems, the function $f(\boldsymbol{x})$ is **strictly convex**—like $f = x^2$ or $f = x^2 + y^2$. Here is the convexity test: *The matrix of second derivatives is positive definite at all points.* We are in high dimensions, but linear algebra identifies the crucial properties of the second derivative matrix.

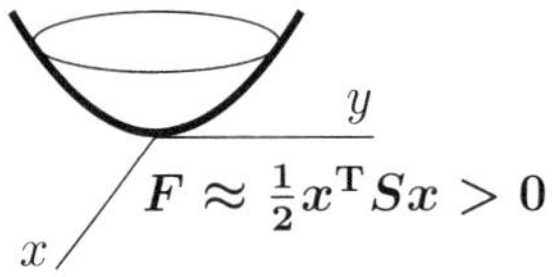

The second derivatives of the energy $\frac{1}{2}\,\boldsymbol{x}^{\mathrm{T}} S \boldsymbol{x}$ are in the matrix S

For an ordinary function $f(x)$ of one variable x, the test for a minimum is famous:

$$\textbf{Minimum} \qquad \begin{matrix} \textbf{First}\text{ derivative} \\ \text{is zero} \end{matrix} \quad \frac{df}{dx} = 0 \qquad \begin{matrix} \textbf{Second}\text{ derivative} \\ \text{is positive} \end{matrix} \quad \frac{d^2 f}{dx^2} > 0 \text{ at } x = x_0$$

For $f(x, y)$ with two variables, the second derivatives go into a matrix: positive definite**!**

$$\begin{matrix} \textbf{Minimum} \\ \textbf{at } \boldsymbol{x_0, y_0} \end{matrix} \quad \frac{\partial f}{\partial x} = 0 \text{ and } \frac{\partial f}{\partial y} = 0 \text{ and } \begin{bmatrix} \partial^2 f/\partial x^2 & \partial^2 f/\partial x \partial y \\ \partial^2 f/\partial x \partial y & \partial^2 f/\partial y^2 \end{bmatrix} \begin{matrix} \textbf{is positive definite} \\ \textbf{at } \boldsymbol{x_0, y_0} \end{matrix}$$

The graph of $z = f(x, y)$ is flat at that point x_0, y_0 because $\partial f/\partial x = \partial f/\partial y = 0$. The graph goes upwards because the second derivative matrix is positive definite. So we have a minimum point of the function $f(x, y)$. Similarly for $f(x, y, z)$.

Positive Semidefinite Matrices

Often we are at the edge of positive definiteness. The determinant is zero. The smallest eigenvalue is $\lambda = 0$. The energy in its eigenvector is $\boldsymbol{x}^{\mathrm{T}} S \boldsymbol{x} = \boldsymbol{x}^{\mathrm{T}} 0 \boldsymbol{x} = 0$. These matrices on the edge are "*positive semidefinite*". Here are two examples (not invertible):

$$S = \begin{bmatrix} 1 & 2 \\ 2 & 4 \end{bmatrix} \text{ and } T = \begin{bmatrix} 2 & -1 & -1 \\ -1 & 2 & -1 \\ -1 & -1 & 2 \end{bmatrix} \quad \begin{matrix} \text{are positive semidefinite} \\ \text{but not positive definite} \end{matrix}$$

S has eigenvalues 5 and 0. Its trace is $1 + 4 = 5$. Its upper left determinants are 1 and 0. The rank of S is only 1. This matrix S factors into $A^{\mathrm{T}}A$ with *dependent columns* in A:

$$\textbf{Dependent columns in } \boldsymbol{A} \quad \textbf{Positive semidefinite } \boldsymbol{S} \qquad \begin{bmatrix} 1 & 2 \\ 2 & 4 \end{bmatrix} = \begin{bmatrix} 1 & 0 \\ 2 & 0 \end{bmatrix} \begin{bmatrix} 1 & 2 \\ 0 & 0 \end{bmatrix} = A^{\mathrm{T}}A.$$

If 4 is increased by any small number, the matrix S will become positive definite.

The cyclic T also has zero determinant. The eigenvector $\boldsymbol{x} = (1,1,1)$ has $T\boldsymbol{x} = \mathbf{0}$ and energy $\boldsymbol{x}^{\mathrm{T}}T\boldsymbol{x} = 0$. Vectors $\boldsymbol{x}$ in all other directions do give positive energy.

$$\begin{array}{l} \textbf{Second differences } \boldsymbol{T} \\ \textbf{from first differences } \boldsymbol{A} \\ \textbf{Columns add to } (\mathbf{0},\mathbf{0},\mathbf{0}) \end{array} \quad \begin{bmatrix} 2 & -1 & -1 \\ -1 & 2 & -1 \\ -1 & -1 & 2 \end{bmatrix} = \begin{bmatrix} 1 & -1 & 0 \\ 0 & 1 & -1 \\ -1 & 0 & 1 \end{bmatrix} \begin{bmatrix} 1 & 0 & -1 \\ -1 & 1 & 0 \\ 0 & -1 & 1 \end{bmatrix}.$$

Positive semidefinite matrices have all $\lambda \geq 0$ and all $\boldsymbol{x}^{\mathrm{T}}S\boldsymbol{x} \geq 0$. Those weak inequalities ($\geq$ **instead of** $>$) include positive definite S along with the singular matrices at the edge.

If S is positive semidefinite, so is every matrix $A^{\mathrm{T}}SA$:

$$\textbf{If } \boldsymbol{x}^{\mathrm{T}}\boldsymbol{S}\boldsymbol{x} \geq \mathbf{0} \textbf{ for every vector } \boldsymbol{x}, \textbf{ then } (\boldsymbol{A}\boldsymbol{x})^{\mathrm{T}}\boldsymbol{S}(\boldsymbol{A}\boldsymbol{x}) \geq \mathbf{0} \textbf{ for every } \boldsymbol{x}.$$

We can tighten this proof to show that $A^{\mathrm{T}}SA$ is actually positive definite. But we have to guarantee that $A\boldsymbol{x}$ is not the zero vector—to be sure that $(A\boldsymbol{x})^{\mathrm{T}}S(A\boldsymbol{x})$ is not zero.

Suppose $\boldsymbol{x}^{\mathrm{T}}S\boldsymbol{x} > 0$ and $A\boldsymbol{x} \neq 0$ whenever $\boldsymbol{x}$ is not zero. Then $A^{\mathrm{T}}SA$ is positive definite.

Again we use the energy test. For every $\boldsymbol{x} \neq 0$ we have $A\boldsymbol{x} \neq \mathbf{0}$. The energy in $A\boldsymbol{x}$ is strictly positive: $(A\boldsymbol{x})^{\mathrm{T}}S(A\boldsymbol{x}) > 0$. The matrix $A^{\mathrm{T}}SA$ is called "**congruent**" to S.

Example 1 The identity matrix $S = I$ is positive definite. Then we have proved:

1. $A^{\mathrm{T}}A$ is positive semidefinite. 2. If A is invertible, then $A^{\mathrm{T}}A$ is positive definite.

This was *Test* 3 for a positive definite matrix. It is mentioned again because $A^{\mathrm{T}}SA$ is such an important matrix in applied mathematics. We want to be sure it is positive definite (not just semidefinite). Then the equations $A^{\mathrm{T}}SA\boldsymbol{x} = \boldsymbol{f}$ in engineering can be solved.

Here is an extension called the **Law of Inertia**.

If $S^{\mathrm{T}} = S$ has P positive eigenvalues and N negative eigenvalues and Z zero eigenvalues, then the same is true for $A^{\mathrm{T}}SA$—provided A is invertible.

The Ellipse $ax^2 + 2bxy + cy^2 = 1$

Think of a tilted ellipse $\boldsymbol{x}^{\mathrm{T}}S\boldsymbol{x} = 1$. Its center is $(0,0)$, as in Figure 6.2a. Turn it to line up with the coordinate axes (X and Y axes). That is Figure 6.2b. These two pictures show the geometry behind the eigenvalues in Λ and the eigenvectors in Q and in $S = Q\Lambda Q^{\mathrm{T}}$. **The eigenvector matrix Q lines up the ellipse.**

The tilted ellipse has $\boldsymbol{x}^{\mathrm{T}}S\boldsymbol{x} = 1$. The lined-up ellipse has $\boldsymbol{X}^{\mathrm{T}}\boldsymbol{\Lambda}\boldsymbol{X} = 1$ with $\boldsymbol{X} = Q\boldsymbol{x}$.

Example 2 Find the axes of this tilted ellipse $\mathbf{5x^2 + 8xy + 5y^2 = 1}$.

Solution Start with the positive definite matrix S that matches this equation :

$$\text{The equation is } \begin{bmatrix} x & y \end{bmatrix} \begin{bmatrix} 5 & 4 \\ 4 & 5 \end{bmatrix} \begin{bmatrix} x \\ y \end{bmatrix} = 1. \qquad \text{The matrix is } \boldsymbol{S} = \begin{bmatrix} \mathbf{5} & \mathbf{4} \\ \mathbf{4} & \mathbf{5} \end{bmatrix}.$$

The eigenvectors are $\begin{bmatrix} 1 \\ 1 \end{bmatrix}$ and $\begin{bmatrix} 1 \\ -1 \end{bmatrix}$. Divide by $\sqrt{2}$ for unit vectors. Then $S = Q\Lambda Q^{\mathrm{T}}$:

$$\begin{array}{l}\textbf{Eigenvectors in } \boldsymbol{Q} \\ \textbf{Eigenvalues 9 and 1}\end{array} \qquad \begin{bmatrix} \mathbf{5} & \mathbf{4} \\ \mathbf{4} & \mathbf{5} \end{bmatrix} = \frac{1}{\sqrt{2}} \begin{bmatrix} 1 & 1 \\ 1 & -1 \end{bmatrix} \begin{bmatrix} \mathbf{9} & 0 \\ 0 & \mathbf{1} \end{bmatrix} \frac{1}{\sqrt{2}} \begin{bmatrix} 1 & 1 \\ 1 & -1 \end{bmatrix}.$$

Now multiply by $\begin{bmatrix} x & y \end{bmatrix}$ on the left and $\begin{bmatrix} x \\ y \end{bmatrix}$ on the right to get $\boldsymbol{x}^{\mathrm{T}} S \boldsymbol{x} = (\boldsymbol{x}^{\mathrm{T}} Q)\Lambda(Q^{\mathrm{T}} \boldsymbol{x})$:

$$\boldsymbol{x}^{\mathrm{T}} S \boldsymbol{x} = \textbf{ sum of squares} \quad 5x^2 + 8xy + 5y^2 = 9\left(\frac{x+y}{\sqrt{2}}\right)^2 + 1\left(\frac{x-y}{\sqrt{2}}\right)^2. \tag{6}$$

The coefficients are the eigenvalues 9 and 1 from Λ. Inside the squares are the eigenvectors $\boldsymbol{q}_1 = (1,1)/\sqrt{2}$ and $\boldsymbol{q}_2 = (1,-1)/\sqrt{2}$.

The axes of the tilted ellipse point along those eigenvectors. This explains why $S = Q\Lambda Q^{\mathrm{T}}$ is called the "principal axis theorem"—it displays the axes. Not only the axis directions (from the eigenvectors) but also the axis lengths (from the eigenvalues).

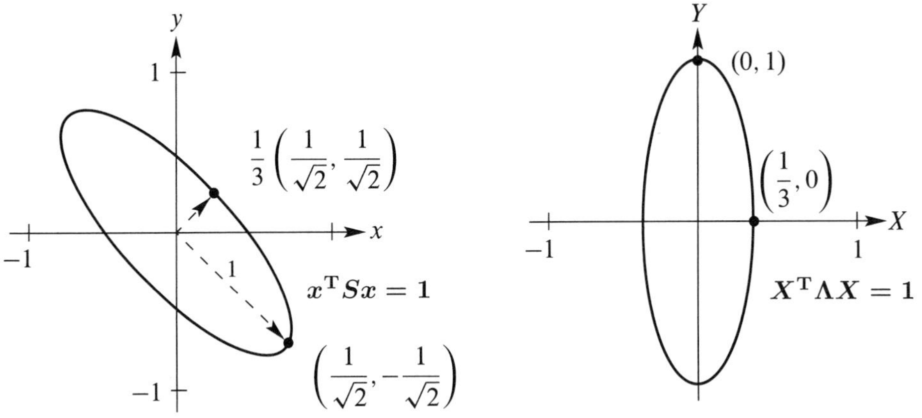

Figure 6.2: The ellipse $\boldsymbol{x}^{\mathrm{T}} S \boldsymbol{x} = 5x^2 + 8xy + 5y^2 = 1$. Lined up it is $9X^2 + Y^2 = 1$.

To see it all, use capital letters for the new coordinates that line up the ellipse :

$$\textbf{Lined up} \qquad \frac{x+y}{\sqrt{2}} = X \qquad \text{and} \qquad \frac{x-y}{\sqrt{2}} = Y \qquad \text{and} \qquad \mathbf{9}\boldsymbol{X}^{\mathbf{2}} + \boldsymbol{Y}^{\mathbf{2}} = \mathbf{1}.$$

The largest value of X^2 is $1/9$. The endpoint of the shorter axis has $X = 1/3$ and $Y = 0$. Notice: The *bigger* eigenvalue λ_1 gives the *shorter* axis, of half-length $1/\sqrt{\lambda_1} = 1/3$. The smaller eigenvalue $\lambda_2 = 1$ gives the greater length $1/\sqrt{\lambda_2} = 1$.

In the xy system, the axes are along the eigenvectors of S. In the XY system, the **axes are along the eigenvectors of $\boldsymbol{\Lambda}$**—the coordinate axes. All comes from $S = Q\Lambda Q^{\mathrm{T}}$.

Optimization and Machine Learning

This book will end with **gradient descent** to minimize $f(\boldsymbol{x})$. Each step to $\boldsymbol{x}_{k+1}$ takes the steepest direction at the current point $\boldsymbol{x}_k$. But that steepest direction changes as we descend. This is where calculus meets linear algebra, at the minimum point $\boldsymbol{x}^*$.

Calculus	The partial derivatives of f are all zero at $\boldsymbol{x}^*$:	$\dfrac{\partial f}{\partial x_i} = 0$
Linear algebra	The matrix S of second derivatives $\dfrac{\partial^2 f}{\partial x_i\, \partial x_j}$	is positive definite

If S is positive definite (or semidefinite) at all points $\boldsymbol{x} = (x_1, \ldots, x_n)$, then **the function $f(x)$ is convex**. If the eigenvalues of S stay above some positive number δ, then **$f(x)$ is strictly convex**. These are the best functions to optimize. They have only one minimum, and gradient descent will find it.

Machine learning produces "loss functions" with hundreds of thousands of variables. They measure the error—which we minimize. But computing all the second derivatives is barely possible. We use first derivatives to tell us a direction to move—the error drops fastest in the steepest direction. Then we take another descent step in a new direction. **This is the central computation in least squares and neural nets and deep learning.**

All Symmetric Matrices are Diagonalizable

This section ends by returning to the proof of the (real) spectral theorem $S = Q\Lambda Q^{\mathrm{T}}$. *Every real symmetric matrix* $\boldsymbol{S}$ *can be diagonalized by a real orthogonal matrix* $\boldsymbol{Q}$.

When no eigenvalues of A are repeated, the eigenvectors are sure to be independent. Then A can be diagonalized. But a repeated eigenvalue can produce a shortage of eigenvectors. This *sometimes* happens for nonsymmetric matrices. It *never* happens for symmetric matrices. ***There are always enough eigenvectors to diagonalize*** $\boldsymbol{S} = \boldsymbol{S}^{\mathrm{T}}$.

The proof comes from **Schur's Theorem**: Every real square matrix factors into $A = QTQ^{-1} = QTQ^{\mathrm{T}}$ for some triangular matrix T. If A is symmetric, then $T = Q^{\mathrm{T}}AQ$ is also symmetric. *But a symmetric triangular matrix T is actually diagonal.*

Thus Schur has found the diagonal matrix we want. If $A = S$ is symmetric, then $T = \Lambda$ and $S = Q\Lambda Q^{\mathrm{T}}$ as required. The website **math.mit.edu/everyone** has the proof of Schur's Theorem. Here we just note that if A can be diagonalized ($A = X\Lambda X^{-1}$) then we can see that triangular matrix T. Use Gram-Schmidt from Section 4.4 to factor X into QR:

$$A = X\Lambda X^{-1} = QR\Lambda R^{-1}Q^{-1} = QTQ^{-1} \quad \text{with triangular } T = R\Lambda R^{-1}.$$

Complex Vectors and Matrices

This page allows for complex numbers like $3+4i$ in $\boldsymbol{x}$ and S. The complex conjugate of $x_1 = 3+4i$ is $\overline{x}_1 = 3-4i$. Then x_1 times $\overline{x}_1$ is $3^2+4^2 = 25$ (real). The magnitude is $|x_1| = \sqrt{25} = 5$. Now suppose we have a complex vector $\boldsymbol{x} = (x_1, x_2) = (3+4i, 1+i)$:

Length squared **$||\boldsymbol{x}||^2$ is real** $\quad ||\boldsymbol{x}||^2 = \overline{\boldsymbol{x}}^{\mathrm{T}}\boldsymbol{x} = \begin{bmatrix} 3-4i & 1-i \end{bmatrix}\begin{bmatrix} 3+4i \\ 1+\ i \end{bmatrix} = 25+2 = 27$

Before we move to matrices, here are the key facts about complex numbers $\boldsymbol{z = a + ib}$.

The conjugate is $\overline{z} = a - ib$	$z + \overline{z} = 2a$ is real	z times $\overline{z} = \|z\|^2 = a^2 + b^2$ is real
$\overline{z}_1 + \overline{z}_2$ equals $\overline{z_1 + z_2}$	$\overline{z}_1$ times $\overline{z}_2$ equals $\overline{z_1 z_2}$	$\dfrac{1}{z} = \dfrac{\overline{z}}{\|z\|^2} = \dfrac{a-ib}{a^2+b^2}$

A real symmetric matrix has $S^{\mathrm{T}} = S$. With complex numbers, we must change S^{T} to $\overline{S}^{\mathrm{T}}$. Here is a matrix that has $\overline{S}^{\mathrm{T}} = S$ (**Hermitian matrix**).

$$\overline{\boldsymbol{S}}^{\mathrm{T}} = \boldsymbol{S} \qquad S = \begin{bmatrix} 2 & 3-3i \\ 3+3i & 5 \end{bmatrix} \qquad \begin{matrix} \overline{\boldsymbol{x}}^{\mathrm{T}}\boldsymbol{S}\boldsymbol{x} = \textbf{real number for complex } \boldsymbol{x} \\ \textbf{8 and } -\textbf{1} = \textbf{ real eigenvalues of } \boldsymbol{S} \end{matrix}$$

$$\det(S - \lambda I) = \det\begin{bmatrix} 2-\lambda & 3-3i \\ 3+3i & 5-\lambda \end{bmatrix} = 10 - 7\lambda + \lambda^2 - 18 = (\lambda - 8)(\lambda + 1)$$

$$\overline{x}^{\mathrm{T}}S\boldsymbol{x} = \begin{bmatrix} \overline{x}_1 & \overline{x}_2 \end{bmatrix}\begin{bmatrix} 2 & 3-3i \\ 3+3i & 5 \end{bmatrix}\begin{bmatrix} x_1 \\ x_2 \end{bmatrix} = \begin{matrix} 2\,\overline{x}_1 x_1 + 5\,\overline{x}_2 x_2 + \\ \overline{x}_1(3-3i)x_2 + \overline{x}_2(3+3i)x_1 \end{matrix} = \textbf{real} \quad (7)$$

$\overline{\boldsymbol{x}}^{\mathrm{T}}S\boldsymbol{x}$ **is real**: Equation (7) ends with complex numbers $z + \overline{z} = a+ib+a-ib = 2a$ (*real*).
Hermitian matrix $\overline{S}^{\mathrm{T}} = S$ is the complex analog of a real symmetric matrix $S^{\mathrm{T}} = S$.
Unitary matrix $\overline{Q}^{\mathrm{T}} = Q^{-1}$ is the complex analog of an orthogonal matrix with $Q^{\mathrm{T}} = Q^{-1}$.

> The eigenvalues of $S = \overline{S}^{\mathrm{T}}$ are real. The eigenvalues of a unitary Q have $|\lambda| = 1$.
>
> The eigenvectors $\boldsymbol{q}_1, \ldots, \boldsymbol{q}_n$ can be complex. *Those eigenvectors are still orthogonal.*

The command S' in MATLAB or Julia automatically returns $\overline{S}^{\mathrm{T}}$ when S is complex. The dot product of complex vectors is $\overline{\boldsymbol{x}}^{\mathrm{T}}\boldsymbol{y} = \overline{x}_1 y_1 + \cdots + \overline{x}_n y_n$.

Example to show two complex dot products $\quad \boldsymbol{x} = \begin{bmatrix} 1 \\ i \end{bmatrix} \quad \boldsymbol{y} = \begin{bmatrix} 1 \\ -i \end{bmatrix} \quad \overline{\boldsymbol{x}}^{\mathrm{T}}\boldsymbol{x} = 2 \quad \overline{\boldsymbol{x}}^{\mathrm{T}}\boldsymbol{y} = \mathbf{1 - 1 = 0}$

Those orthogonal vectors $\boldsymbol{x}, \boldsymbol{y}$ are the eigenvectors of $\boldsymbol{Q} = \begin{bmatrix} \mathbf{0} & \mathbf{1} \\ -\mathbf{1} & \mathbf{0} \end{bmatrix}$ and $S = \begin{bmatrix} \mathbf{0} & \boldsymbol{i} \\ -\boldsymbol{i} & \mathbf{0} \end{bmatrix}$.

■ WORKED EXAMPLES ■

6.3 A Test these symmetric matrices S and T for positive definiteness:

$$S = \begin{bmatrix} 2 & -1 & 0 \\ -1 & 2 & -1 \\ 0 & -1 & 2 \end{bmatrix} \quad \text{and} \quad T = \begin{bmatrix} 2 & -1 & b \\ -1 & 2 & -1 \\ b & -1 & 2 \end{bmatrix}.$$

Solution The pivots of S are 2 and $\frac{3}{2}$ and $\frac{4}{3}$, all positive. Its upper left determinants are 2 and 3 and 4, all positive. The eigenvalues of S are $2-\sqrt{2}$ and 2 and $2+\sqrt{2}$, all positive. That completes three tests. Any one test is decisive!

I have three candidates $\boldsymbol{A_1}, \boldsymbol{A_2}, \boldsymbol{A_3}$ to show that $S = A^{\mathrm{T}}A$ is positive definite. $\boldsymbol{A_1}$ is a first difference matrix with 1 and -1, to produce the second difference $-1, 2, -1$ in S:

$$\boldsymbol{S = A_1^{\mathrm{T}} A_1} \qquad \begin{bmatrix} 2 & -1 & 0 \\ -1 & 2 & -1 \\ 0 & -1 & 2 \end{bmatrix} = \begin{bmatrix} 1 & -1 & 0 & 0 \\ 0 & 1 & -1 & 0 \\ 0 & 0 & 1 & -1 \end{bmatrix} \begin{bmatrix} 1 & 0 & 0 \\ -1 & 1 & 0 \\ 0 & -1 & 1 \\ 0 & 0 & -1 \end{bmatrix}$$

The three columns of A_1 are independent. Therefore S is positive definite.

$\boldsymbol{A_2}$ comes from $S = LDL^{\mathrm{T}}$ (the symmetric version of $S = LU$). Elimination gives the pivots $2, \frac{3}{2}, \frac{4}{3}$ in D and the multipliers $-\frac{1}{2}, 0, -\frac{2}{3}$ in L. **Just put $\boldsymbol{A_2 = L\sqrt{D}}$.**

$$LDL^{\mathrm{T}} = \begin{bmatrix} 1 & & \\ -\frac{1}{2} & 1 & \\ 0 & -\frac{2}{3} & 1 \end{bmatrix} \begin{bmatrix} 2 & & \\ & \frac{3}{2} & \\ & & \frac{4}{3} \end{bmatrix} \begin{bmatrix} 1 & -\frac{1}{2} & 0 \\ & 1 & -\frac{2}{3} \\ & & 1 \end{bmatrix} = (L\sqrt{D})(L\sqrt{D})^{\mathrm{T}} = A_2^{\mathrm{T}} A_2.$$

This triangular matrix A_2 has square roots (not so beautiful). A_2 is the "Cholesky factor" of S and the MATLAB command is $A = \mathsf{chol}(S)$. In applications, the rectangular A_1 is how we build S and this A_2 is how elimination breaks it apart.

Eigenvalues give the symmetric choice $\boldsymbol{A_3 = Q\sqrt{\Lambda}Q^{\mathrm{T}}}$. This succeeds because $A_3^{\mathrm{T}} A_3 = Q\Lambda Q^{\mathrm{T}} = S$. All tests show that the $-1, 2, -1$ matrix S is positive definite.

The three choices A_1, A_2, A_3 give three different ways to split up the energy $\boldsymbol{x}^{\mathrm{T}} S \boldsymbol{x}$:

$$\begin{aligned}
\boldsymbol{x}^{\mathrm{T}} S \boldsymbol{x} &= 2x_1^2 - 2x_1x_2 + 2x_2^2 - 2x_2x_3 + 2x_3^2 && \textbf{Rewrite with squares} \\
\|A_1\boldsymbol{x}\|^2 &= x_1^2 + (x_2 - x_1)^2 + (x_3 - x_2)^2 + x_3^2 && \boldsymbol{S = A_1^{\mathrm{T}} A_1} \\
\|A_2\boldsymbol{x}\|^2 &= 2(x_1 - \tfrac{1}{2}x_2)^2 + \tfrac{3}{2}(x_2 - \tfrac{2}{3}x_3)^2 + \tfrac{4}{3}\,x_3^2 && \boldsymbol{S = LDL^{\mathrm{T}} = A_2^{\mathrm{T}} A_2} \\
\|A_3\boldsymbol{x}\|^2 &= \lambda_1(\boldsymbol{q}_1^{\mathrm{T}}\boldsymbol{x})^2 + \lambda_2(\boldsymbol{q}_2^{\mathrm{T}}\boldsymbol{x})^2 + \lambda_3(\boldsymbol{q}_3^{\mathrm{T}}\boldsymbol{x})^2 && \boldsymbol{S = Q\Lambda Q^{\mathrm{T}} = A_3^{\mathrm{T}} A_3}
\end{aligned}$$

For the second matrix T, the determinant test is easiest.

$$\textbf{Test on } \boldsymbol{T} \qquad \det \boldsymbol{T} = 4 + 2b - 2b^2 = (\boldsymbol{1+b})\,(\boldsymbol{4-2b}) \quad \text{must be positive.}$$

At $b = -1$ and $b = 2$ we get $\det T = 0$. *Between $b = -1$ and $b = 2$ this matrix T is positive definite.* The corner entry $b = 0$ in the matrix S was safely between -1 and 2.

Problem Set 6.3

1 Suppose $S = S^{\mathrm{T}}$. When is ASB also symmetric with the same eigenvalues as S?

(a) Transpose ASB to see that it stays symmetric when $B =$ ______.

(b) ASB is similar to S (same eigenvalues) when $B =$ ______.

Put (a) and (b) together. The symmetric matrices similar to S look like (__)S(__).

2 For S and T, find the eigenvalues and eigenvectors and the factors for $Q\Lambda Q^{\mathrm{T}}$:

$$S = \begin{bmatrix} 2 & 2 & 2 \\ 2 & 0 & 0 \\ 2 & 0 & 0 \end{bmatrix} \quad \text{and} \quad T = \begin{bmatrix} 1 & 0 & 2 \\ 0 & -1 & -2 \\ 2 & -2 & 0 \end{bmatrix}.$$

3 Find *all* orthogonal matrices that diagonalize $S = \begin{bmatrix} 9 & 12 \\ 12 & 16 \end{bmatrix}$.

4 (a) Find a symmetric matrix $\begin{bmatrix} 1 & b \\ b & 1 \end{bmatrix}$ that has a negative eigenvalue.

(b) How do you know it must have a negative pivot?

(c) How do you know it can't have two negative eigenvalues?

5 If C is symmetric prove that $A^{\mathrm{T}}CA$ is also symmetric. (Transpose it.) When A is 6 by 3, what are the shapes of C and $A^{\mathrm{T}}CA$?

6 Find an orthogonal matrix Q that diagonalizes $S = \begin{bmatrix} -2 & 6 \\ 6 & 7 \end{bmatrix}$. What is Λ?

If $A^3 = 0$ then the eigenvalues of A must be ______. Give an example that has $A \neq 0$. But if A is symmetric, diagonalize it to prove that A must be a zero matrix.

7 Write S and T in the form $\lambda_1 \boldsymbol{x}_1 \boldsymbol{x}_1^{\mathrm{T}} + \lambda_2 \boldsymbol{x}_2 \boldsymbol{x}_2^{\mathrm{T}}$ of the spectral theorem $Q\Lambda Q^T$:

$$S = \begin{bmatrix} 3 & 1 \\ 1 & 3 \end{bmatrix} \qquad T = \begin{bmatrix} 9 & 12 \\ 12 & 16 \end{bmatrix} \qquad (\text{keep } \|\boldsymbol{x}_1\| = \|\boldsymbol{x}_2\| = 1).$$

8 Every 2 by 2 symmetric matrix is $\lambda_1 \boldsymbol{x}_1 \boldsymbol{x}_1^{\mathrm{T}} + \lambda_2 \boldsymbol{x}_2 \boldsymbol{x}_2^{\mathrm{T}} = \lambda_1 P_1 + \lambda_2 P_2$. Explain $P_1 + P_2 = \boldsymbol{x}_1 \boldsymbol{x}_1^{\mathrm{T}} + \boldsymbol{x}_2 \boldsymbol{x}_2^{\mathrm{T}} = I$ from columns times rows of Q. Why is $P_1 P_2 = 0$?

9 What are the eigenvalues of $A = \begin{bmatrix} 0 & b \\ -b & 0 \end{bmatrix}$? Create a 4 by 4 antisymmetric matrix ($A^{\mathrm{T}} = -A$) and verify that all its eigenvalues are imaginary.

10 (Recommended) This matrix M is antisymmetric and also ______. Then all its eigenvalues are pure imaginary and they also have $|\lambda| = 1$. ($\|M\boldsymbol{x}\| = \|\boldsymbol{x}\|$ for every $\boldsymbol{x}$ so $\|\lambda \boldsymbol{x}\| = \|\boldsymbol{x}\|$ for eigenvectors.) Find all four eigenvalues from the trace of M:

$$M = \frac{1}{\sqrt{3}} \begin{bmatrix} 0 & 1 & 1 & 1 \\ -1 & 0 & -1 & 1 \\ -1 & 1 & 0 & -1 \\ -1 & -1 & 1 & 0 \end{bmatrix} \quad \text{can only have eigenvalues } i \text{ or } -i.$$

11 Show that this A (**symmetric but complex**) has only one line of eigenvectors:

$$A = \begin{bmatrix} i & 1 \\ 1 & -i \end{bmatrix} \text{ is not even diagonalizable: eigenvalues } \lambda = 0, 0.$$

$A^{\mathrm{T}} = A$ is not such a special property for complex matrices. The good property is $\overline{A}^{\mathrm{T}} = A$. Then all λ's are real and the eigenvectors are orthogonal.

12 Find the eigenvector matrices Q for S and X for B. Show that X doesn't collapse at $d = 1$, even though $\lambda = 1$ is repeated. Are those eigenvectors perpendicular?

$$S = \begin{bmatrix} 0 & d & 0 \\ d & 0 & 0 \\ 0 & 0 & 1 \end{bmatrix} \quad B = \begin{bmatrix} -d & 0 & 1 \\ 0 & 1 & 0 \\ 0 & 0 & d \end{bmatrix} \quad \text{have} \quad \lambda = 1, d, -d.$$

13 Write a 2 by 2 *complex* matrix with $\overline{S}^{\mathrm{T}} = S$ (a "Hermitian matrix"). Find λ_1 and λ_2 for your complex matrix. Check that $\overline{\boldsymbol{x}}_1^{\mathrm{T}} \boldsymbol{x}_2 = 0$ (this is complex orthogonality).

14 ***True*** (with reason) ***or false*** (with example).

(a) A matrix with n real eigenvalues and n real eigenvectors is symmetric.

(b) A matrix with n real eigenvalues and n orthonormal eigenvectors is symmetric.

(c) The inverse of an invertible symmetric matrix is symmetric.

(d) The eigenvector matrix Q of a symmetric matrix is symmetric.

15 (A paradox for instructors) If $AA^{\mathrm{T}} = A^{\mathrm{T}}A$ then A and A^{T} share the same eigenvectors (true). A and A^{T} always share the same eigenvalues. Find the flaw in this conclusion: A and A^{T} must have the same X and same Λ. Therefore A equals A^{T}.

16 Are A and B invertible, orthogonal, projections, permutations, diagonalizable? Which of these factorizations are possible: $\boldsymbol{LU}$, $\boldsymbol{QR}$, $\boldsymbol{X\Lambda X^{-1}}$, $\boldsymbol{Q\Lambda Q^{\mathrm{T}}}$?

$$A = \begin{bmatrix} 0 & 0 & 1 \\ 0 & 1 & 0 \\ 1 & 0 & 0 \end{bmatrix} \qquad B = \frac{1}{3}\begin{bmatrix} 1 & 1 & 1 \\ 1 & 1 & 1 \\ 1 & 1 & 1 \end{bmatrix}.$$

17 What number b in $A = \left[\begin{smallmatrix} 2 & b \\ 1 & 0 \end{smallmatrix}\right]$ makes $A = Q\Lambda Q^{\mathrm{T}}$ possible? What number will make it impossible to diagonalize A? What number makes A singular?

18 Find all 2 by 2 matrices that are orthogonal and also symmetric. Which two numbers can be eigenvalues of those two matrices?

19 This A is nearly symmetric. But what is the angle between the eigenvectors?

$$A = \begin{bmatrix} 1 & 10^{-15} \\ 0 & 1 + 10^{-15} \end{bmatrix} \quad \text{has eigenvectors} \quad \begin{bmatrix} 1 \\ 0 \end{bmatrix} \quad \text{and} \quad [?]$$

20 If $\lambda_{\max}$ is the largest eigenvalue of a symmetric matrix S, no diagonal entry can be larger than $\lambda_{\max}$. What is the first entry a_{11} of $S = Q\Lambda Q^{\mathrm{T}}$? Show why $a_{11} \leq \lambda_{\max}$.

21 Suppose $A^{\mathrm{T}} = -A$ (real *antisymmetric* matrix). Explain these facts about A:

(a) $\boldsymbol{x}^{\mathrm{T}} A\boldsymbol{x} = 0$ for every real vector $\boldsymbol{x}$.

(b) The eigenvalues of A are pure imaginary.

(c) The determinant of A is positive or zero (not negative).

For (a), multiply out an example of $\boldsymbol{x}^{\mathrm{T}} A\boldsymbol{x}$ and watch terms cancel. Or reverse $\boldsymbol{x}^{\mathrm{T}}(A\boldsymbol{x})$ to $-(A\boldsymbol{x})^{\mathrm{T}}\boldsymbol{x}$. For (b), $A\boldsymbol{z} = \lambda\boldsymbol{z}$ leads to $\overline{\boldsymbol{z}}^{\mathrm{T}} A\boldsymbol{z} = \lambda\overline{\boldsymbol{z}}^{\mathrm{T}}\boldsymbol{z} = \lambda\|\boldsymbol{z}\|^2$. Part (a) shows that $\overline{\boldsymbol{z}}^{\mathrm{T}} A\boldsymbol{z} = (\boldsymbol{x} - i\boldsymbol{y})^{\mathrm{T}} A(\boldsymbol{x} + i\boldsymbol{y})$ has zero real part. Then (b) helps with (c).

22 If S is symmetric and all its eigenvalues are $\lambda = 2$, how do you know that S must be $2I$? Key point: Symmetry guarantees that $S = Q\Lambda Q^{\mathrm{T}}$. What is that Λ?

23 *Which symmetric matrices S are also orthogonal*? Show why $S^2 = I$. What are the possible eigenvalues λ ? Then S must be $Q\Lambda Q^{\mathrm{T}}$ for which Λ ?

Problems 24–49 are about tests for positive definiteness.

24 Suppose the 2 by 2 tests $a > 0$ and $ac - b^2 > 0$ are passed by $S = \left[\begin{smallmatrix} a & b \\ c & d \end{smallmatrix}\right]$.

(i) λ_1 and λ_2 have the *same sign* because their product $\lambda_1\lambda_2$ equals ______.

(i) That sign is positive because $\lambda_1 + \lambda_2$ equals ______. So $\lambda_1 > 0$ and $\lambda_2 > 0$.

25 Which of S_1, S_2, S_3, S_4 has two positive eigenvalues? Use a test, don't compute λ's. Also find an $\boldsymbol{x}$ so that $\boldsymbol{x}^{\mathrm{T}} S_1\boldsymbol{x} < 0$, so S_1 is not positive definite.

$$S_1 = \begin{bmatrix} 5 & 6 \\ 6 & 7 \end{bmatrix} \quad S_2 = \begin{bmatrix} -1 & -2 \\ -2 & -5 \end{bmatrix} \quad S_3 = \begin{bmatrix} 1 & 10 \\ 10 & 100 \end{bmatrix} \quad S_4 = \begin{bmatrix} 1 & 10 \\ 10 & 101 \end{bmatrix}.$$

26 For which numbers b and c is positive definite ? Factor S into LDL^{T}.

$$S = \begin{bmatrix} 1 & b \\ b & 9 \end{bmatrix} \qquad S = \begin{bmatrix} 2 & 4 \\ 4 & c \end{bmatrix} \qquad S = \begin{bmatrix} c & b \\ b & c \end{bmatrix}.$$

27 Write $f(x, y) = x^2 + 4xy + 3y^2$ as a *difference* of squares and find a point (x, y) where f is negative. No minimum at $(0, 0)$ even though f has positive coefficients.

28 The function $f(x, y) = 2xy$ certainly has a saddle point and not a minimum at $(0, 0)$. What symmetric matrix S produces $\begin{bmatrix} x & y \end{bmatrix} S \begin{bmatrix} x \\ y \end{bmatrix} = 2xy$? What are its eigenvalues ?

29 Test to see if $A^{\mathrm{T}} A$ is positive definite in each case: A needs independent columns.

$$A = \begin{bmatrix} 1 & 2 \\ 0 & 3 \end{bmatrix} \quad \text{and} \quad A = \begin{bmatrix} 1 & 1 \\ 1 & 2 \\ 2 & 1 \end{bmatrix} \quad \text{and} \quad A = \begin{bmatrix} 1 & 1 & 2 \\ 1 & 2 & 1 \end{bmatrix}.$$

30 Which 3 by 3 symmetric matrices S and T produce these quadratics?

$$\boldsymbol{x}^{\mathrm{T}}S\boldsymbol{x} = 2\left(x_1^2 + x_2^2 + x_3^2 - x_1x_2 - x_2x_3\right). \quad \text{Why is } S \text{ positive definite?}$$
$$\boldsymbol{x}^{\mathrm{T}}T\boldsymbol{x} = 2\left(x_1^2 + x_2^2 + x_3^2 - x_1x_2 - x_1x_3 - x_2x_3\right). \quad \text{Why is } T \text{ semidefinite?}$$

31 Compute the three upper left determinants of S to establish positive definiteness. Verify that their ratios give the second and third pivots in elimination.

$$\textbf{Pivots = ratios of determinants} \qquad S = \begin{bmatrix} 2 & 2 & 0 \\ 2 & 5 & 3 \\ 0 & 3 & 8 \end{bmatrix}.$$

32 For what numbers c and d are S and T positive definite? Test their 3 determinants:

$$S = \begin{bmatrix} c & 1 & 1 \\ 1 & c & 1 \\ 1 & 1 & c \end{bmatrix} \quad \text{and} \quad T = \begin{bmatrix} 1 & 2 & 3 \\ 2 & d & 4 \\ 3 & 4 & 5 \end{bmatrix}.$$

33 Find a matrix with $a > 0$ and $c > 0$ and $a + c > 2b$ that has a negative eigenvalue.

34 *If S is positive definite then S^{-1} is positive definite.* Best proof: The eigenvalues of S^{-1} are positive because ______. Can you use another test?

35 A positive definite matrix cannot have a zero (or even worse, a negative number) on its main diagonal. Show that this matrix fails the energy test $\boldsymbol{x}^{\mathrm{T}}S\boldsymbol{x} > 0$:

$$\begin{bmatrix} x_1 & x_2 & x_3 \end{bmatrix} \begin{bmatrix} 4 & 1 & 1 \\ 1 & 0 & 2 \\ 1 & 2 & 5 \end{bmatrix} \begin{bmatrix} x_1 \\ x_2 \\ x_3 \end{bmatrix} \text{ is not positive when } (x_1, x_2, x_3) = (\quad,\quad,\quad).$$

36 A diagonal entry s_{jj} of a symmetric matrix cannot be smaller than all the λ's. If it were, then $S - s_{jj}I$ would have ______ eigenvalues and would be positive definite. But $S - s_{jj}I$ has a ______ on the main diagonal.

37 Give a quick reason why each of these statements is true:

(a) Every positive definite matrix is invertible.

(b) The only positive definite projection matrix is $P = I$.

(c) A diagonal matrix with positive diagonal entries is positive definite.

(d) A symmetric matrix with a positive determinant might not be positive definite!

38 For which s and t do S and T have all $\lambda > 0$ (therefore positive definite)?

$$S = \begin{bmatrix} s & -4 & -4 \\ -4 & s & -4 \\ -4 & -4 & s \end{bmatrix} \quad \text{and} \quad T = \begin{bmatrix} t & 3 & 0 \\ 3 & t & 4 \\ 0 & 4 & t \end{bmatrix}.$$

39 From $S = Q\Lambda Q^{\mathrm{T}}$ compute the positive definite symmetric square root $Q\sqrt{\Lambda}Q^{\mathrm{T}}$ of each matrix. Check that this square root gives $A^{\mathrm{T}}A = S$:

$$S = \begin{bmatrix} 5 & 4 \\ 4 & 5 \end{bmatrix} \quad \text{and} \quad S = \begin{bmatrix} 10 & 6 \\ 6 & 10 \end{bmatrix}.$$

40 Draw the tilted ellipse $x^2 + xy + y^2 = 1$ and find the half-lengths of its axes from the eigenvalues of the corresponding matrix S.

41 With positive pivots in D, the factorization $S = LDL^{\mathrm{T}}$ becomes $L\sqrt{D}\sqrt{D}L^{\mathrm{T}}$. (Square roots of the pivots give $D = \sqrt{D}\sqrt{D}$.) Then $C = \sqrt{D}L^{\mathrm{T}}$ yields the ***Cholesky factorization*** $A = C^{\mathrm{T}}C$ which is "symmetrized LU":

$$\text{From} \quad C = \begin{bmatrix} 3 & 1 \\ 0 & 2 \end{bmatrix} \quad \text{find } S. \qquad \text{From} \quad S = \begin{bmatrix} 4 & 8 \\ 8 & 25 \end{bmatrix} \quad \text{find } C = \textbf{chol}(S).$$

42 In the Cholesky factorization $S = C^{\mathrm{T}}C$, with $C = \sqrt{D}L^{\mathrm{T}}$, the square roots of the pivots are on the diagonal of C. Find C (upper triangular) for

$$S = \begin{bmatrix} 9 & 0 & 0 \\ 0 & 1 & 2 \\ 0 & 2 & 8 \end{bmatrix} \quad \text{and} \quad S = \begin{bmatrix} 1 & 1 & 1 \\ 1 & 2 & 2 \\ 1 & 2 & 7 \end{bmatrix}.$$

43 Without multiplying $S = \begin{bmatrix} \cos\theta & -\sin\theta \\ \sin\theta & \cos\theta \end{bmatrix} \begin{bmatrix} 2 & 0 \\ 0 & 5 \end{bmatrix} \begin{bmatrix} \cos\theta & \sin\theta \\ -\sin\theta & \cos\theta \end{bmatrix}$, find

(a) the determinant of S (b) the eigenvalues of S
(c) the eigenvectors of S (d) a reason why S is symmetric positive definite.

44 The graph of $z = x^2 + y^2$ is a bowl opening upward. *The graph* of $z = x^2 - y^2$ *is a saddle*. The graph of $z = -x^2 - y^2$ is a bowl opening downward. What is a test on a, b, c for $z = ax^2 + 2bxy + cy^2$ to have a saddle point at $(x, y) = (0, 0)$?

45 Which values of c give a bowl and which c give a saddle point for the graph of $z = 4x^2 + 12xy + cy^2$? Describe this graph at the borderline value of c.

46 When S and T are symmetric positive definite, ST might not even be symmetric. But start from $ST\boldsymbol{x} = \lambda\boldsymbol{x}$ and take dot products with $T\boldsymbol{x}$. Then prove $\lambda > 0$.

47 Suppose C is positive definite (so $\boldsymbol{y}^{\mathrm{T}}C\boldsymbol{y} > 0$ whenever $\boldsymbol{y} \neq \boldsymbol{0}$) and A has independent columns (so $A\boldsymbol{x} \neq \boldsymbol{0}$ whenever $\boldsymbol{x} \neq \boldsymbol{0}$). Apply the energy test to $\boldsymbol{x}^{\mathrm{T}}A^{\mathrm{T}}CA\boldsymbol{x}$ to show that $\boldsymbol{S = A^{\mathrm{T}}CA}$ ***is positive definite: the crucial matrix in engineering***.

48 **Important**! Suppose S is positive definite with eigenvalues $\lambda_1 \geq \lambda_2 \geq \ldots \geq \lambda_n$.

(a) What are the eigenvalues of the matrix $\lambda_1 I - S$? Is it positive semidefinite?

(b) How does it follow that $\lambda_1\boldsymbol{x}^{\mathrm{T}}\boldsymbol{x} \geq \boldsymbol{x}^{\mathrm{T}}S\boldsymbol{x}$ for every $\boldsymbol{x}$?

(c) Draw this conclusion: **The maximum value of $\boldsymbol{x}^{\mathrm{T}}S\boldsymbol{x}/\boldsymbol{x}^{\mathrm{T}}\boldsymbol{x}$ is** _____.

49 For which a and c is this matrix positive definite? For which a and c is it positive semidefinite (this includes definite)?

$$S = \begin{bmatrix} a & a & a \\ a & a+c & a-c \\ a & a-c & a+c \end{bmatrix}$$

All 5 tests are possible.
The energy $\boldsymbol{x}^{\mathrm{T}}S\boldsymbol{x}$ equals
$a\,(x_1 + x_2 + x_3)^2 + c\,(x_2 - x_3)^2$.

6.4 Systems of Differential Equations

1 If $A\boldsymbol{x} = \lambda\boldsymbol{x}$ then $\boldsymbol{u}(t) = e^{\lambda t}\boldsymbol{x}$ will solve $\dfrac{d\boldsymbol{u}}{dt} = A\boldsymbol{u}$. Each λ and $\boldsymbol{x}$ give a solution $e^{\lambda t}\boldsymbol{x}$.

2 If $A = X\Lambda X^{-1}$ then $\boxed{\boldsymbol{u}(t) = e^{At}\boldsymbol{u}(0) = Xe^{\Lambda t}X^{-1}\boldsymbol{u}(0) = c_1 e^{\lambda_1 t}\boldsymbol{x}_1 + \cdots + c_n e^{\lambda_n t}\boldsymbol{x}_n.}$

3 Matrix exponential $e^{\boldsymbol{At}} = I + At + \cdots + (At)^n/n! + \cdots = Xe^{\Lambda t}X^{-1}$ if $A = X\Lambda X^{-1}$.

4 A is stable and $\boldsymbol{u}(t) \to \boldsymbol{0}$ and $e^{At} \to 0$ when all eigenvalues of A have real part < 0.

5 **Second order eqn** / **First order system** $u'' + Bu' + Cu = 0$ is equivalent to $\begin{bmatrix} u \\ u' \end{bmatrix}' = \begin{bmatrix} 0 & 1 \\ -C & -B \end{bmatrix} \begin{bmatrix} u \\ u' \end{bmatrix}$.

Eigenvalues and eigenvectors and $A = X\Lambda X^{-1}$ are perfect for matrix powers A^k. They are also perfect for differential equations $d\boldsymbol{u}/dt = A\boldsymbol{u}$. This section is mostly linear algebra, but to read it you need one fact from calculus: *The derivative of* $e^{\lambda t}$ *is* $\lambda e^{\lambda t}$. The whole point of the section is this: **Constant coefficient differential equations can be converted into linear algebra.**

The ordinary equations $\dfrac{du}{dt} = u$ and $\dfrac{du}{dt} = \lambda u$ are solved by exponentials:

$$\boxed{\frac{d\boldsymbol{u}}{d\boldsymbol{t}} = \boldsymbol{u} \text{ produces } \boldsymbol{u(t) = Ce^t} \qquad \frac{d\boldsymbol{u}}{d\boldsymbol{t}} = \boldsymbol{\lambda u} \text{ produces } \boldsymbol{u(t) = Ce^{\lambda t}}} \tag{1}$$

At time $t = 0$ those solutions start from $u(0) = C$ because $e^0 = 1$. This "initial value" tells us C. **The solutions $\boldsymbol{u(t) = u(0)e^t}$ and $\boldsymbol{u(t) = u(0)e^{\lambda t}}$ start from $\boldsymbol{u(0)}$.**

We just solved a 1 by 1 problem. Linear algebra moves to n by n. The unknown is a vector $\boldsymbol{u}$ (now boldface). It starts from the initial vector $\boldsymbol{u}(0)$. The n equations contain a square matrix A. We expect n solutions $\boldsymbol{u}(t) = e^{\lambda t}\boldsymbol{x}$ from n eigenvalues / eigenvectors.

System of $\boldsymbol{n}$ equations $\quad \boxed{\dfrac{d\boldsymbol{u}}{dt} = A\boldsymbol{u}}$ starting from the vector $\boldsymbol{u}(0) = \begin{bmatrix} u_1(0) \\ \cdots \\ u_n(0) \end{bmatrix}$ at $t = 0$. (2)

These differential equations are *linear*. If $\boldsymbol{u}(t)$ and $\boldsymbol{v}(t)$ are solutions, so is $C\boldsymbol{u}(t) + D\boldsymbol{v}(t)$. We will need n constants like C and D to match the n components of $\boldsymbol{u}(0)$. Our first job is to find n "pure exponential solutions" $\boldsymbol{u} = e^{\lambda t}\boldsymbol{x}$ by using $A\boldsymbol{x} = \lambda\boldsymbol{x}$.

Notice that A is a *constant* matrix. In other linear equations, A changes as t changes. In nonlinear equations, A changes as $\boldsymbol{u}$ changes. We don't have those difficulties, $d\boldsymbol{u}/dt = A\boldsymbol{u}$ is "linear with constant coefficients". Those and only those are the differential equations that we will convert directly to linear algebra. Here is the key:

Solve linear constant coefficient equations by exponentials $e^{\lambda t}\boldsymbol{x}$ ***when*** $A\boldsymbol{x} = \lambda\boldsymbol{x}$.

Solution of $du/dt = Au$

Our pure exponential solution will be $e^{\lambda t}$ times a fixed vector $\boldsymbol{x}$. You may guess that λ is an eigenvalue of A, and $\boldsymbol{x}$ *is the eigenvector*. Substitute $\boldsymbol{u}(t) = e^{\lambda t}\boldsymbol{x}$ into the equation $d\boldsymbol{u}/dt = A\boldsymbol{u}$ to prove you are right. The factor $e^{\lambda t}$ will cancel to leave $\lambda\boldsymbol{x} = A\boldsymbol{x}$:

$$\textbf{Choose } \boldsymbol{u} = e^{\lambda t}\boldsymbol{x} \textbf{ when } \boldsymbol{Ax} = \boldsymbol{\lambda x} \qquad \boxed{\frac{d\boldsymbol{u}}{dt} = \lambda e^{\lambda t}\boldsymbol{x}} \quad \text{agrees } \textbf{with} \quad \boxed{\boldsymbol{Au} = \boldsymbol{A}e^{\boldsymbol{\lambda t}}\boldsymbol{x}} \qquad (3)$$

All components of this special solution $\boldsymbol{u} = e^{\lambda t}\boldsymbol{x}$ share the same $e^{\lambda t}$. The solution grows when $\lambda > 0$. It decays when $\lambda < 0$. If λ is a complex number, its real part decides growth or decay. The imaginary part ω gives oscillation $e^{i\omega t}$ like a sine wave.

Example 1 Solve $\dfrac{d\boldsymbol{u}}{dt} = A\boldsymbol{u} = \begin{bmatrix} \mathbf{0} & \mathbf{1} \\ \mathbf{1} & \mathbf{0} \end{bmatrix}\boldsymbol{u}$ starting from $\boldsymbol{u}(0) = \begin{bmatrix} \mathbf{4} \\ \mathbf{2} \end{bmatrix}$.

This is a vector equation for $\boldsymbol{u}$. It contains two scalar equations for the components y and z. They are "coupled together" because the matrix A is not diagonal:

$$\frac{d\boldsymbol{u}}{dt} = A\boldsymbol{u} \qquad \frac{d}{dt}\begin{bmatrix} y \\ z \end{bmatrix} = \begin{bmatrix} 0 & 1 \\ 1 & 0 \end{bmatrix}\begin{bmatrix} y \\ z \end{bmatrix} \quad \text{means that} \quad \frac{\boldsymbol{dy}}{\boldsymbol{dt}} = \boldsymbol{z} \text{ and } \frac{\boldsymbol{dz}}{\boldsymbol{dt}} = \boldsymbol{y}.$$

The idea of eigenvectors is to combine those equations in a way that gets back to 1 by 1 problems. The combinations $y + z$ and $y - z$ will do it. Add and subtract equations:

$$\frac{d}{dt}(y + z) = z + y \qquad \text{and} \qquad \frac{d}{dt}(y - z) = -(y - z).$$

The combination $y + z$ grows like e^t, because it has $\lambda = 1$. The combination $y - z$ decays like e^{-t}, because it has $\lambda = -1$. Here is the point: We don't have to juggle the original equations $d\boldsymbol{u}/dt = A\boldsymbol{u}$, looking for these special combinations. The eigenvectors and eigenvalues of A will do it for us.

This matrix A has eigenvalues 1 and -1. The eigenvectors $\boldsymbol{x}$ are $(1, 1)$ and $(1, -1)$. The pure exponential solutions $\boldsymbol{u}_1$ and $\boldsymbol{u}_2$ take the form $e^{\lambda t}\boldsymbol{x}$ with $\lambda_1 = 1$ and $\lambda_2 = -1$:

$$\boxed{\boldsymbol{u}_1(t) = e^{\lambda_1 t}\boldsymbol{x}_1 = e^t\begin{bmatrix} 1 \\ 1 \end{bmatrix}} \quad \text{and} \quad \boxed{\boldsymbol{u}_2(t) = e^{\lambda_2 t}\boldsymbol{x}_2 = e^{-t}\begin{bmatrix} 1 \\ -1 \end{bmatrix}.} \qquad (4)$$

$$\begin{matrix}\textbf{Complete solution } \boldsymbol{u}(t) \\ \textbf{Combine } \boldsymbol{u}_1 \textbf{ and } \boldsymbol{u}_2\end{matrix} \qquad \boldsymbol{u}(t) = \boldsymbol{C}e^{\boldsymbol{t}}\begin{bmatrix} \mathbf{1} \\ \mathbf{1} \end{bmatrix} + \boldsymbol{D}e^{-\boldsymbol{t}}\begin{bmatrix} \mathbf{1} \\ \mathbf{-1} \end{bmatrix} = \begin{bmatrix} Ce^t + De^{-t} \\ Ce^t - De^{-t} \end{bmatrix}. \qquad (5)$$

With these two constants C and D, we can match any starting vector $\boldsymbol{u}(0) = (u_1(0), u_2(0))$. Set $t = 0$ and $e^0 = 1$. Example 1 asked for the initial value to be $\boldsymbol{u}(0) = (4, 2)$:

$$\begin{matrix}\boldsymbol{u}(0) \textbf{ decides} \\ \boldsymbol{C} \textbf{ and } \boldsymbol{D}\end{matrix} \qquad C\begin{bmatrix} 1 \\ 1 \end{bmatrix} + D\begin{bmatrix} 1 \\ -1 \end{bmatrix} = \begin{bmatrix} 4 \\ 2 \end{bmatrix} \quad \text{yields} \quad \boldsymbol{C = 3} \quad \text{and} \quad \boldsymbol{D = 1}.$$

With $C = 3$ and $D = 1$ in the solution (5), the initial value problem is completely solved.

For n by n matrices we look for n eigenvectors. Then C and D become $c_1, c_2, \ldots, c_n$.

1. Write $\boldsymbol{u}(0)$ as a **combination** $c_1\boldsymbol{x}_1 + \cdots + c_n\boldsymbol{x}_n$ **of the eigenvectors of** $\boldsymbol{A}$.
2. Multiply each eigenvector $\boldsymbol{x}_i$ by **its growth factor** $e^{\lambda_i t}$.
3. The solution to $d\boldsymbol{u}/dt = A\boldsymbol{u}$ is the same combination of those pure solutions $e^{\lambda t}\boldsymbol{x}$:

$$\boldsymbol{u}(t) = c_1 e^{\lambda_1 t}\boldsymbol{x}_1 + \cdots + c_n e^{\lambda_n t}\boldsymbol{x}_n. \tag{6}$$

Not included: If two λ's are equal, with only one eigenvector, another solution is needed. (It will be $te^{\lambda t}\boldsymbol{x}$.) Step 1 needs a basis of n eigenvectors to diagonalize $A = X\Lambda X^{-1}$.

Example 2 Solve $d\boldsymbol{u}/dt = A\boldsymbol{u}$ knowing the eigenvalues $\lambda = 1, 2, 3$ of A:

Typical example
Equation for $\boldsymbol{u}$
Initial condition $\boldsymbol{u}(0)$

$$\frac{d\boldsymbol{u}}{dt} = \begin{bmatrix} 1 & 1 & 1 \\ 0 & 2 & 1 \\ 0 & 0 & 3 \end{bmatrix} \boldsymbol{u} \quad \text{starting from} \quad \boldsymbol{u}(0) = \begin{bmatrix} 9 \\ 7 \\ 4 \end{bmatrix}.$$

The eigenvector matrix X has $\boldsymbol{x}_1 = (1, 0, 0)$ and $\boldsymbol{x}_2 = (1, 1, 0)$ and $\boldsymbol{x}_3 = (1, 1, 1)$.

Step 1 The vector $\boldsymbol{u}(0) = (9, 7, 4)$ is $\mathbf{2}\boldsymbol{x}_1 + \mathbf{3}\boldsymbol{x}_2 + \mathbf{4}\boldsymbol{x}_3$. Then $(c_1, c_2, c_3) = (\mathbf{2}, \mathbf{3}, \mathbf{4})$.

Step 2 The factors $e^{\lambda t}$ give exponential solutions $e^t\boldsymbol{x}_1$ and $e^{2t}\boldsymbol{x}_2$ and $e^{3t}\boldsymbol{x}_3$.

Step 3 The combination that starts from $\boldsymbol{u}(0)$ is $\boldsymbol{u}(t) = \mathbf{2}e^t\boldsymbol{x}_1 + \mathbf{3}e^{2t}\boldsymbol{x}_2 + \mathbf{4}e^{3t}\boldsymbol{x}_3$.

The coefficients $2, 3, 4$ came from solving the linear equation $c_1\boldsymbol{x}_1 + c_2\boldsymbol{x}_2 + c_3\boldsymbol{x}_3 = \boldsymbol{u}(0)$:

$$\begin{bmatrix} & & \\ \boldsymbol{x}_1 & \boldsymbol{x}_2 & \boldsymbol{x}_3 \\ & & \end{bmatrix} \begin{bmatrix} c_1 \\ c_2 \\ c_3 \end{bmatrix} = \begin{bmatrix} 1 & 1 & 1 \\ 0 & 1 & 1 \\ 0 & 0 & 1 \end{bmatrix} \begin{bmatrix} 2 \\ 3 \\ 4 \end{bmatrix} = \begin{bmatrix} 9 \\ 7 \\ 4 \end{bmatrix} \quad \text{which is} \quad X\boldsymbol{c} = \boldsymbol{u}(0). \tag{7}$$

You now have the basic idea—how to solve $d\boldsymbol{u}/dt = A\boldsymbol{u}$. The rest of this section goes further. We solve equations that contain *second* derivatives, because they arise so often in applications. We also decide whether $\boldsymbol{u}(t)$ approaches zero or blows up or just oscillates.
At the end comes the ***matrix exponential*** e^{At}. The short formula $e^{At}\boldsymbol{u}(0)$ solves the equation $d\boldsymbol{u}/dt = A\boldsymbol{u}$ in the same way that $A^k\boldsymbol{u}_0$ solves the equation $\boldsymbol{u}_{k+1} = A\boldsymbol{u}_k$. Example 3 will show how "difference equations" help to solve differential equations.

All these steps use the λ's and the $\boldsymbol{x}$'s. This section solves the constant coefficient problems that turn into linear algebra. Those are the simplest but most important differential equations—whose solution is completely based on growth factors $e^{\lambda t}$.

Second Order Equations

The most important equation in mechanics is $\boldsymbol{my'' + by' + ky = 0}$. The first term is the mass m times the acceleration $a = y''$. Then by' is damping and ky is force.

This is a second-order equation (it is *Newton's Law* $F = ma$). It contains the second derivative $y'' = d^2y/dt^2$. It is still linear with constant coefficients m, b, k.

In a differential equations course, the method of solution is to substitute $y = e^{\lambda t}$. Each derivative of y brings down a factor λ. We want $y = e^{\lambda t}$ to solve the equation:

$$\boxed{m\frac{d^2y}{dt^2} + b\frac{dy}{dt} + ky = 0 \quad \text{becomes} \quad (m\lambda^2 + b\lambda + k)\, e^{\lambda t} = 0.} \tag{8}$$

Everything depends on $m\lambda^2 + b\lambda + k = 0$. This equation for λ has two roots λ_1 and λ_2. Then the equation for y has two pure solutions $y_1 = e^{\lambda_1 t}$ and $y_2 = e^{\lambda_2 t}$. Their combinations $c_1y_1 + c_2y_2$ give the complete solution. *This is not true if* $\lambda_1 = \lambda_2$.

In a linear algebra course we expect matrices and eigenvalues. Therefore we turn the scalar equation (with y'') into a *vector equation for* y *and* y': First derivative only! Suppose the mass is $m = 1$. Two equations for $\boldsymbol{u} = (y,\ y')$ give $d\boldsymbol{u}/dt = A\boldsymbol{u}$:

$$\begin{aligned} dy/dt &= y' \\ dy'/dt &= -ky - by' \end{aligned} \qquad \text{converts to} \qquad \frac{d}{dt}\begin{bmatrix} y \\ y' \end{bmatrix} = \begin{bmatrix} 0 & 1 \\ -k & -b \end{bmatrix}\begin{bmatrix} y \\ y' \end{bmatrix} = A\boldsymbol{u}. \tag{9}$$

The first equation $dy/dt = y'$ is trivial (but true). The second is equation (8) connecting y'' to y' and y. Together they connect $\boldsymbol{u}'$ to $\boldsymbol{u}$. Now we solve $\boldsymbol{u}' = A\boldsymbol{u}$ by eigenvalues of A:

$$A - \lambda I = \begin{bmatrix} -\lambda & 1 \\ -k & -b-\lambda \end{bmatrix} \quad \text{has determinant} \quad \lambda^2 + b\lambda + k = 0.$$

The equation for the λ's is the same as (8)! It is still $\boldsymbol{\lambda^2 + b\lambda + k = 0}$, since $\boldsymbol{m = 1}$.

The roots λ_1 and λ_2 are now *eigenvalues of* A. The eigenvectors and the solution are

$$\boldsymbol{x}_1 = \begin{bmatrix} 1 \\ \lambda_1 \end{bmatrix} \qquad \boldsymbol{x}_2 = \begin{bmatrix} 1 \\ \lambda_2 \end{bmatrix} \qquad \boldsymbol{u}(t) = c_1e^{\lambda_1 t}\begin{bmatrix} 1 \\ \lambda_1 \end{bmatrix} + c_2e^{\lambda_2 t}\begin{bmatrix} 1 \\ \lambda_2 \end{bmatrix}.$$

The first component of $\boldsymbol{u}(t)$ has $y = c_1e^{\lambda_1 t} + c_2e^{\lambda_2 t}$—the same solution as before. It can't be anything else. In the second component of $\boldsymbol{u}(t)$ you see the velocity dy/dt. The 2 by 2 matrix A in (9) is called a *companion matrix*—a companion to the equation (8).

Example 3 ***Motion around a circle with $y'' + y = 0$ and $y = \cos t$***

This is our master equation with mass $m = 1$ and stiffness $k = 1$ and $d = 0$: no damping. Substitute $y = e^{\lambda t}$ into $y'' + y = 0$ to reach $\boldsymbol{\lambda^2 + 1 = 0}$. *The roots are* $\boldsymbol{\lambda = i}$ *and* $\boldsymbol{\lambda = -i}$. Then half of $e^{it} + e^{-it}$ gives the solution $y = \cos t$.

As a first-order system, the initial values $y(0) = 1$, $y'(0) = 0$ go into $\boldsymbol{u}(0) = (1, 0)$:

$$\textbf{Use } y'' = -y \qquad \frac{d\boldsymbol{u}}{dt} = \frac{d}{dt}\begin{bmatrix} y \\ y' \end{bmatrix} = \begin{bmatrix} 0 & 1 \\ -1 & 0 \end{bmatrix}\begin{bmatrix} y \\ y' \end{bmatrix} = A\boldsymbol{u}. \tag{10}$$

The eigenvalues of A are again the same $\lambda = i$ and $\lambda = -i$ (no surprise). A is antisymmetric with eigenvectors $\boldsymbol{x}_1 = (1, i)$ and $\boldsymbol{x}_2 = (1, -i)$. The combination that matches $\boldsymbol{u}(0) = (1, 0)$ is $\frac{1}{2}(\boldsymbol{x}_1 + \boldsymbol{x}_2)$. Step 2 multiplies the $\boldsymbol{x}$'s by e^{it} and e^{-it}.

Step 3 combines the pure oscillations e^{it} and e^{-it} to find $y = \cos t$ as expected:

$$\boldsymbol{u}(t) = \frac{1}{2}e^{it}\begin{bmatrix}1\\ i\end{bmatrix} + \frac{1}{2}e^{-it}\begin{bmatrix}1\\ -i\end{bmatrix} = \begin{bmatrix}\cos t\\ -\sin t\end{bmatrix}. \qquad \text{This is } \begin{bmatrix}y(t)\\ y'(t)\end{bmatrix}.$$

All good. The vector $\boldsymbol{u} = (\cos t, -\sin t)$ goes around a circle (Figure 6.3). The radius is 1 because $\cos^2 t + \sin^2 t = 1$. The "period" is 2π when $\boldsymbol{u}$ completes a circle.

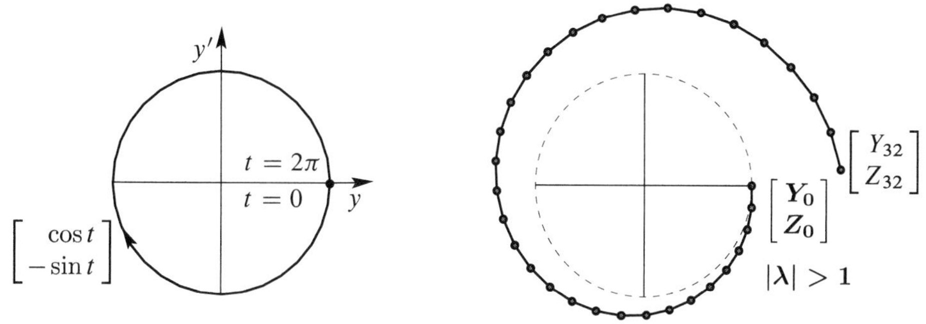

Figure 6.3: The exact solution $\boldsymbol{u} = (\cos t, -\sin t)$ stays on a circle.
Forward differences $\boldsymbol{Y_1}, \boldsymbol{Y_2}, \ldots$ spiral out from the circle in 32 steps.

Difference Equations

To display a circle on a screen, replace $y'' = -y$ by a ***difference equation***. Here are three choices using $\boldsymbol{Y}(t+\Delta t) - 2\boldsymbol{Y}(t) + \boldsymbol{Y}(t-\Delta t)$. Divide by $(\Delta t)^2$ to approximate d^2y/dt^2.

F	**Forward from $n-1$**	$\dfrac{\boldsymbol{Y}_{n+1} - 2\boldsymbol{Y}_n + \boldsymbol{Y}_{n-1}}{(\Delta t)^2} =$	$-\boldsymbol{Y}_{n-1}$	(11 **F**)
C	**Centered at time n**		$-\boldsymbol{Y}_n$	(11 **C**)
B	**Backward from $n+1$**		$-\boldsymbol{Y}_{n+1}$	(11 **B**)

Figure 6.3 shows the exact $y(t) = \cos t$ completing a circle at $t = 2\pi$. The three difference methods *don't* complete a perfect circle in 32 time steps of length $\Delta t = 2\pi/32$. The spirals in those pictures will be explained by eigenvalues λ for **11 F, 11 B, 11 C**.
Forward $|\lambda| > 1$ (spiral out) Centered $|\lambda| = 1$ (best) Backward $|\lambda| < 1$ (spiral in)

The 2-step equations (11) reduce to 1-step systems $\boldsymbol{U}_{n+1} = A\boldsymbol{U}_n$. Instead of $\boldsymbol{u} = (y, y')$ the unknown is $\boldsymbol{U}_n = (Y_n, Z_n)$. We take n time steps of size Δt starting from $\boldsymbol{U}_0$:

$$\textbf{Forward (11F)} \quad \begin{aligned} Y_{n+1} &= Y_n + \Delta t\, Z_n \\ Z_{n+1} &= Z_n - \Delta t\, Y_n \end{aligned} \quad \text{becomes} \quad \boldsymbol{U}_{n+1} = \begin{bmatrix}1 & \Delta t\\ -\Delta t & 1\end{bmatrix}\begin{bmatrix}Y_n\\ Z_n\end{bmatrix} = A\boldsymbol{U}_n. \qquad (12)$$

Those are like $Y' = Z$ and $Z' = -Y$. They are first order equations. Now we have a matrix. Eliminating Z would bring back the second order equation (11 **F**).

My question is simple. *Do the points* (Y_n, Z_n) *stay on the circle* $Y^2 + Z^2 = 1$? No, they are growing to infinity in Figure 6.3. ***We are taking powers*** $\boldsymbol{A^n}$ ***and not*** $\boldsymbol{e^{At}}$, ***so we test the magnitude*** $|\boldsymbol{\lambda}|$ ***and not the real parts of the eigenvalues*** $\boldsymbol{\lambda}$.

Eigenvalues of A $\quad \boldsymbol{\lambda = 1 \pm i\Delta t}$ $\quad$ **Then** $\boldsymbol{|\lambda| > 1}$ **and** $\boldsymbol{(Y_n, Z_n)}$ **spirals out**

The backward choice in (11 **B**) will do the opposite in Figure 6.4. Notice the new A:

$$\textbf{Backward} \quad \begin{aligned} Y_{n+1} &= Y_n + \Delta t\, Z_{n+1} \\ Z_{n+1} &= Z_n - \Delta t\, \mathbf{Y_{n+1}} \end{aligned} \quad \text{is} \quad \begin{bmatrix} 1 & -\Delta t \\ \Delta t & 1 \end{bmatrix} \begin{bmatrix} Y_{n+1} \\ Z_{n+1} \end{bmatrix} = \begin{bmatrix} Y_n \\ Z_n \end{bmatrix} = \boldsymbol{U}_n. \quad (13)$$

That matrix has eigenvalues $1 \pm i\Delta t$. But we invert it to reach $\boldsymbol{U}_{n+1}$ from $\boldsymbol{U}_n$. Then $|\lambda| < 1$ explains why *the solution spirals in* to $(0, 0)$ for backward differences.

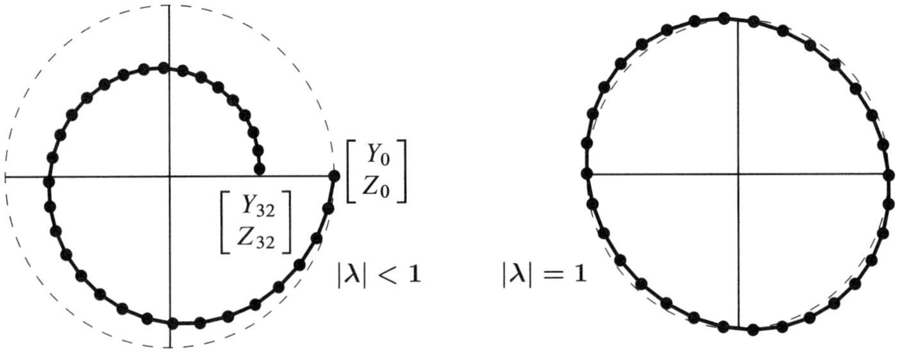

Figure 6.4: Backward differences (**11B**) spiral in. Leapfrog (**11C**) stays near the circle.

On the right side of Figure 6.4 you see 32 steps with the *centered* choice. The solution stays close to the circle (Problem 28) if $\Delta t < 2$. This is the **leapfrog method**, constantly used. The second difference $Y_{n+1} - 2Y_n + Y_{n-1}$ "leaps over" the center value Y_n in (11).

This is the way a chemist follows the motion of molecules (molecular dynamics leads to giant computations). Computational science is lively because one differential equation can be replaced by many difference equations—some unstable, some stable, some neutral. Problem 26 has a fourth (very good) method that exactly completes the circle.

Real engineering and real physics deal with systems (not just a single mass at one point). The unknown $\boldsymbol{y}$ is a vector. The coefficient of $\boldsymbol{y}''$ is a *mass matrix* M, with n masses. The coefficient of $\boldsymbol{y}$ is a *stiffness matrix* K, not a number k. The coefficient of $\boldsymbol{y}'$ is a damping matrix which might be zero.

The vector equation $\boldsymbol{My'' + Ky = f}$ is a major part of computational mechanics.

Stability of 2 by 2 Matrices

For the solution of $du/dt = Au$, there is a fundamental question. *Does the solution approach* $\boldsymbol{u} = \mathbf{0}$ *as* $t \to \infty$? Is the problem *stable*, by dissipating energy? A solution that includes e^t is unstable. Stability depends on the eigenvalues of A.

The complete solution $\boldsymbol{u}(t)$ is built from pure solutions $e^{\lambda t}\boldsymbol{x}$. If the eigenvalue λ is real, we know exactly when $e^{\lambda t}$ will approach zero: *The number* λ *must be negative.* If the eigenvalue is a complex number $\lambda = r + is$, *the real part* r *must be negative.* When $e^{\lambda t}$ splits into $e^{rt}e^{ist}$, the factor e^{ist} has absolute value fixed at 1:

$$e^{ist} = \cos st + i \sin st \quad \text{has} \quad |e^{ist}|^2 = \cos^2 st + \sin^2 st = 1.$$

Then $|e^{\lambda t}| = e^{rt}$ controls the growth ($r > 0$) or the decay ($r < 0$).

The question is: *Which matrices have negative eigenvalues*? More accurately, when are the ***real parts of the*** $\boldsymbol{\lambda}$***'s all negative*** ? The answer is not clear for 2 by 2 matrices.

> A is ***stable*** and $\boldsymbol{u}(t) \to \mathbf{0}$ when all eigenvalues λ of A have ***negative real parts***.
>
> The 2 by 2 matrix $A = \begin{bmatrix} a & b \\ c & d \end{bmatrix}$ must pass two tests :
>
> | | | |
> |---|---|---|
> | $\boldsymbol{\lambda_1 + \lambda_2 < 0}$ | The trace $T = a + d$ must be negative. | (14T) |
> | $\boldsymbol{\lambda_1 \lambda_2 > 0}$ | The determinant $D = ad - bc$ must be positive. | (14D) |

Reason If the λ's are real and negative, their sum is negative. This is the trace $T = a + d$. Their product is positive. This is the determinant D. The argument also goes in reverse.

If $D = \lambda_1\lambda_2$ is positive, then λ_1 and λ_2 have the same sign.

If $T = \lambda_1 + \lambda_2$ is negative, that sign will be negative.

If the λ's are complex numbers, they must have the form $r + is$ and $r - is$. Otherwise T and D will not be real. The determinant D is automatically positive, since $(r + is)(r - is) = r^2 + s^2$. The trace T is $r + is + r - is = 2r$. So a negative trace T means that $r < 0$ and the matrix is stable. The two tests in (14) are correct.

The Exponential of a Matrix

We want to write the solution $\boldsymbol{u}(t)$ ***in a new form*** $e^{At}\boldsymbol{u}(0)$. First we have to say what e^{At} means, with a matrix in the exponent. To define e^{At} for matrices, we copy e^x for numbers.

The direct definition of e^x is by the infinite series $1 + x + \frac{1}{2}x^2 + \frac{1}{6}x^3 + \cdots$. When you change x to a square matrix At, this infinite series defines the matrix exponential e^{At} :

Matrix exponential $\boldsymbol{e^{At}}$	$\boxed{e^{At} = I + At + \frac{1}{2}(At)^2 + \frac{1}{6}(At)^3 + \cdots}$	(15)
Its $\boldsymbol{t}$ **derivative is** $\boldsymbol{Ae^{At}}$	$A + A^2t + \frac{1}{2}A^3t^2 + \cdots = Ae^{At}$	
Its eigenvalues are $\boldsymbol{e^{\lambda t}}$	$(I + At + \frac{1}{2}(At)^2 + \cdots)\boldsymbol{x} = (1 + \lambda t + \frac{1}{2}(\lambda t)^2 + \cdots)\boldsymbol{x}$	

The number that divides $(At)^n$ is "n factorial". This is $n! = (1)(2)\cdots(n-1)(n)$. The factorials after $1, 2, 6$ are $4! = 24$ and $5! = 120$. They grow quickly. The series always converges and its derivative is always Ae^{At}. Therefore $e^{At}\boldsymbol{u}(0)$ solves the differential equation with one quick formula—*even if there is a shortage of eigenvectors.*

This chapter emphasizes how to find $\boldsymbol{u}(t) = e^{At}\boldsymbol{u}(0)$ by diagonalization. Assume A does have n independent eigenvectors, so we can substitute $A = X\Lambda X^{-1}$ into the series for e^{At}. Whenever $X\Lambda X^{-1}X\Lambda X^{-1}$ appears, cancel $X^{-1}X$ in the middle:

Use the series $$\boldsymbol{e^{At}} = I + X\Lambda X^{-1}t + \tfrac{1}{2}(X\Lambda X^{-1}t)(X\Lambda X^{-1}t) + \cdots$$

Factor out X and X^{-1} $$= \boldsymbol{X}\left[I + \Lambda t + \tfrac{1}{2}(\Lambda t)^2 + \cdots\right]\boldsymbol{X^{-1}}$$

e^{At} is diagonalized! $$\boxed{\boldsymbol{e^{At} = X\, e^{\Lambda t}\, X^{-1}}.}$$

e^{At} has the same eigenvector matrix X as A. Then Λ is a diagonal matrix and so is $e^{\Lambda t}$. The numbers $e^{\lambda_i t}$ are on the diagonal. Multiply $Xe^{\Lambda t}X^{-1}\boldsymbol{u}(0)$ to recognize $\boldsymbol{u}(t)$:

$$e^{At}\boldsymbol{u}(0) = Xe^{\Lambda t}X^{-1}\boldsymbol{u}(0) = \begin{bmatrix} \\ \boldsymbol{x}_1 & \cdots & \boldsymbol{x}_n \\ \\ \end{bmatrix}\begin{bmatrix} e^{\lambda_1 t} & & \\ & \ddots & \\ & & e^{\lambda_n t} \end{bmatrix}\begin{bmatrix} c_1 \\ \vdots \\ c_n \end{bmatrix}. \qquad (16)$$

This solution $e^{At}\boldsymbol{u}(0)$ is the same answer that came in equation (6) from three steps:

1. $\boldsymbol{u}(0) = c_1\boldsymbol{x}_1 + \cdots + c_n\boldsymbol{x}_n = X\boldsymbol{c}$. Here we need n independent eigenvectors.
2. Multiply each $\boldsymbol{x}_i$ by its growth factor $e^{\lambda_i t}$ to follow it forward in time.
3. The best form of $e^{At}\boldsymbol{u}(0)$ is $\quad \boldsymbol{u}(t) = \boldsymbol{c_1 e^{\lambda_1 t}x_1 + \cdots + c_n e^{\lambda_n t}x_n}$. (17)

Example 4 Use the infinite series to find e^{At} for $A = \left[\begin{smallmatrix} 0 & 1 \\ -1 & 0 \end{smallmatrix}\right]$. Notice that $A^4 = I$:

$$A = \begin{bmatrix} & 1 \\ -1 & \end{bmatrix} \quad A^2 = \begin{bmatrix} -1 & \\ & -1 \end{bmatrix} \quad A^3 = \begin{bmatrix} & -1 \\ 1 & \end{bmatrix} \quad A^4 = \begin{bmatrix} 1 & \\ & 1 \end{bmatrix}.$$

A^5, A^6, A^7, A^8 will be a repeat of A, A^2, A^3, A^4. The top right corner has $1, 0, -1, 0$ repeating over and over in powers of A. Then $t - \frac{1}{6}t^3$ starts the infinite series for e^{At} in that top right corner, and $1 - \frac{1}{2}t^2$ starts the top left corner:

$$e^{At} = I + At + \tfrac{1}{2}(At)^2 + \tfrac{1}{6}(At)^3 + \cdots = \begin{bmatrix} 1 - \frac{1}{2}t^2 + \cdots & t - \frac{1}{6}t^3 + \cdots \\ -t + \frac{1}{6}t^3 - \cdots & 1 - \frac{1}{2}t^2 + \cdots \end{bmatrix}.$$

That matrix e^{At} shows the infinite series for $\cos t$ and $\sin t$!

$$\boldsymbol{A} = \begin{bmatrix} \boldsymbol{0} & \boldsymbol{1} \\ \boldsymbol{-1} & \boldsymbol{0} \end{bmatrix} \qquad \boldsymbol{e^{At}} = \begin{bmatrix} \boldsymbol{\cos t} & \boldsymbol{\sin t} \\ \boldsymbol{-\sin t} & \boldsymbol{\cos t} \end{bmatrix}. \qquad (18)$$

A is an antisymmetric matrix ($A^{\mathrm{T}} = -A$). Its exponential e^{At} is an orthogonal matrix. The eigenvalues of A are i and $-i$. Then the eigenvalues of e^{At} are e^{it} and e^{-it}.

1 ***The inverse of*** $\boldsymbol{e^{At}}$ ***is always*** $\boldsymbol{e^{-At}}$**.**

2 ***The eigenvalues of*** $\boldsymbol{e^{At}}$ ***are always*** $\boldsymbol{e^{\lambda t}}$**.**

3 ***When A is antisymmetric,*** $\boldsymbol{e^{At}}$ ***is orthogonal. Inverse = transpose =*** $\boldsymbol{e^{-At}}$**.**

Antisymmetric is the same as "skew-symmetric". Those matrices have pure imaginary eigenvalues like i and $-i$. Then e^{At} has eigenvalues like e^{it} and e^{-it}. Their absolute value is 1: neutral stability, pure oscillation, energy is conserved. So $||\boldsymbol{u}(t)|| = ||\boldsymbol{u}(0)||$.

If A is triangular, the eigenvector matrix X is also triangular. So are X^{-1} and e^{At}. The solution $\boldsymbol{u}(t)$ is a combination of eigenvectors. Its short form is $e^{At}\boldsymbol{u}(0)$.

Example 5 Solve $\dfrac{d\boldsymbol{u}}{dt} = A\boldsymbol{u} = \begin{bmatrix} 1 & 1 \\ 0 & 2 \end{bmatrix} \boldsymbol{u}$ starting from $\boldsymbol{u}(0) = \begin{bmatrix} 2 \\ 1 \end{bmatrix}$ at $t = 0$.

Solution The eigenvalues 1 and 2 are on the diagonal of A (since A is triangular). The eigenvectors are $(1,0)$ and $(1,1)$. Then e^{At} produces $\boldsymbol{u}(t)$ for every $\boldsymbol{u}(0)$:

$$\boldsymbol{u}(t) = Xe^{\Lambda t}X^{-1}\boldsymbol{u}(0) \text{ is } \begin{bmatrix} 1 & 1 \\ 0 & 1 \end{bmatrix} \begin{bmatrix} e^t & \\ & e^{2t} \end{bmatrix} \begin{bmatrix} 1 & -1 \\ 0 & 1 \end{bmatrix} \boldsymbol{u}(0) = \begin{bmatrix} e^t & e^{2t} + e^t \\ 0 & e^{2t} \end{bmatrix} \boldsymbol{u}(0).$$

That last matrix is $\boldsymbol{e^{At}}$**.** It is nice because A is triangular. The situation is the same as for $A\boldsymbol{x} = \boldsymbol{b}$ and inverses. We don't need A^{-1} to find $\boldsymbol{x}$, and we don't need e^{At} to solve $d\boldsymbol{u}/dt = A\boldsymbol{u}$. But as quick formulas for the answers, $A^{-1}\boldsymbol{b}$ and $e^{At}\boldsymbol{u}(0)$ are unbeatable.

Example 6 Solve $y'' + 4y' + 3y = 0$ by substituting $e^{\lambda t}$ and also by linear algebra.

Solution Substituting $y = e^{\lambda t}$ yields $(\lambda^2 + 4\lambda + 3)e^{\lambda t} = 0$. That quadratic factors into $\lambda^2+4\lambda+3 = (\lambda+1)(\lambda+3) = 0$. Therefore $\boldsymbol{\lambda_1 = -1}$ and $\boldsymbol{\lambda_2 = -3}$. The pure solutions are $y_1 = e^{-t}$ and $y_2 = e^{-3t}$. The complete solution $\boldsymbol{y} = c_1y_1 + c_2y_2$ approaches zero.

To use linear algebra we set $\boldsymbol{u} = (y, y')$. Then the vector equation is $\boldsymbol{u}' = A\boldsymbol{u}$:

$$\begin{aligned} dy/dt &= y' \\ dy'/dt &= -3y - 4y' \end{aligned} \quad \text{converts to} \quad \frac{d\boldsymbol{u}}{dt} = \begin{bmatrix} 0 & 1 \\ -3 & -4 \end{bmatrix} \boldsymbol{u}.$$

This A is a "companion matrix" and its eigenvalues are again $\lambda_1 = -1$ and $\lambda_2 = -3$:

$$\textbf{Same quadratic} \qquad \det(A - \lambda I) = \begin{vmatrix} -\lambda & 1 \\ -3 & -4 - \lambda \end{vmatrix} = \lambda^2 + 4\lambda + 3 = 0.$$

The eigenvectors of A are $(1, \lambda_1)$ and $(1, \lambda_2)$. Either way, the decay in $y(t)$ comes from e^{-t} and e^{-3t}. With constant coefficients, calculus leads to linear algebra $A\boldsymbol{x} = \lambda\boldsymbol{x}$.

The eigenvectors are *orthogonal* (proved in Section 6.3 for all symmetric matrices). All three λ_i are negative. This A is *negative definite* and e^{At} decays to zero (stability). The starting $\boldsymbol{u}(0) = (0, 2\sqrt{2}, 0)$ is $\boldsymbol{x}_3 - \boldsymbol{x}_2$. The solution is $\boldsymbol{u}(t) = \boldsymbol{e^{\lambda_3 t}x_3 - e^{\lambda_2 t}x_2}$.

Heat equation In Figure 6.5a, the temperature at the center starts at $2\sqrt{2}$. Heat diffuses into the neighboring boxes and then to the outside boxes (frozen at $0°$). The rate of heat flow between boxes is the temperature difference. From box 2, heat flows left and right at the rate $u_1 - u_2$ and $u_3 - u_2$. So the flow out is $u_1 - 2u_2 + u_3$ in the second row of $A\boldsymbol{u}$.

Wave equation $d^2\boldsymbol{u}/dt^2 = A\boldsymbol{u}$ has the same eigenvectors $\boldsymbol{x}$. But now the eigenvalues λ lead to **oscillations** $e^{i\omega t}\boldsymbol{x}$ and $e^{-i\omega t}\boldsymbol{x}$. The frequencies come from $\omega^2 = -\lambda$:

$$\frac{d^2}{dt^2}(e^{i\omega t}\boldsymbol{x}) = A(e^{i\omega t}\boldsymbol{x}) \qquad \text{becomes} \qquad (i\omega)^2 e^{i\omega t}\boldsymbol{x} = \lambda e^{i\omega t}\boldsymbol{x} \quad \text{and} \quad \boldsymbol{\omega^2 = -\lambda}.$$

There are two square roots of $-\lambda$, so we have $e^{i\omega t}\boldsymbol{x}$ and $e^{-i\omega t}\boldsymbol{x}$. With three eigenvectors this makes *six* solutions to $\boldsymbol{u}'' = A\boldsymbol{u}$. A combination will match the six components of $\boldsymbol{u}(0)$ and $\boldsymbol{u}'(0)$. Since $\boldsymbol{u}' = \boldsymbol{0}$ in this problem, $e^{i\omega t}\boldsymbol{x}$ and $e^{-i\omega t}\boldsymbol{x}$ produce $2\cos\omega t\,\boldsymbol{x}$.

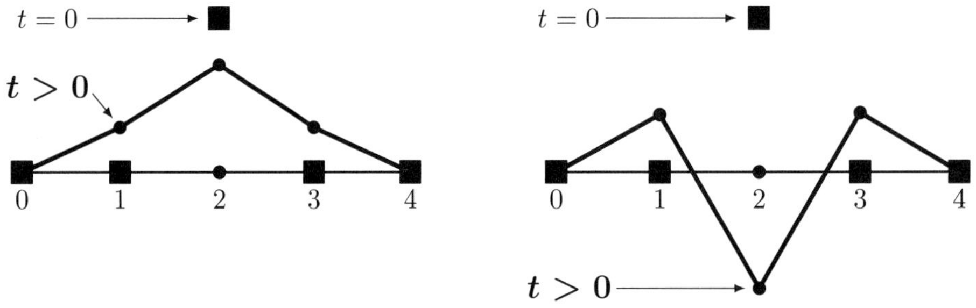

Figure 6.5: Heat diffuses away from box 2 (left). Waves travel from box 2 (right).

Example 7 Substituting $y = e^{\lambda t}$ into $y'' - 2y' + y = 0$ gives an equation with **repeated roots**: $\lambda^2 - 2\lambda + 1 = 0$ is $(\lambda - 1)^2 = 0$ with $\boldsymbol{\lambda = 1, 1}$. A differential equations course would propose e^t and te^t as two independent solutions. Here we discover why.

Linear algebra reduces $y'' - 2y' + y = 0$ to a vector equation for $\boldsymbol{u} = (y, y')$:

$$\frac{d}{dt}\begin{bmatrix} y \\ y' \end{bmatrix} = \begin{bmatrix} y' \\ 2y' - y \end{bmatrix} \quad \text{is} \quad \frac{d\boldsymbol{u}}{dt} = A\boldsymbol{u} = \begin{bmatrix} \mathbf{0} & \mathbf{1} \\ \mathbf{-1} & \mathbf{2} \end{bmatrix} \boldsymbol{u}. \tag{20}$$

A has a **repeated eigenvalue** $\boldsymbol{\lambda = 1, 1}$ (with trace $= 2$ and $\det A = 1$). The only eigenvectors are multiples of $\boldsymbol{x} = (1, 1)$. *Diagonalization is not possible.* This matrix A has only one line of eigenvectors. So we compute e^{At} from its definition as a series:

$$\textbf{Short series} \qquad \boldsymbol{e^{At}} = e^{It}\,e^{(A-I)t} = \boldsymbol{e^t\,[I + (A - I)t]}. \tag{21}$$

That "infinite" series for $e^{(A-I)t}$ ended quickly because $(A - I)^2$ is the zero matrix. You can see $\boldsymbol{te^t}$ in equation (21). The first component of $\boldsymbol{e^{At}}\,\boldsymbol{u}(0)$ is our answer $y(t)$:

$$\begin{bmatrix} y \\ y' \end{bmatrix} = e^t \left[I + \begin{bmatrix} -1 & 1 \\ -1 & 1 \end{bmatrix} t \right] \begin{bmatrix} y(0) \\ y'(0) \end{bmatrix} \qquad \boldsymbol{y(t) = e^t y(0) - te^t y(0) + te^t y'(0)}.$$

Problem Set 6.4

Note In linear algebra the serious danger is a shortage of eigenvectors. Our eigenvectors $(1, \lambda_1)$ and $(1, \lambda_2)$ are the same if $\lambda_1 = \lambda_2$. Then we can't diagonalize A. In this case we don't yet have two independent solutions to $d\boldsymbol{u}/dt = A\boldsymbol{u}$.

In differential equations the danger is also a repeated λ. After $y = e^{\lambda t}$, a second solution has to be found. It turns out to be $y = \boldsymbol{t}e^{\boldsymbol{\lambda t}}$. This "impure" solution (with an extra $\boldsymbol{t}$) appears in the matrix exponential e^{At}. Example 7 showed how.

1 Find two λ's and $\boldsymbol{x}$'s so that $\boldsymbol{u} = e^{\lambda t}\boldsymbol{x}$ solves $\dfrac{d\boldsymbol{u}}{dt} = \begin{bmatrix} 4 & 3 \\ 0 & 1 \end{bmatrix} \boldsymbol{u}$.

What combination $\boldsymbol{u} = c_1 e^{\lambda_1 t}\boldsymbol{x}_1 + c_2 e^{\lambda_2 t}\boldsymbol{x}_2$ starts from $\boldsymbol{u}(0) = (5, -2)$?

2 Solve Problem 1 for $\boldsymbol{u} = (y, z)$ by back substitution, z before y:

$$\text{Solve } \frac{dz}{dt} = z \text{ from } z(0) = -2. \qquad \text{Then solve } \frac{dy}{dt} = 4y + 3z \text{ from } y(0) = 5.$$

The solution for y will be a combination of e^{4t} and e^t. The λ's are 4 and 1.

3 (a) If every column of A adds to zero, why is $\lambda = 0$ an eigenvalue?

(b) With negative diagonal and positive off-diagonal adding to zero, $\boldsymbol{u}' = A\boldsymbol{u}$ will be a "continuous" Markov equation. Find the eigenvalues and eigenvectors, and the *steady state* as $t \to \infty$

$$\text{Solve } \frac{d\boldsymbol{u}}{dt} = \begin{bmatrix} -2 & 3 \\ 2 & -3 \end{bmatrix} \boldsymbol{u} \text{ with } \boldsymbol{u}(0) = \begin{bmatrix} 4 \\ 1 \end{bmatrix}. \quad \text{What is } \boldsymbol{u}(\infty)?$$

4 A door is opened between rooms that hold $v(0) = 30$ people and $w(0) = 10$ people. The movement between rooms is proportional to the difference $v - w$:

$$\frac{dv}{dt} = w - v \qquad \text{and} \qquad \frac{dw}{dt} = v - w.$$

Show that the total $v + w$ is constant (40 people). Find the matrix in $d\boldsymbol{u}/dt = A\boldsymbol{u}$ and its eigenvalues and eigenvectors. What are v and w at $t = 1$ and $t = \infty$?

5 Reverse the diffusion of people in Problem 4 to $d\boldsymbol{u}/dt = -A\boldsymbol{u}$:

$$\frac{dv}{dt} = v - w \qquad \text{and} \qquad \frac{dw}{dt} = w - v.$$

The total $v+w$ still remains constant. How are the λ's changed now that A is changed to $-A$? But show that $v(t)$ grows to infinity from $v(0) = 30$.

6 $A = \begin{bmatrix} a & 1 \\ 1 & a \end{bmatrix}$ has real eigenvalues but $B = \begin{bmatrix} b & -1 \\ 1 & b \end{bmatrix}$ has complex eigenvalues:

Find the conditions on a and b (real) so that all solutions of $d\boldsymbol{u}/dt = A\boldsymbol{u}$ and $d\boldsymbol{v}/dt = B\boldsymbol{v}$ approach zero as $t \to \infty$: Re $\lambda < 0$ for all eigenvalues.

7 Suppose P is the projection matrix onto the 45° line $y = x$ in $\mathbf{R}^2$. What are its eigenvalues? If $d\boldsymbol{u}/dt = -P\boldsymbol{u}$ (notice minus sign) can you find the limit of $\boldsymbol{u}(t)$ at $t = \infty$ starting from $\boldsymbol{u}(0) = (3, 1)$?

8 The rabbit population shows fast growth (from $6r$) but loss to wolves (from $-2w$). The wolf population always grows in this model ($-w^2$ would control wolves):

$$\frac{dr}{dt} = 6r - 2w \qquad \text{and} \qquad \frac{dw}{dt} = 2r + w.$$

Find the eigenvalues and eigenvectors. If $r(0) = w(0) = 30$ what are the populations at time t? After a long time, what is the ratio of rabbits to wolves?

9 (a) Write $(4, 0)$ as a combination $c_1\boldsymbol{x}_1 + c_2\boldsymbol{x}_2$ of these two eigenvectors of A:

$$\begin{bmatrix} 0 & 1 \\ -1 & 0 \end{bmatrix}\begin{bmatrix} 1 \\ i \end{bmatrix} = i\begin{bmatrix} 1 \\ i \end{bmatrix} \qquad \begin{bmatrix} 0 & 1 \\ -1 & 0 \end{bmatrix}\begin{bmatrix} 1 \\ -i \end{bmatrix} = -i\begin{bmatrix} 1 \\ -i \end{bmatrix}.$$

(b) The solution to $d\boldsymbol{u}/dt = A\boldsymbol{u}$ starting from $(4, 0)$ is $c_1e^{it}\boldsymbol{x}_1 + c_2e^{-it}\boldsymbol{x}_2$. Substitute $e^{it} = \cos t + i\sin t$ and $e^{-it} = \cos t - i\sin t$ to find $\boldsymbol{u}(t)$.

10 Find A to change $y'' = 5y' + 4y$ into a vector equation for $\boldsymbol{u} = (y, y')$:

$$\frac{d\boldsymbol{u}}{dt} = \begin{bmatrix} y' \\ y'' \end{bmatrix} = \begin{bmatrix} & \\ & \end{bmatrix}\begin{bmatrix} y \\ y' \end{bmatrix} = A\boldsymbol{u}.$$

What are the eigenvalues of A? Find them also from $y'' = 5y' + 4y$ with $y = e^{\lambda t}$.

11 The solution to $y'' = 0$ is a straight line $y = C + Dt$. Convert to a matrix equation:

$$\frac{d}{dt}\begin{bmatrix} y \\ y' \end{bmatrix} = \begin{bmatrix} 0 & 1 \\ 0 & 0 \end{bmatrix}\begin{bmatrix} y \\ y' \end{bmatrix} \text{ has the solution } \begin{bmatrix} y \\ y' \end{bmatrix} = e^{At}\begin{bmatrix} y(0) \\ y'(0) \end{bmatrix}.$$

This matrix A has $\lambda = 0, 0$ and it cannot be diagonalized. Find A^2 and compute $e^{At} = I + At + \frac{1}{2}A^2t^2 + \cdots$. Multiply your e^{At} times $\big(y(0), y'(0)\big)$ to check the straight line $y(t) = y(0) + y'(0)t$.

12 Substitute $y = e^{\lambda t}$ into $y'' = 6y' - 9y$ to show that $\lambda = 3$ is a repeated root. This is trouble; we need a second solution after e^{3t}. The matrix equation is

$$\frac{d}{dt}\begin{bmatrix} y \\ y' \end{bmatrix} = \begin{bmatrix} 0 & 1 \\ -9 & 6 \end{bmatrix}\begin{bmatrix} y \\ y' \end{bmatrix}.$$

Show that this matrix has $\lambda = 3, 3$ and only one line of eigenvectors. *Trouble here too.* Check that $y = te^{3t}$ is a second solution to $y'' = 6y' - 9y$.

13 (a) Write down two familiar functions that solve the equation $d^2y/dt^2 = -9y$. Which one starts with $y(0) = 3$ and $y'(0) = 0$?

(b) This second-order equation $y'' = -9y$ produces a vector equation $\boldsymbol{u}' = A\boldsymbol{u}$:

$$\boldsymbol{u} = \begin{bmatrix} y \\ y' \end{bmatrix} \qquad \frac{d\boldsymbol{u}}{dt} = \begin{bmatrix} y' \\ y'' \end{bmatrix} = \begin{bmatrix} 0 & 1 \\ -9 & 0 \end{bmatrix} \begin{bmatrix} y \\ y' \end{bmatrix} = A\boldsymbol{u}.$$

Find $\boldsymbol{u}(t)$ by using the eigenvalues and eigenvectors of A: $\boldsymbol{u}(0) = (3, 0)$.

14 The matrix in this question is skew-symmetric ($A^{\mathrm{T}} = -A$):

$$\frac{d\boldsymbol{u}}{dt} = \begin{bmatrix} 0 & c & -b \\ -c & 0 & a \\ b & -a & 0 \end{bmatrix} \boldsymbol{u} \qquad \text{or} \qquad \begin{aligned} u_1' &= cu_2 - bu_3 \\ u_2' &= au_3 - cu_1 \\ u_3' &= bu_1 - au_2. \end{aligned}$$

(a) The derivative of $\|\boldsymbol{u}(t)\|^2 = u_1^2 + u_2^2 + u_3^2$ is $2u_1u_1' + 2u_2u_2' + 2u_3u_3'$. Substitute u_1', u_2', u_3 to get *zero*. Then $\|\boldsymbol{u}(t)\|^2$ stays equal to $\|\boldsymbol{u}(0)\|^2$.

(b) $A^{\mathrm{T}} = -A$ *makes* $Q = e^{At}$ *orthogonal*. Prove $Q^{\mathrm{T}} = e^{-At}$ from the series for Q.

15 A particular solution to $d\boldsymbol{u}/dt = A\boldsymbol{u} - \boldsymbol{b}$ is $\boldsymbol{u}_p = A^{-1}\boldsymbol{b}$, if A is invertible. The usual solutions to $d\boldsymbol{u}/dt = A\boldsymbol{u}$ give $\boldsymbol{u}_n$. Find the complete solution $\boldsymbol{u} = \boldsymbol{u}_p + \boldsymbol{u}_n$:

(a) $\dfrac{du}{dt} = u - 4$ (b) $\dfrac{d\boldsymbol{u}}{dt} = \begin{bmatrix} 1 & 0 \\ 1 & 1 \end{bmatrix} \boldsymbol{u} - \begin{bmatrix} 4 \\ 6 \end{bmatrix}$.

Questions 16–25 are about the matrix exponential e^{At}.

16 Write five terms of the infinite series for e^{At}. Take the t derivative of each term. Show that you have four terms of Ae^{At}. Conclusion: $e^{At}\boldsymbol{u}_0$ solves $\boldsymbol{u}' = A\boldsymbol{u}$.

17 The matrix $B = \begin{bmatrix} 0 & -4 \\ 0 & 0 \end{bmatrix}$ has $B^2 = 0$. Find e^{Bt} from a (short) infinite series. Check that the derivative of e^{Bt} is Be^{Bt}.

18 Starting from $\boldsymbol{u}(0)$ the solution at time T is $e^{AT}\boldsymbol{u}(0)$. Go an additional time t to reach $e^{At}\, e^{AT}\boldsymbol{u}(0)$. This solution at time $t + T$ can also be written as _____. Conclusion: e^{At} times e^{AT} equals _____.

19 If $A^2 = A$ show that the infinite series produces $e^{At} = I + (e^t - 1)A$.

20 Generally $e^A e^B \neq e^B e^A$. They are both different from e^{A+B}. Check this for

$$A = \begin{bmatrix} 1 & 4 \\ 0 & 0 \end{bmatrix} \qquad B = \begin{bmatrix} 0 & -4 \\ 0 & 0 \end{bmatrix} \qquad A + B = \begin{bmatrix} 1 & 0 \\ 0 & 0 \end{bmatrix}.$$

21 Put $A = \begin{bmatrix} 1 & 3 \\ 0 & 0 \end{bmatrix}$ into the infinite series to find e^{At}. First compute A^2 and A^n :

$$e^{At} = \begin{bmatrix} 1 & 0 \\ 0 & 1 \end{bmatrix} + \begin{bmatrix} t & 3t \\ 0 & 0 \end{bmatrix} + \frac{1}{2}\begin{bmatrix} & \\ & \end{bmatrix} + \cdots = \begin{bmatrix} e^t & \\ 0 & \end{bmatrix}.$$

22 (Recommended) Give two reasons why the matrix exponential e^{At} is never singular:
(a) Write down its inverse. (b) Why are its eigenvalues $e^{\lambda t}$ nonzero?

23 Find a solution $x(t), y(t)$ that gets large as $t \to \infty$. To avoid this instability a scientist exchanged the two equations to get $\lambda < 0$. How can this be?

24 $Y_{n+1} - 2Y_n + Y_{n-1} = -(\Delta t)^2 Y_n$ can be written as a one-step difference equation:

$$\begin{aligned} Y_{n+1} &= Y_n + \Delta t\, Z_n \\ Z_{n+1} &= Z_n - \Delta t\, Y_{n+1} \end{aligned} \qquad \begin{bmatrix} 1 & 0 \\ \Delta t & 1 \end{bmatrix} \begin{bmatrix} Y_{n+1} \\ Z_{n+1} \end{bmatrix} = \begin{bmatrix} 1 & \Delta t \\ 0 & 1 \end{bmatrix} \begin{bmatrix} Y_n \\ Z_n \end{bmatrix}$$

Invert the matrix on the left side to write this as $\boldsymbol{U}_{n+1} = A\boldsymbol{U}_n$. Show that $\det A = 1$. Choose the large time step $\Delta t = 1$ and find the eigenvalues λ_1 and $\lambda_2 = \overline{\lambda}_1$ of A:

$$\boldsymbol{A} = \begin{bmatrix} \mathbf{1} & \mathbf{1} \\ \mathbf{-1} & \mathbf{0} \end{bmatrix} \textbf{ has } |\boldsymbol{\lambda_1}| = |\boldsymbol{\lambda_2}| = \mathbf{1}\textbf{. Show that } \boldsymbol{A^6} = \boldsymbol{I} \text{ so } u_6 = u_0 \textbf{ exactly}.$$

25 That *leapfrog method* in Problem 24 is very successful for small time steps Δt. But find the eigenvalues of A for $\Delta t = \sqrt{2}$ and 2. Any time step $\Delta t > 2$ will lead to $|\lambda| > 1$, and the powers in $\boldsymbol{U}_n = A^n \boldsymbol{U}_0$ will explode.

$$A = \begin{bmatrix} 1 & \sqrt{2} \\ -\sqrt{2} & -1 \end{bmatrix} \quad \text{and} \quad A = \begin{bmatrix} 1 & 2 \\ -2 & -3 \end{bmatrix} \quad \begin{matrix} \textbf{borderline} \\ \textbf{unstable} \end{matrix}$$

26 A very good idea for $y'' = -y$ is the trapezoidal method (half forward/half back). ***This may be the best way to keep (Y_n, Z_n) exactly on a circle.***

$$\textbf{Trapezoidal} \quad \begin{bmatrix} 1 & -\Delta t/2 \\ \Delta t/2 & 1 \end{bmatrix} \begin{bmatrix} Y_{n+1} \\ Z_{n+1} \end{bmatrix} = \begin{bmatrix} 1 & \Delta t/2 \\ -\Delta t/2 & 1 \end{bmatrix} \begin{bmatrix} Y_n \\ Z_n \end{bmatrix}.$$

(a) Invert the left matrix to write this equation as $\boldsymbol{U}_{n+1} = A\boldsymbol{U}_n$. *Show that A is an orthogonal matrix*: $A^{\mathrm{T}}A = I$. **These points $\boldsymbol{U}_n$ never leave the circle.** $A = (I - B)^{-1}(I + B)$ is always an orthogonal matrix if $B^{\mathrm{T}} = -B$.

(b) (Optional MATLAB) Take 32 steps from $\boldsymbol{U}_0 = (1, 0)$ to $\boldsymbol{U}_{32}$ with $\Delta t = 2\pi/32$. Is $\boldsymbol{U}_{32} = \boldsymbol{U}_0$? I think there is a small error.

27 Explain one of these three proofs that the square of e^A is e^{2A}.

1. Solving with e^A from $t = 0$ to 1 and then 1 to 2 agrees with e^{2A} from 0 to 2.
2. The squared series $(I + A + \frac{A^2}{2} + \cdots)^2$ matches $I + 2A + \frac{(2A)^2}{2} + \cdots = e^{2A}$.
3. If A can be diagonalized then $(Xe^{\Lambda}X^{-1})(Xe^{\Lambda}X^{-1}) = Xe^{2\Lambda}X^{-1}$.

Notes on a Differential Equations Course

Constant coefficient linear equations are the simplest to solve. This Section 6.4 shows you part of a differential equations course, but there is more:

1. The second order equation $mu'' + bu' + ku = 0$ has major importance in applications. The exponents λ in the solutions $u = e^{\lambda t}$ solve $m\lambda^2 + b\lambda + k = 0$.

   Underdamping $b^2 < 4mk$ **Critical damping** $b^2 = 4mk$ **Overdamping** $b^2 > 4mk$

   This decides whether λ_1 and λ_2 are real roots or repeated roots or complex roots. With complex $\lambda = a + i\omega$ the solution $u(t)$ oscillates from $e^{i\omega t}$ as it decays from e^{at}.

2. Our equations had no forcing term $f(t)$. We were finding the "nullspace solution". To $\boldsymbol{u_n(t)}$ we need to add a particular solution $\boldsymbol{u_p(t)}$ that balances the force $f(t)$. This solution can also be discovered and studied by *Laplace transform*:

   Input $\boldsymbol{f(s)}$ at time $\boldsymbol{s}$
   Growth factor $\boldsymbol{e^{A(t-s)}}$
   Add up outputs at time $\boldsymbol{t}$

   $$u_{\textbf{particular}} = \int_0^t e^{A(t-s)} f(s)\, ds.$$

In real applications, nonlinear differential equations are solved numerically. A method with good accuracy is "Runge-Kutta". The constant solutions to $du/dt = f(u)$ are $u(t) = Y$ with $f(Y) = 0$ and $du/dt = 0$: *no movement*. Far from Y, the computer takes over.

This basic course is the subject of my textbook (a companion to this one) on **Differential Equations and Linear Algebra**: **math.mit.edu / dela**.

The individual sections of the book are described in a series of short videos, and a parallel series about numerical solutions was prepared by Cleve Moler:
ocw.mit.edu/resources/res-18-009-learn-differential-equations-up-close-with-gilbert-strang-and-cleve-moler-fall-2015/
www.mathworks.com/academia/courseware/learn-differential-equations.html

7 The Singular Value Decomposition (SVD)

This chapter develops one idea. That idea applies to every matrix, square or rectangular. It is an extension of eigenvectors, and now we need *two sets of orthogonal vectors*: input vectors $\boldsymbol{v}_1$ to $\boldsymbol{v}_n$ and output vectors $\boldsymbol{u}_1$ to $\boldsymbol{u}_m$. This is completely natural for an m by n matrix. The vectors $\boldsymbol{v}_1$ to $\boldsymbol{v}_r$ are in the row space and $\boldsymbol{u}_1$ to $\boldsymbol{u}_r$ are in the column space. Then we recover A from r pieces of rank one, with $r = \text{rank}(A)$ and positive singular values $\sigma_1 \geq \sigma_2 \geq \cdots \geq \sigma_r > 0$ in the diagonal matrix Σ.

$$\textbf{SVD} \qquad A = \boldsymbol{U\Sigma V}^{\mathbf{T}} = \sigma_1 \boldsymbol{u}_1 \boldsymbol{v}_1^{\mathrm{T}} + \sigma_2 \boldsymbol{u}_2 \boldsymbol{v}_2^{\mathrm{T}} + \cdots + \sigma_r \boldsymbol{u}_r \boldsymbol{v}_r^{\mathrm{T}}.$$

The right singular vectors $\boldsymbol{v}_i$ are eigenvectors of $A^{\mathrm{T}}A$. They give bases for the row space and nullspace of A. The left singular vectors $\boldsymbol{u}_i$ are eigenvectors of AA^{T}. They give bases for the column space and left nullspace of A. **Then $\boldsymbol{A v_i}$ equals $\boldsymbol{\sigma_i u_i}$ for $\boldsymbol{i \leq r}$. The matrix $\boldsymbol{A}$ is diagonalized by these two bases: $\boldsymbol{AV} = \boldsymbol{U\Sigma}$.**

Each $\boldsymbol{u}_i = \boldsymbol{v}_i$ when A is a symmetric positive definite matrix. Those singular vectors will be the eigenvectors $\boldsymbol{q}_i$. And the singular values σ_i become the eigenvalues of A. If A is not square or not symmetric, then $A = U\Sigma V^{\mathrm{T}}$ replaces $S = Q\Lambda Q^{\mathrm{T}}$.

The SVD is a valuable way to understand a matrix of data. In that case AA^{T} is the **sample covariance matrix**, after centering the data and dividing by $n-1$ (Section 7.3). Its eigenvalues are σ_1^2 to σ_r^2. Its eigenvectors are the $\boldsymbol{u}$'s in the SVD. Principal Component Analysis (**PCA**) is totally based on the singular vectors of the data matrix A.

The SVD allows wonderful projects, by separating a photograph = matrix of pixels into its rank-one components. Each time you include one more piece $\sigma_i \boldsymbol{u}_i \boldsymbol{v}_i^{\mathrm{T}}$, the picture becomes clearer. Section 7.2 shows examples and a link to an excellent website. Section 7.3 describes PCA and its connection to the covariance matrix in statistics.

Section 7.4 shows how it all develops from and depends on one idea: *orthogonality*.

7.1 Singular Values and Singular Vectors

1 The Singular Value Decomposition of any matrix is $\boldsymbol{A} = \boldsymbol{U\Sigma V}^{\mathrm{T}}$ or $\boldsymbol{AV} = \boldsymbol{U\Sigma}$.

2 Singular vectors in $A\boldsymbol{v}_i = \sigma_i \boldsymbol{u}_i$ are orthonormal: $\boldsymbol{V}^{\mathrm{T}}\boldsymbol{V} = I$ and $\boldsymbol{U}^{\mathrm{T}}\boldsymbol{U} = I$.

3 The diagonal matrix $\boldsymbol{\Sigma}$ contains the singular values $\sigma_1 \geq \sigma_2 \geq \cdots \geq \sigma_r > 0$.

4 The squares of those singular values are eigenvalues of $A^{\mathrm{T}}A$ and AA^{T}.

The eigenvectors of symmetric matrices are orthogonal. We want to go beyond symmetric matrices to all matrices—without giving up orthogonality. To make this possible for every m by n matrix (even if $m \neq n$) we need one set of orthonormal vectors $\boldsymbol{v}_1, \ldots, \boldsymbol{v}_n$ in $\mathbf{R}^n$ and a second set $\boldsymbol{u}_1, \ldots, \boldsymbol{u}_m$ in $\mathbf{R}^m$. **Instead of $\boldsymbol{Sx} = \boldsymbol{\lambda x}$ we want $\boldsymbol{Av} = \boldsymbol{\sigma u}$.**

Here is a 2 by 2 unsymmetric example with orthogonal inputs and orthogonal outputs:

$$A\boldsymbol{v}_1 = \begin{bmatrix} 3 & 0 \\ 4 & 5 \end{bmatrix}\begin{bmatrix} 1 \\ 1 \end{bmatrix} = \begin{bmatrix} 3 \\ 9 \end{bmatrix} \quad \text{and} \quad A\boldsymbol{v}_2 = \begin{bmatrix} 3 & 0 \\ 4 & 5 \end{bmatrix}\begin{bmatrix} -1 \\ 1 \end{bmatrix} = \begin{bmatrix} -3 \\ 1 \end{bmatrix}. \tag{1}$$

$(1,1)$ is orthogonal to $(-1,1)$, and $(3,9)$ is orthogonal to $(-3,1)$. Those are not unit vectors but that is easily fixed. The inputs $(1,1)$ and $(-1,1)$ need to be divided by $\sqrt{2}$. The outputs need to be divided by $\sqrt{10}$. That leaves the singular values $3\sqrt{5}$ and $\sqrt{5}$:

$$\boldsymbol{Av} = \boldsymbol{\sigma u} \qquad \begin{bmatrix} 3 & 0 \\ 4 & 5 \end{bmatrix}\boldsymbol{v}_1 = \mathbf{3}\sqrt{\mathbf{5}}\,\boldsymbol{u}_1 \quad \text{and} \quad \begin{bmatrix} 3 & 0 \\ 4 & 5 \end{bmatrix}\boldsymbol{v}_2 = \sqrt{\mathbf{5}}\,\boldsymbol{u}_2. \tag{2}$$

Multiply the singular values $\sigma_1 = 3\sqrt{5}$ and $\sigma_2 = \sqrt{5}$ to get $\sigma_1\sigma_2 = \det A = 15$. We can move from vector formulas to a matrix formula. $SQ = Q\Lambda$ becomes $\boldsymbol{AV} = \boldsymbol{U\Sigma}$.

$$\boldsymbol{AV} = \begin{bmatrix} 3 & 0 \\ 4 & 5 \end{bmatrix}\begin{bmatrix} 1 & -1 \\ 1 & 1 \end{bmatrix}\frac{1}{\sqrt{2}} = \begin{bmatrix} 1 & -3 \\ 3 & 1 \end{bmatrix}\frac{1}{\sqrt{10}}\begin{bmatrix} \mathbf{3}\sqrt{\mathbf{5}} & 0 \\ 0 & \sqrt{\mathbf{5}} \end{bmatrix} = \boldsymbol{U\Sigma}. \tag{3}$$

V and U are orthogonal matrices. So if we multiply equation (3) by V^{T}, A will be alone:

$$\boxed{\boldsymbol{AV} = \boldsymbol{U\Sigma} \quad \text{becomes} \quad \boldsymbol{A} = \boldsymbol{U\Sigma V}^{\mathrm{T}} = \sigma_1\boldsymbol{u}_1\boldsymbol{v}_1^{\mathrm{T}} + \sigma_2\boldsymbol{u}_2\boldsymbol{v}_2^{\mathrm{T}}} \tag{4}$$

This says everything except how to find $\boldsymbol{V}$ and $\boldsymbol{U}$ and $\boldsymbol{\Sigma}$, and what they mean.

When equation (4) multiplies $\boldsymbol{v}_i$, orthogonality produces $A\boldsymbol{v}_i = \sigma_i\boldsymbol{u}_i$. Key point: *Every matrix A is diagonalized by* two *sets of singular vectors*, not one set of eigenvectors.

In the 2 by 2 example, the first piece is more important than the second piece because $\sigma_1 = 3\sqrt{5}$ is greater than $\sigma_2 = \sqrt{5}$. To recover A, add the pieces $\sigma_1\boldsymbol{u}_1\boldsymbol{v}_1^{\mathrm{T}} + \sigma_2\boldsymbol{u}_2\boldsymbol{v}_2^{\mathrm{T}}$:

$$\frac{\mathbf{3}\sqrt{\mathbf{5}}}{\sqrt{10}\sqrt{2}}\begin{bmatrix} \mathbf{1} \\ \mathbf{3} \end{bmatrix}\begin{bmatrix} \mathbf{1} & \mathbf{1} \end{bmatrix} + \frac{\sqrt{\mathbf{5}}}{\sqrt{10}\sqrt{2}}\begin{bmatrix} \mathbf{-3} \\ \mathbf{1} \end{bmatrix}\begin{bmatrix} \mathbf{-1} & \mathbf{1} \end{bmatrix} = \frac{3}{2}\begin{bmatrix} \mathbf{1} & \mathbf{1} \\ \mathbf{3} & \mathbf{3} \end{bmatrix} + \frac{1}{2}\begin{bmatrix} \mathbf{3} & \mathbf{-3} \\ \mathbf{-1} & \mathbf{1} \end{bmatrix} = \begin{bmatrix} \mathbf{3} & \mathbf{0} \\ \mathbf{4} & \mathbf{5} \end{bmatrix}$$

The Reduced Form of the SVD

That full form $AV = U\Sigma$ can have a lot of zeros in Σ when the rank of A is small and the nullspace of A is large. Those zeros contribute nothing to matrix multiplication. The heart of the SVD is in the first r $\boldsymbol{v}$'s and $\boldsymbol{u}$'s and σ's. We can change $AV = U\Sigma$ to $\boldsymbol{AV_r = U_r\Sigma_r}$ by removing the parts that are sure to produce zeros. This leaves the reduced SVD where $\boldsymbol{\Sigma_r}$ **is now square:** $(\boldsymbol{m \times n})\,(\boldsymbol{n \times r}) = (\boldsymbol{m \times r})\,(\boldsymbol{r \times r})$.

Reduced SVD
$\boldsymbol{AV_r = U_r\Sigma_r}$
$\boldsymbol{Av_i = \sigma_i u_i}$

$$\boldsymbol{A}\begin{bmatrix} \boldsymbol{v}_1 & .. & \boldsymbol{v}_r \\ & \text{row space} & \end{bmatrix} = \begin{bmatrix} \boldsymbol{u}_1 & .. & \boldsymbol{u}_r \\ & \text{column space} & \end{bmatrix}\begin{bmatrix} \sigma_1 & & \\ & \ddots & \\ & & \sigma_r \end{bmatrix} \quad (5)$$

We still have $V_r^{\mathrm{T}} V_r = I_r$ and $U_r^{\mathrm{T}} U_r = I_r$ from those orthogonal unit vectors $\boldsymbol{v}$'s and $\boldsymbol{u}$'s. When V_r and U_r are not square, we can't have full inverses: $V_r V_r^{\mathrm{T}} \neq I$ and $U_r U_r^{\mathrm{T}} \neq I$. **But** $\boldsymbol{A = U_r\Sigma_r V_r^{\mathrm{T}}}$ **is true**. The other multiplications in $A = U\Sigma V^{\mathrm{T}}$ give only zeros.

Example 2 $\quad \boldsymbol{A} = \begin{bmatrix} \mathbf{1} & \mathbf{2} & \mathbf{2} \end{bmatrix} = \begin{bmatrix} \mathbf{1} \end{bmatrix}\begin{bmatrix} \mathbf{3} \end{bmatrix}\begin{bmatrix} \mathbf{1} & \mathbf{2} & \mathbf{2} \end{bmatrix}/3 = \boldsymbol{U_r\Sigma_r V_r}^{\mathrm{T}}$ has $r = 1$ and $\sigma_1 = 3$.

The rest of $U\Sigma V^{\mathrm{T}}$ contributes nothing to A, because of all the zeros in Σ. The key separation of A into $\sigma_1\boldsymbol{u}_1\boldsymbol{v}_1^{\mathrm{T}} + \cdots + \sigma_r\boldsymbol{u}_r\boldsymbol{v}_r^{\mathrm{T}}$ stops at $\sigma_1\boldsymbol{u}_1\boldsymbol{v}_1^{\mathrm{T}}$ because the rank is $r = 1$.

The Important Fact for Data Science

Why is the SVD so important for this subject and this book? Like the other factorizations $A = LU$ and $A = QR$ and $S = Q\Lambda Q^{\mathrm{T}}$, it separates the matrix into rank one pieces. A special property of the SVD is that **those pieces come in order of importance**. The first piece $\sigma_1\boldsymbol{u}_1\boldsymbol{v}_1^{\mathrm{T}}$ when $\sigma_1 > \sigma_2$ is the closest rank one matrix to A. More is true: **For every** $\boldsymbol{k}$**, the sum of the first** $\boldsymbol{k}$ **pieces is the rank** $\boldsymbol{k}$ **matrix that is closest to** $\boldsymbol{A}$.

$$\boldsymbol{A_k = \sigma_1 u_1 v_1^{\mathrm{T}} + \cdots + \sigma_k u_k v_k^{\mathrm{T}}} \quad \textbf{is the best rank } \boldsymbol{k} \textbf{ approximation to } \boldsymbol{A}$$

"Eckart-Young" $\quad$ **If** $\boldsymbol{B}$ **has rank** $\boldsymbol{k}$ **then** $\boldsymbol{||A - A_k|| \leq ||A - B||}$. $\quad$ (6)

To interpret that statement you need to know the meaning of the symbol $\boldsymbol{||A - B||}$. This is the "**norm**" of the matrix $A - B$, a measure of its size (like the absolute value of a number). The norm could be σ_1 or $\sigma_1 + \cdots + \sigma_r$ or the square root of $\sigma_1^2 + \cdots + \sigma_r^2$.

Our first job is to discover how U and Σ and V^{T} can be computed. For a small matrix they come from eigenvectors and eigenvalues of $A^{\mathrm{T}}A$ and AA^{T}.

For a large matrix, multiplying A by A^{T} is not wise. Two steps are much better: Reduce A to two nonzero diagonals and modify the QR algorithm that finds eigenvalues.

First Proof of the SVD

The goal is $A = U\Sigma V^{\mathrm{T}}$. We want to identify the two sets of singular vectors, the $\boldsymbol{u}$'s and the $\boldsymbol{v}$'s. One way to find those vectors is to form the symmetric matrices $A^{\mathrm{T}}A$ and AA^{T}:

$$\boldsymbol{A}^{\mathrm{T}}\boldsymbol{A} = (V\Sigma^{\mathrm{T}}U^{\mathrm{T}})\ (U\Sigma V^{\mathrm{T}}) = \boldsymbol{V\Sigma^{\mathrm{T}}\Sigma V^{\mathrm{T}}} \tag{7}$$

$$\boldsymbol{A}\boldsymbol{A}^{\mathrm{T}} = (U\Sigma V^{\mathrm{T}})\ (V\Sigma^{\mathrm{T}}U^{\mathrm{T}}) = \boldsymbol{U\Sigma\Sigma^{\mathrm{T}}U^{\mathrm{T}}} \tag{8}$$

Both (7) and (8) produced symmetric matrices. Usually $A^{\mathrm{T}}A$ and AA^{T} are different. Both right hand sides have the special form $Q\Lambda Q^{\mathrm{T}}$. **Eigenvectors are in $\boldsymbol{Q = V}$ or $\boldsymbol{Q = U}$.** So we know from (7) and (8) how V and U and Σ connect to those symmetric matrices $A^{\mathrm{T}}A$ and AA^{T}.

> **V contains orthonormal eigenvectors of $A^{\mathrm{T}}A$**
>
> **U contains orthonormal eigenvectors of AA^{T}**
>
> **σ_1^2 to σ_r^2 are the nonzero eigenvalues of both $A^{\mathrm{T}}A$ and AA^{T}**

We are not quite finished, for this reason. **The SVD requires that $\boldsymbol{Av_k = \sigma_k u_k}$.** It connects each right singular vector $\boldsymbol{v}_k$ to a left singular vector $\boldsymbol{u}_k$, for $k = 1, \ldots, r$. When I choose the $\boldsymbol{v}$'s, that choice will decide the signs of the $\boldsymbol{u}$'s. If $S\boldsymbol{u} = \lambda\boldsymbol{u}$ then also $S(-\boldsymbol{u}) = \lambda(-\boldsymbol{u})$ and I have to know the sign to choose. More than that, there is a whole plane of eigenvectors when λ is a double eigenvalue. When I choose two $\boldsymbol{v}$'s in that plane, then $A\boldsymbol{v} = \sigma\boldsymbol{u}$ will tell me both $\boldsymbol{u}$'s. This is in equation (9).

The plan is to start with the $\boldsymbol{v}$'s. **Choose orthonormal eigenvectors** $\boldsymbol{v}_1, \ldots, \boldsymbol{v}_r$ of $A^{\mathrm{T}}A$. **Then choose $\boldsymbol{\sigma_k = \sqrt{\lambda_k}}$.** To determine the $\boldsymbol{u}$'s we require that $\boldsymbol{Av = \sigma u}$:

$$\boldsymbol{v}\text{'s then }\boldsymbol{u}\text{'s} \qquad \boldsymbol{A}^{\mathrm{T}}\boldsymbol{A}\boldsymbol{v}_k = \boldsymbol{\sigma_k^2}\boldsymbol{v}_k \quad \text{and then} \quad \boldsymbol{u}_k = \frac{\boldsymbol{A}\boldsymbol{v}_k}{\boldsymbol{\sigma_k}} \quad \text{for} \quad k = 1, \ldots, r \tag{9}$$

This produces the SVD ! Let me check that those vectors $\boldsymbol{u}_1$ to $\boldsymbol{u}_r$ are eigenvectors of AA^{T}.

$$\boldsymbol{AA^{\mathrm{T}}}\boldsymbol{u}_k = AA^{\mathrm{T}}\left(\frac{A\boldsymbol{v}_k}{\sigma_k}\right) = A\left(\frac{A^{\mathrm{T}}A\,\boldsymbol{v}_k}{\sigma_k}\right) = A\,\frac{\sigma_k^2\,\boldsymbol{v}_k}{\sigma_k} = \boldsymbol{\sigma}_k^2\,\boldsymbol{u}_k \tag{10}$$

The $\boldsymbol{v}$'s were chosen to be orthonormal. I must check that the $\boldsymbol{u}$'s are also orthonormal:

$$\boldsymbol{u}_j^{\mathrm{T}}\boldsymbol{u}_k = \left(\frac{A\boldsymbol{v}_j}{\sigma_j}\right)^{\mathrm{T}}\left(\frac{A\boldsymbol{v}_k}{\sigma_k}\right) = \frac{\boldsymbol{v}_j^{\mathrm{T}}(A^{\mathrm{T}}A\boldsymbol{v}_k)}{\sigma_j\,\sigma_k} = \frac{\sigma_k}{\sigma_j}\,\boldsymbol{v}_j^{\mathrm{T}}\,\boldsymbol{v}_k = \begin{cases} 1 & \text{if } j = k \\ 0 & \text{if } j \neq k \end{cases} \tag{11}$$

Notice that $(AA^{\mathrm{T}})A = A(A^{\mathrm{T}}A)$ was the key to equation (10). The law $(AB)C = A(BC)$ is the key to a great many proofs in linear algebra. Moving the parentheses is a powerful idea. This is the *associative law*.

Finally we have to choose the last $n-r$ vectors $\boldsymbol{v}_{r+1}$ to $\boldsymbol{v}_n$ and the last $m-r$ vectors $\boldsymbol{u}_{r+1}$ to $\boldsymbol{u}_m$. This is easy. **These $\boldsymbol{v}$'s and $\boldsymbol{u}$'s are in the nullspaces of A and A^{T}.** We can choose any orthonormal bases for those nullspaces. They will automatically be orthogonal to the first $\boldsymbol{v}$'s in the row space of A and the first $\boldsymbol{u}$'s in the column space. This is because the whole spaces are orthogonal: $\mathbf{N}(A) \perp \mathbf{C}(A^{\mathrm{T}})$ and $\mathbf{N}(A^{\mathrm{T}}) \perp \mathbf{C}(A)$. *The proof of the SVD is complete* by that Fundamental Theorem of Linear Algebra.

Now we have U and V and Σ in the full size SVD of equation (1): m $\boldsymbol{u}$'s, n $\boldsymbol{v}$'s. You may have noticed that the eigenvalues of $A^{\mathrm{T}}A$ are in $\Sigma^{\mathrm{T}}\Sigma$, and *the same numbers σ_1^2 to σ_r^2 are also eigenvalues of AA^{T} in $\Sigma\Sigma^{\mathrm{T}}$*. An amazing fact: **$BA$ always has the same nonzero eigenvalues as AB.** If B is invertible, $BA = B(AB)B^{-1}$ is similar to AB.

AB and BA: Equal Nonzero Eigenvalues

If A is m by n and B is n by m, AB and BA have the same nonzero eigenvalues.

Start with $AB\boldsymbol{x} = \lambda\boldsymbol{x}$ and $\lambda \neq 0$. Multiply both sides by B, to get $BAB\boldsymbol{x} = \lambda B\boldsymbol{x}$. This says that $B\boldsymbol{x}$ is an eigenvector of BA with the same eigenvalue λ—exactly what we wanted. We needed $\lambda \neq 0$ to be sure that $B\boldsymbol{x}$ is truly a nonzero eigenvector.

Notice that if B is square and invertible, then $B^{-1}(BA)B = AB$. This says that BA is *similar* to AB: *same eigenvalues*. But our first proof allows A and B to be m by n and n by m. This covers the important example of the SVD when $B = A^{\mathrm{T}}$. In that case $A^{\mathrm{T}}A$ and AA^{T} both lead to the r nonzero singular values of A.

If m is larger than n, then AB has $m-n$ extra zero eigenvalues compared to BA.

Example 1 (completed) Find the matrices U and Σ and V for $A = \begin{bmatrix} \mathbf{3} & \mathbf{0} \\ \mathbf{4} & \mathbf{5} \end{bmatrix}$.

With rank 2, this A has two positive singular values σ_1 and σ_2. We will see that σ_1 is larger than $\lambda_{\max} = 5$, and σ_2 is smaller than $\lambda_{\min} = 3$. Begin with $A^{\mathrm{T}}A$ and AA^{T}:

$$A^{\mathrm{T}}A = \begin{bmatrix} 25 & 20 \\ 20 & 25 \end{bmatrix} \qquad AA^{\mathrm{T}} = \begin{bmatrix} 9 & 12 \\ 12 & 41 \end{bmatrix}$$

Those have the same trace (50) and the same eigenvalues $\sigma_1^2 = 45$ and $\sigma_2^2 = 5$. The square roots are $\boldsymbol{\sigma_1} = \sqrt{\mathbf{45}}$ and $\boldsymbol{\sigma_2} = \sqrt{\mathbf{5}}$. Then $\sigma_1\sigma_2 = 15$ and this is the determinant of A.

The key step for V is to find the eigenvectors of $A^{\mathrm{T}}A$ (with eigenvalues 45 and 5):

$$\begin{bmatrix} 25 & 20 \\ 20 & 25 \end{bmatrix}\begin{bmatrix} 1 \\ 1 \end{bmatrix} = \mathbf{45}\begin{bmatrix} \mathbf{1} \\ \mathbf{1} \end{bmatrix} \qquad \begin{bmatrix} 25 & 20 \\ 20 & 25 \end{bmatrix}\begin{bmatrix} -1 \\ 1 \end{bmatrix} = \mathbf{5}\begin{bmatrix} \mathbf{-1} \\ \mathbf{1} \end{bmatrix}$$

Then $\boldsymbol{v}_1$ and $\boldsymbol{v}_2$ are those orthogonal eigenvectors rescaled to length 1. Divide by $\sqrt{2}$.

Right singular vectors $\boldsymbol{v}_1 = \dfrac{1}{\sqrt{2}}\begin{bmatrix} 1 \\ 1 \end{bmatrix}$ $\quad \boldsymbol{v}_2 = \dfrac{1}{\sqrt{2}}\begin{bmatrix} -1 \\ 1 \end{bmatrix}$ **Left singular vectors** $\boldsymbol{u}_i = \dfrac{A\boldsymbol{v}_i}{\sigma_i}$

Now compute Av_1 and Av_2 which must equal $\sigma_1 u_1 = \sqrt{45}\, u_1$ and $\sigma_2 u_2 = \sqrt{5}\, u_2$:

$$Av_1 = \frac{3}{\sqrt{2}}\begin{bmatrix}1\\3\end{bmatrix} = \sqrt{45}\frac{\mathbf{1}}{\sqrt{\mathbf{10}}}\begin{bmatrix}\mathbf{1}\\\mathbf{3}\end{bmatrix} = \sigma_1\, u_1$$

$$Av_2 = \frac{1}{\sqrt{2}}\begin{bmatrix}-3\\1\end{bmatrix} = \sqrt{5}\,\frac{\mathbf{1}}{\sqrt{\mathbf{10}}}\begin{bmatrix}\mathbf{-3}\\\mathbf{1}\end{bmatrix} = \sigma_2\, u_2$$

The division by $\sqrt{10}$ makes u_1 and u_2 orthonormal. Then $\sigma_1 = \sqrt{45}$ and $\sigma_2 = \sqrt{5}$ as expected. The Singular Value Decomposition of A is U times Σ times V^{T}.

$$\boxed{U = \frac{1}{\sqrt{10}}\begin{bmatrix}1 & -3\\3 & 1\end{bmatrix} \quad \Sigma = \begin{bmatrix}\sqrt{45} & \\ & \sqrt{5}\end{bmatrix} \quad V = \frac{1}{\sqrt{2}}\begin{bmatrix}1 & -1\\1 & 1\end{bmatrix}} \qquad (12)$$

U and V contain orthonormal bases for the column space and the row space of A (both spaces are just $\mathbf{R}^2$). The real achievement is that those two bases diagonalize A: AV equals $U\Sigma$. The matrix $A = U\Sigma V^{\mathrm{T}}$ splits into two rank-one matrices, columns times rows, with $\sqrt{2}\,\sqrt{10} = \sqrt{20}$. *Their sum is* A, with $\sqrt{5}/\sqrt{20} = \frac{1}{2}$.

$$\sigma_1 u_1 v_1^{\mathrm{T}} + \sigma_2 u_2 v_2^{\mathrm{T}} = \frac{\sqrt{\mathbf{45}}}{\sqrt{\mathbf{20}}}\begin{bmatrix}\mathbf{1} & \mathbf{1}\\\mathbf{3} & \mathbf{3}\end{bmatrix} + \frac{\sqrt{\mathbf{5}}}{\sqrt{\mathbf{20}}}\begin{bmatrix}\mathbf{3} & \mathbf{-3}\\\mathbf{-1} & \mathbf{1}\end{bmatrix} = \begin{bmatrix}\mathbf{3} & \mathbf{0}\\\mathbf{4} & \mathbf{5}\end{bmatrix} = A.$$

Every matrix is a sum of rank one matrices with orthogonal v's and orthogonal u's. Orthogonal rows $\begin{bmatrix}\mathbf{1} & \mathbf{1}\end{bmatrix}$ and $\begin{bmatrix}\mathbf{3} & \mathbf{-3}\end{bmatrix}$, orthogonal columns $(\mathbf{1},\mathbf{3})$ and $(\mathbf{3},\mathbf{-1})$.

To say again: **Good codes do not start with $A^{\mathrm{T}}A$ and AA^{T}.** Instead we produce zeros in A by rotations that leave only two diagonals (and don't affect the singular values). The last page of this section describes a successful way to compute the SVD.

Question: If $S = Q\Lambda Q^{\mathrm{T}}$ is symmetric positive definite, what is its SVD?

Answer: The SVD is exactly $U\Sigma V^{\mathrm{T}} = Q\Lambda Q^{\mathrm{T}}$. The matrix $U = V = Q$ is orthogonal. And the eigenvalue matrix Λ becomes the singular value matrix Σ.

Question: If $S = Q\Lambda Q^{\mathrm{T}}$ has a negative eigenvalue ($Sx = -\alpha\, x$), what is the singular value and what are the vectors v and u?

Answer: The singular value will be $\sigma = +\alpha$ (positive). One singular vector (either u or v) must be $-x$ (reverse the sign). Then $Sx = -\alpha\, x$ is the same as $Sv = \sigma u$. The two sign changes cancel.

Question: If $A = Q$ is an orthogonal matrix, why does every singular value equal 1?

Answer: All singular values are $\sigma = 1$ because $A^{\mathrm{T}}A = Q^{\mathrm{T}}Q = I$. **Then $\Sigma = I$.** But $U = Q$ and $V = I$ is only one of many choices for the singular vectors u and v:

$$Q = U\Sigma V^{\mathrm{T}} \text{ can be } Q = QII^{\mathrm{T}} \text{ or any } Q = (QQ_1)IQ_1^{\mathrm{T}}.$$

Question : Why are all eigenvalues of a square matrix A less than or equal to σ_1 ?

Answer : Multiplying by orthogonal matrices U and V^{T} does not change vector lengths :

$$||A\boldsymbol{x}|| = ||U\Sigma V^{\mathrm{T}}\boldsymbol{x}|| = ||\Sigma V^{\mathrm{T}}\boldsymbol{x}|| \leq \sigma_1||V^{\mathrm{T}}\boldsymbol{x}|| = \boldsymbol{\sigma_1}||\boldsymbol{x}|| \text{ for all } x. \qquad (13)$$

An eigenvector has $||A\boldsymbol{x}|| = |\lambda|\,||\boldsymbol{x}||$. Then (13) gives $|\lambda|\,||\boldsymbol{x}|| \leq \sigma_1\,||\boldsymbol{x}||$ and $|\boldsymbol{\lambda}| \leq \boldsymbol{\sigma_1}$.

Question : If $A = \boldsymbol{x}\boldsymbol{y}^{\mathrm{T}}$ has rank 1, what are $\boldsymbol{u}_1$ and $\boldsymbol{v}_1$ and σ_1 ? **Check that $|\boldsymbol{\lambda_1}| \leq \boldsymbol{\sigma_1}$.**

Answer : The singular vectors $\boldsymbol{u}_1 = \boldsymbol{x}/||\boldsymbol{x}||$ and $\boldsymbol{v}_1 = \boldsymbol{y}/||\boldsymbol{y}||$ have length 1. Then $\boldsymbol{\sigma_1} = ||\boldsymbol{x}||\,||\boldsymbol{y}||$ is the only nonzero number in the singular value matrix Σ. Here is the SVD :

$$\textbf{Rank 1 matrix} \qquad \boldsymbol{x}\boldsymbol{y}^{\mathrm{T}} = \frac{\boldsymbol{x}}{||\boldsymbol{x}||}\left(||\boldsymbol{x}||\,||\boldsymbol{y}||\right)\frac{\boldsymbol{y}^{\mathrm{T}}}{||\boldsymbol{y}||} = \boldsymbol{u}_1\sigma_1\boldsymbol{v}_1^{\mathrm{T}}.$$

Observation The only nonzero eigenvalue of $A = \boldsymbol{x}\boldsymbol{y}^{\mathrm{T}}$ is $\lambda_1 = \boldsymbol{y}^{\mathrm{T}}\boldsymbol{x}$. The eigenvector is $\boldsymbol{x}$ because $(\boldsymbol{x}\boldsymbol{y}^{\mathrm{T}})\,\boldsymbol{x} = \boldsymbol{x}\,(\boldsymbol{y}^{\mathrm{T}}\boldsymbol{x}) = \lambda_1\boldsymbol{x}$. Then the key inequality $|\lambda_1| \leq \sigma_1$ becomes exactly the Schwarz inequality $|\boldsymbol{y}^{\mathrm{T}}\boldsymbol{x}| \leq ||\boldsymbol{x}||\,||\boldsymbol{y}||$.

The Geometry of the SVD

The SVD separates a matrix into $A = U\Sigma V^{\mathrm{T}}$: (**orthogonal**) $\times$ (**diagonal**) $\times$ (**orthogonal**). In two dimensions we can draw those steps. The orthogonal matrices U and V rotate the plane. The diagonal matrix Σ stretches it along the axes. Figure 7.1 shows **rotation** times **stretching** times **rotation**. **Vectors x on the unit circle go to Ax on an ellipse.**

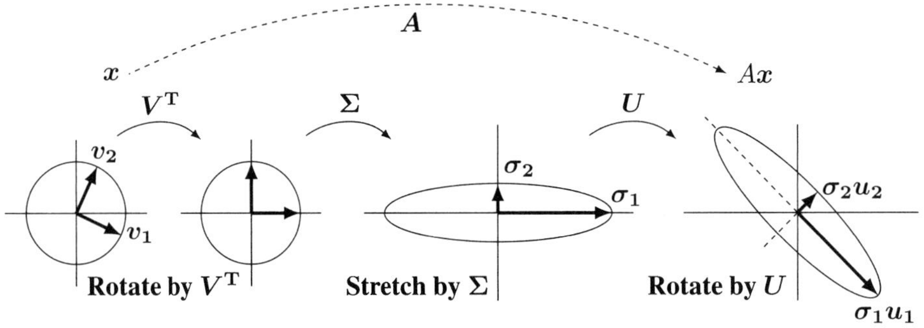

Figure 7.1: U and V are rotations and possible reflections. Σ stretches circle to ellipse.

This picture applies to a 2 by 2 invertible matrix (because $\sigma_1 > 0$ and $\sigma_2 > 0$). First is a rotation of any $\boldsymbol{x}$ to $V^{\mathrm{T}}\boldsymbol{x}$. Then Σ stretches that vector to $\Sigma V^{\mathrm{T}}\boldsymbol{x}$. Then U rotates to $A\boldsymbol{x} = U\Sigma V^{\mathrm{T}}\boldsymbol{x}$. We kept all determinants positive to avoid reflections. The four numbers a, b, c, d in the matrix connect to *two rotation angles* θ, ϕ and *two numbers* σ_1, σ_2 *in* Σ.

$$A = \begin{bmatrix} a & b \\ c & d \end{bmatrix} = \begin{bmatrix} \cos\theta & -\sin\theta \\ \sin\theta & \cos\theta \end{bmatrix} \begin{bmatrix} \sigma_1 & \\ & \sigma_2 \end{bmatrix} \begin{bmatrix} \cos\phi & \sin\phi \\ -\sin\phi & \cos\phi \end{bmatrix}. \qquad (14)$$

Question. If the matrix is symmetric then $b = c$. Now A has only 3 (not 4) parameters. *How do the* 4 *numbers* θ, ϕ, σ_1, σ_2 *reduce to* 3 *numbers for a symmetric matrix*?

Question 2 *If* $\theta = 30^\circ$ *and* $\sigma_1 = 2$ *and* $\sigma_2 = 1$ *and* $\phi = 60^\circ$, *what is* A?

The First Singular Vector $\boldsymbol{v}_1$

This page will establish a new way to look at $\boldsymbol{v}_1$. The previous pages chose the $\boldsymbol{v}$'s as eigenvectors of $A^{\mathrm{T}}A$. Certainly that remains true. But there is a valuable way to understand these singular vectors **one at a time instead of all at once**. We start with $\boldsymbol{v}_1$ and the singular value σ_1. The length of $\boldsymbol{x}$ comes from $||\boldsymbol{x}||^2 = x_1^2 + \cdots + x_n^2 = \boldsymbol{x}^{\mathrm{T}}\boldsymbol{x}$.

$$\textbf{Maximize the ratio } \frac{||A\boldsymbol{x}||}{||\boldsymbol{x}||}. \textbf{ The maximum is } \sigma_1 \textbf{ at the vector } \boldsymbol{x} = \boldsymbol{v}_1. \tag{15}$$

The ellipse in Figure 7.1 showed why the maximizing $\boldsymbol{x}$ is $\boldsymbol{v}_1$. When you follow $\boldsymbol{v}_1$ across the page, it ends at $A\boldsymbol{v}_1 = \sigma_1\boldsymbol{u}_1$. The longer axis of the ellipse has length $||A\boldsymbol{v}_1|| = \sigma_1$.

But we aim for an independent approach to the SVD ! We are not assuming that we already know U or Σ or V. How do we recognize that the ratio $||A\boldsymbol{x}||/||\boldsymbol{x}||$ is a maximum when $\boldsymbol{x} = \boldsymbol{v}_1$? Calculus tells us that the first derivatives must be zero. The derivatives will be easier to compute if we square our function and work with $S = A^{\mathrm{T}}A$:

$$\textbf{Problem: Find the maximum value } \lambda \textbf{ of } \frac{||A\boldsymbol{x}||^2}{||\boldsymbol{x}||^2} = \frac{\boldsymbol{x}^{\mathrm{T}}A^{\mathrm{T}}A\boldsymbol{x}}{\boldsymbol{x}^{\mathrm{T}}\boldsymbol{x}} = \frac{\boldsymbol{x}^{\mathrm{T}}S\boldsymbol{x}}{\boldsymbol{x}^{\mathrm{T}}\boldsymbol{x}}. \tag{16}$$

This "Rayleigh quotient" depends on $x_1, \ldots, x_n$. Calculus uses the quotient rule, so we need the partial derivatives $2\boldsymbol{x}$ and $2S\boldsymbol{x}$ of $\boldsymbol{x}^{\mathrm{T}}\boldsymbol{x}$ and $\boldsymbol{x}^{\mathrm{T}}S\boldsymbol{x}$:

$$\frac{\partial}{\partial x_i}\left(\boldsymbol{x}^{\mathrm{T}}\boldsymbol{x}\right) = \frac{\partial}{\partial x_i}\left(x_1^2 + \cdots + x_i^2 + \cdots + x_n^2\right) = 2\boldsymbol{x}_i \tag{17}$$

$$\frac{\partial}{\partial x_i}\left(\boldsymbol{x}^{\mathrm{T}}S\boldsymbol{x}\right) = \frac{\partial}{\partial x_i}\left(\sum_i\sum_j S_{ij}x_ix_j\right) = 2\sum_j S_{ij}x_j = 2\left(S\boldsymbol{x}\right)_i \tag{18}$$

Use the quotient rule for $\dfrac{\partial}{\partial x_i}\left(\dfrac{\boldsymbol{x}^{\mathrm{T}}S\boldsymbol{x}}{\boldsymbol{x}^{\mathrm{T}}\boldsymbol{x}}\right)$ and set those n partial derivatives of (16) to zero :

$$\left(\boldsymbol{x}^{\mathrm{T}}\boldsymbol{x}\right)2\left(S\boldsymbol{x}\right)_i - \left(\boldsymbol{x}^{\mathrm{T}}S\boldsymbol{x}\right)2x_i = 0 \text{ for } i = 1, \ldots, n \tag{19}$$

Equation (19) says that $(S\boldsymbol{x})_i = \lambda x_i$. The number λ is $\boldsymbol{x}^{\mathrm{T}}S\boldsymbol{x}/\boldsymbol{x}^{\mathrm{T}}\boldsymbol{x}$. **Then $S\boldsymbol{x} = \lambda\boldsymbol{x}$ and the best $\boldsymbol{x}$ to maximize the ratio in (16) is an eigenvector of S !**

$$\boldsymbol{2S\boldsymbol{x} = 2\lambda\boldsymbol{x}} \textbf{ and the maximum value of } \frac{\boldsymbol{x}^{\mathrm{T}}S\boldsymbol{x}}{\boldsymbol{x}^{\mathrm{T}}\boldsymbol{x}} = \frac{||A\boldsymbol{x}||^2}{||\boldsymbol{x}||^2} \textbf{ is an eigenvalue } \lambda \textbf{ of } S.$$

The search is narrowed to eigenvectors of $S = A^{\mathrm{T}}A$. The eigenvector with largest λ is $\boldsymbol{x} = \boldsymbol{v}_1$. The eigenvalue is $\lambda_1 = \sigma_1^2$. Calculus has confirmed the solution (15) of the maximum problem. That problem has led to σ_1 and $\boldsymbol{v}_1$ in the SVD.

For the full SVD, we need *all* the singular vectors and singular values. To find $\boldsymbol{v}_2$ and σ_2, we adjust the maximum problem so it looks only at vectors $\boldsymbol{x}$ orthogonal to $\boldsymbol{v}_1$.

Maximize $\dfrac{||Ax||}{||x||}$ **under the condition** $v_1^{\mathrm{T}}x = 0$**. The maximum is** σ_2 **at** $x = v_2$**.**

"Lagrange multipliers" were invented to deal with constraints on $\boldsymbol{x}$ like $\boldsymbol{v}_1^{\mathrm{T}}\boldsymbol{x} = 0$. And Problem 9 gives a simple direct way to work with this condition $\boldsymbol{v}_1^{\mathrm{T}}\boldsymbol{x} = 0$.

Every singular vector $\boldsymbol{v}_{k+1}$ gives the maximum ratio $||A\boldsymbol{x}||/||\boldsymbol{x}||$ over all vectors $\boldsymbol{x}$ that are perpendicular to the first $\boldsymbol{v}_1, \ldots, \boldsymbol{v}_k$. The left singular vectors would come from maximizing $||A^{\mathrm{T}}\boldsymbol{y}||/||\boldsymbol{y}||$. We are always finding the axes of an ellipsoid and the eigenvectors of symmetric matrices $A^{\mathrm{T}}A$ or AA^{T} : all at once or separately.

Computing Eigenvalues and Singular Values

What is the main difference between the symmetric eigenvalue problem $Sx = \lambda x$ and $Av = \sigma u$? How much can we simplify S and A before computing λ's and σ's ?

Eigenvalues are the same for S and $Q^{-1}SQ = Q^{\mathrm{T}}SQ$ when Q is orthogonal.

So we have limited freedom to create zeros in $Q^{-1}SQ$ (which stays symmetric). If we try for too many zeros in $Q^{-1}S$, the final Q will destroy them. The good $Q^{-1}SQ$ will be **tridiagonal**: we can reduce S to three nonzero diagonals.

Singular values are the same for A and $Q_1^{-1}AQ_2$ even if Q_1 is different from Q_2.

We have more freedom to create zeros in $Q_1^{-1}AQ_2$. With the right Q's, this will be **bidiagonal** (two nonzero diagonals). We can quickly find Q and Q_1 and Q_2 so that

$$\boldsymbol{Q^{-1}SQ} = \begin{bmatrix} a_1 & b_1 & & \\ b_1 & a_2 & b_2 & \\ & b_2 & \cdot & \cdot \\ & & \cdot & a_n \end{bmatrix} \begin{matrix} \leftarrow \text{for } \lambda\text{'s} \\ \boldsymbol{Q_1^{-1}AQ_2} = \\ \text{for } \sigma\text{'s} \rightarrow \end{matrix} \begin{bmatrix} c_1 & d_1 & & \\ 0 & c_2 & d_2 & \\ & 0 & \cdot & \cdot \\ & & 0 & c_n \end{bmatrix} \tag{20}$$

The reader will know that the singular values of A are the square roots of the eigenvalues of $S = A^{\mathrm{T}}A$. And the singular values of $Q_1^{-1}AQ_2$ are the same as the singular values of A. **Multiply (bidiagonal)^{T}(bidiagonal) to see tridiagonal**.

This offers an option that we should not take. Don't multiply $A^{\mathrm{T}}A$ and find its eigenvalues. This is unnecessary work and the condition of the problem will be unnecessarily squared. The Golub-Kahan algorithm for the SVD works directly on A, in two steps:

1. Find Q_1 and Q_2 so that $Q_1^{-1}AQ_2$ is bidiagonal as in (20).

2. Adjust the shifted QR algorithm to preserve singular values of this bidiagonal matrix.

Step 1 requires $O(mn^2)$ multiplications to put an m by n matrix A into bidiagonal form. Then later steps will work only with bidiagonal matrices. Normally it then takes $O(n^2)$ multiplications to find singular values (correct to nearly machine precision). The full algorithm is described on pages 489–492 in the 4th edition of Golub-Van Loan.

When A is truly large, we turn to **random sampling**. With very high probability, *randomized linear algebra gives accurate results.* Most gamblers would say that a good outcome from careful random sampling is certain.

Problem Set 7.1

1 Find $A^{\mathrm{T}}A$ and AA^{T} and the singular vectors $\boldsymbol{v}_1, \boldsymbol{v}_2, \boldsymbol{u}_1, \boldsymbol{u}_2$ for A:

$$A = \begin{bmatrix} 0 & 1 & 0 \\ 0 & 0 & 8 \\ 0 & 0 & 0 \end{bmatrix} \quad \text{has rank } r = 2. \quad \text{The eigenvalues are } 0, 0, 0.$$

Check the equations $A\boldsymbol{v}_1 = \sigma_1 \boldsymbol{u}_1$ and $A\boldsymbol{v}_2 = \sigma_2 \boldsymbol{u}_2$ and $A = \sigma_1 \boldsymbol{u}_1 \boldsymbol{v}_1^{\mathrm{T}} + \sigma_2 \boldsymbol{u}_2 \boldsymbol{v}_2^{\mathrm{T}}$. If you remove row 3 of A (all zeros), show that σ_1 and σ_2 don't change.

2 Find the singular values and also the eigenvalues of B:

$$B = \begin{bmatrix} 0 & 1 & 0 \\ 0 & 0 & 8 \\ \frac{1}{1000} & 0 & 0 \end{bmatrix} \quad \text{has rank } r = 3 \text{ and determinant } \frac{8}{1000}.$$

Compared to A above, eigenvalues have changed much more than singular values.

3 The SVD connects $\boldsymbol{v}$'s in the row space to $\boldsymbol{u}$'s in the column space. Transpose $A = U\Sigma V^{\mathrm{T}}$ to see that $\boldsymbol{A}^{\mathrm{T}} = \boldsymbol{V}\boldsymbol{\Sigma}^{\mathrm{T}}\boldsymbol{U}^{\mathrm{T}}$ goes the opposite way, from $\boldsymbol{u}$'s to $\boldsymbol{v}$'s:

$$\boldsymbol{A}^{\mathrm{T}}\boldsymbol{u}_k = \sigma_k \boldsymbol{v}_k \ \text{ for } k = 1, \ldots, r \qquad \boldsymbol{A}^{\mathrm{T}}\boldsymbol{u}_k = \boldsymbol{0} \ \text{ for } k = r+1, \ldots, m$$

Multiply $A\boldsymbol{v}_k = \sigma_k \boldsymbol{u}_k$ by A^{T}. Divide $A^{\mathrm{T}}A\boldsymbol{v}_k = \sigma_k^2 \boldsymbol{v}_k$ in equation (9) by σ_k. What are the $\boldsymbol{u}$'s and $\boldsymbol{v}$'s for the transpose $\begin{bmatrix} 3 & 4 & ; & 0 & 5 \end{bmatrix}$ of our example matrix?

4 When $A\boldsymbol{v}_k = \sigma_k \boldsymbol{u}_k$ and $A^{\mathrm{T}}\boldsymbol{u}_k = \sigma_k \boldsymbol{v}_k$, show that S has eigenvalues σ_k and $-\sigma_k$:

$$S = \begin{bmatrix} 0 & \boldsymbol{A} \\ \boldsymbol{A}^{\mathrm{T}} & 0 \end{bmatrix} \quad \text{has eigenvectors} \quad \begin{bmatrix} \boldsymbol{u}_k \\ \boldsymbol{v}_k \end{bmatrix} \quad \text{and} \quad \begin{bmatrix} -\boldsymbol{u}_k \\ \boldsymbol{v}_k \end{bmatrix} \quad \text{and trace} = 0.$$

The eigenvectors of this symmetric S tell us the singular vectors of A.

5 Find the eigenvalues and the singular values of this 2 by 2 matrix A.

$$A = \begin{bmatrix} 2 & 1 \\ 4 & 2 \end{bmatrix} \quad \text{with} \quad A^{\mathrm{T}}A = \begin{bmatrix} 20 & 10 \\ 10 & 5 \end{bmatrix} \quad \text{and} \quad AA^{\mathrm{T}} = \begin{bmatrix} 5 & 10 \\ 10 & 20 \end{bmatrix}.$$

The eigenvectors $(1, 2)$ and $(1, -2)$ of A are not orthogonal. How do you know the eigenvectors $\boldsymbol{v}_1, \boldsymbol{v}_2$ of $A^{\mathrm{T}}A$ will be orthogonal? Notice that $A^{\mathrm{T}}A$ and AA^{T} have the same eigenvalues $\lambda_1 = 25$ and $\lambda_2 = 0$.

6 The two columns of $AV = U\Sigma$ are $A\boldsymbol{v}_1 = \sigma_1 \boldsymbol{u}_1$ and $A\boldsymbol{v}_2 = \sigma_2 \boldsymbol{u}_2$. So we hope that

$$A\boldsymbol{v}_1 = \begin{bmatrix} 1 & 0 \\ 1 & 1 \end{bmatrix} \begin{bmatrix} \sigma_1 \\ 1 \end{bmatrix} = \sigma_1 \begin{bmatrix} 1 \\ \sigma_1 \end{bmatrix} \quad \text{and} \quad \begin{bmatrix} 1 & 0 \\ 1 & 1 \end{bmatrix} \begin{bmatrix} 1 \\ -\sigma_1 \end{bmatrix} = \sigma_2 \begin{bmatrix} \sigma_1 \\ -1 \end{bmatrix}$$

The first needs $\sigma_1 + 1 = \sigma_1^2$ and the second needs $1 - \sigma_1 = -\sigma_2$. Are those true?

7 The MATLAB commands $A = \text{rand}\,(20, 40)$ and $B = \text{randn}\,(20, 40)$ produce 20 by 40 random matrices. The entries of A are between 0 and 1 with uniform probability. The entries of B have a normal "bell-shaped" probability distribution. Using an svd command, find and graph their singular values σ_1 to σ_{20}. Why do they have 20 σ's?

8 A symmetric matrix $S = A^{\mathrm{T}}A$ has eigenvalues λ_1 to λ_n and eigenvectors $\boldsymbol{v}_1$ to $\boldsymbol{v}_n$. Then any vector has the form $\boldsymbol{x} = c_1\boldsymbol{v}_1 + \cdots + c_n\boldsymbol{v}_n$. The Rayleigh quotient is

$$R(\boldsymbol{x}) = \frac{\boldsymbol{x}^{\mathrm{T}}S\boldsymbol{x}}{\boldsymbol{x}^{\mathrm{T}}\boldsymbol{x}} = \frac{\lambda_1 c_1^2 + \cdots + \lambda_n c_n^2}{c_1^2 + \cdots + c_n^2}.$$

Which vector $\boldsymbol{x}$ gives the maximum of R? What are the numbers c_1 to c_n for that maximizing vector $\boldsymbol{x}$? Which $\boldsymbol{x}$ gives the minimum of R?

9 To find σ_2 and $\boldsymbol{v}_2$ from maximizing that ratio $R(\boldsymbol{x})$, we must rule out the first singular vectors $\boldsymbol{v}_1$ by requiring $\boldsymbol{x}^{\mathrm{T}}\boldsymbol{v}_1 = 0$. What does this mean for c_1? Which c's give the new maximum σ_2 at the second eigenvector $\boldsymbol{x} = \boldsymbol{v}_2$?

10 Find $A^{\mathrm{T}}A$ and the singular vectors in $A\boldsymbol{v}_1 = \sigma_1\boldsymbol{u}_1$ and $A\boldsymbol{v}_2 = \sigma_2\boldsymbol{u}_2$:

$$A = \begin{bmatrix} 2 & 2 \\ -1 & 1 \end{bmatrix} \quad \text{and} \quad A = \begin{bmatrix} 3 & 3 \\ 4 & 4 \end{bmatrix}.$$

11 For this rectangular matrix find $\boldsymbol{v}_1, \boldsymbol{v}_2, \boldsymbol{v}_3$ and $\boldsymbol{u}_1, \boldsymbol{u}_2$ and σ_1, σ_2. Then write the SVD for A as $U\Sigma V^{\mathrm{T}} = (2 \times 2)\,(2 \times 3)\,(3 \times 3)$.

$$A = \begin{bmatrix} 1 & 1 & 0 \\ 0 & 1 & 1 \end{bmatrix}$$

12 If $(A^{\mathrm{T}}A)\boldsymbol{v} = \sigma^2\boldsymbol{v}$, multiply by A. *Move the parentheses to get* $(AA^{\mathrm{T}})A\boldsymbol{v} = \sigma^2(A\boldsymbol{v})$. **If $\boldsymbol{v}$ is an eigenvector of $A^{\mathrm{T}}A$, then _____ is an eigenvector of AA^{T}.**

13 Find the eigenvalues and unit eigenvectors $\boldsymbol{v}_1, \boldsymbol{v}_2$ of $A^{\mathrm{T}}A$. Then find $\boldsymbol{u}_1 = A\boldsymbol{v}_1/\sigma_1$:

$$A = \begin{bmatrix} 1 & 2 \\ 3 & 6 \end{bmatrix} \text{ and } A^{\mathrm{T}}A = \begin{bmatrix} 10 & 20 \\ 20 & 40 \end{bmatrix} \text{ and } AA^{\mathrm{T}} = \begin{bmatrix} 5 & 15 \\ 15 & 45 \end{bmatrix}.$$

Verify that $\boldsymbol{u}_1$ is a unit eigenvector of AA^{T}. Complete the matrices U, Σ, V.

$$\textbf{SVD} \qquad \begin{bmatrix} 1 & 2 \\ 3 & 6 \end{bmatrix} = \begin{bmatrix} \boldsymbol{u}_1 & \boldsymbol{u}_2 \end{bmatrix} \begin{bmatrix} \sigma_1 & \\ & 0 \end{bmatrix} \begin{bmatrix} \boldsymbol{v}_1 & \boldsymbol{v}_2 \end{bmatrix}^{\mathbf{T}}.$$

14 (a) Why is the trace of $A^{\mathrm{T}}A$ equal to the sum of all a_{ij}^2?

(b) For every rank-one matrix, why is σ_1^2 = sum of all a_{ij}^2?

15 If $A = U\Sigma V^{\mathrm{T}}$ is a square invertible matrix then $A^{-1} =$ _____ _____ _____.
The largest singular value of A^{-1} is therefore $1/\sigma_{\min}(A)$. The largest eigenvalue is $1/|\lambda(A)|_{\min}$. Then equation (13) says that $\boldsymbol{\sigma_{\min}(A) \leq |\lambda(A)|_{\min}}$.

16 Suppose $A = U\Sigma V^{\mathrm{T}}$ is 2 by 2 with $\sigma_1 > \sigma_2 > 0$. Change A by ***as small a matrix as possible*** to produce a singular matrix A_0. Hint: U and V do not change.

17 Why doesn't the SVD for $A + I$ just use $\Sigma + I$?

7.2 Compressing Images by the SVD

1 An image is a large matrix of grayscale values, one for each pixel and color.

2 When nearby pixels are correlated (not random) the image can be compressed.

3 Flags often give simple images. Photographs can be compressed by the SVD.

Image processing and compression are major consumers of linear algebra. Recognizing images often uses convolutional neural nets in deep learning. These topics in Chapters 7-8 represent part of the enormous intersection of computational engineering and mathematics.

This section will begin with stylized images like flags: often with *low complexity*. Then we move to photographs with many more pixels. In all cases we want efficient ways to process and transmit signals. The image is effectively a large matrix (!) that can represent light/dark and red/green/blue for every small pixel.

The SVD offers one approach to matrix approximation: *Replace A by A_k.* The sum A of r rank one matrices $\sigma_j \boldsymbol{u}_j \boldsymbol{v}_j^{\mathrm{T}}$ can be reduced to the sum A_k of the first k terms. This section (plus online help) will consider the effect on an image such as a flag or a photograph. Section 7.3 will explore more examples in which we need to approximate and understand a matrix of data.

Start with flags. More than 30 countries chose flags with three stripes. Those flags have a particularly simple form: easy to compress. I found a book called "Flags of the World" and the pictures range from one solid color (Libya's flag was entirely green during the Gaddafi years) to very complicated images. How would those pictures be compressed with minimum loss ?

The linear algebra answer is: *Use the* SVD. Notice that 3 stripes still produce rank 1. France has blue-white-red vertical stripes and $b\ w\ r$ in its columns. By coincidence the German flag is nearly its transpose with the colors Black-White-Red:

$$\begin{bmatrix} b\,b\,w\,w\,r\,r \\ b\,b\,w\,w\,r\,r \\ b\,b\,w\,w\,r\,r \\ b\,b\,w\,w\,r\,r \\ b\,b\,w\,w\,r\,r \\ b\,b\,w\,w\,r\,r \end{bmatrix} = \begin{bmatrix} 1 \\ 1 \\ 1 \\ 1 \\ 1 \\ 1 \end{bmatrix} \begin{bmatrix} b\,b\,w\,w\,r\,r \end{bmatrix} \quad \textbf{France} \qquad \begin{bmatrix} B & B & B & B & B & B \\ B & B & B & B & B & B \\ W & W & W & W & W & W \\ W & W & W & W & W & W \\ R & R & R & R & R & R \\ R & R & R & R & R & R \end{bmatrix} = \begin{bmatrix} B \\ B \\ W \\ W \\ R \\ R \end{bmatrix} \begin{bmatrix} 1\,1\,1\,1\,1\,1 \end{bmatrix} \quad \textbf{Germany}$$

Each matrix reduces to two vectors. To transmit those images we can replace N^2 pixels by $2N$ pixels. Similarly, Italy is green-white-red and Iceland is green-white-orange. But many many countries make the problem infinitely more difficult by adding a small badge on top of those stripes. Japan has a red sun on a white background and the Maldives have an elegant white moon on a green rectangle on a white rectangle. Those curved images have infinite rank—compression is still possible and necessary, but not to rank one.

A few flags stay with finite rank but they add a cross to increase the rank. Here are two flags (Greece and Tonga) with rank 3 and rank 4.

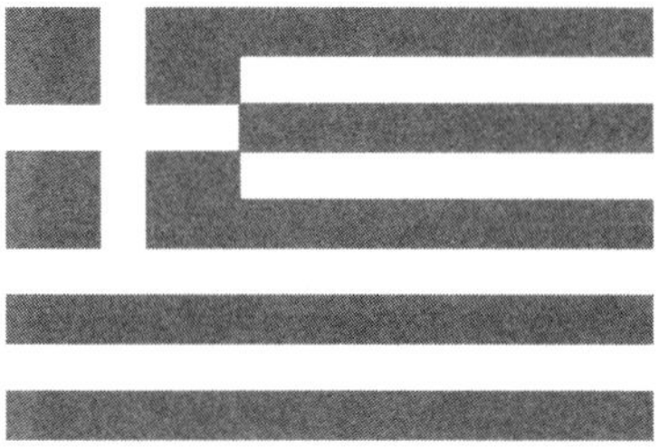

Greece

Tonga

I see four different rows in the Greek flag, but only three columns. Mistakenly, I thought the rank was 4. But now I think that row $2 +$ row $3 =$ row 1 and the rank of Greece is 3.

On the other hand, Tonga's flag does seem to have rank 4. The left half has four rows: all white-short red-longer red-all red. We can't produce any row from a linear combination of the other rows. The island kingdom of Tonga has the champion flag of finite rank!

Singular Values with Diagonals

Three countries have flags with only two diagonal lines: Bahamas, Czech Republic, and Kuwait. Many countries have added in stars and multiple diagonals. From my book I can't be sure whether Canada also has small curves. It is interesting to find the SVD of this matrix with lower triangular 1's—including the main diagonal—and upper triangular 0's.

Flag with a triangle $$A = \begin{bmatrix} \mathbf{1} & 0 & 0 & 0 \\ \mathbf{1} & \mathbf{1} & 0 & 0 \\ \mathbf{1} & \mathbf{1} & \mathbf{1} & 0 \\ \mathbf{1} & \mathbf{1} & \mathbf{1} & \mathbf{1} \end{bmatrix} \quad \text{has} \quad A^{-1} = \begin{bmatrix} \mathbf{1} & 0 & 0 & 0 \\ \mathbf{-1} & \mathbf{1} & 0 & 0 \\ 0 & \mathbf{-1} & \mathbf{1} & 0 \\ 0 & 0 & \mathbf{-1} & \mathbf{1} \end{bmatrix}.$$

A has full rank $r = N$. All eigenvalues are 1, on the main diagonal. Then A has N singular values (all positive, but not equal to 1). The SVD will produce n pieces $\sigma_i \boldsymbol{u}_i \boldsymbol{v}_i^{\mathrm{T}}$ of rank one. Perfect reproduction needs all n pieces.

In compression the small σ's can be discarded with no serious loss in image quality. We want to understand the singular values for $n = 4$ and also to plot all σ's for large n. The graph on the next page will decide if A is compressed by the SVD.

Working by hand, we begin with AA^{T} (a computer would proceed differently):

$$AA^{\mathrm{T}} = \begin{bmatrix} 1 & 1 & 1 & 1 \\ 1 & 2 & 2 & 2 \\ 1 & 2 & 3 & 3 \\ 1 & 2 & 3 & 4 \end{bmatrix} \quad \text{and} \quad (AA^{\mathrm{T}})^{-1} = (A^{-1})^{\mathrm{T}} A^{-1} = \begin{bmatrix} \mathbf{2} & -1 & 0 & 0 \\ -1 & \mathbf{2} & -1 & 0 \\ 0 & -1 & \mathbf{2} & -1 \\ 0 & 0 & -1 & \mathbf{1} \end{bmatrix}. \quad (1)$$

That $-1, 2, -1$ inverse matrix is included because all its eigenvalues have the form $2 - 2\cos\theta$. We know those eigenvalues! So we know the singular values of A.

$$\lambda(AA^{\mathrm{T}}) = \frac{1}{2 - 2\cos\theta} = \frac{1}{4\sin^2(\theta/2)} \quad \text{gives} \quad \sigma(A) = \sqrt{\lambda} = \frac{1}{2\sin(\theta/2)}. \qquad (2)$$

The n different angles θ are equally spaced, which makes this example so exceptional:

$$\theta = \frac{\pi}{2n+1}, \frac{3\pi}{2n+1}, \ldots, \frac{(2n-1)\,\pi}{2n+1} \qquad \left(n = 4 \text{ includes } \theta = \frac{3\pi}{9} \text{ with } 2\sin\frac{\theta}{2} = 1\right).$$

The important point is to graph the n singular values of A. Those numbers drop off (unlike the eigenvalues of A, which are all 1). But the dropoff is not steep. So the SVD gives only moderate compression of this triangular flag. *Great compression for Hilbert.*

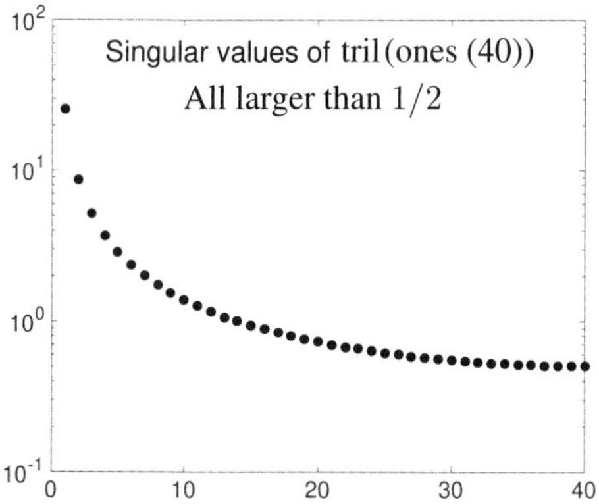

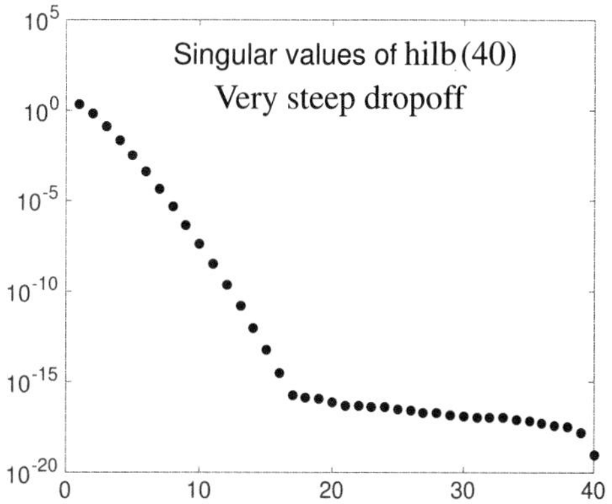

Figure 7.2: Singular values of the 40 by 40 triangle of 1's (it is not compressible). The evil Hilbert matrix $H(i,j) = (i+j-1)^{-1}$ has low effective rank: we must compress it.

The striking point about the graph is that the singular values for the triangular matrix never go below $\frac{1}{2}$. Working with Alex Townsend, we have seen this phenomenon for 0-1 matrices with the 1's in other shapes (such as circular). This has not yet been explained.

Image Compression by the SVD

Compressing photographs and images is an exceptional way to see the SVD in action. The action comes by varying the number of rank one pieces $\sigma \boldsymbol{u}\boldsymbol{v}^{\mathrm{T}}$ in the display. By keeping more terms the image improves.

We hoped to find a website that would show this improvement. By good fortune, Tim Baumann has achieved exactly what we hoped for. He has given all of us permission to use his work: **https://timbaumann.info/svd-image-compression-demo/**

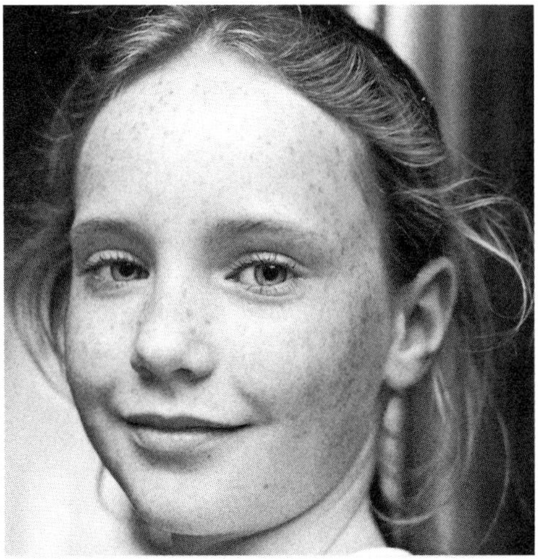

Uncompressed image. Slider at 300.

IMAGE SIZE 600×600
#PIXELS $= 360000$

UNCOMPRESSED SIZE
proportional to number of pixels

COMPRESSED SIZE
approximately proportional to
$600 \times 300 + 300 + 300 \times 600$
$= 360300$

COMPRESSION RATIO
$360000/360300 = 1.00$

Show singular values

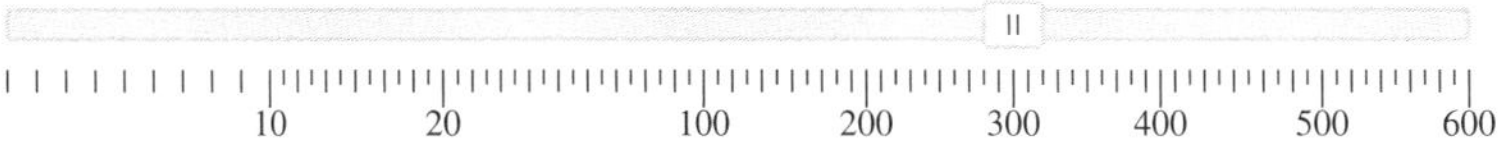

Compressed image. Slider at 20.

IMAGE SIZE 600×600
#PIXELS $= 360000$

UNCOMPRESSED SIZE
proportional to number of pixels

COMPRESSED SIZE
approximately proportional to
$600 \times 20 + 20 + 20 \times 600$
$= 24020$

COMPRESSION RATIO
$360000/24020 = 14.99$

Show singular values

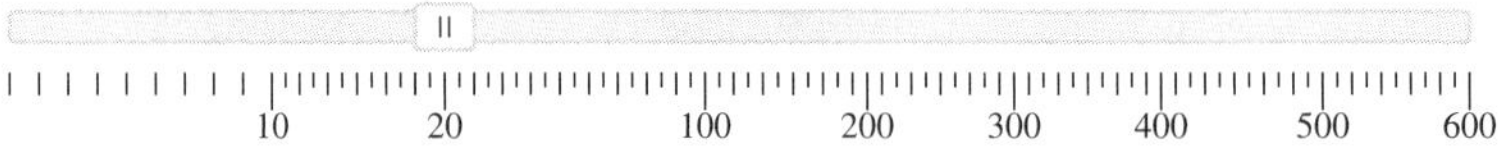

Change the number of singular values using the slider. Click on one of these images to compress it :

You can compress your own images by using the file picker or by dropping them on this page.

This is one of the five images directly available. The position of the slider determines the compression ratio. The best ratio depends on the complexity of the image—the girl and the Mondrian painting are less complex and allow higher compression than the city or the tree or the cats. When the computation of compressed size adds $600 \times 80 + 80 + 80 \times 600$, we have 80 terms $\sigma \boldsymbol{u}\boldsymbol{v}^{\mathrm{T}}$ with vectors $\boldsymbol{u}$ and $\boldsymbol{v}$ of dimension 600. The slider is set at 80.

You can compress your own images by using the "file picker" button below the six sample images provided on the site, or by dropping them directly onto the page.

One beautiful feature of Tim Baumann's site is that it operates in the browser, with instant results. This book's website **math.mit.edu/everyone** can include ideas from readers. Please see that edited site for questions and comments and suggestions.

Problem Set 7.2

1 We usually think that the identity matrix I is as simple as possible. But why is I difficult to compress? *Create the matrix for a rank 3 flag with a horizontal-vertical cross.*

2 These flags have rank 2. Write A and B in any way as $\boldsymbol{u}_1\boldsymbol{v}_1^{\mathrm{T}} + \boldsymbol{u}_2\boldsymbol{v}_2^{\mathrm{T}}$.

$$\boldsymbol{A}_{\text{Sweden}} = \boldsymbol{A}_{\text{Finland}} = \begin{bmatrix} 1 & 2 & 1 & 1 \\ 2 & 2 & 2 & 2 \\ 1 & 2 & 1 & 1 \end{bmatrix} \qquad \boldsymbol{B}_{\text{Benin}} = \begin{bmatrix} 1 & 2 & 2 \\ 1 & 3 & 3 \end{bmatrix}$$

3 Now find the trace and determinant of BB^{T} and also $B^{\mathrm{T}}B$ in Problem 2. The singular values of B are close to $\sigma_1^2 = 28 - \frac{1}{14}$ and $\sigma_2^2 = \frac{1}{14}$. Is B compressible or not?

4 Use $[U, S, V] = \mathsf{svd}\,(A)$ to find two orthogonal pieces $\sigma \boldsymbol{u}\boldsymbol{v}^{\mathrm{T}}$ of $\boldsymbol{A}_{\text{Sweden}}$.

5 A matrix for the Japanese flag has a circle of ones surrounded by all zeros. Suppose the center line of the circle (the diameter) has $2N$ ones. Then the circle will contain about πN^2 ones. We think of the flag as a 1-0 matrix. Its rank will be approximately CN, proportional to N. *What is that number C?*

Hint: Remove a big square submatrix of ones, with corners at $\pm 45°$ and $\pm 135°$. The rows above the square and the columns to the right of the square are independent. Draw a picture and estimate the number cN of those rows. Then $C = 2c$.

6 Here is one way to start with a function $F(x, y)$ and construct a matrix A. Set $A_{ij} = \boldsymbol{F(i/N, j/N)}$. (The indices i and j go from 0 to N or from $-N$ to N.) The rank of A as N increases should reflect the simplicity or complexity of F. Find the ranks of the matrices for the functions $F_1 = \boldsymbol{xy}$ and $F_2 = \boldsymbol{x} + \boldsymbol{y}$ and $F_3 = \boldsymbol{x^2} + \boldsymbol{y^2}$. Then find three singular values and singular vectors for F_3.

7 In Problem 6, what conditions on $F(x, y)$ will produce a symmetric matrix S? An antisymmetric matrix A? A singular matrix M? A matrix of rank 2?

7.3 Principal Component Analysis

The "principal components" of A are its singular vectors, the orthogonal columns $\boldsymbol{u}_j$ and $\boldsymbol{v}_j$ of the matrices U and V. This section aims to apply the Singular Value Decomposition $A = U\Sigma V^{\mathrm{T}}$. **Principal Component Analysis (PCA)** uses the largest σ's connected to the first $\boldsymbol{u}$'s and $\boldsymbol{v}$'s to understand the information in a matrix of data.

We are given a matrix A, and we extract its most important part A_k (**largest $\boldsymbol{\sigma}$'s**):

$$\boldsymbol{A_k = \sigma_1 u_1 v_1^{\mathrm{T}} + \cdots + \sigma_k u_k v_k^{\mathrm{T}}} \quad \textbf{with rank } (\boldsymbol{A_k}) = \boldsymbol{k}.$$

A_k solves a matrix optimization problem—and we start there. **The closest rank $\boldsymbol{k}$ matrix to $\boldsymbol{A}$ is $\boldsymbol{A_k}$.** In statistics we are identifying the rank one pieces of A with largest variance. This puts the SVD at the center of data science.

In that world, PCA is "unsupervised" learning. Our only instructor is linear algebra—the SVD tells us to choose A_k. When the learning is supervised, we have a big set of training data. Deep Learning constructs a (nonlinear!) function F to correctly classify most of that data. Then we apply this F to new data, as you will see in Chapter 8.

Principal Component Analysis is based on matrix approximation by A_k. The proof that A_k is the best choice was begun by Schmidt (1907). He wrote about operators in function space; his ideas extend directly to matrices. Eckart and Young gave a new proof (using the Frobenius norm to measure $A - A_k$). Then Mirsky allowed any norm $||A||$ that depends only on the singular values—as in the definitions (2), (3), and (4) below.

Here is that key property of the special rank k matrix $A_k = \sigma_1 \boldsymbol{u}_1 \boldsymbol{v}_1^{\mathrm{T}} + \cdots + \sigma_k \boldsymbol{u}_k \boldsymbol{v}_k^{\mathrm{T}}$:

$$\boxed{\boldsymbol{A_k} \textbf{ is closest to } \boldsymbol{A} \qquad \textbf{If } \boldsymbol{B} \textbf{ has rank } \boldsymbol{k} \textbf{ then } ||\boldsymbol{A} - \boldsymbol{A_k}|| \leq ||\boldsymbol{A} - \boldsymbol{B}||.} \tag{1}$$

Three choices for the matrix norm $||A||$ have special importance and their own names:

Spectral norm	$\lVert\boldsymbol{A}\rVert = \max \dfrac{\lVert A\boldsymbol{x}\rVert}{\lVert\boldsymbol{x}\rVert} = \boldsymbol{\sigma_1}$	(often called the $\boldsymbol{\ell^2}$ **norm**)	(2)
Frobenius norm	$\lVert\boldsymbol{A}\rVert_{\boldsymbol{F}} = \sqrt{\sigma_1^2 + \cdots + \sigma_r^2}$	(7) also defines $\lVert\boldsymbol{A}\rVert_{\boldsymbol{F}}$	(3)
Nuclear norm	$\lVert\boldsymbol{A}\rVert_{\boldsymbol{N}} = \boldsymbol{\sigma_1 + \sigma_2 + \cdots + \sigma_r}$	(the trace norm)	(4)

These norms have different values already for the n by n identity matrix:

$$||I||_2 = 1 \qquad ||I||_F = \sqrt{n} \qquad ||I||_N = n. \tag{5}$$

Replace I by any orthogonal matrix Q and the norms stay the same (because all $\sigma_i = 1$):

$$||Q||_2 = 1 \qquad ||Q||_F = \sqrt{n} \qquad ||Q||_N = n. \tag{6}$$

More than this, the spectral and Frobenius and nuclear norms of any matrix stay the same when A is multiplied (on either side) by an orthogonal matrix. So the norm of $A = U\Sigma V^{\mathrm{T}}$ equals the norm of Σ: $||\boldsymbol{A}|| = ||\boldsymbol{\Sigma}||$ because U and V are orthogonal matrices.

Norm of a Matrix

We need a way to measure the size of a vector or a matrix. For a vector, the most important norm is the usual length $||\boldsymbol{v}||$. For a matrix, Frobenius extended the sum of squares idea to include all the entries in A. This norm $||A||_F$ is also named Hilbert-Schmidt.

$$||\boldsymbol{v}||^2 = v_1^2 + \cdots + v_n^2 \qquad ||A||_F^2 = a_{11}^2 + \cdots + a_{1n}^2 + \cdots + a_{m1}^2 + \cdots + a_{mn}^2 \qquad (7)$$

Clearly $||\boldsymbol{v}|| \geq 0$ and $||c\boldsymbol{v}|| = |c|\ ||\boldsymbol{v}||$. Similarly $||A||_F \geq 0$ and $||cA||_F = |c|\ ||A||_F$. Equally essential is the triangle inequality for $\boldsymbol{v} + \boldsymbol{w}$ and $A + B$:

Triangle inequalities $\quad ||\boldsymbol{v} + \boldsymbol{w}|| \leq ||\boldsymbol{v}|| + ||\boldsymbol{w}||$ and $||A + B||_F \leq ||A||_F + ||B||_F \quad (8)$

We use one more fact when we meet dot products $\boldsymbol{v}^{\mathrm{T}}\boldsymbol{w}$ or matrix products AB :

Schwarz inequalities $\quad |\boldsymbol{v}^{\mathrm{T}}\boldsymbol{w}| \leq ||\boldsymbol{v}||\ ||\boldsymbol{w}|| \quad$ and $\quad ||AB||_F \leq ||A||_F\ ||B||_F \quad (9)$

That Frobenius matrix inequality comes directly from the Schwarz vector inequality:

$$|(AB)_{ij}|^2 \leq ||\text{row } i \text{ of } A||^2\ ||\text{column } j \text{ of } B||^2. \text{ Add for all } i \text{ and } j \text{ to see } ||AB||_F^2.$$

This suggests that there could be a dot product of matrices. It is $A \cdot B = \text{trace}(A^{\mathrm{T}}B)$.

Note. The largest size $|\lambda|$ of the eigenvalues of A is *not an acceptable norm*! We know that a nonzero matrix could have all zero eigenvalues—but its norm $||A||$ is not allowed to be zero. In this respect singular values are superior to eigenvalues.

The Eckart-Young Theorem

The theorem was in equation (1): **If B has rank k then $||A - A_k|| \leq ||A - B||$.** In all three norms $||A||$ and $||A||_F$ and $||A||_N$, we come closest to A by cutting off the SVD after k terms. The closest matrix is $A_k = \sigma_1 \boldsymbol{u}_1 \boldsymbol{v}_1^{\mathrm{T}} + \cdots + \sigma_k \boldsymbol{u}_k \boldsymbol{v}_k^{\mathrm{T}}$. This is the fact to use in approximating A by a low rank matrix!

We need an example and it can look extremely simple: a diagonal matrix A.

$$\text{The rank 2 matrix closest to } A = \begin{bmatrix} \mathbf{4} & 0 & 0 & 0 \\ 0 & \mathbf{3} & 0 & 0 \\ 0 & 0 & \mathbf{2} & 0 \\ 0 & 0 & 0 & \mathbf{1} \end{bmatrix} \text{ is } A_2 = \begin{bmatrix} \mathbf{4} & 0 & 0 & 0 \\ 0 & \mathbf{3} & 0 & 0 \\ 0 & 0 & \mathbf{0} & 0 \\ 0 & 0 & 0 & \mathbf{0} \end{bmatrix}$$

The difference $A - A_2$ is all zero except for the 2 and 1. Then $||A - A_2||_F = \sqrt{2^2 + 1^2}$. How could any other rank 2 matrix be closer to A than this A_2?

Please realize that this deceptively simple example includes *all matrices of the form* $Q_1 A Q_2$. The norms and the rank are not changed by the orthogonal matrices Q_1, Q_2. So this example includes all matrices with singular values $4, 3, 2, 1$. The best approximation A_2 keeps 4 and 3. Several proofs are collected in my 2019 book *Linear Algebra and Learning from Data* (Wellesley-Cambridge Press). Chi-Kwong Li has simplified Mirsky's proof that A_k is closest to A, for all norms that depend only on the σ's.

Principal Component Analysis

Now we start using the SVD. *The matrix A is full of data.* We have n samples. For each sample we measure m variables (like height and weight). The data matrix A_0 has n columns and m rows. In many applications it is a very large matrix.

The first step is to find the average (the sample mean) along each row of A_0. Subtract that mean from all m entries in the row. Now each row of the centered matrix A has *mean zero*. The columns of A are n points in $\mathbf{R}^m$. Because of centering, the sum of the n column vectors is zero. So the average column is the zero vector.

Often those n points are clustered near a line or a plane or another low-dimensional subspace of $\mathbf{R}^m$. Figure 7.3 shows a typical set of data points clustered along a line in $\mathbf{R}^2$ (after centering A_0 to shift the points left-right and up-down to have mean $(0,0)$ in A).

How will linear algebra find that closest line through $(0,0)$? **It is in the direction of the first singular vector $\boldsymbol{u}_1$ of $\boldsymbol{A}$.** This is the key point of PCA !

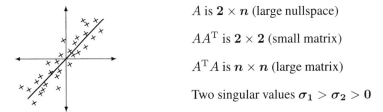

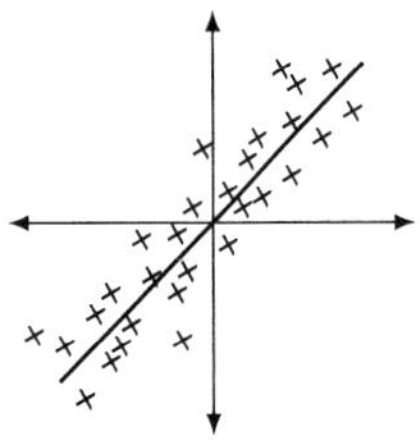

A is $\mathbf{2} \times \boldsymbol{n}$ (large nullspace)

AA^{T} is $\mathbf{2} \times \mathbf{2}$ (small matrix)

$A^{\mathrm{T}}A$ is $\boldsymbol{n} \times \boldsymbol{n}$ (large matrix)

Two singular values $\boldsymbol{\sigma_1} > \boldsymbol{\sigma_2} > \mathbf{0}$

Figure 7.3: Data points (columns of A) are often close to a line in $\mathbf{R}^2$ or a subspace in $\mathbf{R}^m$.

The Geometry Behind PCA

The best line in Figure 7.3 solves a problem in **perpendicular least squares**. This is also called *orthogonal regression*. It is different from the standard least squares fit to n data points, or the least squares solution to a linear system $A\boldsymbol{x} = \boldsymbol{b}$. That classical problem in Section 4.3 minimizes $||A\boldsymbol{x} - \boldsymbol{b}||^2$. It measures distances up and down to the best line. Our problem minimizes *perpendicular* distances. The older problem leads to a linear equation $A^{\mathrm{T}}A\widehat{\boldsymbol{x}} = A^{\mathrm{T}}\boldsymbol{b}$ for the best $\widehat{\boldsymbol{x}}$. Our problem leads to singular vectors $\boldsymbol{u}_i$ (eigenvectors of AA^{T}). Those are the two sides of linear algebra: not the same side.

The sum of squared distances from the data points to the $\boldsymbol{u}_1$ line is a minimum.

To see this, separate each column $\boldsymbol{a}_j$ of A into its components along $\boldsymbol{u}_1$ and $\boldsymbol{u}_2$:

$$\sum_1^n ||\boldsymbol{a}_j||^2 = \sum_1^n |\boldsymbol{a}_j^{\mathrm{T}}\boldsymbol{u}_1|^2 + \sum_1^n |\boldsymbol{a}_j^{\mathrm{T}}\boldsymbol{u}_2|^2. \tag{10}$$

The sum on the left is fixed by the data. The first sum on the right has terms $\boldsymbol{u}_1^{\mathrm{T}}\boldsymbol{a}_j\boldsymbol{a}_j^{\mathrm{T}}\boldsymbol{u}_1$. It adds to $\boldsymbol{u}_1^{\mathrm{T}}(AA^{\mathrm{T}})\boldsymbol{u}_1$. So when we maximize that sum in PCA by choosing the top eigenvector $\boldsymbol{u}_1$ of AA^{T}, we minimize the second sum. That second sum of squared distances from data points to the best line (or best subspace) is the smallest possible.

The Geometric Meaning of Eckart-Young

Figure 7.3 was in two dimensions and it led to the closest line. Now suppose our data matrix A_0 is 3 by n: Three measurements like age, height, weight for each of n samples. Again we center each row of the matrix, so all the rows of A add to zero. And the points move into three dimensions.

We can still look for the nearest line. It will be revealed by the first singular vector $\boldsymbol{u}_1$ of A. The best line will go through $(0,0,0)$. But if the data points fan out compared to Figure 7.3, *we really need to look for the best plane.* The meaning of "best" is still this: The sum of perpendicular distances squared to the best plane is a minimum.

That plane will be spanned by the singular vectors $\boldsymbol{u}_1$ and $\boldsymbol{u}_2$. This is the meaning of Eckart-Young. *It leads to a neat conclusion: The best plane contains the best line.*

The Statistics Behind PCA

The key numbers in probability and statistics are the **mean** and **variance**. The "mean" is an average of the data (in each row of A_0). Subtracting those means from each row of A_0 produced the centered A. The crucial quantities are the "variances" and "covariances". The variances are sums of squares of distances from the mean—along each row of A.

The variances are the diagonal entries of the matrix AA^{T}.

Suppose the columns of A correspond to a child's age on the x-axis and its height on the y-axis. (Those ages and heights are measured from the average age and height.) We are looking for the straight line that stays closest to the data points in the figure. And we have to account for the *joint age-height distribution* of the data.

The covariances are the off-diagonal entries of the matrix AA^{T}.

Those are dot products (row i of A) $\cdot$ (row j of A). High covariance means that increased height goes with increased age. (Negative covariance means that one variable increases when the other decreases.) Our first example has only two rows from age and height: the symmetric matrix AA^{T} is 2 by 2. As the number n of sample children increases, we divide by $n-1$ to give AA^{T} its statistically correct scale.

The sample covariance matrix is defined by $S = \dfrac{AA^{\mathrm{T}}}{n-1}$.

The factor is $n-1$ because one degree of freedom has already been used for mean $=0$. This example with six ages and heights is already centered to make each row add to zero:

Example $$A = \begin{bmatrix} 3 & -4 & 7 & 1 & -4 & -3 \\ 7 & -6 & 8 & -1 & -1 & -7 \end{bmatrix}$$

For this data, the sample covariance matrix S is easily computed. It is positive definite.

Variances and covariances $$S = \frac{1}{6-1}AA^{\mathrm{T}} = \begin{bmatrix} 20 & 25 \\ 25 & 40 \end{bmatrix}.$$

The two orthogonal eigenvectors of S are $\boldsymbol{u}_1$ and $\boldsymbol{u}_2$. Those are the left singular vectors (often called the *principal components*) of A. The Eckart-Young theorem says that **the vector $\boldsymbol{u}_1$ points along the closest line in Figure 7.3**.

The second singular vector $\boldsymbol{u}_2$ will be perpendicular to that closest line.

Important note PCA can be described using the symmetric $S = AA^{\mathrm{T}}/(n-1)$ or the rectangular A. No doubt S is the nicer matrix. But given the data in A, computing S can be a computational mistake. For large matrices, a direct SVD of A is faster and more accurate. By going to AA^{T} we square σ_1 and σ_r and the condition number σ_1/σ_r.

In the example, S has eigenvalues near 57 and 3. Their sum is $20 + 40 = 60$, the trace of S. The first rank one piece $\sqrt{57}\boldsymbol{u}_1\boldsymbol{v}_1^{\mathrm{T}}$ is much larger than the second piece $\sqrt{3}\boldsymbol{u}_2\boldsymbol{v}_2^{\mathrm{T}}$. **The leading eigenvector $\boldsymbol{u}_1 \approx (0.6, 0.8)$ tells us that the closest line in the scatter plot has slope near $8/6$.** The direction in the graph nearly produces a $6-8-10$ right triangle.

The Linear Algebra Behind PCA

Principal Component Analysis is a way to understand n sample points $\boldsymbol{a}_1, \ldots, \boldsymbol{a}_n$ in m-dimensional space—the data. That data plot is centered: all rows of A add to zero. The crucial connection to linear algebra is in the singular values and the left singular vectors $\boldsymbol{u}_i$ of A. Those come from the eigenvalues $\lambda_i = \sigma_i^2$ and the eigenvectors of the sample covariance matrix $S = AA^{\mathrm{T}}/(n-1)$.

The total variance in the data comes from the squared Frobenius norm of A:

$$\textbf{Total variance } \boldsymbol{T} = ||\boldsymbol{A}||_F^2/(\boldsymbol{n}-\boldsymbol{1}) = (||\boldsymbol{a}_1||^2 + \cdots + ||\boldsymbol{a}_n||^2)/(\boldsymbol{n}-\boldsymbol{1}). \tag{11}$$

This is the trace of S—the sum down the diagonal. Linear algebra tells us that the trace equals the **sum of the eigenvalues of the sample covariance matrix $\boldsymbol{S}$.**

The SVD is producing orthogonal singular vectors $\boldsymbol{u}_i$ that separate the data into uncorrelated pieces (with zero covariance). They come in order of decreasing variance, and the first pieces tell us what we need to know.

The trace of S connects the total variance to the sum of variances of the principal components $\boldsymbol{u}_1, \ldots, \boldsymbol{u}_r$:

$$\textbf{Total variance} \qquad \boldsymbol{T} = \boldsymbol{\sigma}_1^2 + \cdots + \boldsymbol{\sigma}_r^2. \tag{12}$$

The first principal component $\boldsymbol{u}_1$ accounts for (or "*explains*") a fraction σ_1^2/T of the total variance. The next singular vector $\boldsymbol{u}_2$ of A explains the next largest fraction σ_2^2/T. Each singular vector is doing its best to capture the meaning in a matrix—and all together they succeed.

The point of the Eckart-Young Theorem is that k singular vectors (acting together) explain *more of the data than any other set of k vectors*. So we are justified in choosing $\boldsymbol{u}_1$ to $\boldsymbol{u}_k$ as a basis for the k-dimensional subspace closest to the n data points.

The "effective rank" of A and S is the number of singular values above the point where noise drowns the true signal in the data. Often this point is visible on a "**scree plot**" showing the dropoff in the singular values (or their squares σ_i^2). Look for the "elbow" in the scree plot (Figure 7.2) where the signal ends and noise takes over.

Problem Set 7.3

1 Suppose A_0 holds these 2 measurements of 5 samples:

$$A_0 = \begin{bmatrix} 5 & 4 & 3 & 2 & 1 \\ -1 & 1 & 0 & 1 & -1 \end{bmatrix}$$

Find the average of each row and subtract it to produce the centered matrix A. Compute the sample covariance matrix $S = AA^{\mathrm{T}}/(n-1)$ and find its eigenvalues λ_1 and λ_2. What line through the origin is closest to the 5 samples in columns of A?

2 Take the steps of Problem 1 for this 2 by 6 matrix A_0:

$$A_0 = \begin{bmatrix} 1 & 0 & 1 & 0 & 1 & 0 \\ 1 & 2 & 3 & 3 & 2 & 1 \end{bmatrix}$$

3 The sample variances s_1^2, s_2^2 and the sample covariance s_{12} are the entries of S. Find S after subtracting row averages from $A_0 = \begin{bmatrix} 1 & 2 & 3 \\ 5 & 2 & 2 \end{bmatrix}$. What is σ_1 ?

4 From the eigenvectors of $S = AA^{\mathrm{T}}$, find the line (the $\boldsymbol{u}_1$ direction through the center point) and then the plane ($\boldsymbol{u}_1$ and $\boldsymbol{u}_2$ directions) closest to these four points in three-dimensional space:

$$A = \begin{bmatrix} 1 & -1 & 0 & 0 \\ 0 & 0 & 2 & -2 \\ 1 & 1 & -1 & -1 \end{bmatrix}.$$

5 Compare ordinary least squares (Section 4.3) with PCA (perpendicular least squares). They both give a closest line $C + Dt$ to the symmetric data $\boldsymbol{b} = -1, 0, 1$ at times $t = -3, 1, 2$.

$$A = \begin{bmatrix} 1 & -3 \\ 1 & 1 \\ 1 & 2 \end{bmatrix} \qquad \boldsymbol{b} = \begin{bmatrix} -1 \\ 0 \\ 1 \end{bmatrix}$$

Least squares : $A^{\mathrm{T}}A\widehat{\boldsymbol{x}} = A^{\mathrm{T}}\boldsymbol{b}$
PCA : eigenvector of AA^{T}
(singular vector $\boldsymbol{u}_1$ of A)

6 The idea of **eigenfaces** begins with N images : same size and alignment. Subtract the average image from each of the N images. Create the sample covariance matrix $S = \Sigma A_i A_i^{\mathrm{T}}/N - 1$ and find the eigenvectors ($=$ eigenfaces) with largest eigenvalues. They don't look like faces but their combinations come close to faces. Wikipedia gives a code for this *dimension reduction* pioneered by Turk and Pentland.

7 What are the singular values of $A - A_3$ if A has singular values $5, 4, 3, 2, 1$ and A_3 is the closest matrix of rank 3 to A ?

8 If A has $\sigma_1 = 9$ and B has $\sigma_1 = 4$, what are upper and lower bounds to σ_1 for $A + B$? Why is this true ?

7.4 The Victory of Orthogonality (and a Revolution)

If I look back at the linear algebra in this book, orthogonal matrices have won. You could say that they deserved to win. The key to their success goes all the way back to Section 1.2 on lengths and dot products. Let me recall some of their victories and add new ones.

1 The length of $Q\boldsymbol{x}$ equals the length of $\boldsymbol{x}$: $(Q\boldsymbol{x})^{\mathrm{T}}(Q\boldsymbol{x}) = \boldsymbol{x}^{\mathrm{T}}Q^{\mathrm{T}}Q\boldsymbol{x} = \boldsymbol{x}^{\mathrm{T}}\boldsymbol{x} = ||\boldsymbol{x}||^2$.

2 The dot product $(Q\boldsymbol{x})^{\mathrm{T}}(Q\boldsymbol{y})$ equals the dot product $\boldsymbol{x}^{\mathrm{T}}\boldsymbol{y}$: $\boldsymbol{x}^{\mathrm{T}}Q^{\mathrm{T}}Q\boldsymbol{y}=\boldsymbol{x}^{\mathrm{T}}\boldsymbol{y}$.

3 All powers Q^N and products $Q_1Q_2\dots Q_N$ of orthogonal matrices are orthogonal.

4 The projection matrix onto the column space of Q (m by n) is $QQ^{\mathrm{T}} = (QQ^{\mathrm{T}})^2$.

5 The least squares solution to $Q\boldsymbol{x} = \boldsymbol{b}$ $(m > n)$ is $\widehat{\boldsymbol{x}} = Q^{\mathrm{T}}\boldsymbol{b} = Q^{+}\boldsymbol{b}$ (pseudoinverse).

6 The eigenvectors of a symmetric matrix S can be chosen orthonormal : $S = Q\Lambda Q^{\mathrm{T}}$.

7 The singular vectors of every matrix are orthonormal : $A = Q_1\Sigma\, Q_2^{\mathrm{T}} = U\Sigma V^{\mathrm{T}}$.

8 The pseudoinverse of $U\Sigma V^{\mathrm{T}}$ is $V\Sigma^{+}U^{\mathrm{T}}$. The nonzeros in Σ^{+} (n by m) are $\frac{1}{\sigma_1}, \dots, \frac{1}{\sigma_r}$.

That list shows something important. The success of orthogonal matrices is tied to the **sum of squares definition of length:** $||\boldsymbol{x}||^2 = \boldsymbol{x}^{\mathrm{T}}\boldsymbol{x}$. In least squares, the derivative of $||A\boldsymbol{x} - \boldsymbol{b}||^2$ leads to a symmetric matrix $S = A^{\mathrm{T}}A$. Then S is diagonalized by an orthogonal matrix Q (the eigenvectors). A is diagonalized by two orthogonal matrices U and V. And here is more about orthogonal matrices : $\boldsymbol{A} = \boldsymbol{QS}$ and $\boldsymbol{A} = \boldsymbol{QR}$.

9 Every invertible matrix equals an orthogonal $\boldsymbol{Q}$ times a positive definite $\boldsymbol{S}$.

$$\textbf{Polar Decomposition} \qquad \boldsymbol{A} = \boldsymbol{U\Sigma\, V^{\mathrm{T}}} = (\boldsymbol{UV^{\mathrm{T}}})\,(\boldsymbol{V\Sigma V^{\mathrm{T}}}) = \boldsymbol{QS} \tag{1}$$

S is like a positive number r, and Q is like a complex number $e^{i\theta}$ of magnitude 1. Every complex number $x + iy$ can be written as $e^{i\theta}$ times r. Every matrix factors into Q times S. The square root of $(x-iy)\,(x+iy)$ is r. The square root of $A^{\mathrm{T}}A=V\Sigma^2V^{\mathrm{T}}$ is $S=V\Sigma V^{\mathrm{T}}$.

Example $Q = UV^{\mathrm{T}}$ and $S = V\Sigma V^{\mathrm{T}}$ come from the SVD of A in Section 7.1 :

$$A = \begin{bmatrix} 3 & 0 \\ 4 & 5 \end{bmatrix} = \frac{1}{\sqrt{5}}\begin{bmatrix} 2 & -1 \\ 1 & 2 \end{bmatrix}\sqrt{5}\begin{bmatrix} 2 & 1 \\ 1 & 2 \end{bmatrix} = \boldsymbol{QS}$$

10 Every invertible matrix equals an orthogonal $\boldsymbol{Q}$ times an upper triangular $\boldsymbol{R}$.

Example Q and R come from the Gram-Schmidt algorithm in Section 4.4 :

$$A = \begin{bmatrix} 3 & 0 \\ 4 & 5 \end{bmatrix} = \frac{1}{5}\begin{bmatrix} 3 & -4 \\ 4 & 3 \end{bmatrix}\begin{bmatrix} 5 & 4 \\ 0 & 3 \end{bmatrix} = \boldsymbol{QR}$$

Householder Reflection Matrices

Here is a neat construction of orthogonal matrices. Instead of rotations (determinant $= 1$) these are reflections (determinant $= -1$). Each matrix H is determined by a unit vector $\boldsymbol{u}$:

Reflection matrix $\quad \boldsymbol{H} = \boldsymbol{I} - 2\boldsymbol{u}\boldsymbol{u}^{\mathrm{T}}$ $\quad$ Notice that $\quad H\boldsymbol{u} = \boldsymbol{u} - 2\boldsymbol{u}\boldsymbol{u}^{\mathrm{T}}\boldsymbol{u} = -\boldsymbol{u}$

Clearly $H^{\mathrm{T}} = H$. Verify that H is an orthogonal matrix: $\boldsymbol{H}^{\mathrm{T}}\boldsymbol{H} = \boldsymbol{I}$ because $\boldsymbol{u}^{\mathrm{T}}\boldsymbol{u} = 1$:

$$H^{\mathrm{T}}H = (I - 2\boldsymbol{u}\boldsymbol{u}^{\mathrm{T}})(I - 2\boldsymbol{u}\boldsymbol{u}^{\mathrm{T}}) = I - 4\boldsymbol{u}\boldsymbol{u}^{\mathrm{T}} + 4\boldsymbol{u}\boldsymbol{u}^{\mathrm{T}}\boldsymbol{u}\boldsymbol{u}^{\mathrm{T}} = I \qquad (2)$$

One eigenvector of H is $\boldsymbol{u}$ with eigenvalue $\lambda = -1$: $H\boldsymbol{u} = \boldsymbol{u} - 2\boldsymbol{u}\boldsymbol{u}^{\mathrm{T}}\boldsymbol{u}$ simplifies to $\boldsymbol{u} - 2\boldsymbol{u} = -\boldsymbol{u}$. The other eigenvectors $\boldsymbol{x}$ fill the plane $\boldsymbol{u}^{\mathrm{T}}\boldsymbol{x} = 0$ that is orthogonal to $\boldsymbol{u}$.

$$\boldsymbol{u}^{\mathrm{T}}\boldsymbol{x} = 0 \quad \text{leads to} \quad H\boldsymbol{x} = \boldsymbol{x} - 2\boldsymbol{u}\boldsymbol{u}^{\mathrm{T}}\boldsymbol{x} = \boldsymbol{x} \quad \text{so } \lambda = 1$$

Notice: The eigenvalues 1 and -1 are real, since H is symmetric.
The eigenvalues have $|\lambda| = 1$, since H is orthogonal.
The eigenvalues are $\lambda = 1$ ($n-1$ times) and $\lambda = -1$ (one time).
The determinant of H is -1, from multiplying the λ's

Examples $\boldsymbol{u} = \dfrac{1}{\sqrt{2}}\begin{bmatrix} 1 \\ -1 \end{bmatrix}$ leads to the permutation $H = I - \dfrac{2}{2}\begin{bmatrix} 1 & -1 \\ -1 & 1 \end{bmatrix} = \begin{bmatrix} \mathbf{0} & \mathbf{1} \\ \mathbf{1} & \mathbf{0} \end{bmatrix}$

$\begin{bmatrix} \cos\theta \\ \sin\theta \end{bmatrix}$ leads to the reflection $H = I - 2\begin{bmatrix} \cos^2\theta & \cos\theta\sin\theta \\ \cos\theta\sin\theta & \sin^2\theta \end{bmatrix} = \begin{bmatrix} -\cos 2\theta & -\sin 2\theta \\ -\sin 2\theta & \cos 2\theta \end{bmatrix}$

Both examples have determinant -1. The neatest formula is the answer to this question: *If* $||\boldsymbol{a}|| = ||\boldsymbol{r}||$, *which matrix reflects* $\boldsymbol{a}$ *into* $H\boldsymbol{a} = \boldsymbol{r}$?

Choose the unit vector $\boldsymbol{u} = \dfrac{\boldsymbol{v}}{\lVert\boldsymbol{v}\rVert}$ with $\boldsymbol{v} = \boldsymbol{a} - \boldsymbol{r}$. $\quad$ Then $H\boldsymbol{a} = \left(I - 2\boldsymbol{u}\boldsymbol{u}^{\mathrm{T}}\right)\boldsymbol{a} = \boldsymbol{r}$

This leads to an error-free algorithm that factors A into QR: orthogonal times triangular. Q will be a product $H_n \ldots H_2 H_1$ of Householder reflections. One column at a time, we choose H_j to produce the desired column $\boldsymbol{r}_j$ in R. We keep a record of the vectors $\boldsymbol{u}_j$ that lead to each H_j (storing vectors not matrices). When we need the triangular R, we just use the $\boldsymbol{u}_j$ to recover those matrices H_j. This idea can replace Gram-Schmidt for $A = QR$.

Long ago Euler found another way to produce all orthogonal matrices. He rotated in some order around the x and y and z axes (three plane rotations). To an airline pilot those three rotations are roll and pitch and yaw.

Now we show that orthogonality also wins in "function space". The vectors $\boldsymbol{q}$ become functions $q(x)$. The dot products become integrals $\int f\, g\, dx$. The dimension becomes ∞.

Calculus : Vectors Become Functions

This is a book about linear algebra (for matrices). Orthogonality is just as important in calculus (for functions). Linear combinations of functions produce a function space. Lengths $||\boldsymbol{f}||$ are still the square roots of inner products $\boldsymbol{f}^{\mathrm{T}}\boldsymbol{f}$. Where the inner product of two vectors is a sum, *the inner product of two functions* $f(x)$ *and* $g(x)$ *is an integral*:

$$\textbf{Inner product} \quad (\boldsymbol{f},\boldsymbol{g})=\int f(x)\,g(x)\,dx \quad \textbf{Length } ||\boldsymbol{f}||^2=(f,f)=\int |f(x)|^2\,dx \tag{3}$$

The two great inequalities of mathematics extend from vectors to functions :

$$\textbf{Schwarz inequality} \quad \left|\int f(x)\,g(x)\,dx\right| \le ||f||\,||g|| \qquad \textbf{Triangle inequality} \quad ||f+g|| \le ||f||+||g|| \tag{4}$$

The orthogonal basis from Gram-Schmidt now contains functions $q_k(x)$ instead of vectors :

$$\textbf{Basis functions } \boldsymbol{q(x)} \qquad \boldsymbol{f(x)} = c_1q_1(x) + c_2q_2(x) + \cdots \text{ (infinite series)} \tag{5}$$

In a Fourier Series, those q's are sines or cosines. Other series use Chebyshev functions : **Chebfun.org** computes with very high accuracy. It is orthogonality $\int q_i\,q_k\,dx = 0$ that allows us to find each Fourier coefficient c_k. *Just multiply the series by* q_k *and integrate*:

$$\int f(x)\,q_k(x)\,dx = c_1\int q_1\,q_k\,dx + c_2\int q_2\,q_k\,dx + \cdots + c_k\int (q_k)^2 dx + \cdots \tag{6}$$

By orthogonality, all terms on the right side are zero except the kth term :

$$\textbf{Find } \boldsymbol{c_k} \qquad \int f(x)\,q_k(x)\,dx = c_k\int \Big(q_k(x)\Big)^2 dx \tag{7}$$

Basis functions like $q_k = \sin kx$ and $\cos kx$ are guaranteed to be orthogonal because they are eigenfunctions for a symmetric differential equation $A^{\mathrm{T}}A\,q_k = \lambda_k q_k$:

$$A = \frac{d}{dx} \qquad A^{\mathrm{T}} = -\frac{d}{dx} \qquad \boldsymbol{A^{\mathrm{T}}A\sin qx} = -\frac{d^2}{dx^2}\sin qx = \boldsymbol{q^2\sin qx}. \tag{8}$$

Symmetric matrix equations $A^{\mathrm{T}}A\boldsymbol{x} = \boldsymbol{b}$ become symmetric differential equations. Here is Newton's Law using $-A^{\mathrm{T}}A\boldsymbol{y}$ for acceleration (the second derivative) :

$$A = \frac{d}{dt} \text{ and } A^{\mathrm{T}} = -\frac{d}{dt} \qquad - A^{\mathrm{T}}A\boldsymbol{y} = \frac{d^2y}{dt^2} = \frac{\text{force}}{\text{mass}}.$$

The equations of physics and mechanics tell us about minimizing the energy $\frac{1}{2}\boldsymbol{y}^{\mathrm{T}}S\boldsymbol{y}$— just as we saw for positive definite matrices. The important point is that the basic laws of physics (as presented by Feynman) produce equations from minimum principles. They lead to symmetry and positive definiteness. Then the eigenfunctions are orthogonal.

The Counterrevolution from Sparsity

The message from classical mechanics was unmistakable: Symmetric equations have orthogonal eigenfunctions. But the world of data is not classical. We have large rectangular matrices and all kinds of training data. We want to understand that data in a simple way. Smaller sums of squares are not necessarily the overriding goal. *Sparse vectors* with few nonzeros are the easiest to understand.

We would like to build that goal of sparsity into the minimization. Right away a sum of squares is inappropriate. If we minimize $x_1^2 + \cdots + x_N^2$ subject to $x_1 + \cdots + x_N = 1$, the best vector $\boldsymbol{x}^*$ has N components all equal to $1/N$. This is the opposite of sparse! So we add a constraint that will push the solution $\boldsymbol{x}^*$ toward few nonzero components.

The difficulty is that the **cardinality of** $\boldsymbol{x}$ (the number of nonzero components) is not a convex function of $\boldsymbol{x}$. The set of vectors satisfying "card $(\boldsymbol{x}) \leq 3$" is not convex, because the halfway vector $\frac{1}{2}(\boldsymbol{x} + X)$ can have cardinality 6 (six nonzeros). We look for a different convex function whose minimization encourages sparsity. It was not at all sure that a good convex function would be found.

In fact there is an excellent substitute for cardinality. It is the $\boldsymbol{\ell}^1$ **norm of** $\boldsymbol{x}$:

$$\boldsymbol{\ell}^1\textbf{norm} \qquad ||\boldsymbol{x}||_1 = |x_1| + \cdots + |x_n| \text{ has } ||\boldsymbol{x} + \boldsymbol{y}||_1 \leq ||\boldsymbol{x}||_1 + ||\boldsymbol{y}||_1. \tag{9}$$

$\boldsymbol{\ell}^1$ is the first in a sequence of norms. The exponents in $\boldsymbol{\ell}^p$ go between $p = 1$ and $p = \infty$:

$$\boldsymbol{\ell}^p \textbf{ norm} \qquad ||x||_p = \left[|x_1|^p + \cdots + |x_n|^p\right]^{1/p} \text{ has } ||\boldsymbol{x} + \boldsymbol{y}||_p \leq ||\boldsymbol{x}||_p + ||\boldsymbol{y}||_p. \tag{10}$$

$p = 2$ gives our usual sum of squares (the $\boldsymbol{\ell}^2$ norm). At the top end $p = \infty$ is the "maximum norm" $||\boldsymbol{x}||_\infty = \max|x_i|$. If we go below $p = 1$, the triangle inequality fails: convexity is lost. As $p \to 0$, $||\boldsymbol{x}||_p$ approaches the cardinality of $\boldsymbol{x}$: not a norm.

Let me show how adding an $\boldsymbol{\ell}^1$ penalty to the $\boldsymbol{\ell}^2$ norm produces a sparser solution $\boldsymbol{x}^*$.

$$\textbf{No penalty} \quad 2E = (x-1)^2 + (y-1)^2 + (x+y)^2 \text{ has } \begin{array}{l} \partial E/\partial x = 2x + \ \ y = 1 \\ \partial E/\partial y = \ \ x + 2y = 1 \end{array}.$$

The minimum at $x = y = \frac{1}{3}$ (not sparse) is $E = \frac{2}{3}$. Now add a penalty $||\boldsymbol{x}||_1 = |x| + |y|$:

$$2E = (x-1)^2 + (y-1)^2 + (x+y)^2 + \mathbf{2}|\boldsymbol{x}| + \mathbf{2}|\boldsymbol{y}| \text{ has } \begin{array}{l} \partial E/\partial x = 2x + \ \ y = 0 \\ \partial E/\partial y = \ \ x + 2y = 0 \end{array}.$$

The minimizing solution $(\boldsymbol{x}^*, \boldsymbol{y}^*)$ moves from $\left(\frac{\mathbf{1}}{\mathbf{3}}, \frac{\mathbf{1}}{\mathbf{3}}\right)$ to $(\mathbf{0}, \mathbf{0})$: totally sparse.

There is a geometric way to see why an $\boldsymbol{\ell}^1$ minimization picks out a sparse solution. The vectors with $|v_1| + |v_2| \leq 1$ fill a diamond in the $v_1 - v_2$ plane. The ℓ^2 norm gives a circle $v_1^2 + v_2^2 \leq 1$ and the ℓ^∞ norm $|v_1| \leq 1, |v_2| \leq 1$ gives a square.

It is the corners of the diamond that touch a line at a sparse point. One of those sharp corners will lead to the minimum in the following typical optimization.

The Minimum of $||v||$ on the line $a_1v_1 + a_2v_2 = 1$

Which point on a diagonal line like $3v_1 + 4v_2 = 1$ is *closest to* $(0,0)$? The answer (and the meaning of "closest") will depend on the norm—the measure of distance. This is a good way to see important differences between ℓ^1 and ℓ^2 and ℓ^∞.

To find the closest point to $(0,0)$, rescale the ℓ^1 diamond and ℓ^2 circle and ℓ^∞ square (where the vectors have $||v||_1 \leq 1$ and $||v||_2 \leq 1$ and $||v||_\infty \leq 1$). When they touch the line $3v_1 + 4v_2 = 1$, the touching point v^* will solve our optimization problem:

> Minimize $||v||$ among vectors (v_1, v_2) on the line $3v_1 + 4v_2 = 1$

$\left(0, \frac{1}{4}\right)$ has $||v^*||_1 = \frac{1}{4}$ $\left(\frac{3}{25}, \frac{4}{25}\right)$ has $||v^*||_2 = \frac{1}{5}$ $\left(\frac{1}{7}, \frac{1}{7}\right)$ has $||v^*||_\infty = \frac{1}{7}$

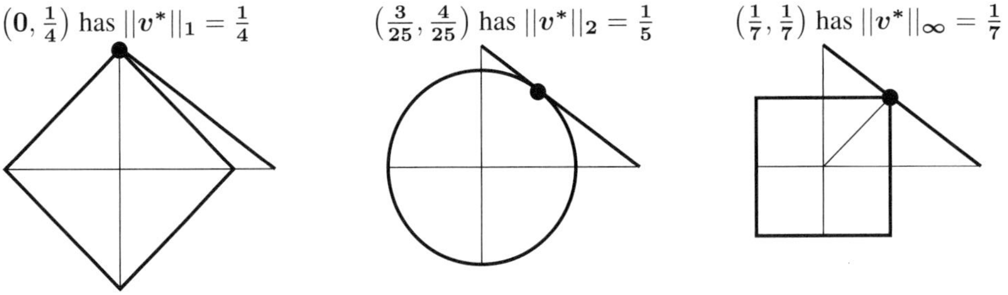

Figure 7.4: The solutions v^* to the ℓ^1 and ℓ^2 and ℓ^∞ minimizations. The first is **sparse**.

The first figure displays a highly important property of the minimizing solution to the ℓ^1 problem: **That solution v^* has a zero component**. The vector v^* is "sparse". To repeat, this happened because **a diamond will touch a line at a sharp point**. The line (or the hyperplane in high dimensions) contains the vectors that satisfy the constraints $Av = b$. The diamond expands to meet the line at a sharp corner!

The essential point is that **the solutions to ℓ^1 problems are sparse**. They have few nonzero components, and those components have meaning. By contrast the least squares solution (using ℓ^2) has many small and non-interesting components. By squaring, those small components hardly affect the ℓ^2 distance and they turn up in the ℓ^2 solution.

Minimizing with the ℓ^1 norm

The point of these pages is that computations are not all based on minimum energy. When sparsity is desirable, ℓ^1 comes in. We need new methods for new problems like these:

Basis Pursuit	Minimize $||x||_1$ subject to $Ax = b$
LASSO with Penalty	Minimize $||Ax - b||_2^2 + \lambda ||x||_1$
LASSO with Constraint	Minimize $||Ax - b||_2^2$ with $||x||_1 \leq T$

LASSO was invented by Tibshirani to improve on ordinary regression (= least squares).

Numerical methods for $\ell^2 + \ell^1$ minimization are always iterative: Step by step improvement in $\boldsymbol{x}$. We have learned to break each step into an ℓ^2 piece and separately an ℓ^1 piece. Lagrange's idea builds the constraints into the function to be minimized (by introducing Lagrange multipliers as unknowns). Those multipliers have meanings—they are derivatives of the minimum cost with respect to the constraints.

In mathematical finance the multipliers are the Greeks: represented by Greek letters. They measure the risks in buying an option—the right to buy or sell an asset when its value equals a designated strike price.

Problem Set 7.4

1 If $\boldsymbol{v}$ is a complex vector $(\boldsymbol{v}_1, \boldsymbol{v}_2)$, its length has $||\boldsymbol{v}||^2 = \overline{\boldsymbol{v}}^{\mathrm{T}}\boldsymbol{v} = |v_1|^2 + |v_2|^2$. For $\boldsymbol{v} = \begin{bmatrix} 1 \\ i \end{bmatrix}$ find $\overline{v}$ and $||\boldsymbol{v}||^2$ and $\overline{\boldsymbol{v}}^{\mathrm{T}}$ and $||\overline{\boldsymbol{v}}||^2$.

2 Find the eigenvalues and eigenvectors of a rotation matrix and a reflection matrix:

$$Q = \begin{bmatrix} \cos\theta & -\sin\theta \\ \sin\theta & \cos\theta \end{bmatrix} \qquad H = \frac{1}{\sqrt{2}}\begin{bmatrix} 1 & 1 \\ 1 & -1 \end{bmatrix}$$

3 A **permutation matrix** has the same columns as the identity matrix (in some order). *Explain why this permutation matrix and every permutation matrix is orthogonal*:

$$P = \begin{bmatrix} 0 & 1 & 0 & 0 \\ 0 & 0 & 1 & 0 \\ 0 & 0 & 0 & 1 \\ 1 & 0 & 0 & 0 \end{bmatrix} \text{ has orthonormal columns so } P^{\mathrm{T}}P = \_\_\_\_ \text{ and } P^{-1} = \_\_\_\_ .$$

When a matrix is symmetric or orthogonal, **it will have orthogonal eigenvectors**. This is the most important source of orthogonal vectors in applied mathematics.

4 Four eigenvectors of that matrix P are $\boldsymbol{x}_1 = (1,1,1,1), \boldsymbol{x}_2 = (1, i, i^2, i^3)$, $\boldsymbol{x}_3 = (1, i^2, i^4, i^6)$, and $\boldsymbol{x}_4 = (1, i^3, i^6, i^9)$. Multiply P times each vector to find $\lambda_1, \lambda_2, \lambda_3, \lambda_4$. The eigenvectors are the columns of the 4 by 4 **Fourier matrix** $\boldsymbol{F}$.

$$\text{Show that } Q = \frac{F}{2} = \frac{1}{2}\begin{bmatrix} 1 & 1 & 1 & 1 \\ 1 & i & -1 & -i \\ 1 & i^2 & 1 & -1 \\ 1 & i^3 & -1 & i \end{bmatrix} \text{ has orthonormal columns } \overline{\boldsymbol{Q}}^{\mathrm{T}}\boldsymbol{Q} = \boldsymbol{I}$$

5 **Haar wavelets are orthogonal vectors (columns of W) using only $1, -1$, and 0.**

$n = 4$

$$W = \begin{bmatrix} 1 & 1 & 1 & 0 \\ 1 & 1 & -1 & 0 \\ 1 & -1 & 0 & 1 \\ 1 & -1 & 0 & -1 \end{bmatrix}$$

Find $W^{\mathrm{T}}W$ and W^{-1} and the eight Haar wavelets for $n = 8$.

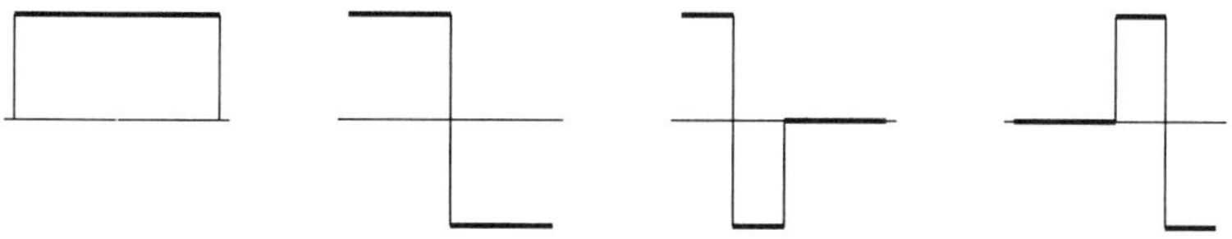

8 Learning from Data

This chapter describes a combination of linear algebra and calculus and machine learning. They produce an algorithm called "deep learning" that approximates a nonlinear function of many variables. That unknown function classifies the data, and recognizes the image, and translates the sentence, and finds the best move in Go. The learning function $F(\boldsymbol{x}, \boldsymbol{v})$ has to combine complexity with simplicity.

Simplicity comes from two key steps in each layer F_k of the overall learning function F :

$$\textbf{Layer } \boldsymbol{k-1} \textbf{ to layer } \boldsymbol{k} \qquad \boldsymbol{v}_k = \boldsymbol{F}_k(\boldsymbol{v}_{k-1}) = \text{ReLU}(A_k \boldsymbol{v}_{k-1} + \boldsymbol{b}_k) \tag{1}$$

That function F_k begins with a linear step to the vector $\boldsymbol{w}_k = A_k \boldsymbol{v}_{k-1} + \boldsymbol{b}_k$. Then comes a fixed nonlinear function like ReLU (my pronunciation is RayLoo). That function acts on every component of every vector $A_k \boldsymbol{v}_{k-1} + \boldsymbol{b}_k$ to give $\boldsymbol{v}_k$:

$$\boxed{\text{ReLU (any number } x) = \max(0, x) = \begin{matrix} 0 & \text{if } x \leq 0 \\ x & \text{if } x \geq 0 \end{matrix}}$$

It is amazing that this nonlinear function can achieve so much. The key is **composition**: functions of functions of functions. We have $L+1$ layers $\ell = 0, 1, \ldots, L$ (layer 0 is input, layer L is output). Composition produces $\boldsymbol{v}_L$ from $\boldsymbol{v}_{L-1}$ and eventually from the input $\boldsymbol{v}_0$:

$$\boldsymbol{v}_L = F_L(\boldsymbol{v}_{L-1}) = F_L(F_{L-1}(\ldots(F_1(\boldsymbol{v}_0)))) = \text{chain of nonlinear functions } F_k. \tag{2}$$

The "weights" $\boldsymbol{x}$ are all the entries in the matrices A_1 to A_L and the vectors $\boldsymbol{b}_1$ to $\boldsymbol{b}_L$. A deep network will have many weights from dense matrices A_k and fewer weights from convolution matrices (Section 8.2). The big computation is to choose those A's and $\boldsymbol{b}$'s in $\boldsymbol{x}$ so that $\boldsymbol{v}_L = F(\boldsymbol{x}, \boldsymbol{v}_0)$ is close to the known outputs $\boldsymbol{w}$ from the training data $\boldsymbol{v}_0$.

More training data should give more accurate weights A_k and $\boldsymbol{b}_k$—at a cost of extra computations. Those computations aim for weights that minimize the *loss*—the difference between $\boldsymbol{v}_L$ and $\boldsymbol{w}$. *Stochastic gradient descent* is a favorite algorithm to find those weights. *Backpropagation* computes derivatives of F from derivatives of every F_k by the chain rule. The design of F is a balance between computing cost and learning power.

Amazingly, F can achieve accurate outputs on new test data that it has never seen.

Reference: *Linear Algebra and Learning from Data*, Gilbert Strang, Wellesley-Cambridge (2019).

The Functions of Deep Learning

Suppose one of the digits $0, 1, \ldots, 9$ is drawn in a square. How does a person recognize which digit it is ? That neuroscience question is not answered here. How can a computer recognize which digit it is ? This is a machine learning question. Probably both answers begin with the same idea : *Learn from examples.*

So we start with M different images (the training set). An image will be a set of p small pixels—or a vector $\boldsymbol{v} = (v_1, \ldots, v_p)$. The component v_i tells us the "grayscale" of the ith pixel in the image : how dark or light it is. So we have M images each with p features : M vectors $\boldsymbol{v}$ in p-dimensional space. For every $\boldsymbol{v}$ in that training set we know the digit it represents.

In a way, we know a function. We have M inputs in $\mathbf{R}^p$ each with an output from 0 to 9. But we don't have a "rule". We are helpless with a new input. Machine learning proposes to create a rule that succeeds on (most of) the training images. But "succeed" means much more than that : The rule should give the correct digit for a much wider set of test images, taken from the same population. This essential requirement is called *generalization.*

What form shall the rule take ? Here we meet the fundamental question. Our first answer might be : $F(\boldsymbol{v})$ could be a linear function from $\mathbf{R}^p$ to $\mathbf{R}^{10}$ (a 10 by p matrix). The 10 outputs would be probabilities of the numbers 0 to 9. We would have $10p$ entries and M training samples to get mostly right.

The difficulty is : Linearity is far too limited. Artistically, two zeros could make an 8. 1 and 0 could combine into a handwritten 9 or possibly 6. Images don't add. In recognizing faces instead of numbers, we will need a lot of pixels—and the input-output rule is nowhere near linear.

Artificial intelligence languished for a generation, waiting for new ideas. There is no claim that the absolutely best class of functions has now been found. That class needs to allow a great many parameters (called weights). And it must remain feasible to compute all those weights (in a reasonable time) from knowledge of the training set.

The choice that has succeeded beyond expectation—and has turned shallow learning into deep learning—is *Continuous Piecewise Linear (CPL) functions.* **Linear** for simplicity, **continuous** to model an unknown but reasonable rule, and **piecewise** to achieve the nonlinearity that is an absolute requirement for real images and data.

This leaves the crucial question of computability. What parameters will quickly describe a large family of CPL functions ? Linear finite elements start with a triangular mesh. But specifying many individual nodes in $\mathbf{R}^p$ is expensive. Much better if those nodes are the *intersections* of a smaller number of lines (or hyperplanes). Please know that a regular grid is too simple.

Here is a first construction of a piecewise linear function of the data vector $\boldsymbol{v}$. Choose a matrix A_1 and vector $\boldsymbol{b}_1$. Then set to zero (this is the nonlinear step called ReLU) all negative components of $A_1\boldsymbol{v}+\boldsymbol{b}_1$. Then multiply by a matrix A_2 to produce 10 outputs in $\boldsymbol{w} = F(\boldsymbol{v}) = A_2(A_1\boldsymbol{v} + \boldsymbol{b}_1)_+$. That vector $(A_1\boldsymbol{v} + \boldsymbol{b}_1)_+$ forms a "hidden layer" between the input $\boldsymbol{v}$ and the output $\boldsymbol{w}$.

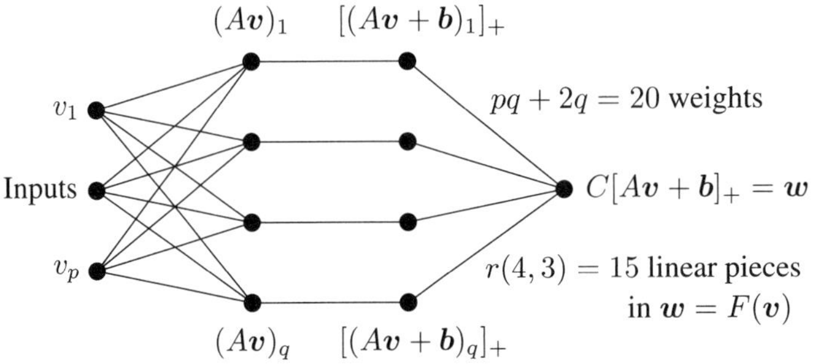

Actually the nonlinear function ReLU $(x) = x_+ = \max(x, 0)$ was originally smoothed into a logistic curve like $1/(1+e^{-x})$. It was reasonable to think that continuous derivatives would help in optimizing the weights $A_1, \boldsymbol{b}_1, A_2$. That proved to be wrong.

The graph of each component of $(A_1\boldsymbol{v} + \boldsymbol{b}_1)_+$ has two halfplanes (one is flat, from the zeros where $A_1\boldsymbol{v} + \boldsymbol{b}_1$ is negative). If A_1 is q by p, the input space $\mathbf{R}^p$ is sliced by q hyperplanes into r pieces. We can count those pieces ! This measures the "expressivity" of the overall function $F(\boldsymbol{v})$. The formula from combinatorics uses the binomial coefficients (see Section 8.1):

$$r(q,p) = \binom{q}{0} + \binom{q}{1} + \cdots + \binom{q}{p}$$

This number gives an impression of the graph of F with a hidden layer. But our function is not yet sufficiently expressive, and one more idea is needed.

Here is the indispensable ingredient in the learning function F. The best way to create complex functions from simple functions is by **composition**. Each F_i is linear (or affine) followed by the nonlinear ReLU : $F_i(\boldsymbol{v}) = (A_i\boldsymbol{v} + \boldsymbol{b}_i)_+$. Their composition is $F(\boldsymbol{v}) = F_L(F_{L-1}(\ldots F_2(F_1(\boldsymbol{v}))))$. We now have $L - 1$ hidden layers before the final output layer. The network becomes deeper as L increases. That depth can grow quickly for convolutional nets (with banded Toeplitz matrices A: many zeros).

The great optimization problem of deep learning is to compute weights A_i and $\boldsymbol{b}_i$ that will make the outputs $F(\boldsymbol{v})$ nearly correct—close to the digit $w(\boldsymbol{v})$ that the image $\boldsymbol{v}$ represents. This problem of minimizing some measure of $F(\boldsymbol{v}) - w(\boldsymbol{v})$ is solved by following a gradient downhill. The derivatives of this complicated function are computed by *backpropagation*—the workhorse of deep learning that executes the chain rule.

A historic competition in 2012 was to identify the 1.2 million images collected in ImageNet. The breakthrough neural network in AlexNet had 60 million weights. Its accuracy (after 5 days of stochastic gradient descent) cut in half the next best error rate. Deep learning had arrived.

Our goal here was to identify continuous piecewise linear functions as powerful approximators. That family is also convenient—closed under addition and maximization and composition. The magic is that the learning function $F(A_i, \boldsymbol{b}_i, \boldsymbol{v})$ gives accurate results on images $\boldsymbol{v}$ that F has never seen.

8.1 Piecewise Linear Learning Functions

Deep neural networks have evolved into a major force in machine learning. Step by step, the structure of the network has become more resilient and powerful—and more easily adapted to new applications. One way to begin is to describe essential pieces in the structure. Those pieces come together into a **learning function** $\boldsymbol{F(x, v)}$ with weights $\boldsymbol{x}$ that capture information from the training data $\boldsymbol{v}$—to prepare for use with new test data.

Here are important steps in creating that function F:

1	Key operation	**Composition** $\boldsymbol{F = F_3(F_2(F_1(x, v)))}$
2	Key rule	**Matrix chain rule for** x**-derivatives of** $\boldsymbol{F}$
3	Key algorithm	**Stochastic gradient descent to find the best weights** $\boldsymbol{x}$
4	Key subroutine	**Backpropagation to execute the chain rule**
5	Key nonlinearity	**ReLU**$\boldsymbol{(y)}$ **= max**$\boldsymbol{(y, 0)}$ **= ramp function**

Our first step is to describe the pieces $F_1, F_2, F_3, \ldots$ for one layer of neurons at a time. The weights $\boldsymbol{x}$ that connect the layers $\boldsymbol{v}$ are optimized in creating $\boldsymbol{F}$. The vector $\boldsymbol{v} = \boldsymbol{v}_0$ comes from the training set, and the function F_k produces the vector $\boldsymbol{v}_k$ at layer k. The whole success is to build the power of F from those pieces F_k in equation (1).

$\boldsymbol{F_k}$ is a Piecewise Linear Function of $\boldsymbol{v_{k-1}}$

The input to F_k is a vector $\boldsymbol{v}_{k-1}$ of length N_{k-1}. The output is a vector $\boldsymbol{v}_k$ of length N_k, ready for input to F_{k+1}. This function F_k has two parts, first linear and then nonlinear:

1. The linear part of F_k yields $A_k\boldsymbol{v}_{k-1} + \boldsymbol{b}_k$ (that bias vector $\boldsymbol{b}_k$ makes this "affine")
2. A fixed nonlinear function like ReLU is applied to *each component* of $A_k\boldsymbol{v}_{k-1} + \boldsymbol{b}_k$

$$\textbf{Layer } \boldsymbol{k} \qquad \boxed{\boldsymbol{v}_k = F_k(\boldsymbol{v}_{k-1}) = \text{ReLU}\,(A_k\boldsymbol{v}_{k-1} + \boldsymbol{b}_k)} \tag{1}$$

The training data for each sample is a feature vector $\boldsymbol{v}_0$. The matrices A_k and the column vectors $\boldsymbol{b}_k$ are "weights" to be chosen—so that the final output $\boldsymbol{v}_L$ is close to the correct output $\boldsymbol{w}$. Frequently *stochastic gradient descent* computes optimal weights $\boldsymbol{x} = (A_1, \boldsymbol{b}_1, \ldots, A_L)$ in the central computation of deep learning. Minimizing a loss function of $\boldsymbol{v}_L - \boldsymbol{w}$ relies on "*backpropagation*" to find the $\boldsymbol{x}$-derivatives.

The activation function $\text{ReLU}(y) = \max(y, 0)$ gives flexibility and adaptability. Linear steps alone were of limited power and ultimately they were unsuccessful.

ReLU is applied to every "neuron" in every internal layer. There are N_k neurons in layer k, containing the N_k outputs from equation (1). Notice that ReLU itself is continuous and piecewise linear, as its graph shows. (The graph is just a ramp with slopes 0 and 1. Its derivative is the usual step function.) When we choose ReLU, the composite function $\boldsymbol{F} = F_L \cdot\cdot (F_2(F_1(\boldsymbol{x}, \boldsymbol{v})))$ has an important and attractive property:

The learning function $\boldsymbol{F}$ **is continuous and piecewise linear in** $\boldsymbol{v}$**.**

One Internal Layer ($L = 2$)

Suppose we have measured $m = 3$ features of one sample point in the training set. Those features are the 3 components of the input vector $\boldsymbol{v} = \boldsymbol{v}_0$. Then the first function F_1 in the chain multiplies $\boldsymbol{v}_0$ by a matrix A_1 and adds an offset vector $\boldsymbol{b}_1$ (bias vector). If A_1 is 4 by 3 and the vector $\boldsymbol{b}_1$ is 4 by 1, we have 4 components of $A_0\boldsymbol{v}_0 + \boldsymbol{b}_0$.

That step found 4 combinations of the 3 original features in $\boldsymbol{v} = \boldsymbol{v}_0$. The 12 weights in the matrix A_1 were optimized over many feature vectors $\boldsymbol{v}_0$ in the training set, to choose a 4 by 3 matrix (and a 4 by 1 bias vector) that would find 4 insightful combinations.

The final step to reach $\boldsymbol{v}_1$ is to apply the nonlinear "activation function" to each of the 4 components of $A_1\boldsymbol{v}_0 + \boldsymbol{b}_1$. Historically, the graph of that nonlinear function was often given by a smooth "*S-curve*". Particular choices then and now are in Figure 8.1.

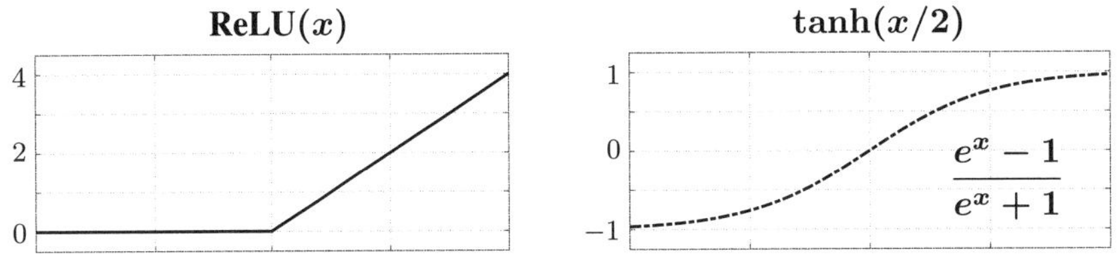

Figure 8.1: The **Re**ctified **L**inear **U**nit and a sigmoid option for nonlinearity.

Previously it was thought that a sudden change of slope would be dangerous and possibly unstable. But large scale numerical experiments indicated otherwise! A better result was achieved by the **ramp function** $\text{ReLU}(y) = \max(y, 0)$. We will work with ReLU:

$$\boxed{\textbf{Substitute } A_1\boldsymbol{v}_0 + \boldsymbol{b}_1 \textbf{ into ReLU to find } \boldsymbol{v}_1 \qquad (\boldsymbol{v}_1)_k = \max((A_1\boldsymbol{v}_0 + \boldsymbol{b}_1)_k, 0).} \quad (2)$$

Now we have the components of $\boldsymbol{v}_1$ at the four "neurons" in layer 1. The input layer held the three components of this particular sample of training data. We may have thousands of samples. The optimization algorithm found A_1 and $\boldsymbol{b}_1$, probably by gradient descent.

Suppose our neural net is shallow instead of deep. *It only has this first layer of* 4 *neurons.* Then the final step will multiply the 4-component vector $\boldsymbol{v}_1$ by a 1 by 4 matrix A_2 (a row vector). A vector $\boldsymbol{b}_2$ and the nonlinear ReLU are not applied to the output.

$$\boxed{\begin{array}{l}\text{Overall we compute } \boldsymbol{v}_2 = F(\boldsymbol{x}, \boldsymbol{v}_0) \text{ for each feature vector } \boldsymbol{v}_0 \text{ in the training set.} \\ \text{The steps are } \boldsymbol{v}_2 = A_2\,(\text{ReLU}\,(A_1\boldsymbol{v}_0 + \boldsymbol{b}_1)) = F(\boldsymbol{x}, \boldsymbol{v}_0).\end{array}} \quad (3)$$

The goal in optimizing $\boldsymbol{x} = A_1, \boldsymbol{b}_1, A_2$ is that the output values $\boldsymbol{v}_L = \boldsymbol{v}_2$ at the last layer $L = 2$ should correctly capture the important features of the training data $\boldsymbol{v}_0$.

At the beginning of machine learning the function F was *linear*—a severe limitation. Now F is certainly nonlinear. Just the inclusion of ReLU at each neuron in each layer has made a dramatic difference. It is the processing power of the computer that makes for fast operations on the data. For a deep network we depend on the speed of GPU's (the Graphical Processing Units that were developed for computer games).

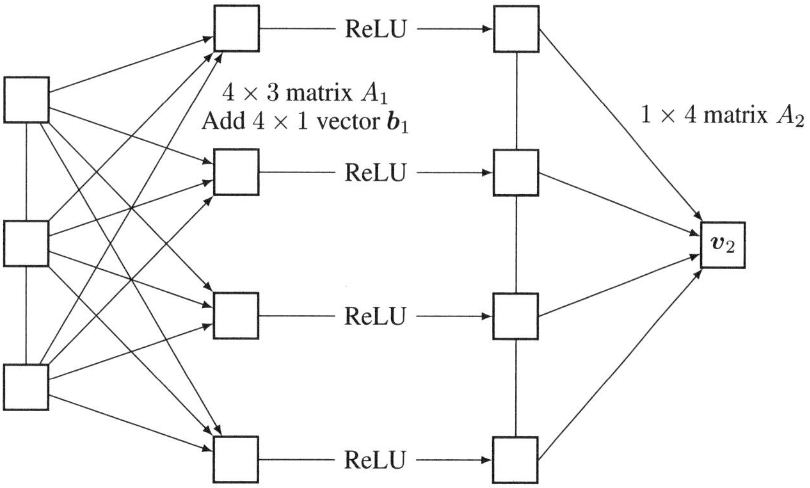

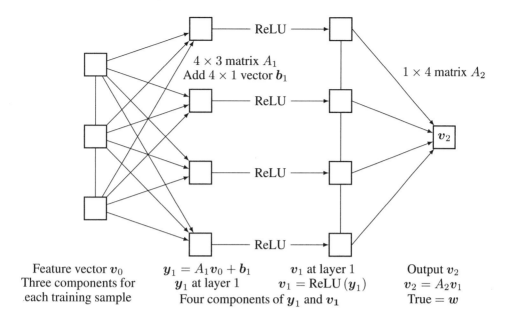

Figure 8.2: A feed-forward neural net with 4 neurons on **one internal layer**. The output $\boldsymbol{v}_2$ (plus or minus) classifies the input $\boldsymbol{v}_0$ (dog or cat). Then $\boldsymbol{v}_2$ is a composite measure of the 3-component feature vector $\boldsymbol{v}_0$. This net has 20 weights in A_k and $\boldsymbol{b}_k$.

For a **classification problem** each sample $\boldsymbol{v}_0$ of the training data is assigned **1 or −1**. We want the output $\boldsymbol{v}_2$ to have that correct sign (most of the time). For a **regression problem** we use the numerical value (**not just the sign**) of $\boldsymbol{v}_2$. Depending on our choice of loss function of $\boldsymbol{v}_2 - \boldsymbol{w}$, this problem can be like least squares or entropy minimization. We are choosing $\boldsymbol{x}$ = (weight matrices A_k and vectors $\boldsymbol{b}_k$) to minimize the loss. Here are 3 possible loss functions:

1 Square loss $$L(\boldsymbol{x}) = \frac{1}{N}\sum_{1}^{N} ||F(\boldsymbol{x},\boldsymbol{v}_2) - \text{true}||^2 \ : \ N \text{ training samples}$$

2 Hinge loss $$L(\boldsymbol{x}) = \frac{1}{N}\sum_{1}^{N} \textbf{max}\,(0, 1 - y\,F(\boldsymbol{x})) \text{ for } \textbf{classification } y = 1 \text{ or } -1$$

3 Cross-entropy loss $$L(\boldsymbol{x}) = -\frac{1}{N}\sum_{1}^{N} [y_i \log \widehat{y_i} + (1-y_i)\log(1-\widehat{y_i})] \text{ for } y_i = 0 \text{ or } 1$$

Our hope is that **the function *F* has "learned" the data**. This is machine learning. We aim for enough weights so that F has discovered what is important in recognizing *dog versus cat—or identifying an oncoming car versus a turning car.*

Machine learning doesn't aim to capture every detail of the numbers $0, 1, 2\ldots, 9$. It just aims to capture enough information to decide correctly *which number it is.*

The Initial Weights x_0 in Gradient Descent

The architecture in a neural net decides the form of the learning function $F(\boldsymbol{x}, \boldsymbol{v})$. The training data goes into $\boldsymbol{v}$. Then we *initialize* the weights $\boldsymbol{x}$ in the matrices A and vectors $\boldsymbol{b}$. Starting from $\boldsymbol{x}_0$, the optimization algorithm (see Section 8.3) computes weights $\boldsymbol{x}_1$ and $\boldsymbol{x}_2$ and onward, aiming to find weights $\boldsymbol{x}$ that minimize the total loss.

The question is : *What weights $\boldsymbol{x}_0$ to start with* ? Choosing $\boldsymbol{x}_0 = 0$ would be a disaster. Poor initialization is an important cause of failure in deep learning. A proper choice of the net and the initial $\boldsymbol{x}_0$ has random (and independent) weights that meet two requirements:

1. $\boldsymbol{x}_0$ has a carefully chosen variance σ^2.

2. The hidden layers in the neural net have enough neurons.

Hanin and Rolnick show that the initial variance σ^2 controls the mean of the computed weights. The layer widths control the variance of the weights. The key point is this: *Many-layered depth can reduce the loss on the training set. But if σ^2 is wrong or if width is sacrificed, then gradient descent can lose control of the weights.* The computation of $\boldsymbol{x}$ can explode to infinity or implode to zero.

The danger controlled by the variance σ^2 of $\boldsymbol{x}_0$ is exponentially large or exponentially small weights. The good choice is $\sigma^2 = 2$/fan-in. The fan-in is the maximum number of inputs to neurons (Figure 8.2 has fan-in $= 4$ at the output). Software like Keras makes a good choice of σ^2.

Max-Pooling to Reduce Dimensions

An image can have many pixels; the input $\boldsymbol{v}_0$ can include many features. Then the size of our computations can grow out of control. We have to reduce the number of components (sometimes called neurons) in a typical layer. If you look at the architecture of AlexNet, you see "pooling layers" to reduce dimension.

The most popular choice is simply **max-pooling**. Divide the image into blocks of 4 pixels. Replace each 2×2 block by 1×1: usually the largest of the four numbers. (Average pooling would keep the average of those numbers.) Here is AlexNet.

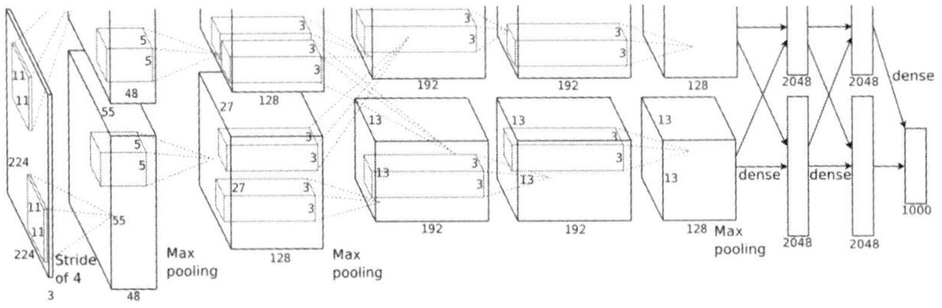

Figure 8.3: AlexNet uses two GPU's linked at certain layers. Pooling simply reduces the image dimensions. Most layers connect by convolution matrices A_k (Section 8.2) but the final layers are fully connected by dense matrices A_k. The input dimension is $150,528$ and AlexNet had $60,000,000$ weights—it won the 2012 competition to classify ImageNet.

The Graph of the Learning Function $F(v)$

The graph of $F(\boldsymbol{v})$ is a surface made up of many, many flat pieces—they are planes or hyperplanes that fit together along all the folds where ReLU produced a change of slope. This is like origami except that this graph has flat pieces going to infinity. And the graph might not be in $\mathbf{R}^3$—the feature vector $\boldsymbol{v} = \boldsymbol{v}_0$ has $N_0 = m$ components.

Part of the *mathematics of deep learning* is to estimate the number r of flat pieces and to visualize how they fit into one piecewise linear surface. That estimate comes after an example of a neural net with one internal layer. Each feature vector $\boldsymbol{v}_0$ contains m measurements like height, weight, age of a sample in the training set.

In Figure 8.2, F had three inputs in $\boldsymbol{v}_0$ and one output $\boldsymbol{v}_2$. Its graph will be a piecewise flat surface in 4-dimensional space. The height of the graph is $\boldsymbol{v}_2 = F(\boldsymbol{v}_0)$, over the point $\boldsymbol{v}_0$ in 3-dimensional space. Limitations of space in the book (and severe limits on the author's imagination) prevent us from drawing that graph in $\mathbf{R}^4$. Nevertheless we can try to count the flat pieces, based on 3 inputs and 4 neurons and 1 output.

Note You actually see points on the graph of F when you run examples on **playground.tensorflow.org**. This is a very instructive website.

That website offers four options for the training set of points $\boldsymbol{v}_0$. You choose the number of layers and neurons. Please choose the ReLU activation function ! Then the program counts epochs as gradient descent optimizes the weights. (An *epoch* sees all samples on average once.) If you have allowed enough layers and neurons to correctly classify the blue and orange training samples, you will see a polygon separating them. **That polygon shows where $\boldsymbol{F} = \mathbf{0}$.** It is the cross-section of the graph of $z = F(\boldsymbol{v})$ at height $z = 0$.

That polygon separating blue from orange (or *plus* from *minus* : this is classification) is the analog of a separating hyperplane in a Support Vector Machine. If we were limited to linear functions and a straight line between a blue ball and an orange ring around it, separation would be impossible. But for the deep learning function F this is not difficult...

We will discuss experiments on this **playground.tensorflow** site in the Problem Set.

Important Note : Fully Connected versus Convolutional

We don't want to mislead the reader. "Fully connected" nets are often not the most effective. If the weights around one pixel in an image can be repeated around all pixels (why not ?), then one row of A is all we need. The row can assign zero weights to faraway pixels. Local **convolutional neural nets (CNN's)** are in AlexNet and Section 8.2. The website **math.mit.edu/ENNUI** allows the reader to create a CNN with pooling.

You will see that the count r grows exponentially with the number of neurons and layers. That is a useful insight into the power of deep learning. We badly need insight because the size and depth of the neural network make it difficult to visualize in full detail.

Counting Flat Pieces in the Graph : One Internal Layer

It is easy to count entries in the weight matrices A_k and the bias vectors $\boldsymbol{b}_k$. Those numbers determine the function F. But it is far more interesting to count the number of flat pieces in the graph of F. This number $\boldsymbol{r}$ measures the **expressivity** of the neural network. $F(\boldsymbol{x}, \boldsymbol{v})$ is a more complicated function than we fully understand (at least so far). The system is deciding and acting on its own, without explicit approval of its "thinking". For driverless cars we will see the consequences fairly soon.

Suppose $\boldsymbol{v}_0$ has m components and $A_1\boldsymbol{v}_0+\boldsymbol{b}_1$ has N components. We have N functions of $\boldsymbol{v}_0$. Each of those linear functions is zero along a hyperplane (dimension $m-1$) in $\mathbf{R}^m$. When we apply ReLU to that linear function it becomes piecewise linear, with a *fold along that hyperplane.* On one side of the fold its graph is sloping, on the other side this component of $\boldsymbol{v}_1$ is zero from ReLU.

Then the next matrix A_2 combines those N piecewise linear functions of $\boldsymbol{v}_0$, so $\boldsymbol{v}_2$ now has *folds along N different hyperplanes in* $\mathbf{R}^m$. Those words describe each piecewise linear component of the next layer $A_2(\text{ReLU}(A_1\boldsymbol{v}_0 + \boldsymbol{b}_1))$: the output in our case.

You could think of N straight folds in a plane (the folds are actually along N hyperplanes in m-dimensional space). The first fold separates the plane in two pieces. The next fold from ReLU will leave us with four pieces. The third fold is more difficult to visualize, but the following Figure 8.4 shows that there are seven (*not eight*) pieces.

In combinatorial theory, we have a **hyperplane arrangement**—and a theorem of Tom Zaslavsky counts the pieces. The proof is presented in Richard Stanley's great textbook on *Enumerative Combinatorics* (2001). But that theorem is more complicated than we need, because it allows the fold lines to meet in all possible ways. We assume that the fold lines are in "general position"—$m+1$ folds don't meet. A neat way to count linear pieces is given by *On the Expressive Power of Deep Neural Networks* (arXiv : 1606.05336).

Theorem For $\boldsymbol{v}$ in m dimensions $\mathbf{R}^m$, suppose the graph of $F(\boldsymbol{v})$ has folds along N hyperplanes $H_1, \ldots, H_N$. Those come from N linear equations $\boldsymbol{a}_i^{\mathrm{T}}\boldsymbol{v} + b_i = 0$, in other words from ReLU at N neurons. Then the number of linear pieces of F and regions bounded by the N hyperplanes is $r(N, m)$:

$$\boldsymbol{r}(\boldsymbol{N}, \boldsymbol{m}) = \binom{\boldsymbol{N}}{0} + \binom{\boldsymbol{N}}{1} + \cdots + \binom{\boldsymbol{N}}{\boldsymbol{m}}. \tag{4}$$

These binomial coefficients are

$$\binom{\boldsymbol{N}}{i} = \frac{N!}{i!(N-i)!} \quad \text{with } 0! = 1 \text{ and } \binom{\boldsymbol{N}}{0} = 1 \text{ and } \binom{\boldsymbol{N}}{i} = 0 \text{ for } i > N.$$

Example The function $F(x, y, z) = \text{ReLU}(x) + \text{ReLU}(y) + \text{ReLU}(z)$ has 3 folds along the 3 planes $x = 0, y = 0, z = 0$. Those planes divide $\mathbf{R}^3$ into $r(3, 3) = 8$ pieces where $F = x + y + z$ and $x + z$ and x and 0 (and 4 more). Adding ReLU $(x + y + z - 1)$ gives a fourth fold and $r(4, 3) = 15$ pieces of $\mathbf{R}^3$. Not 16 because the new fold plane $x + y + z = 1$ does not meet the 8th original piece where $x < 0, y < 0, z < 0$.

George Polya's famous YouTube video *Let Us Teach Guessing* cut a cake by 5 planes. He helps the class to find $r(5,3) = 26$ pieces. Formula (4) allows m-dimensional cakes.

One hyperplane in $\mathbf{R}^m$ produces $\binom{1}{0} + \binom{1}{1} = 2$ regions. And $N = 2$ hyperplanes will produce $r(2,m) = 1+2+1 = 4$ regions provided $m > 1$. When $m = 1$ we have two folds in a line, which only separates the line into $r(2,1) = 3$ pieces.

The count $\boldsymbol{r}$ of linear pieces from N folds will follow from the recursive formula

$$\boxed{r(N,m) = r(N-1,m) + r(N-1,m-1).} \tag{5}$$

To understand that recursion, start with $N-1$ hyperplanes in $\mathbf{R}^m$ and $r(N-1,m)$ regions. *Add one more hyperplane* H (dimension $m-1$). The established $N-1$ hyperplanes cut H into $r(N-1,m-1)$ regions. Each of those pieces of H divides one existing region into two, adding $r(N-1,m-1)$ regions to the original $r(N-1,m)$; see Figure 8.4. So the recursion is correct, and we now apply equation (5) to compute $r(N,m)$.

The count starts at $r(1,0) = r(0,1) = \mathbf{1}$. Then (4) is proved by induction on $N+m$:

$$\begin{aligned} r(N-1,m) + r(N-1,m-1) &= \sum_0^m \binom{N-1}{i} + \sum_0^{m-1} \binom{N-1}{i} \\ &= \binom{N-1}{0} + \sum_0^{m-1} \left[\binom{N-1}{i} + \binom{N-1}{i+1}\right] \\ &= \binom{N}{0} + \sum_0^{m-1} \binom{N}{i+1} = \sum_0^m \binom{\boldsymbol{N}}{\boldsymbol{i}}. \end{aligned} \tag{6}$$

The two terms in brackets (second line) became one term because of a useful identity:

$\binom{N-1}{i} + \binom{N-1}{i+1} = \binom{N}{i+1}$ and the induction is complete.

Mike Giles made that presentation clearer, and he suggested Figure 8.4 to show the effect of the last hyperplane H. There are $r = 2^N$ linear pieces of $F(\boldsymbol{v})$ for $N \leq m$ and $r \approx N^m/m!$ pieces for $N >> m$, when the hidden layer has many neurons.

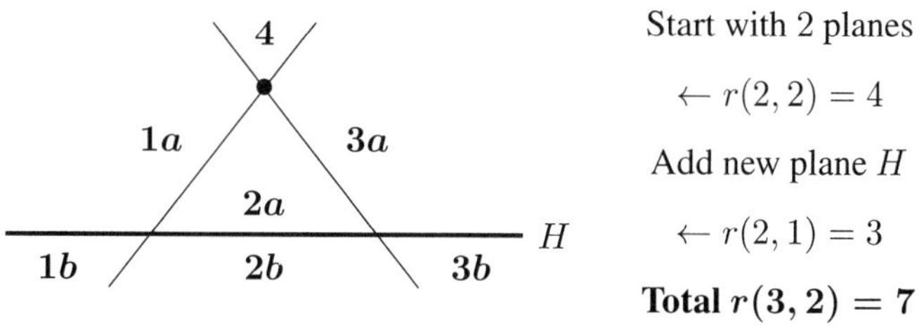

Figure 8.4: The $r(2,1) = 3$ pieces of H create 3 new regions. Then the count becomes $r(3,2) = 4+3 = 7$ flat regions in the continuous piecewise linear surface $\boldsymbol{v}_2 = F(\boldsymbol{v}_0)$. A fourth fold would cross all 3 existing folds and create 4 new regions, so $r(4,2) = 11$.

Flat Pieces of $F(v)$ with More Hidden Layers

Counting the linear pieces of $F(\boldsymbol{v})$ is much harder with 2 internal layers in the network. Again $\boldsymbol{v}_0$ and $\boldsymbol{v}_1$ have m and N_1 components. Now $A_1\boldsymbol{v}_1 + \boldsymbol{b}_1$ will have N_2 components before ReLU. Each one is like the function F for one layer, described above. Then application of ReLU will create new folds in its graph. Those folds are along the lines where a component of $A_1\boldsymbol{v}_1 + \boldsymbol{b}_1$ is zero.

Remember that each component of $A_1\boldsymbol{v}_1 + \boldsymbol{b}_1$ is piecewise linear, not linear. So it crosses zero (if it does) along a piecewise linear surface, not a hyperplane. The straight lines in Figure 8.4 for the folds in $\boldsymbol{v}_1$ will change to *piecewise straight lines* for the folds in $\boldsymbol{v}_2$. So the count becomes variable, depending on the details of $\boldsymbol{v}_0, A_1, \boldsymbol{b}_1, A_2$, and $\boldsymbol{b}_2$.

Still we can estimate the number of linear pieces. We have N_2 piecewise straight lines (or piecewise hyperplanes in $\mathbf{R}^m$) from N_2 ReLU's at the second hidden layer. If those lines were actually straight, we would have a total of $N_1 + N_2$ folds in each component of $\boldsymbol{v}_3 = F(\boldsymbol{v}_0)$. *Then the formula* (4) *to count the pieces would have* $N_1 + N_2$ *in place of* N. This estimate is confirmed by Hanin and Rolnick (*arXiv*: 1906.00904). So the count of neurons, not layers and depth, decides the number of linear pieces in $F(\boldsymbol{x}, \boldsymbol{v})$.

Composition $F_3(F_2(F_1(v)))$

The word "composition" would simply represent "matrix multiplication" if all our functions were linear: $F_k(\boldsymbol{v}) = A_k\boldsymbol{v}$. Then $F(\boldsymbol{v}_0) = A_3A_2A_1\boldsymbol{v}_0$: just one matrix. For nonlinear F_k the meaning is the same: Compute $\boldsymbol{v}_1 = F_1(\boldsymbol{v}_0)$, then $\boldsymbol{v}_2 = F_2(\boldsymbol{v}_1)$, and finally $\boldsymbol{v}_3 = F_3(\boldsymbol{v}_2)$. Now we get remarkable functions. This operation of **composition $F_3(F_2(F_1(\boldsymbol{v}_0)))$** is far more powerful in creating functions than addition !

For a neural network, composition produces continuous piecewise linear functions $F(\boldsymbol{v}_0)$. The 13th problem on Hilbert's list of 23 unsolved problems in 1900 asked a question about *all* continuous functions. A famous generalization of his question was this :

> Is every continuous function $F(x, y, z)$ of three variables the composition of continuous functions $G_1, \ldots, G_N$ of two variables ? The answer is *yes*.

Hilbert seems to have expected the answer *no*. But a positive answer was given in 1957 by Vladimir Arnold (age 19). His teacher Andrey Kolmogorov had previously created multivariable functions out of 3-variable functions. The 2-variable functions xy and x^y use 1-variable functions exp and log, and you must allow *addition*.

$$\boldsymbol{xy} = \mathbf{exp}(\mathbf{log}\, \boldsymbol{x} + \mathbf{log}\, \boldsymbol{y}) \quad \text{and} \quad \boldsymbol{x^y} = \mathbf{exp}(\mathbf{exp}(\mathbf{log}\, \boldsymbol{y} + \mathbf{log}\,\mathbf{log}\, \boldsymbol{x})). \tag{7}$$

So much to learn from the Web. A chapter of *Kolmogorov's Heritage in Mathematics* (Springer, 2007) connects these questions explicitly to neural networks.

Is the answer to Hilbert still yes for piecewise linear continuous functions ?

With enough layers and neurons, F can approximate any continuous $f(v)$. This is **universality**. New research by Telgarsky and Townsend shows the power of *rational functions*.

Problem Set 8.1

1 In the example $F = \text{ReLU}(x) + \text{ReLU}(y) + \text{ReLU}(z)$ that follows formula (4) for $r(N, m)$, suppose the 4th fold comes from $\text{ReLU}(x + y + z)$. Its fold plane $x + y + z = 0$ now meets the 3 original fold planes $x = 0, y = 0, z = 0$ at a single point $(0, 0, 0)$—*an exceptional case*. Describe the 16 (not 15) linear pieces of F = sum of these four ReLU's.

2 Suppose we have $m = 2$ inputs and N neurons on a hidden layer, so $F(x, y)$ is a linear combination of N ReLU's. Write out the formula for $r(N, 2)$ to show that the count of linear pieces of F has leading term $\frac{1}{2}N^2$.

3 Suppose we have $N = 18$ lines in a plane. If 9 are vertical and 9 are horizontal, how many pieces of the plane ? Compare with $r(18, 2)$ when the lines are in general position and no three lines meet.

4 What weight matrix A_1 and bias vector $\boldsymbol{b}_1$ will produce $\text{ReLU}(x + 2y - 4)$ and $\text{ReLU}(3x - y + 1)$ and $\text{ReLU}(2x + 5y - 6)$ as the $N = 3$ components of the first hidden layer ? (The input layer has 2 components x and y.) If the output w is the sum of those three ReLU's, how many flat pieces in the piecewise linear $w(x, y)$?

5 Folding a line four times gives $r(4, 1) = 5$ pieces. Folding a plane four times gives $r(4, 2) = 11$ pieces. According to formula (4), how many flat subsets come from folding $\mathbf{R}^3$ four times ? The flat subsets of $\mathbf{R}^3$ meet at 2D planes (like a door frame).

6 The binomial theorem finds the coefficients $\binom{N}{k}$ in $(\boldsymbol{a} + \boldsymbol{b})^N = \sum_0^N \binom{N}{k} a^k b^{N-k}$.

For $a = b = 1$ what does this reveal about those coefficients and $r(N, m)$ for $m \geq N$?

7 In Figure 8.4, one more fold will produce 11 flat pieces in the graph of $z = F(x, y)$. Check that formula (4) gives $r(4, 2) = 11$. How many pieces after five folds ?

8 Explain with words or show with graphs why each of these statements about Continuous Piecewise Linear functions (CPL functions) is true :

M	The maximum $M(x, y)$ of two CPL functions $F_1(x, y)$ and $F_2(x, y)$ is CPL.
S	The sum $S(x, y)$ of two CPL functions $F_1(x, y)$ and $F_2(x, y)$ is CPL.
C	If the one-variable functions $y = F_1(x)$ and $z = F_2(y)$ are CPL, so is the composition $C(x) = z = (F_2(F_1(x))$.

9 How many weights and biases are in a network with $m = N_0 = 4$ inputs in each feature vector $\boldsymbol{v}_0$ and $N = 6$ neurons on each of the 3 hidden layers ? How many activation functions (ReLU) are in this network, before the final output ?

10 (Experimental) In a neural network with two internal layers and a total of 10 neurons, should you put more of those neurons in layer **1** or layer **2** ?

Problems 11–13 use the blue ball, orange ring example on playground.tensorflow.org with one hidden layer and activation by ReLU (not Tanh). When learning succeeds, a white polygon separates blue from orange in the figure that the code creates.

11 Does learning succeed for $N = 4$? What is the count $r(N, 2)$ of flat pieces in $F(\boldsymbol{x})$? The white polygon shows where flat pieces in the graph of $F(\boldsymbol{x})$ change sign as they go through the base plane $z = 0$. How many sides in the polygon ?

12 Reduce to $N = 3$ neurons in one layer. Does F still classify blue and orange correctly ? How many flat pieces $r(3, 2)$ in the graph of $F(\boldsymbol{v})$ and how many sides in the separating polygon ?

13 Reduce further to $N = 2$ neurons in one layer. Does learning still succeed ? What is the count $r(2, 2)$ of flat pieces ? How many folds in the graph of $F(\boldsymbol{v})$? How many sides in the white separator ?

14 Example 2 has blue and orange in two quadrants each. With one layer, do $N = 3$ neurons and even $N = 2$ neurons classify that training data correctly ? How many flat pieces are needed for success ? Describe the unusual graph of $F(\boldsymbol{v})$ when $N = 2$.

15 Example 4 with blue and orange spirals is much more difficult ! With one hidden layer, can the network learn this training data ? Describe the results as N increases.

16 Try that difficult example with two hidden layers. Start with $4 + 4$ and $6 + 2$ and $2 + 6$ neurons. Is $2 + 6$ better or worse or more unusual than $6 + 2$?

17 How many neurons bring complete separation of the spirals with two hidden layers ? Can three layers succeed with fewer neurons than two layers ?

I found that $4 + 4 + 2$ and $4 + 4 + 4$ neurons give very unstable iterations for that spiral graph. There were spikes in the training loss until the algorithm stopped trying. playground.tensorflow.org was created by Daniel Smilkov.

18 What is the smallest number of pieces that 20 fold lines can produce in a plane ?

19 How many pieces are produced from 10 vertical and 10 horizontal folds ?

20 What is the maximum number of pieces from 20 fold lines in a plane ?

8.2 Convolutional Neural Nets

This section is about networks with a different architecture. An m by n matrix still connects a layer with n neurons to the next layer with m neurons. Up to now, the layers were *fully connected: A had mn independent weights.* Now we might have only $E = 3$ or $E^2 = 9$ independent weights in A.

The fully connected "dense net" will be extremely inefficient for image recognition. First, the weight matrices A will be huge. If one image has 200 by 300 pixels, then its input layer has $60,000$ components. The weight matrix A_1 for the first hidden layer has $60,000$ columns. The problem is : We are looking for connections between faraway pixels.

Almost always, the important connections in an image are **local**.

Music has a 1D local structure

Images have a 2D local structure (3 copies for red-green-blue)

Video has a 3D local structure : Images in a time series

More than this, the search for structure is essentially *the same everywhere in the image.* There is normally no reason to process one part of a text or image or video differently from its other parts. We can use the same weights in all parts : *Share the weights.* The neural net of local connections between pixels is **shift-invariant**: the same everywhere.

The result is a big reduction in the number of independent weights. Suppose each neuron is connected to only E neurons on the next layer, and those connections are the same for all neurons. Then the matrix A between those layers has only E independent weights x. The optimization of those weights becomes enormously faster. In reality we have time to create several different channels with their own E or E^2 weights. They can look for edges in different directions (horizontal, vertical, and diagonal).

In one dimension, a banded shift-invariant matrix is a **Toeplitz matrix** or a **filter**. Multiplication by that matrix A is a **convolution** $\boldsymbol{x} * \boldsymbol{v}$. The network of connections between the layers is a **Convolutional Neural Net** (**CNN** or **ConvNet**). Here $E = 3$.

$$A = \begin{bmatrix} x_1 & x_0 & x_{-1} & 0 & 0 & 0 \\ 0 & x_1 & x_0 & x_{-1} & 0 & 0 \\ 0 & 0 & x_1 & x_0 & x_{-1} & 0 \\ 0 & 0 & 0 & x_1 & x_0 & x_{-1} \end{bmatrix} \qquad \begin{array}{l} \boldsymbol{v} = (v_0, v_1, v_2, v_3, v_4, v_5) \\ \boldsymbol{y} = A\boldsymbol{v} = \quad (y_1, y_2, y_3, y_4) \\ N+2 \text{ inputs } \boldsymbol{v} \text{ and } N \text{ outputs } \boldsymbol{y} \end{array}$$

It is valuable to see A as *a combination of* **shift matrices $\boldsymbol{L}, \boldsymbol{C}, \boldsymbol{R}$**: Left, Center, Right.

$$\textbf{Each shift has a diagonal of 1's} \qquad \boldsymbol{A = x_1 L + x_0 C + x_{-1} R}$$

Then the derivatives of $\boldsymbol{y} = A\boldsymbol{v} = \boldsymbol{x_1 L v + x_0 C v + x_{-1} R v}$ are exceptionally simple :

$$\frac{\partial(\text{output})}{\partial(\text{weight})} \qquad \boxed{\frac{\partial \boldsymbol{y}}{\partial x_1} = L\boldsymbol{v} \qquad \frac{\partial \boldsymbol{y}}{\partial x_0} = C\boldsymbol{v} \qquad \frac{\partial \boldsymbol{y}}{\partial x_{-1}} = R\boldsymbol{v}} \tag{1}$$

Convolution of Vectors $x * v$

The convolution of two vectors is written $\boldsymbol{x} * \boldsymbol{v} = (2,1,2) * (3,3,1)$. Computing the result $\boldsymbol{x} * \boldsymbol{v} = (6,9,11,7,2)$ is like multiplying the numbers 212 and 331, without carrying.

$$\begin{array}{rrrrr} & & 3 & 3 & 1 \\ & & 2 & 1 & 2 \\ \hline & & 6 & 6 & 2 \\ & 3 & 3 & 1 & \\ 6 & 6 & 2 & & \\ \hline \mathbf{6} & \mathbf{9} & \mathbf{11} & \mathbf{7} & \mathbf{2} \end{array}$$

Notice that we leave
$6+3+2=11$ as is (no carrying)
Same steps for multiplying
$2x^2+x+2$ times $3x^2+3x+1$
That answer would be
$\mathbf{6x^4+9x^3+11x^2+7x+2}$

The previous page just put the numbers $(x_1, x_0, x_{-1}) = (2,1,2)$ on three diagonals of A. Then ordinary multiplication 212 times 331 converts to matrix-vector multiplication $A\boldsymbol{v}$. When $\boldsymbol{x}$ has length $j+1$ and $\boldsymbol{v}$ has length $k+1$, convolution $\boldsymbol{x} * \boldsymbol{v}$ has length $j+k+1$.

Convolution as a Moving Window

Suppose we average each number with the next number in $\boldsymbol{v} = (1,3,5)$. The result is $\boldsymbol{y} = (2,4)$. This is a typical convolution of $\boldsymbol{v}$ with the averaging vector $\boldsymbol{x} = \left(\frac{1}{2}, \frac{1}{2}\right)$:

$$\boldsymbol{y} = A\boldsymbol{v} = \begin{bmatrix} \frac{1}{2} & \frac{1}{2} & 0 \\ 0 & \frac{1}{2} & \frac{1}{2} \end{bmatrix} \begin{bmatrix} 1 \\ 3 \\ 5 \end{bmatrix} = \begin{bmatrix} 2 \\ 4 \end{bmatrix}$$

Notice the decision not to pad $\boldsymbol{v}$ with a zero at each end (and extend A to be 4 by 5). That would lead to 4 outputs $\boldsymbol{y}$ instead of 2. It would be consistent with multiplying numbers: 11 times 135 is 1485 and dividing by 2 gives $\frac{1}{2}, 2, 4, \frac{5}{2}$.

Python and MATLAB offer both versions of convolution, padded or not (and a third option with three outputs). We will choose not to pad the input with zeros. Each row of A is a perfect shift of the previous row, as above.

Often the convolution process $A\boldsymbol{v}$ is seen as a moving window. The window starts at 1 3 and moves to 3 5. Averaging produces 2 in the first window and 4 in the second window. The whole point of "shift invariance" is that *a convolution does the same thing in each window.*

Windows in Two Dimensions

This approach is helpful in two dimensions where the window is a square or a rectangle. It is easy to see 2 by 2 overlapping windows filling an n by n square. There would be $(n-1)^2$ windows and an average over each window. The matrix A has $(n-1)^2$ outputs from n^2 inputs. Each row of A has $\left(\frac{1}{4}, \frac{1}{4}, \frac{1}{4}, \frac{1}{4}\right)$: four nonzeros to average over a 2 by 2 window.

To produce that two-dimensional averaging convolution, first average neighboring pairs *in every row*. Then average every pair of neighbors *in every column*. The combination will average over every square. It works because the 2D averaging matrix has rank 1:

$$\begin{matrix}\text{average over a}\\ \text{2 by 2 window}\end{matrix} = \begin{bmatrix}\frac{1}{4} & \frac{1}{4}\\ \frac{1}{4} & \frac{1}{4}\end{bmatrix} = \begin{bmatrix}\frac{1}{2}\\ \frac{1}{2}\end{bmatrix}\begin{bmatrix}\frac{1}{2} & \frac{1}{2}\end{bmatrix} = \begin{matrix}\text{column averages}\\ \text{of row averages}\end{matrix}$$

Experiments have pointed to $E = 3$ as a good choice for convolutions in deep learning. Deep learning could look for an E by E matrix like this of rank 1 — and apply the two averages along rows and then down columns. The rank 1 matrix for 2D convolution from an E by $E = 3$ by 3 matrix would have $2E - 1 = 5$ instead of $E^2 = 9$ unknown weights.

For clarity here are *nine* 3 *by* 3 *windows that fill a* 5 *by* 5 *square*.

3 × 3 windows in a 5 × 5 square

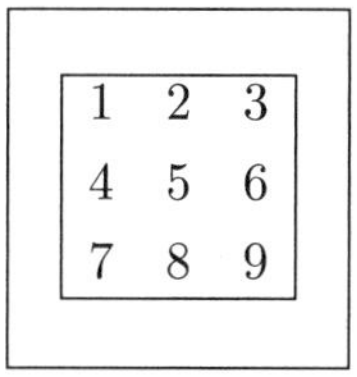

Move this window left / right / up / down / left up / right up / left down / right down to produce 9 windows centered in these nine positions

2D Convolution by One Large Matrix

When the input $\boldsymbol{v}$ is an image, convolution becomes two-dimensional. $E = 3$ numbers x_{-1}, x_0, x_1 change to $E^2 = 3^2$ independent weights. The inputs v_{ij} have two indices and $\boldsymbol{v}$ represents $(N+2)^2$ pixels. The outputs have only N^2 pixels unless we pad $\boldsymbol{v}$ with zeros at the boundary. The 2D convolution $A\boldsymbol{v}$ is a linear combination of **9 shifts of** $\boldsymbol{v}$.

	x_{11}	x_{01}	x_{-11}	**Input image**	v_{ij}	i, j from $(0,0)$ to $(N+1, N+1)$
Weights	x_{10}	x_{00}	x_{-10}	**Output image**	y_{ij}	i, j from $(1,1)$ to (N,N)
	x_{1-1}	x_{0-1}	x_{-1-1}	**Shifts L, C, R, U, D**		= **L**eft, **C**enter, **R**ight, **U**p, **D**own

$$A = x_{11}LU + x_{01}CU + x_{-11}RU + x_{10}L + x_{00}C + x_{-10}R + x_{1-1}LD + x_{0-1}CD + x_{-1-1}RD$$

This expresses the convolution matrix A as a combination of 9 shifts. The derivatives of the output $\boldsymbol{y} = A\boldsymbol{v}$ are again exceptionally simple. We use the nine derivatives in (2) to create the gradients $\boldsymbol{\nabla F}$ and $\boldsymbol{\nabla L}$ (learning function, loss function) that are needed in gradient descent to improve the weights $\boldsymbol{x}_k$. The next iteration $\boldsymbol{x}_{k+1} = \boldsymbol{x}_k - s\boldsymbol{\nabla L}_k$ has weights that better match the correct outputs from the training data.

Backpropagation finds these 9 derivatives of $\boldsymbol{y} = A\boldsymbol{v}$ with respect to 9 weights:

$$\frac{\partial \boldsymbol{y}}{\partial x_{11}} = LU\boldsymbol{v} \quad \frac{\partial \boldsymbol{y}}{\partial x_{01}} = CU\boldsymbol{v} \quad \frac{\partial \boldsymbol{y}}{\partial x_{-11}} = RU\boldsymbol{v} \quad \ldots \quad \frac{\partial \boldsymbol{y}}{\partial x_{-1-1}} = RD\boldsymbol{v} \qquad (2)$$

CNN's can readily afford to have B **parallel channels** (and that number B can vary as we go deeper into the net). The count of weights in $\boldsymbol{x}$ is so much reduced by weight sharing and weight locality, that we don't need and we can't expect one set of $E^2 = 9$ weights to do all the work of a convolutional net. *B convolutions give the next layer.*

A convolution is a combination of shift matrices (producing a filter or Toeplitz matrix)
In deep learning, the coefficients in the combination will be the "weights" to be learned.
Several convolutions in parallel will extract more information from the image.

Two-dimensional Convolutional Nets

Now we come to the real success of CNN's : **Image recognition**. ConvNets and deep learning have produced a small revolution in computer vision. The applications are to self-driving cars, drones, medical imaging, security, robotics—there is nowhere to stop. Our interest is in the algebra and geometry and intuition that makes all this possible.

In two dimensions (for images) the matrix A is **block Toeplitz**. Each small block is E by E. This is a familiar structure in computational engineering. The count E^2 of independent weights to be optimized is far smaller than for a fully connected network. The same weights are used around all pixels (*shift-invariance*). The matrix A produces a 2D convolution $\boldsymbol{x} * \boldsymbol{v}$. Frequently A is called a **filter**.

To understand an image, look to see where it changes. *Find the edges.* Our eyes look for sharp cutoffs and steep gradients. The computer can do the same by creating a filter. The difficulty with two or more dimensions is that edges can have many directions. We will need horizontal and vertical and diagonal filters for the test images. And filters have many purposes, including smoothing and gradient detection and edge detection.

Smoothing For functions, one smoother is *convolution with a Gaussian* $e^{-x^2/2\sigma^2}$. For vectors, we could convolve $v * G$ with $G = \frac{1}{17}(1, 4, 7, 4, 1)$.

Gradient detection Image processing (as distinct from learning by a CNN) needs filters that detect the gradient. They contain specially chosen weights. We mention some simple filters just to indicate how they can find first derivatives of $\boldsymbol{f}$.

One dimension $E = 3$ $\quad (x_1, x_0, x_{-1}) = \left(-\frac{\mathbf{1}}{\mathbf{2}}, \mathbf{0}, \frac{\mathbf{1}}{\mathbf{2}}\right)$ Then $(A\boldsymbol{v})_i = \frac{1}{2}v_{i+1} - \frac{1}{2}v_{i-1}$.

Two dimensions These 3×3 *Sobel operators* approximate $\partial/\partial x$ and $\partial/\partial y$:

$$E = 3 \qquad \frac{\boldsymbol{\partial}}{\boldsymbol{\partial x}} \approx \frac{1}{2}\begin{bmatrix} -1 & 0 & 1 \\ -2 & 0 & 2 \\ -1 & 0 & 1 \end{bmatrix} \qquad \frac{\boldsymbol{\partial}}{\boldsymbol{\partial y}} \approx \frac{1}{2}\begin{bmatrix} -1 & -2 & -1 \\ 0 & 0 & 0 \\ 1 & 2 & 1 \end{bmatrix} \tag{3}$$

Edge detection Those weights were created for image processing, to locate the most important features of a typical image : *its edges*. These would be candidates for E by E filters inside a 2D convolutional matrix A. But remember that in deep learning, weights like $\frac{1}{2}$ and $-\frac{1}{2}$ are not chosen by the user. They are created from the training data.

Important The filters described so far all have a **stride** $S = 1$. For a larger stride, the *moving window takes longer steps* as it moves across the image. Here is the matrix A for a 1-dimensional 3-weight filter with a stride of 2. Notice especially that the length of the output $\boldsymbol{y} = A\boldsymbol{v}$ is reduced by that factor of 2 (previously four outputs and now two):

$$\textbf{Stride } S = \mathbf{2} \qquad A = \begin{bmatrix} \boldsymbol{x_1} & \boldsymbol{x_0} & \boldsymbol{x_{-1}} & 0 & 0 \\ 0 & 0 & \boldsymbol{x_1} & \boldsymbol{x_0} & \boldsymbol{x_{-1}} \end{bmatrix} \tag{4}$$

Now the nonzero weights like x_1 are two columns apart (S columns apart for stride S). In 2D, a stride $S = 2$ reduces each direction by 2 and the size of the output by 4.

Counting the Number of Inputs and Outputs

In a one-dimensional problem, suppose a layer has N neurons. We apply a convolutional matrix with E nonzero weights. The stride is S, and we pad the input signal by P zeros at each end. How many outputs (M numbers) does this filter produce?

$$\boxed{\textbf{Karpathy's formula} \qquad \boldsymbol{M = \frac{N - E + 2P}{S} + 1}} \tag{5}$$

In a 2D or 3D problem, this 1D formula applies in each direction.

Suppose $E = 3$ and the stride is $S = 1$. If we add one zero ($P = 1$) at each end, then

$$M = N - 3 + 2 + 1 = N \qquad \text{(input length = output length)}$$

This case $2P = E - 1$ with stride $S = 1$ is the most common architecture for CNN's.

If we don't pad the ends of the input with zeros, then $P = 0$ and $M = N - 2$ (as in the 4 by 6 matrix A at the start of this section). In 2 dimensions this becomes $M^2 = (N-2)^2$. We lose neurons this way, but we avoid any artificial zero-padding.

Now suppose the stride is $S = 2$. Then $N - E$ must be an even number. Otherwise the formula (5) produces a fraction. Here are two examples of success for stride $S = 2$, with $N - E = 5 - 3$ and padding $P = 0$ or $P = 1$ at both ends of the $N = 5$ inputs:

$$\begin{matrix}\textbf{Stride} \\ \mathbf{2}\end{matrix} \quad \begin{bmatrix} x_{-1} & x_0 & x_1 & 0 & 0 \\ 0 & 0 & x_{-1} & x_0 & x_1 \end{bmatrix} \begin{bmatrix} x_{-1} & x_0 & x_1 & 0 & 0 & 0 & 0 \\ 0 & 0 & x_{-1} & x_0 & x_1 & 0 & 0 \\ 0 & 0 & 0 & 0 & x_{-1} & x_0 & x_1 \end{bmatrix}$$

A Deep Convolutional Network

Recognizing images is a major application of deep learning (and a major success). The success came with the creation of AlexNet and the development of convolutional nets. This page will describe a deep network of local convolutional matrices for image recognition. We follow the prize-winning paper of Simonyan and Zisserman from ICLR 2015. That paper recommends a deep architecture of $L = 16$–19 layers with small (3×3) filters. The network has a breadth of B parallel channels (B images on each layer).

If the breadth B were to stay the same at all layers, and all filters had E by E local weights, a straightforward formula would estimate the number W of weights in the net:

$$\boxed{\boldsymbol{W \approx LBE^2} \qquad \textbf{\textit{L} layers, \textit{B} channels, \textit{E} by \textit{E} local convolutions}} \qquad (6)$$

Notice that W does not depend on the count of neurons on each layer. This is because the E^2 weights are shared. Pooling will reduce the count of neurons.

It is very common to end a CNN with fully-connected layers. You see the last layers in AlexNet (Section 8.1). Those dense layers radically increased the count of weights to $W \approx 135{,}000{,}000$.

Softmax Outputs for Multiclass Networks

In recognizing digits, we have 10 possible outputs. For letters and other symbols, 26 or more. With multiple output classes, we need an appropriate way to decide the very last layer (the output layer $\boldsymbol{w}$ in the neural net that started with $\boldsymbol{v}$). "Softmax" replaces the two-output case of logistic regression. **We are turning n numbers into probabilities.**

The outputs $w_1, \ldots, w_n$ are converted to probabilities $p_1, \ldots, p_n$ that add to 1.

$$\boxed{\textbf{Softmax} \qquad \boldsymbol{p_j = \frac{1}{S}\, e^{w_j}} \quad \text{where} \quad \boldsymbol{S = \sum_{k=1}^{n} e^{w_k}}} \qquad (7)$$

Certainly softmax assigns the largest probability p_j to the largest output w_j. But e^w is a nonlinear function of w. So the softmax assignment is not invariant to scale: If we double all the outputs w_j, softmax will produce different probabilities p_j. For small w's softmax actually deemphasizes the largest number $w_{\max}$.

In the CNN example of **teachyourmachine.com** that recognizes digits, you will see how softmax produces the probabilities displayed in a pie chart—an excellent visual aid.

Residual Networks (ResNets)

Networks are becoming seriously deeper with more and more hidden layers. Mostly these are convolutional layers with a moderate number of independent weights. But depth brings dangers. Information can jam up and never reach the output. The problem of "vanishing gradients" can be serious: so many multiplications in propagating so far, with the result that computed gradients are exponentially small. When it is well designed, depth is a good thing—**but you must create paths for learning to move forward**.

The remarkable thing is that those fast paths can be very simple: "*skip connections*" that go directly to the next layer—bypassing the usual step $\boldsymbol{v}_n = (A_n \boldsymbol{v}_{n-1} + \boldsymbol{b}_n)_+$. L layers could allow 2^L possible routes— fast or normal from each layer to the next.

One result is that entire layers can be removed without significant impact. The nth layer is reached by 2^{n-1} possible paths. Many paths have length well below n, not counting the skips. By sending information far forward, features that are learned early don't get lost before the output. Residual networks have become highly successful deep networks.

A Simple CNN : Learning to Read Letters

One of the class projects at MIT was a convolutional net **teachyourmachine.com**. The user begins by drawing multiple copies (not many) of A and B. On this training set, the correct classification is part of the input from the user. Then comes the mysterious step of *learning this data*—creating a continuous piecewise linear function $F(\boldsymbol{v})$ that gives high probability to the correct answer (the letter that was intended).

For learning to read digits, 10 probabilities appear in a pie chart. You quickly discover that too small a training set leads to frequent errors. If the examples had centered numbers or letters, and the test images are not centered, the user understands why those errors appear. **One purpose of teachyourmachine.com is education in machine learning at all levels.**

The World Championship at the Game of Go

A dramatic achievement by a deep convolutional network was to defeat the (human) world champion at Go. This is a difficult game played on a 19 by 19 board. In turn, two players put down "stones" in attempting to surround those of the opponent. When a group of one color has no open space beside it (left, right, up, or down), those stones are removed from the board. Wikipedia has an animated game to show the steps.

AlphaGo defeated the leading player Lee Sedol by 4 games to 1 in 2016. It had trained on thousands of human games. This was a convincing victory, but not overwhelming. Then the neural network was deepened and improved. Google's new version AlphaGo Zero learned to play without any human intervention—simply by playing against itself. Now it defeated its former self AlphaGo by 100 to 0.

The key point about the new and better version is that **the machine learned by itself**. *It was told the rules and nothing more.* The first version had been fed earlier games, aiming to discover why winners had won and losers had lost. The outcome from the new approach was parallel to the machine translation of languages. To master a language, special cases from grammar seemed essential. How else to learn all those exceptions? The translation team at Google was telling the system what it needed to know.

Meanwhile another small team was taking a different approach: *Let the machine figure it out.* In both cases, playing Go and translating languages, success came with a deeper neural net and more games and no coaching.

It is interesting that the machine often makes opening moves that have seldom or never been chosen by humans. The input to the network is a board position and its history. The output vector gives the probability of selecting each move—and also a scalar that estimates the probability of winning from that position. Every step communicates with a Monte Carlo tree search, to produce *reinforcement learning*.

8.3 Minimizing Loss by Gradient Descent

This section of the final chapter is about a fundamental problem: **Minimize a function** $\boldsymbol{F(x_1,\ldots,x_n)}$. Calculus teaches us that all the first derivatives $\partial F/\partial x_i$ are zero at the minimum (when F is smooth). If we have $n = 20$ unknowns (a small number in deep learning) then minimizing one function F leads to 20 equations $\partial F/\partial x_i = 0$. *"Gradient descent" uses the derivatives $\partial F/\partial x_i$ to find a direction that reduces $F(\boldsymbol{x})$.* The steepest direction, in which $F(\boldsymbol{x})$ decreases fastest, is given by the gradient $-\boldsymbol{\nabla F}$:

$$\textbf{Gradient descent / Learning rate } s_k \qquad \boldsymbol{x}_{k+1} = \boldsymbol{x}_k - s_k \boldsymbol{\nabla F}(\boldsymbol{x}_k) \tag{1}$$

$\boldsymbol{\nabla F}$ represents the vector $(\partial F/\partial x_1,\ldots,\partial F/\partial x_n)$ of the n partial derivatives of F. So (1) is a vector equation for each step $k = 1,2,3,\ldots$ and s_k is the *stepsize* or the *learning rate*. We hope to move toward the point $\boldsymbol{x}^*$ where the graph of $F(\boldsymbol{x})$ hits bottom.

We are willing to assume for now that 20 first derivatives exist and can be computed. We are not willing to assume that those 20 functions also have 20 convenient derivatives $\partial/\partial x_j(\partial F/\partial x_i)$. Those are the 210 **second derivatives** of F which go into a 20 by 20 symmetric matrix H—the Hessian matrix. (Symmetry reduces $n^2 = 400$ to $\frac{1}{2}n^2 + \frac{1}{2}n = 210$ computations.) Those second derivatives would be very useful extra information, but in many problems we have to go without.

You should know that 20 first derivatives and 210 second derivatives don't multiply the computing cost by 20 and 210. The neat idea of **automatic differentiation**—rediscovered and extended as **backpropagation** in machine learning—makes those cost factors much smaller in practice. Backpropagation is a fast way to follow n chain rules at once.

Return for a moment to equation (1). The step $-s_k\boldsymbol{\nabla F}(\boldsymbol{x}_k)$ includes a minus sign (to descend) and a factor s_k (to control the the stepsize) and the gradient vector $\boldsymbol{\nabla F}$ (containing the first derivatives of F computed at the current point $\boldsymbol{x}_k$). A lot of thought and computational experience has gone into the choice of stepsize and search direction.

We start with the main facts about derivatives and gradient vectors $\boldsymbol{\nabla F}$. Please forgive me, this linear algebra book is ending with a touch of calculus.

Multivariable Calculus

Machine learning involves functions $F(x_1,\ldots,x_n)$ of many variables. We need basic facts about the first and second derivatives of F. These are "partial derivatives" when $n > 1$.

$$\begin{array}{l}\textbf{One function } \boldsymbol{F}\\ \boldsymbol{n} \textbf{ variables } \boldsymbol{x}\end{array} \qquad \boldsymbol{F(x+\Delta x) \approx F(x) + (\Delta x)^{\mathrm{T}}\,\nabla F + \frac{1}{2}\,(\Delta x)^{\mathrm{T}} H (\Delta x)} \tag{2}$$

This is the beginning of a Taylor series—and we don't often go beyond that second-order term. The first terms $F(\boldsymbol{x}) + (\Delta\boldsymbol{x})^{\mathrm{T}}\,\boldsymbol{\nabla} F$ give a *first order* approximation to $F(\boldsymbol{x}+\Delta\boldsymbol{x})$, using information at $\boldsymbol{x}$. Then the $(\Delta\boldsymbol{x})^2$ term makes it a *second order* approximation.

Example 1 When S is symmetric, the gradient of $F(\boldsymbol{x}) = \boldsymbol{x}^{\mathrm{T}} \boldsymbol{S} \boldsymbol{x}$ is $\nabla \boldsymbol{F} = \mathbf{2}\, \boldsymbol{S}\boldsymbol{x}$. To see this vector, write out the function $F(x_1, x_2)$ when $n = 2$. The matrix S is 2 by 2:

$$F = \begin{bmatrix} x_1 & x_2 \end{bmatrix} \begin{bmatrix} a & b \\ b & c \end{bmatrix} \begin{bmatrix} x_1 \\ x_2 \end{bmatrix} = \begin{matrix} ax_1^2 + cx_2^2 \\ + 2bx_1x_2 \end{matrix} \qquad \begin{bmatrix} \partial F/\partial x_1 \\ \partial F/\partial x_2 \end{bmatrix} = 2 \begin{bmatrix} ax_1 + bx_2 \\ bx_1 + cx_2 \end{bmatrix} = \mathbf{2}\boldsymbol{S} \begin{bmatrix} \boldsymbol{x_1} \\ \boldsymbol{x_2} \end{bmatrix}.$$

Example 2 For a positive definite symmetric S, the minimum of a quadratic $F(\boldsymbol{x}) = \frac{1}{2}\, \boldsymbol{x}^{\mathrm{T}} \boldsymbol{S} \boldsymbol{x} - \boldsymbol{a}^{\mathrm{T}} \boldsymbol{x}$ is the negative number $\boldsymbol{F}_{\text{min}} = -\frac{1}{2}\, \boldsymbol{a}^{\mathrm{T}} \boldsymbol{S} \boldsymbol{a}$.

This is an important example! The minimum occurs where first derivatives of F are zero:

$$\boxed{\nabla \boldsymbol{F} = \begin{bmatrix} \partial F/\partial x_1 \\ \vdots \\ \partial F/\partial x_n \end{bmatrix} = \boldsymbol{S}\boldsymbol{x} - \boldsymbol{a} = \mathbf{0} \ \text{ at } \ \boldsymbol{x}^* = \boldsymbol{S}^{-1}\boldsymbol{a} = \arg\min \boldsymbol{F}.} \tag{3}$$

As always, that notation **arg min F stands for the point x^* where the minimum of $F(\boldsymbol{x}) = \frac{1}{2}\boldsymbol{x}^{\mathrm{T}} S\boldsymbol{x} - \boldsymbol{a}^{\mathrm{T}}\boldsymbol{x}$ is reached.** Often we are more interested in this minimizing vector $\boldsymbol{x}^* = S^{-1}\boldsymbol{a}$ than in the actual minimum value $\boldsymbol{F}_{\text{min}} = F(\boldsymbol{x}^*)$ at that point:

$$\boldsymbol{F}_{\text{min}} \text{ is } \frac{1}{2}(S^{-1}\boldsymbol{a})^{\mathrm{T}} S (S^{-1}\boldsymbol{a}) - \boldsymbol{a}^{\mathrm{T}}(S^{-1}\boldsymbol{a}) = \frac{1}{2}\boldsymbol{a}^{\mathrm{T}} S^{-1}\boldsymbol{a} - \boldsymbol{a}^{\mathrm{T}} S^{-1}\boldsymbol{a} = -\frac{\mathbf{1}}{\mathbf{2}}\boldsymbol{a}^{\mathrm{T}}\boldsymbol{S}^{-1}\boldsymbol{a}.$$

The graph of F is a bowl passing through zero at $\boldsymbol{x} = \mathbf{0}$ and dipping to its minimum at $\boldsymbol{x}^*$.

The Geometry of the Gradient Vector ∇f

Start with a function $f(x, y)$. It has $n = 2$ variables. Its gradient is $\nabla \boldsymbol{f} = (\partial \boldsymbol{f}/\partial \boldsymbol{x}, \partial \boldsymbol{f}/\partial \boldsymbol{y})$. This vector changes length as we move the point x, y where the derivatives are computed:

$$\nabla \boldsymbol{f} = \left(\frac{\partial f}{\partial x}, \frac{\partial f}{\partial y}\right) \qquad \textbf{Length} = ||\nabla \boldsymbol{f}|| = \sqrt{\left(\frac{\partial f}{\partial x}\right)^2 + \left(\frac{\partial f}{\partial y}\right)^2} = \begin{matrix}\text{slope in the} \\ \text{steepest direction}\end{matrix}$$

That length $||\nabla \boldsymbol{f}||$ tells us the steepness of the graph of $z = f(x, y)$. The graph is normally a curved surface—like a mountain or a valley in xyz space. At each point there is a slope $\partial f/\partial x$ in the x-direction and a slope $\partial f/\partial y$ in the y-direction. **The steepest slope is in the direction of $\nabla \boldsymbol{f}$ = grad $\boldsymbol{f}$.** The magnitude of that steepest slope is $||\nabla \boldsymbol{f}||$.

Example 3 The graph of a linear function $f(x, y) = ax + by$ is the plane $z = ax + by$. The gradient is the vector $\nabla \boldsymbol{f} = \begin{bmatrix} a \\ b \end{bmatrix}$ of partial derivatives. The length of that vector is $||\nabla \boldsymbol{f}|| = \sqrt{a^2 + b^2}$ = slope of the roof. The slope is steepest in the direction of $\nabla \boldsymbol{f}$.

That steepest direction is perpendicular to the level direction. The level direction z = constant has $ax + by$ = constant. It is the safe direction to walk, perpendicular to $\nabla \boldsymbol{f}$. Figure 8.5 shows the perpendicular directions (level and steepest) on the plane $\boldsymbol{z} = \boldsymbol{x} + \mathbf{2}\boldsymbol{y}$.

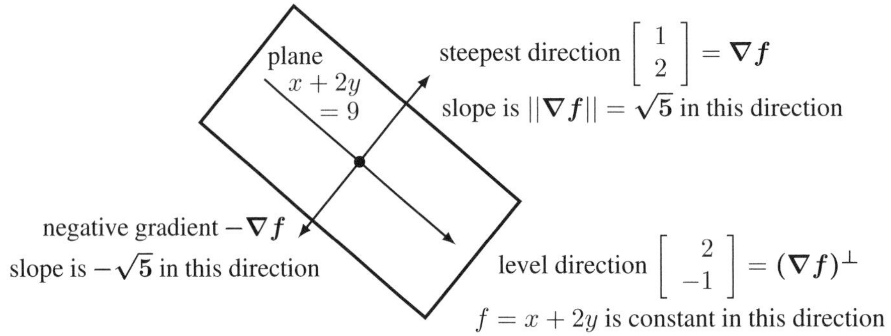

Figure 8.5: The negative gradient $-\nabla f$ gives the direction of steepest descent.

Example 4 The gradient of the quadratic $f = ax^2 + by^2$ is $\nabla f = \begin{bmatrix}\partial f/\partial x\\ \partial f/\partial y\end{bmatrix} = \begin{bmatrix}2ax\\2by\end{bmatrix}$. That tells us the steepest direction, changing from point to point. We are on a curved surface (a bowl opening upward). The bottom of the bowl is at $x = y = 0$ where the gradient vector is zero. The slope in the steepest direction is $||\nabla f||$. At the minimum, $\nabla f = (2ax, 2by) = (0, 0)$ and *slope = zero.*

The level direction has $z = ax^2 + by^2$ = constant height. That plane z = constant cuts through the bowl in a level curve. In this example the level curve $ax^2 + by^2 = c$ is an ellipse. The tangent line to the ellipse (level direction) is perpendicular to the gradient (steepest direction). But there is a serious difficulty for steepest descent:

The steepest direction changes as you go down ! The gradient doesn't point to the bottom !

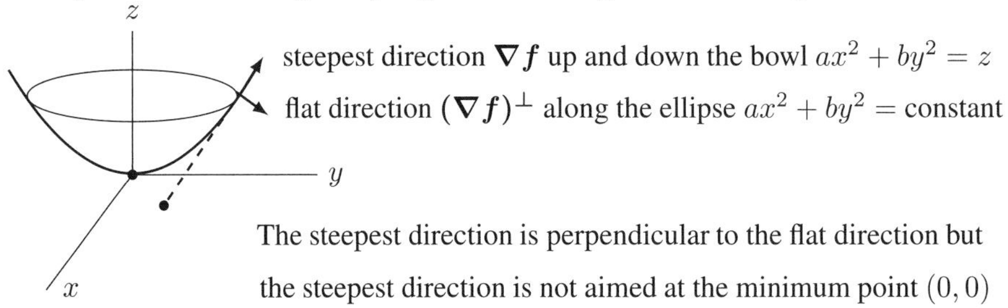

Figure 8.6: **Steepest descent moves down the bowl in the gradient direction** $\begin{bmatrix}-2ax\\-2by\end{bmatrix}$.

Let me repeat. At the point x_0, y_0 the gradient direction for $f = ax^2 + by^2$ is along $\nabla f = (2ax_0, 2by_0)$. The steepest line through x_0, y_0 is $2ax_0(y - y_0) = 2by_0(x - x_0)$. But then the lowest point $(x, y) = (0, 0)$ does not lie on the line ! **We will not find that minimum point in one step of "gradient descent". The steepest direction does not lead to the bottom of the bowl**—except when $b = a$ and the bowl is circular.

Water changes direction as it goes down a mountain. Sooner or later, we must change direction too. In practice we keep going in the gradient direction and stop when our cost function f is not decreasing quickly (or starts upward). At that point Step 1 ends and we recompute the gradient ∇f. This gives a new descent direction for Step 2.

An Important Example with Zig-Zag

The example $f(x,y) = \frac{1}{2}(x^2 + by^2)$ is extremely useful for $0 < b \le 1$. Its gradient ∇f has two components $\partial f/\partial x = x$ and $\partial f/\partial y = by$. The minimum value of f is zero. That minimum is reached at the point $(x^*, y^*) = (0,0)$. Best of all, steepest descent with exact line search produces a simple formula for each point (x_k, y_k) in the slow progress down the bowl toward $(0,0)$. Starting from $(x_0, y_0) = (b, 1)$ we find these points (x_k, y_k).

$$\boxed{x_k = b\left(\frac{b-1}{b+1}\right)^k \quad y_k = \left(\frac{1-b}{1+b}\right)^k \quad f(x_k, y_k) = \left(\frac{1-b}{1+b}\right)^{2k} f(x_0, y_0)} \quad (4)$$

If $b = 1$, you see immediate success in one step. The point (x_1, y_1) is $(0,0)$. The bowl is perfectly circular with $f = \frac{1}{2}(x^2 + y^2)$. The negative gradient gives a direction down the bowl, and it goes exactly through $(0,0)$. Then the first step of gradient descent finds that correct minimizing point where $f = 0$.

The real purpose of this example is seen when b is small. The crucial ratio in equation (4) is $r = (1-b)/(b+1)$. For $b = \frac{1}{10}$ this ratio is $r = 9/11$. For $b = \frac{1}{100}$ the ratio is $99/101$. The ratio is approaching 1 and the progress toward $(0,0)$ has virtually stopped when b is very small.

Figure 8.7 shows the frustrating zig-zag pattern of the steps toward $(0,0)$. Every step is short and progress is very slow. This is a case where the stepsize s_k in $\boldsymbol{x}_{k+1} = \boldsymbol{x}_k - s_k \nabla f(\boldsymbol{x}_k)$ was exactly chosen to minimize f (an exact line search). But the direction of $-\nabla f$, even if steepest, is pointing far from the final answer $(x^*, y^*) = (0,0)$.

The bowl has become a narrow valley when b is small. We are uselessly *crossing the valley* instead of moving down the valley to the bottom.

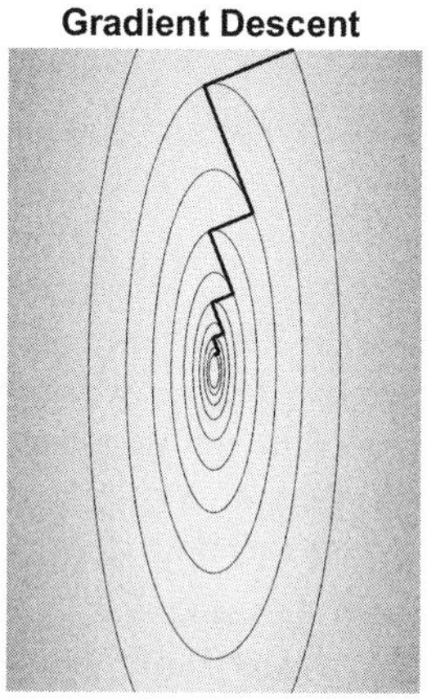

The first descent step starts out perpendicular to the level set. As it crosses through lower level sets, the function $f(x,y)$ is decreasing. **Eventually its path is tangent to a level set L.** Descent has stopped. Going further will increase f. The first step ends. The next step is perpendicular to L. So the zig-zag path took a 90° turn.

Figure 8.7: Slow convergence on a zig-zag path to the minimum of $f = x^2 + by^2$.

For b close to 1, the bowl is rounder and the descent is faster. First-order convergence means that the distance to $(x^*, y^*) = (0,0)$ is reduced by a constant factor at each step. For gradient descent and this f, the convergence factor in (4) is $(1-b)^2/(1+b)^2$.

Momentum and the Path of a Heavy Ball

The slow zig-zag path of steepest descent is a real problem. We have to improve it. Our model example $f = \frac{1}{2}(x^2 + by^2)$ has only two variables x, y and its second derivative matrix H is diagonal—constant entries $f_{xx} = 1$ and $f_{yy} = b$ and $f_{xy} = 0$. But it shows the zig-zag problem very clearly when $\boldsymbol{b = \lambda_{\text{min}}/\lambda_{\text{max}} = b/1}$ **is small**.

Key idea: Zig-zag would not happen for a heavy ball rolling downhill. Its momentum carries it through the narrow valley—bumping the sides but moving mostly forward. So we **add momentum with coefficient β to the gradient** (this is Polyak's important idea).

The direction $\boldsymbol{z}_k$ of the improved step remembers the previous direction $\boldsymbol{z}_{k-1}$.

Descent with momentum $$\boxed{\boldsymbol{x}_{k+1} = \boldsymbol{x}_k - s\boldsymbol{z}_k \quad \text{with} \quad \boldsymbol{z}_k = \boldsymbol{\nabla f}(\boldsymbol{x}_k) + \beta \boldsymbol{z}_{k-1}} \qquad (5)$$

Now we have two coefficients to choose—the stepsize s and also β. Most important, **the step to $\boldsymbol{x}_{k+1}$ in equation (5) involves $\boldsymbol{z}_{k-1}$.** Momentum has turned a one-step method (gradient descent) into a two-step method. To get back to one step, we have to rewrite equation (5) as **two coupled equations** (one vector equation) for the state at time $k+1$:

Vector equation with momentum $$\boxed{\begin{aligned} \boldsymbol{x}_{k+1} \qquad\qquad &= \boldsymbol{x}_k - s\boldsymbol{z}_k \\ \boldsymbol{z}_{k+1} - \boldsymbol{\nabla f}(\boldsymbol{x}_{k+1}) &= \qquad \beta \boldsymbol{z}_k \end{aligned}} \qquad (6)$$

With those two equations, we have recovered a one-step method. This is exactly like reducing a single second order differential equation to a system of two first order equations. Second order reduces to first order when dy/dt becomes a second unknown along with y.

2nd order equation
1st order system $$\frac{d^2y}{dt^2} + b\frac{dy}{dt} + ky = 0 \text{ becomes } \frac{d}{dt}\begin{bmatrix} y \\ dy/dt \end{bmatrix} = \begin{bmatrix} 0 & 1 \\ -k & -b \end{bmatrix}\begin{bmatrix} y \\ dy/dt \end{bmatrix}.$$

Interesting that this b is damping the motion while β adds momentum to encourage it.

The Quadratic Model

When $f(\boldsymbol{x}) = \frac{1}{2}\boldsymbol{x}^{\mathrm{T}}S\boldsymbol{x}$ is quadratic, its gradient $\boldsymbol{\nabla f} = S\boldsymbol{x}$ is linear. This is the model problem to understand: S is symmetric positive definite and $\boldsymbol{\nabla f}(\boldsymbol{x}_{k+1})$ becomes $S\boldsymbol{x}_{k+1}$. Our 2 by 2 supermodel is included, when the matrix S is diagonal with entries 1 and b. For a bigger matrix S, you will see that its largest and smallest eigenvalues determine the best choices for β and the stepsize s—so the 2 by 2 case actually contains the essence of the whole problem.

To understand the steps of accelerated descent, we track each eigenvector $\boldsymbol{q}$ of S. Here we are using a key idea from linear algebra (Chapter 6): *Follow the eigenvectors.*

Suppose $S\boldsymbol{q} = \lambda\boldsymbol{q}$ and $\boldsymbol{x}_k = c_k\,\boldsymbol{q}$ and $\boldsymbol{z}_k = d_k\,\boldsymbol{q}$ and $\boldsymbol{\nabla f}_k = S\boldsymbol{x}_k = \lambda\, c_k\,\boldsymbol{q}$. Then equation (7) connects the numbers c_k and d_k at step k to c_{k+1} and d_{k+1} at step $k+1$.

Following the eigenvector q
$$\begin{aligned} c_{k+1} &= c_k - s\,d_k \\ -\lambda\,c_{k+1} + d_{k+1} &= \beta\,d_k \end{aligned} \qquad \begin{bmatrix} \mathbf{1} & 0 \\ -\boldsymbol{\lambda} & \mathbf{1} \end{bmatrix} \begin{bmatrix} c_{k+1} \\ d_{k+1} \end{bmatrix} = \begin{bmatrix} \mathbf{1} & -\boldsymbol{s} \\ 0 & \boldsymbol{\beta} \end{bmatrix} \begin{bmatrix} c_k \\ d_k \end{bmatrix} \tag{7}$$

Now we invert the first matrix ($-\lambda$ becomes λ) to see each descent step clearly:

Descent step multiplies by R
$$\begin{bmatrix} c_{k+1} \\ d_{k+1} \end{bmatrix} = \begin{bmatrix} 1 & 0 \\ \lambda & 1 \end{bmatrix} \begin{bmatrix} 1 & -s \\ 0 & \beta \end{bmatrix} \begin{bmatrix} c_k \\ d_k \end{bmatrix} = \begin{bmatrix} \mathbf{1} & -\boldsymbol{s} \\ \boldsymbol{\lambda} & \boldsymbol{\beta - \lambda s} \end{bmatrix} \begin{bmatrix} c_k \\ d_k \end{bmatrix} = \boldsymbol{R} \begin{bmatrix} c_k \\ d_k \end{bmatrix} \tag{8}$$

After k steps the starting vector is multiplied by R^k. For fast convergence to zero (which is the minimum of $f = \frac{1}{2}\boldsymbol{x}^{\mathrm{T}}S\boldsymbol{x}$) we want both eigenvalues e_1 and e_2 of R to be as small as possible. Clearly those eigenvalues of R depend on the eigenvalue λ of S. That eigenvalue λ could be anywhere between $\lambda_{\min}(S)$ and $\lambda_{\max}(S)$. Our problem is:

$$\textbf{Choose } s \textbf{ and } \beta \textbf{ to minimize } \max\Big[|e_1(\boldsymbol{\lambda})|,\ |e_2(\boldsymbol{\lambda})|\Big] \quad \text{for } \lambda_{\min}(S) \le \lambda \le \lambda_{\max}(S).$$

It seems a miracle that this problem has a beautiful solution. The optimal s and β are

$$s = \left(\frac{2}{\sqrt{\lambda_{\max}} + \sqrt{\lambda_{\min}}}\right)^2 \quad \text{and} \quad \beta = \left(\frac{\sqrt{\lambda_{\max}} - \sqrt{\lambda_{\min}}}{\sqrt{\lambda_{\max}} + \sqrt{\lambda_{\min}}}\right)^2. \tag{9}$$

Think of the 2 by 2 supermodel, when S has eigenvalues $\lambda_{\max} = 1$ and $\lambda_{\min} = b$:

$$s = \left(\frac{2}{1+\sqrt{b}}\right)^2 \quad \text{and} \quad \beta = \left(\frac{1-\sqrt{b}}{1+\sqrt{b}}\right)^2 \tag{10}$$

These choices of stepsize and momentum give a convergence rate that looks like the rate in equation (4) for ordinary steepest descent (no momentum). But there is a crucial difference between (10) and (4): **b is replaced by $\sqrt{b}$**.

$$\textbf{Ordinary descent factor} \quad \left(\frac{\mathbf{1}-\boldsymbol{b}}{\mathbf{1}+\boldsymbol{b}}\right)^{\mathbf{2}} \qquad \textbf{Accelerated descent factor} \quad \left(\frac{\mathbf{1}-\sqrt{\boldsymbol{b}}}{\mathbf{1}+\sqrt{\boldsymbol{b}}}\right)^{\mathbf{2}} \tag{11}$$

So similar but so different. The real test comes when b is very small. Then the ordinary descent factor is essentially $1 - 4b$, very close to 1. The accelerated descent factor is essentially $1 - 4\sqrt{b}$, *much further below* 1.

To emphasize the improvement that momentum brings, suppose $b = 1/100$. Then $\sqrt{b} = 1/10$ (ten times larger than b). The convergence factors in equation (11) are

$$\textbf{Steepest descent} \quad \left(\frac{.99}{1.01}\right)^2 \approx \mathbf{.96} \qquad \textbf{Accelerated descent} \quad \left(\frac{.9}{1.1}\right)^2 \approx \mathbf{.67}$$

Ten steps of ordinary descent multiply the starting error by 0.67. This is matched by a single momentum step. Ten steps with momentum multiply the error by 0.018. **$\boldsymbol{\lambda_{\max}/\lambda_{\min} = 1/b = \kappa}$ is the condition number of $\boldsymbol{S}$.**

Stochastic Gradient Descent

Gradient descent is fundamental in training a deep neural network. It is based on a step of the form $\boldsymbol{x}_{k+1} = \boldsymbol{x}_k - s_k \boldsymbol{\nabla} L(\boldsymbol{x}_k)$. That step should lead us downhill toward the point $\boldsymbol{x}^*$ where the loss function $L(\boldsymbol{x})$ is minimized for the test data $\boldsymbol{v}$. But for large networks with many samples in the training set, this algorithm (as it stands) is not successful!

It is important to recognize two different problems with classical steepest descent:

1. Computing $\boldsymbol{\nabla} L$ at every descent step—the derivatives of the total loss L with respect to all the weights $\boldsymbol{x}$ in the network—is *too expensive.* That total loss adds the individual losses $\ell(\boldsymbol{x}, \boldsymbol{v}_i)$ *for every sample* $\boldsymbol{v}_i$ *in the training set*—potentially millions of separate losses are computed and added in every computation of $\boldsymbol{L}$.

2. The number of weights is even larger. So $\boldsymbol{\nabla}_{\boldsymbol{x}} \boldsymbol{L} = \boldsymbol{0}$ for many different choices $\boldsymbol{x}^*$ of the weights. **Some of those choices can give poor results on unseen test data.** The learning function $\boldsymbol{F}$ can fail to "generalize". But **stochastic gradient descent (SGD)** does find weights $\boldsymbol{x}^*$ that generalize—weights that will succeed on unseen input vectors $\boldsymbol{v}$ from a similar population.

Stochastic gradient descent uses only a "minibatch" of the training data at each step. B samples will be chosen randomly. Replacing the full batch of all the training data by a minibatch changes $L(\boldsymbol{x}) = \frac{1}{n} \sum \ell_i(\boldsymbol{x})$ to a sum of only B losses. This resolves both difficulties at once. The success of deep learning rests on these two facts:

1. Computing $\boldsymbol{\nabla L}$ by backpropagation on B samples is much faster. Often $B = 1$.

2. The stochastic algorithm produces weights $\boldsymbol{x}^*$ that also succeed on unseen data.

The first point is clear. The calculation per step is greatly reduced. The second point is a miracle. Generalizing well to new data is a gift that researchers work hard to explain.

Stochastic Descent Using One Sample Per Step

To simplify, suppose each minibatch contains only one sample $\boldsymbol{v}_k$ (so $B = 1$). That sample is chosen randomly. The theory of stochastic descent usually assumes that the sample is replaced after use—in principle the sample could be chosen again at step $k+1$. But replacement is expensive compared to starting with a random ordering of the samples. In practice, we often omit replacement and work through samples in a random order.

Each pass through the training data is **one epoch** of the descent algorithm. Ordinary gradient descent computes one epoch per step (batch mode). Stochastic gradient descent needs many steps (for minibatches). The online advice is to choose $B \leq 32$.

Stochastic descent is more sensitive to the stepsizes s_k than full gradient descent. If we randomly choose sample $\boldsymbol{v}_i$ at step k, then the kth descent step sees only one loss:

$$\boxed{\boldsymbol{x}_{k+1} = \boldsymbol{x}_k - s_k \boldsymbol{\nabla}_{\boldsymbol{x}} \boldsymbol{\ell}(\boldsymbol{x}_k, \boldsymbol{v}_i)} \qquad \boxed{\boldsymbol{\nabla}_{\boldsymbol{x}} \boldsymbol{\ell} = \text{derivative of the loss term from sample } \boldsymbol{v}_i}$$

We are doing much less work per step (B inputs instead of all inputs from the training set). But we do not necessarily converge more slowly. A typical feature of stochastic gradient descent is **"semi-convergence"**: *fast convergence at the start.*

Fast Start: Least Squares with $n = 1$

Early steps of SGD often converge more quickly than GD toward the solution x^*. We admit immediately that later iterations of SGD are frequently erratic. **Convergence at the start changes to large oscillations near the solution**. Figure 8.8 will show this. One response is to stop early. And thereby we avoid overfitting the data.

In the following example, the solution $\boldsymbol{x}^*$ is in a specific interval I. If the current approximation $\boldsymbol{x}_k$ is outside I, the next approximation $\boldsymbol{x}_{k+1}$ is closer to I (or inside I). That gives semiconvergence—a good start. *But eventually the $\boldsymbol{x}_k$ bounce around inside I.*

We learned from Suvrit Sra that the simplest example is the best. The vector $\boldsymbol{x}$ has only one component x. The ith loss is $\ell_i = \frac{1}{2}(a_i x - b_i)^2$ with $a_i > 0$. The gradient of ℓ_i is its derivative $a_i(a_i x - b_i)$. It is zero and ℓ_i is minimized at $\boldsymbol{x} = \boldsymbol{b_i/a_i}$. The total loss over all N samples is $L(x) = \frac{1}{2N}\sum(a_i x - b_i)^2$: Least squares with N equations, 1 unknown.

$$\text{The equation to solve is } \boldsymbol{\nabla L} = \frac{1}{N}\sum_{1}^{N} a_i(a_i x - b_i) = 0. \text{ The solution is } \boldsymbol{x}^* = \frac{\sum \boldsymbol{a_i b_i}}{\sum \boldsymbol{a_i^2}}$$

Important If B/A is the largest ratio b_i/a_i, then the true solution $\boldsymbol{x}^*$ **is below $\boldsymbol{B/A}$**. This follows from a row of four inequalities:

$$\text{all } \frac{\boldsymbol{b_i}}{\boldsymbol{a_i}} \leq \frac{\boldsymbol{B}}{\boldsymbol{A}} \qquad A\,a_i\,b_i \leq B\,a_i^2 \qquad A\left(\sum a_i b_i\right) \leq B\left(\sum a_i^2\right) \qquad \boldsymbol{x}^* = \frac{\sum a_i b_i}{\sum a_i^2} \leq \frac{\boldsymbol{B}}{\boldsymbol{A}}.$$

Similarly x^* is above the smallest ratio $\boldsymbol{\beta/\alpha}$. Conclusion: If $\boldsymbol{x}_k$ is outside the interval $\boldsymbol{I}$ from β/α to B/A, then the kth gradient descent step will move *toward that interval $\boldsymbol{I}$* containing $\boldsymbol{x}^*$. Here is what we can expect from stochastic gradient descent:

If x_k is outside I, then x_{k+1} moves toward the interval $\beta/\alpha \leq x \leq B/A$.
If x_k is inside I, then so is x_{k+1}. The iterations can bounce around inside I.

A typical sequence $\boldsymbol{x}_0, \boldsymbol{x}_1, \boldsymbol{x}_2, \ldots$ from minimizing $||A\boldsymbol{x} - \boldsymbol{b}||^2$ by stochastic gradient descent is graphed in Figure 8.8. *You see the fast start and the oscillating finish.* This behavior is a perfect signal to think about early stopping or else averaging.

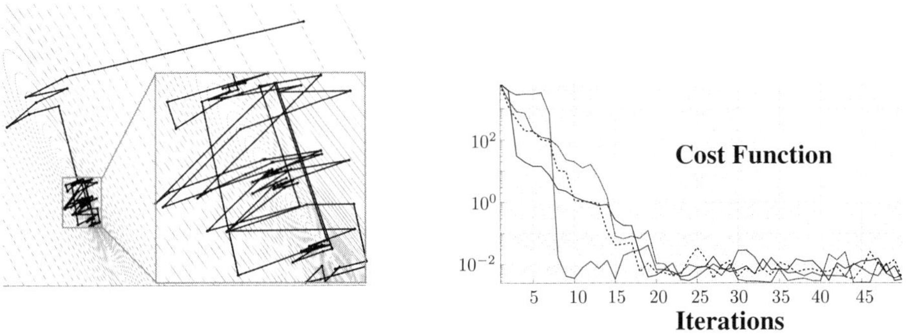

Figure 8.8: Early iterations succeed but later iterations oscillate. For these four SGD paths, the quadratic cost function decreases quickly and then fluctuates instead of converging.

Overfitting

Here is an observation from experience. *We may not want to fit the training data too perfectly.* That could be **overfitting**. The function F becomes oversensitive. It memorized everything but it hasn't learned anything. **Generalization** is the ability to give the correct classification for unseen test data $\boldsymbol{v}$, based on the weights $\boldsymbol{x}$ that were learned from the training data.

I compare overfitting with choosing a polynomial of degree 60 *that fits exactly to* 61 *data points.* Its 61 coefficients a_0 to a_{60} will perfectly learn the data. But that high degree polynomial will oscillate wildly between the data points. For test data at a nearby point, the perfect-fit polynomial gives a totally wrong answer. But see Figure 8.9 for an unexpected result from severe overfitting.

One fundamental strategy in training a neural network (which means finding a function that learns from the training data and generalizes well to test data) is **early stopping**. Machine learning needs to know when to quit ! Possibly this is true of human learning too.

The Double Descent Curve

This section is a report on experiments (including ordinary least squares). Usually we are fitting a large problem with a smaller number of parameters. We may have m equations and $n < m$ unknowns: $A\boldsymbol{x} = \boldsymbol{b}$ with a *tall thin matrix* A. The m measurements $\boldsymbol{b}$ are not exact and too few parameters ($n = 2$ for fitting by a straight line) cannot model the data.

To improve our result, we allow more parameters. As n increases, the model begins to improve. "The bias is reduced." But nothing is perfect, and the first descent in Figure 8.5 turns upward. The model begins to fail from **overfitting**. All this is for $n < m$, the usual situation.

It was fully expected that deep learning would become deep suffering, when the number of layers and matrix weights increased too far. The evidence pointed that way, until computers became so fast and powerful that ***$\boldsymbol{n}$ went past $\boldsymbol{m}$*** : The model is **overparameterized**. Now we have many solutions to choose from—many different weights $\boldsymbol{x}$ minimize the loss function $\boldsymbol{L}(\boldsymbol{x})$.

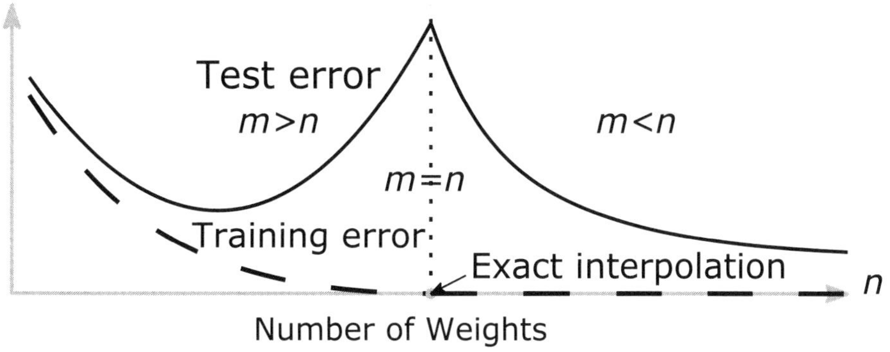

Figure 8.9: This is Belkin's **double descent curve** (arXiv 2003.00307). It shows the test error going down as n increases, and then up for overfitting. The surprise is that the error goes back down for $n > m$. Among many solutions, a good one is chosen.

Gradient descent (full batch or minibatch chosen stochastically) tries to converge to good weights! Apparently it does, if you look at the second descent in Figure 8.9. The method generalizes well to new data—by choosing a particularly good solution among all the possible solutions. That process is not fully understood as this book is being written.

For the linear least squares equation $A^{\mathrm{T}}A\widehat{\boldsymbol{x}} = A^{\mathrm{T}}\boldsymbol{b}$, any solution to $A\boldsymbol{x} = \mathbf{0}$ could be added to $\widehat{\boldsymbol{x}}$. Then $A^{\mathrm{T}}A(\widehat{\boldsymbol{x}} + \boldsymbol{x})$ still equals $A^{\mathrm{T}}\boldsymbol{b}$. Those null solutions $\boldsymbol{x}$ will appear as soon as $n > m$ (more unknowns than equations). But the **minimum norm solution** keeps $\widehat{\boldsymbol{x}}$ in the row space of A. It avoids adding $\boldsymbol{x}$ from the nullspace of A because that would increase the norm: $||\widehat{\boldsymbol{x}} + \boldsymbol{x}||^2 = ||\widehat{\boldsymbol{x}}||^2 + ||\boldsymbol{x}||^2$. And it succeeds!

A neat observation by Poggio (arXiv 1912.06190) looks at the condition number of $A^{\mathrm{T}}A$. As n increases, the graph of that number is very close to Figure 8.9—the error goes down again for large n. Out of many solutions for $n > m$, Belkin and other authors show that gradient descent somehow chooses a solution that generalizes well to new data.

ADAM: Adaptive Methods Using Earlier Gradients

For faster convergence of gradient descent and stochastic gradient descent, adaptive methods have been a major success. The idea is *to use gradients from earlier steps.* "Momentum" went one step back, to $k - 1$. These adaptive methods (ADAM) go all the way back. Memory partly guides the choice of search direction $\boldsymbol{D}_k$ and stepsize s_k. We are searching for the vector $\boldsymbol{x}^*$ that minimizes a specified loss function $L(\boldsymbol{x})$. In the step to $\boldsymbol{x}_{k+1} = \boldsymbol{x}_k - s\boldsymbol{D}_k$, we are free to choose the *direction* $\boldsymbol{D}_k$ *and stepsize* s_k.

$$\boldsymbol{D}_k = \boldsymbol{D}(\boldsymbol{\nabla L}_k, \boldsymbol{\nabla L}_{k-1}, \ldots, \boldsymbol{\nabla L}_0) \quad \text{and} \quad s_k = s(\boldsymbol{\nabla L}_k, \boldsymbol{\nabla L}_{k-1}, \ldots, \boldsymbol{\nabla L}_0). \quad (12)$$

For a standard iteration (not adaptive), $\boldsymbol{D}_k$ depends only on the current gradient $\boldsymbol{\nabla L_k}$ (and s_k can be $s/\sqrt{k}$). That gradient is evaluated only on a random minibatch B of the test data. Now, deep networks often have the option of averaging some or all of the gradients from earlier minibatches: Success or failure will depend on $\boldsymbol{D}_k$ and s_k.

Exponential moving averages in ADAM have become the favorites. Recent gradients $\nabla \boldsymbol{L}$ have greater weight than earlier gradients in both s_k and the step direction $\boldsymbol{D}_k$. The exponential weights in $\boldsymbol{D}$ and s come from $\delta < 1$ and $\beta < 1$. Typical values are $\delta = 0.9$ and $\beta = 0.999$. Small values of δ and β will effectively kill off the moving memory and lose the advantages of adaptive methods.

The actual computation of $\boldsymbol{D}_k$ and s_k will be a recursive combination of old and new :

$$\boxed{\boldsymbol{D}_k = \delta \boldsymbol{D}_{k-1} + (1-\delta)\nabla \boldsymbol{L}(\boldsymbol{x}_k) \qquad s_k^2 = \beta s_{k-1}^2 + (1-\beta)\,||\nabla \boldsymbol{L}\,(\boldsymbol{x}_k)||^2} \tag{13}$$

For several class projects, adaptive methods clearly produced faster convergence.

After fast convergence to weights that nearly solve $\nabla \boldsymbol{L}(\boldsymbol{x}) = \boldsymbol{0}$ there is still the crucial issue : **Why do those weights generalize well to unseen test data ?**

Randomized Kaczmarz for $Ax = b$ is Stochastic Gradient Descent

$$\boxed{\textbf{Kaczmarz for } \boldsymbol{A x = b} \textbf{ with random } \boldsymbol{i(k)} \qquad \boldsymbol{x}_{k+1} = \boldsymbol{x}_k + \frac{b_i - \boldsymbol{a}_i^{\mathrm{T}} \boldsymbol{x}_k}{||\boldsymbol{a}_i||^2}\, \boldsymbol{a}_i} \tag{14}$$

The Kaczmarz idea is simple. Choose row i of A at step k. Adjust $\boldsymbol{x}_{k+1}$ to solve equation i in $A\boldsymbol{x} = \boldsymbol{b}$. (Multiply equation (14) by $\boldsymbol{a}_i^{\mathrm{T}}$ to verify that $\boldsymbol{a}_i^{\mathrm{T}} \boldsymbol{x}_{k+1} = b_i$. This is equation i in $A\boldsymbol{x} = \boldsymbol{b}$.) Geometrically, $\boldsymbol{x}_{k+1}$ is the projection of $\boldsymbol{x}_k$ onto one of the hyperplanes $\boldsymbol{a}_i^{\mathrm{T}} \boldsymbol{x} = \boldsymbol{b}_i$ that meet at $\boldsymbol{x}^* = A^{-1}\boldsymbol{b}$.

This algorithm resisted a close analysis for many years. The equations $\boldsymbol{a}_1^{\mathrm{T}} \boldsymbol{x} = b_1$, $\boldsymbol{a}_2^{\mathrm{T}} \boldsymbol{x} = b_2 \ldots$ were taken in cyclic order 1 to n, 1 to n,... Then Strohmer and Vershynin proved fast convergence for random Kaczmarz. They used SGD with *norm-squared sampling* : Choose row i of A with probability p_i proportional to $||\boldsymbol{a}_i||^2$.

A previous page described the Kaczmarz iterations for $A\boldsymbol{x} = \boldsymbol{b}$ when A is N by 1. The sequence $x_0, x_1, x_2, \ldots$ moved toward the interval I. The least squares solution $\boldsymbol{x}^*$ was in that interval. For an N by K matrix A, we expect the K by 1 vectors $\boldsymbol{x}_i$ to move into a K-dimensional box around $\boldsymbol{x}^*$. Figure 8.8 showed this for $K = 2$.

The next page will present numerical experiments for stochastic gradient descent.

Random Kaczmarz and Iterated Projections

Suppose $A\boldsymbol{x}^* = \boldsymbol{b}$. A typical step of random Kaczmarz projects the current error $\boldsymbol{x}_k - \boldsymbol{x}^*$ onto the hyperplane $\boldsymbol{a}_i^{\mathrm{T}}\boldsymbol{x} = b_i$. Here i is chosen randomly at step k (often with importance sampling using probabilities proportional to $||a_i||^2$). To see that projection matrix $\boldsymbol{a}_i\boldsymbol{a}_i^{\mathrm{T}}/\boldsymbol{a}_i^{\mathrm{T}}\boldsymbol{a}_i$, substitute $b_i = \boldsymbol{a}_i^{\mathrm{T}}\boldsymbol{x}^*$ into the update step :

$$\boldsymbol{x}_{k+1} - \boldsymbol{x}^* = \boldsymbol{x}_k - \boldsymbol{x}^* + \frac{b_i - \boldsymbol{a}_i^{\mathrm{T}}\boldsymbol{x}_k}{||\boldsymbol{a}_i||^2}\,\boldsymbol{a}_i = (\boldsymbol{x}_k - \boldsymbol{x}^*) - \frac{\boldsymbol{a}_i\boldsymbol{a}_i^{\mathrm{T}}}{\boldsymbol{a}_i^{\mathrm{T}}\boldsymbol{a}_i}(\boldsymbol{x}_k - \boldsymbol{x}^*) \tag{15}$$

Orthogonal projection never increases length. The error can only decrease. The error norm $||\boldsymbol{x}_k - \boldsymbol{x}^*||$ decreases steadily, even if the cost function $||A\boldsymbol{x}_k - \boldsymbol{b}||$ does not. *But convergence is usually slow* ! Strohmer-Vershynin estimate the expected error :

$$\mathrm{E}\left[||\boldsymbol{x}_k - \boldsymbol{x}^*||^2\right] \le \left(1 - \frac{1}{c^2}\right)^k ||\boldsymbol{x}_0 - \boldsymbol{x}^*||^2,\ c = \text{ condition number of } A. \tag{16}$$

This is slow compared to gradient descent (there c^2 is replaced by c, and then $\sqrt{c}$ with momentum added). But (16) is independent of the size of A : attractive for large problems.

Our experiments converge slowly ! The 100 by 10 matrix A is random with $c \approx 400$. The figures show random Kaczmarz for $600,000$ steps. We measure convergence by the angle θ_k between $\boldsymbol{x}_k - \boldsymbol{x}^*$ and the row $\boldsymbol{a}_i$ chosen at step k. The error equation (15) is

$$||\boldsymbol{x}_{k+1} - \boldsymbol{x}^*||^2 = (1 - \cos^2\theta_k)\,||\boldsymbol{x}_k - \boldsymbol{x}^*||^2 \tag{17}$$

The graph shows that those numbers $1 - \cos^2\theta_k$ *are very close to* 1 : **slow convergence**. But the second graph confirms that convergence does occur. The Strohmer-Vershynin bound becomes $\mathrm{E}\left[\cos^2\theta_k\right] \ge 1/c^2$. Our example matrix has $1/c^2$ close to 10^{-5} and experimentally $\cos^2\theta_k \approx 2\cdot 10^{-5}$, confirming that bound.

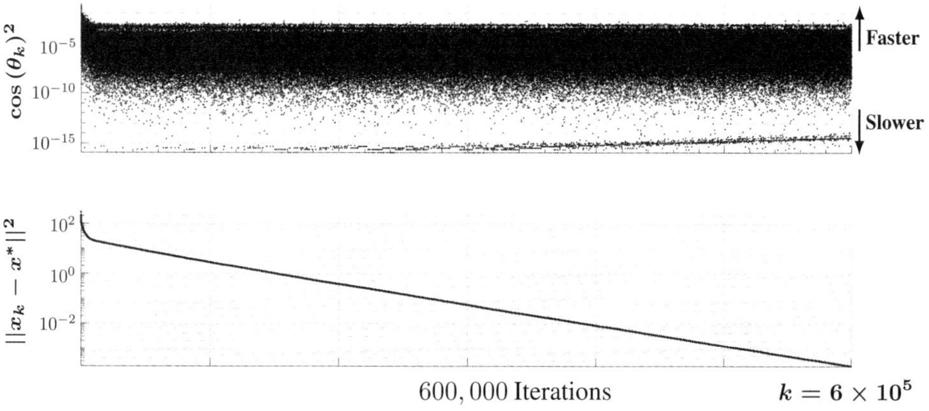

Figure 8.10: Convergence of the squared error for random Kaczmarz. Equation (17) with $1 - \cos^2\theta_k$ close to $1 - 10^{-5}$ produces the slow convergence in the lower graph.

Product of Matrices ABC : Which Order ?

Backpropagation is an incredibly efficient improvement on computing each $\partial F/\partial x_i$ separately. At first it seems unbelievable, that reorganizing the computations can make such an enormous difference. In the end (the doubter might say) you have to compute derivatives for each small step and multiply by the chain rule. But the reordering works—and N derivatives are computed in far less than N times the cost of one derivative $\partial F/\partial x_1$.

Backpropagation is backward-mode automatic differentiation (*reverse mode AD*). A chain of functions in F leads to a chain of derivatives in $\boldsymbol{\nabla F}$. Those are numbers in a basic calculus course, but they are matrices in deep learning—when each layer is connected to the next layer.

It is beautiful that matrix multiplication still gives the correct chain rule (see below). But when we are asked to multiply three matrices ABC, the associative law offers two choices for the multiplication order. The word "*back*" will tell us the better order.

$$\textbf{\textit{AB} first or \textit{BC} first ?} \qquad \textbf{Compute } (\boldsymbol{AB})\,\boldsymbol{C} \textbf{ or } \boldsymbol{A}(\boldsymbol{BC})\textbf{ ?}$$

The result is the same but the number of individual multiplications can be very different. Suppose the matrix A is $\boldsymbol{m}$ by $\boldsymbol{n}$, and B is $\boldsymbol{n}$ by $\boldsymbol{p}$, and C is $\boldsymbol{p}$ by $\boldsymbol{q}$.

First way

$$AB = (m \times n)\,(n \times p) \text{ has } \boldsymbol{mnp} \text{ multiplications}$$
$$(AB)C = (m \times p)\,(p \times q) \text{ has } \boldsymbol{mpq} \text{ multiplications}$$

Second way

$$BC = (n \times p)\,(p \times q) \text{ has } \boldsymbol{npq} \text{ multiplications}$$
$$A(BC) = (m \times n)\,(n \times q) \text{ has } \boldsymbol{mnq} \text{ multiplications}$$

So the comparison is between $\boldsymbol{mp(n+q)}$ and $\boldsymbol{nq(m+p)}$. Divide both by $mnpq$:

The first way is faster when $\dfrac{\mathbf{1}}{\boldsymbol{q}} + \dfrac{\mathbf{1}}{\boldsymbol{n}}$ is smaller than $\dfrac{\mathbf{1}}{\boldsymbol{m}} + \dfrac{\mathbf{1}}{\boldsymbol{p}}$.

Here is an extreme case (extremely important). *Suppose C is a column vector* : $\boldsymbol{p}$ **by 1**. Thus $q = 1$. Should you multiply BC to get another column vector (n by 1) and then $A(BC)$ to find the output (m by 1) ? Or should you multiply matrices AB first ?

The question almost answers itself. The correct $A(BC)$ produces a vector at each step. The matrix-vector multiplication BC has np steps. The next matrix-vector multiplication $A(BC)$ has mn steps. Compare those matrix-vector steps to the cost of starting with the matrix-matrix option AB (mnp steps !). Nobody in their right mind would do that.

In the application of $A(BC)$ to the chain rule, we start with the last layer C is the derivative of the last F_L and we go **back** to the first layer (A is the derivative of F_1).

The Multivariable Chain Rule

Suppose the vector $\boldsymbol{v}$ with n components v_i is a function of the vector $\boldsymbol{u}$ with m components u_j. The derivative of each v_i with respect to each u_j goes into the derivative matrix $\partial \boldsymbol{v}/\partial \boldsymbol{u}$ (often called the Jacobian matrix J). Its shape is n by m (not m by n). Similarly, the derivative matrix $\partial \boldsymbol{w}/\partial \boldsymbol{v}$ of the vector function $\boldsymbol{w} = (w_1, \ldots, w_p)$ with respect to the components of $\boldsymbol{v} = (v_1, \ldots, v_n)$ is a p by n matrix:

$$\frac{\partial \boldsymbol{w}}{\partial \boldsymbol{v}} = \begin{bmatrix} \dfrac{\partial w_1}{\partial v_1} & \cdots & \dfrac{\partial w_1}{\partial v_n} \\ & \cdots & \\ \dfrac{\partial w_p}{\partial v_1} & \cdots & \dfrac{\partial w_p}{\partial v_n} \end{bmatrix} \qquad \frac{\partial \boldsymbol{v}}{\partial \boldsymbol{u}} = \begin{bmatrix} \dfrac{\partial v_1}{\partial u_1} & \cdots & \dfrac{\partial v_1}{\partial u_m} \\ & \cdots & \\ \dfrac{\partial v_n}{\partial u_1} & \cdots & \dfrac{\partial v_n}{\partial u_m} \end{bmatrix} \tag{18}$$

Each w_i depends on the v's and each v_j depends on the u's. Therefore each function $w_1, \ldots, w_p$ depends on $u_1, \ldots, u_m$. The chain rule aims to find the derivatives $\partial w_i/\partial u_k$. And the rule is exactly a dot product: (row i of $\partial \boldsymbol{w}/\partial \boldsymbol{v}$) $\cdot$ (column k of $\partial \boldsymbol{v}/\partial \boldsymbol{u}$).

$$\frac{\partial w_i}{\partial u_k} = \frac{\partial w_i}{\partial v_1}\frac{\partial v_1}{\partial u_k} + \cdots + \frac{\partial w_i}{\partial v_n}\frac{\partial v_n}{\partial u_k} = \left(\frac{\partial w_i}{\partial v_1}, \ldots, \frac{\partial w_i}{\partial v_n}\right) \cdot \left(\frac{\partial v_1}{\partial u_k}, \ldots, \frac{\partial v_n}{\partial u_k}\right) \tag{19}$$

Multivariable chain rules
Multiply the matrices in (18)
$$\frac{\partial \boldsymbol{w}}{\partial \boldsymbol{u}} = \left(\frac{\partial \boldsymbol{w}}{\partial \boldsymbol{v}}\right)\left(\frac{\partial \boldsymbol{v}}{\partial \boldsymbol{u}}\right)$$

The key to matrix calculus is linear algebra—shown again in this chain rule.

Problem Set 8.3

1 The rank-one matrix $P = \boldsymbol{a}\boldsymbol{a}^{\mathrm{T}}/\boldsymbol{a}^{\mathrm{T}}\boldsymbol{a}$ is an orthogonal projection onto the line through $\boldsymbol{a}$. Verify that $P^2 = P$ (projection) and that $P\boldsymbol{x}$ is on that line and that $\boldsymbol{x} - P\boldsymbol{x}$ is always perpendicular to $\boldsymbol{a}$. (*Why is* $\boldsymbol{a}^{\mathrm{T}}\boldsymbol{x} = \boldsymbol{a}^{\mathrm{T}}P\boldsymbol{x}$?)

2 Verify equation (15) which shows that $\boldsymbol{x}_{k+1} - \boldsymbol{x}^*$ is exactly $P(\boldsymbol{x}_k - \boldsymbol{x}^*)$.

3 If A has only two rows $\boldsymbol{a}_1$ and $\boldsymbol{a}_2$, then Kaczmarz will produce the *alternating projections* in this figure. Starting from any error vector $\boldsymbol{e}_0 = \boldsymbol{x}_0 - \boldsymbol{x}^*$, why does $\boldsymbol{e}_k$ approach zero ? How fast—if you know that angle θ at $\mathbf{0}$?

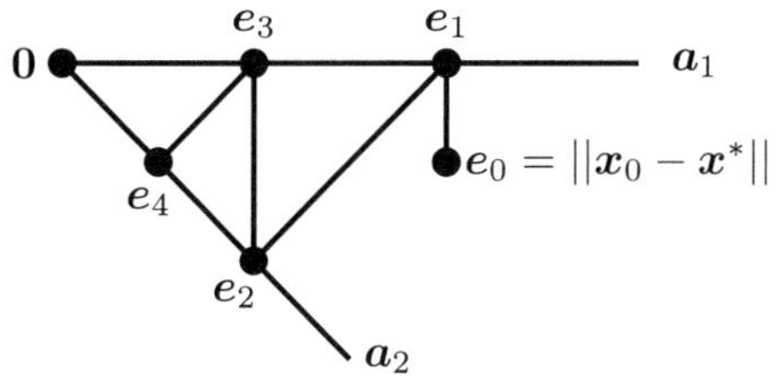

4 Suppose we want to minimize $F(x,y) = y^2 + (y-x)^2$. The actual minimum is $F = 0$ at $(x^*, y^*) = (0,0)$. Find the gradient vector $\boldsymbol{\nabla F}$ at the starting point $(x_0, y_0) = (1,1)$. For full gradient descent (*not stochastic*) with stepsize $s = \frac{1}{2}$, where is (x_1, y_1) ?

5 In minimizing $F(\boldsymbol{x}) = ||A\boldsymbol{x} - \boldsymbol{b}||^2$, stochastic gradient descent with minibatch size $B = 1$ will solve one equation $\boldsymbol{a}_i^{\mathrm{T}}\boldsymbol{x} = b_i$ at each step. Explain the typical step for minibatch size $B = 2$.

6 (Experiment) For a random A and $\boldsymbol{b}$ (20 by 4 and 20 by 1), try stochastic gradient descent with minibatch sizes $B = 1$ and $B = 2$. Compare the convergence rates—the ratios $r_k = ||\boldsymbol{x}_{k+1} - \boldsymbol{x}^*|| \,/\, ||\boldsymbol{x}_k - \boldsymbol{x}^*||$.

7 (Experiment) Try the weight averaging proposed in arXiv : 1803.05407 on page 365. Apply it to the minimization of $||A\boldsymbol{x} - \boldsymbol{b}||^2$ with randomly chosen A (20 by 10) and $\boldsymbol{b}$ (20 by 1), and minibatch $B = 1$.

Do averages in stochastic descent converge faster than the usual iterates $\boldsymbol{x}_k$?

8.4 Mean, Variance, and Covariance

1 The **probabilities** p_1 to p_n of outcomes x_1 to x_n are positive numbers adding to 1.

2 The **mean** m is the *average value* or expected value of the outcome: $m = \sum p_i x_i$.

3 The **variance** σ^2 is the average *squared distance* from the mean $\sum p_i(x_i - m)^2$

The mean is simple and we will start there. Right away we have two different situations. We may have the results (*sample values*) from a completed trial, or we may have the expected results (*expected values*) from future trials. Examples will show the difference:

Sample values Five random freshmen have ages $\mathbf{18, 17, 18, 19, 17}$

Sample mean $\frac{1}{5}(18 + 17 + 18 + 19 + 17) = \mathbf{17.8}$

Probabilities The ages in a freshmen class are 17 (**20%**), 18 (**50%**), or 19 (**30%**)

Expected age E [x] of a random freshman $= (0.2)\,17 + (0.5)\,18 + (0.3)\,19 = \mathbf{18.1}$

Both numbers 17.8 and 18.1 are correct averages. The sample mean starts with N samples x_1 to x_N from a completed trial. Their mean is the *average* of the N observed samples:

$$\textbf{Sample mean} \qquad m = \mu = \frac{1}{N}(x_1 + x_2 + \cdots + x_N) \tag{1}$$

The **expected value of** x starts with the probabilities $p_1, \ldots, p_n$ of the ages $x_1, \ldots, x_n$:

$$\textbf{Expected value} \quad m = \mathrm{E}[x] = p_1x_1 + p_2x_2 + \cdots + p_nx_n. \tag{2}$$

This is $\boldsymbol{p} \cdot \boldsymbol{x}$. The number $m = \mathrm{E}[x]$ tells us what to expect, $m = \mu$ tells us what we got.

A fair coin has probability $p_0 = \frac{1}{2}$ of tails and $p_1 = \frac{1}{2}$ of heads. Then $\mathrm{E}[x] = \left(\frac{1}{2}\right) 0 + \frac{1}{2}(1)$. The fraction of heads in N coin flips is the sample mean. The **"Law of Large Numbers"** says that with probability 1, the sample mean will converge to its expected value $\mathrm{E}[x] = \frac{1}{2}$ as the sample size N increases. This does *not* mean that if we have seen more tails than heads, the next sample is likely to be heads. The odds remain 50-50.

The first 1000 flips do affect the sample mean. *But* 1000 *flips will not affect its limit*— because you are dividing by $N \to \infty$.

Note Probability and statistics are essential for modern applied mathematics. With multiple experiments, the mean m is a vector. The variances/covariances go into a matrix. Probabilities $p(t)$ change with time in a master equation.

Variance (around the mean)

The **variance σ^2** measures expected distance (squared) from the expected mean $\mathrm{E}[x]$. The **sample variance S^2** measures actual distance (squared) from the actual sample mean. The square root is the **standard deviation σ or S**. After an exam, I email the results μ and S to the class. I don't know the expected m and σ^2 because I don't know the probabilities p_1 to p_{100} for each score. (After 60 years, I still have no idea what to expect.)

The distance is always *from the mean*—sample or expected. We are looking for the size of the "spread" around the mean value $x = m$. Start with N samples.

$$\boxed{\textbf{Sample variance} \qquad S^2 = \frac{1}{N-1}\left[(x_1 - m)^2 + \cdots + (x_N - m)^2\right]} \qquad (3)$$

The sample ages $x = 18, 17, 18, 19, 17$ have mean $m = 17.8$. That sample has variance 0.7 :

$$S^2 = \frac{1}{4}\left[(.2)^2 + (-.8)^2 + (.2)^2 + (1.2)^2 + (-.8)^2\right] = \frac{1}{4}(2.8) = \mathbf{0.7}$$

The minus signs disappear when we compute squares. Please notice ! Statisticians divide by $N - 1 = 4$ (and not $N = 5$) so that S^2 is an unbiased estimate of σ^2. One degree of freedom is already accounted for in the sample mean.

An important identity comes from splitting each $(x - m)^2$ into $x^2 - 2mx + m^2$:

$$\begin{aligned} \text{sum of } (x_i - m)^2 &= (\text{sum of } x_i^2) - 2m(\text{sum of } x_i) + (\text{sum of } m^2) \\ &= (\text{sum of } x_i^2) - 2m(Nm) + Nm^2 \\ \textbf{sum of } (\boldsymbol{x_i - m})^2 &= (\textbf{sum of } \boldsymbol{x_i^2}) - \boldsymbol{Nm^2}. \end{aligned} \qquad (4)$$

This is an equivalent way to find $(x_1 - m)^2 + \cdots + (x_N - m^2)$ by adding $x_1^2 + \cdots + x_N^2$.

Now start with probabilities p_i (never negative !) instead of samples. We find expected values instead of sample values. The variance σ^2 is the crucial number in statistics.

$$\boxed{\textbf{Variance} \quad \boldsymbol{\sigma^2} = \mathbf{E}\left[(\boldsymbol{x - m})^2\right] = \boldsymbol{p_1(x_1 - m)^2} + \cdots + \boldsymbol{p_n(x_n - m)^2}.} \qquad (5)$$

We are squaring the distance from the expected value $m = \mathrm{E}[x]$. We don't have samples, only expectations. We know probabilities but we don't know the experimental outcomes. Equation (3) for the sample variance S^2 extends directly to equation (6) for the variance σ^2 :

$$\text{Sum of } p_i(x_i - m)^2 = (\text{Sum of } p_i x_i^2) - (\text{Sum of } p_i x_i)^2 \quad \text{or} \quad \boxed{\boldsymbol{\sigma^2} = \mathrm{E}[\boldsymbol{x^2}] - (\mathrm{E}[\boldsymbol{x}]^2)} \qquad (6)$$

Example 1 Coin flipping has outputs $x = 0$ and 1 with probabilities $p_0 = p_1 = \frac{1}{2}$.

Mean m $= \frac{1}{2}(0) + \frac{1}{2}(1) = \mathbf{\frac{1}{2}}$ = average outcome = $\mathrm{E}[x]$

Variance σ^2 $= \frac{1}{2}\left(0 - \frac{1}{2}\right)^2 + \frac{1}{2}\left(1 - \frac{1}{2}\right)^2 = \frac{1}{8} + \frac{1}{8} = \mathbf{\frac{1}{4}}$ = average of (distance from m)^2

For average distance (not squared) from m, what do we expect ? $\mathbf{E}[\boldsymbol{x - m}]$ **is zero !**

Example 2 Find the variance σ^2 of the ages of college freshmen.

Solution The probabilities of ages $x_i = 17, 18, 19$ were $p_i = 0.2$ and 0.5 and 0.3. The expected value was $\boldsymbol{m} = \sum \boldsymbol{p_i x_i} = \mathbf{18.1}$. The variance uses those same probabilities:

$$\boldsymbol{\sigma^2} = (0.2)(\mathbf{17} - \mathbf{18.1})^2 + (0.5)(\mathbf{18} - \mathbf{18.1})^2 + (0.3)(\mathbf{19} - \mathbf{18.1})^2$$
$$= (0.2)(\mathbf{1.21}) + (0.5)(\mathbf{0.01}) + (0.3)(\mathbf{0.81}) = \mathbf{0.49}. \text{ } \textit{Then } \boldsymbol{\sigma} = \mathbf{0.7}.$$

This measures the spread of $17, 18, 19$ around $\mathrm{E}[x]$, weighted by probabilities $0.2, 0.5, 0.3$.

Continuous Probability Distributions

Up to now we have allowed for ages $17, 18, 19$: 3 outcomes. If we measure age in days instead of years, there will be too many possible ages. Better to allow *every number between* 17 *and* 20—a continuum of possible ages. Then probabilities p_1, p_2, p_3 for ages x_1, x_2, x_3 change to a **probability distribution** $\boldsymbol{p(x)}$ for a range of ages $17 \leq x \leq 20$.

The best way to explain probability distributions is to give you two examples. They will be the **uniform distribution** and the **normal distribution**. The first (uniform) is easy. The normal distribution is all-important.

Uniform distribution Suppose ages are uniformly distributed between 17.0 and 20.0. All those ages are "equally likely". Of course any one exact age has no chance at all. There is zero probability that you will hit the exact number $x = 17.1$ or $x = 17 + \sqrt{2}$. But you can provide **the chance $\boldsymbol{F(x)}$ that a random freshman has age less than** $\leq \boldsymbol{x}$:

The chance of age less than $x = 17$ is $F(17) = \mathbf{0}$	$x \leq 17$ won't happen
The chance of age less than $x = 20$ is $F(20) = \mathbf{1}$	$x \leq 20$ will happen
The chance of age less than x is $\boldsymbol{F(x) = \frac{1}{3}(x - 17)}$	$\mathbf{17}$ **to** $\mathbf{20}$: $\boldsymbol{F}$ **goes from 0 to 1**

From 17 to 20, the **cumulative distribution $\boldsymbol{F(x)}$** increases linearly where p is constant.

You could say that $p(x)\,dx$ is the probability of a sample falling in between x and $x + dx$. This is "infinitesimally true": $p(x)\,dx$ is $F(x + dx) - F(x)$. Here is calculus:

$$\boldsymbol{F} = \textbf{integral of } \boldsymbol{p} \qquad \textbf{Probability of } \boldsymbol{a \leq x \leq b} = \int_a^b \boldsymbol{p(x)\,dx = F(b) - F(a)} \qquad (7)$$

$F(b)$ is the probability of $x \leq b$. Subtract $F(a)$ to keep $x \geq a$. That leaves $a \leq x \leq b$.

Mean and Variance of $p(x)$

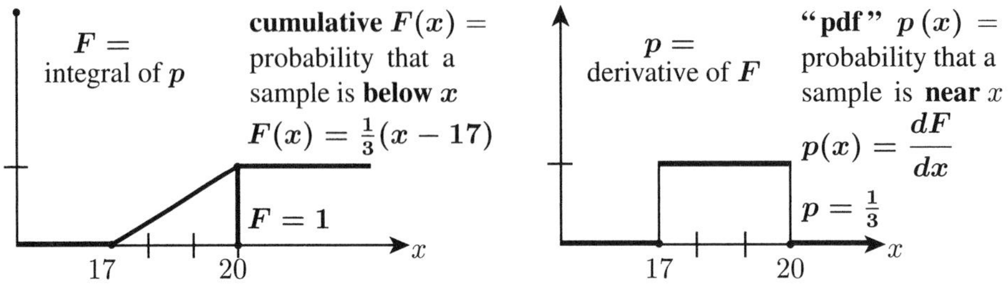

Figure 8.11: $F(x)$ is the cumulative distribution and its derivative $p(x) = dF/dx$ is the **probability density function (pdf)**. The area up to x under the graph of $p(x)$ is $F(x)$.

What are the mean m and variance σ^2 for a probability distribution? Previously we added $p_i x_i$ to get the mean (expected value). With a continuous distribution we **integrate** $\boldsymbol{xp(x)}$:

Mean	$m = \mathrm{E}[x] = \displaystyle\int x\,p(x)\,dx$
Variance	$\sigma^2 = \mathbf{E}\left[(x-m)^2\right] = \displaystyle\int p(x)\,(x-m)^2\,dx$

When ages are uniform between 17 and 20, the mean is $m = 18.5$ with $\sigma^2 = \dfrac{3}{4}$.

$$\sigma^2 = \int_{17}^{20} \frac{1}{3}(x-18.5)^2\,dx = \int_{0}^{3} \frac{1}{3}(x-1.5)^2\,dx = \frac{1}{9}(x-1.5)^3\Bigg]_{x=0}^{x=3} = \frac{2}{9}(1.5)^3 = \frac{3}{4}.$$

That is a typical example, and here is the complete picture for a uniform $p(x)$, 0 to a.

Uniform for $\mathbf{0 \le x \le a}$	**Density** $p(x) = \dfrac{1}{a}$	**Cumulative** $F(x) = \dfrac{x}{a}$
Mean $m = \int x\;p(x)\,dx = \dfrac{a}{2}$	**Variance** $\sigma^2 = \displaystyle\int_0^a \frac{1}{a}\left(x-\frac{a}{2}\right)^2 dx = \frac{a^2}{12}$	(8)

For one random number between 0 and 1 $\left(\text{mean } \frac{1}{2}\right)$ the variance is $\sigma^2 = \frac{1}{12}$.

Normal Distribution : Bell-shaped Curve

The normal distribution is also called the "Gaussian" distribution. It is the most important of all probability density functions $p(x)$. The reason for its overwhelming importance comes from repeating an experiment and averaging the outcomes. The experiments have their own distribution (like heads and tails). *The average approaches a normal distribution.*

Central Limit Theorem (informal) The average of N samples of "any" probability distribution approaches a normal distribution as $N \to \infty$.

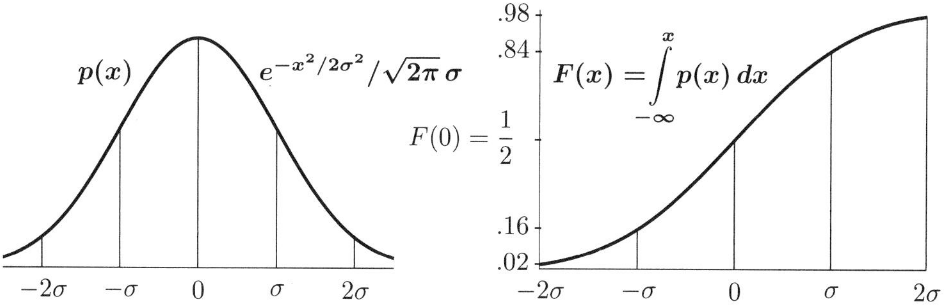

Figure 8.12: The standard normal distribution $p\,(x)$ has mean $m = 0$ and $\sigma = 1$.

The "standard normal distribution" $p(x)$ is symmetric around $x = 0$, so its mean value is $m = 0$. It is chosen to have a standard variance $\sigma^2 = 1$. It is called $\mathbf{N}\,(0, 1)$.

The graph of $\boldsymbol{p(x) = \frac{1}{\sqrt{2\pi}}\, e^{-x^2/2}}$ is the **bell-shaped curve** with variance $\sigma^2 = 1$.

By symmetry the mean is $m = 0$. The integral for σ^2 uses the idea in Problem 11 to reach 1. Figure 8.12 shows a graph of $p(x)$ for $\mathbf{N}\,(0, \sigma)$ and also its cumulative distribution $F(x) =$ integral of $p(x)$. From $F(x)$ you see a very important approximation for opinion polling:

The probability that a random sample falls between $-\sigma$ and σ is $\boldsymbol{F(\sigma) - F(-\sigma) \approx \frac{2}{3}}$.

Similarly, the probability that a random x lies between -2σ and 2σ ("*less than two standard deviations from the mean*") is $F(2\sigma) - F(-2\sigma) \approx 0.95$. If you have an experimental result further than 2σ from the mean, it is fairly sure to be not accidental. The normal distribution with any mean m and standard deviation σ comes by shifting and stretching the standard $\mathbf{N}\,(0, 1)$. **Shift $\boldsymbol{x}$ to $\boldsymbol{x - m}$. Stretch $\boldsymbol{x - m}$ to $\boldsymbol{(x - m)/\sigma}$.**

Gaussian density $\boldsymbol{p(x)}$ **Normal distribution $\mathbf{N}(\boldsymbol{m, \sigma})$**	$\boldsymbol{p(x) = \dfrac{1}{\sigma\sqrt{2\pi}}\, e^{-(x-m)^2/2\sigma^2}}$	(8)

The integral of $p(x)$ is $F(x)$—the probability that a random sample will fall below x. There is no simple formula to integrate $e^{-x^2/2}$, so $F(x)$ is computed very carefully.

N Coin Flips and $N \to \infty$

Example 3 **Suppose x is 1 or -1 with equal probabilities $p_1 = p_{-1} = \frac{1}{2}$.**

The mean value is $m = \frac{1}{2}(1) + \frac{1}{2}(-1) = \mathbf{0}$. The variance is $\sigma^2 = \frac{1}{2}(1)^2 + \frac{1}{2}(-1)^2 = \mathbf{1}$.

The key question is the *average* $A_N = (x_1 + \cdots + x_N)/N$. The independent x_i are ± 1 and we are dividing their sum by N. The expected mean of A_N is still zero. The law of large numbers says that this sample average approaches zero with probability 1. How fast does A_N approach zero? **What is its variance σ_N^2 ?**

$$\text{By linearity} \quad \boldsymbol{\sigma_N^2} = \frac{\sigma^2}{N^2} + \frac{\sigma^2}{N^2} + \cdots + \frac{\sigma^2}{N^2} = N\frac{\sigma^2}{N^2} = \frac{\mathbf{1}}{\boldsymbol{N}} \quad \text{since } \sigma^2 = 1.$$

> Here are the results from three numerical tests: random 0 or 1 averaged over N trials.
>
> [**48** 1's from $N = \mathbf{100}$] [**5035** 1's from $N = \mathbf{10000}$] [**19967** 1's from $N = \mathbf{40000}$].
>
> **The standardized X** $= (x - m)/\sigma = \left(A_N - \frac{1}{2}\right) / \, 2\sqrt{N}$ was [$\mathbf{-.40}$] [$\mathbf{.70}$] [$\mathbf{-.33}$].

The Central Limit Theorem says that the average of many coin flips will approach a normal distribution. Let us begin to see how that happens: **binomial approaches normal.**

The "binomial" probabilities $p_0, \ldots, p_N$ count the number of heads in N coin flips.

For each (fair) flip, the probability of heads is $\frac{1}{2}$. For $N = 3$ flips, the probability of heads all three times is $\left(\frac{1}{2}\right)^3 = \frac{1}{8}$. The probability of heads twice and tails once is $\frac{3}{8}$, from three sequences HHT and HTH and THH. These numbers $\frac{\mathbf{1}}{\mathbf{8}}$ and $\frac{\mathbf{3}}{\mathbf{8}}$ are pieces of $\left(\frac{1}{2} + \frac{1}{2}\right)^3 = \frac{\mathbf{1}}{\mathbf{8}} + \frac{\mathbf{3}}{\mathbf{8}} + \frac{\mathbf{3}}{\mathbf{8}} + \frac{\mathbf{1}}{\mathbf{8}} = 1$. *The average number of heads in* 3 *flips is* 1.5.

$$\textbf{Mean} \;\; m = (3 \text{ heads})\frac{1}{8} + (2 \text{ heads})\frac{3}{8} + (1 \text{ head})\frac{3}{8} + 0 = \frac{3}{8} + \frac{6}{8} + \frac{3}{8} = \mathbf{1.5 \text{ heads}}$$

With N flips, Example 3 (or common sense) gives a mean of $\boldsymbol{m = \Sigma\, x_i p_i = \frac{1}{2}N}$ heads.

The variance σ^2 is based on the *squared distance* from this mean $N/2$. With $N = 3$ the variance is $\sigma^2 = \frac{\mathbf{3}}{\mathbf{4}}$ (*which is* $N/4$). To find σ^2 we add $(x_i - m)^2\, p_i$ with $m = 1.5$:

$$\boldsymbol{\sigma^2} = (3-1.5)^2\,\frac{1}{8} + (2-1.5)^2\,\frac{3}{8} + (1-1.5)^2\,\frac{3}{8} + (0-1.5)^2\,\frac{1}{8} = \frac{9+3+3+9}{32} = \frac{\mathbf{3}}{\mathbf{4}}.$$

For any N, the variance for a binomial distribution is $\boldsymbol{\sigma_N^2 = N/4}$. Then $\sigma_N = \sqrt{N}/2$.

Figure 8.13 shows how the probabilities of 0, 1, 2, 3, 4 heads in $N = 4$ flips come close to a bell-shaped Gaussian. That Gaussian is centered at the mean value $m = N/2 = 2$.

To reach the standard Gaussian (mean 0 and variance 1) we shift and rescale Figure 6.8. If x is the number of heads in N flips—the average of N zero-one outcomes—then x is shifted by its mean $m = N/2$ and rescaled by $\sigma = \sqrt{N}/2$ to produce the standard X:

Shifted and scaled $\quad \boldsymbol{X = \dfrac{x - m}{\sigma} = \dfrac{x - \frac{1}{2}N}{\sqrt{N}/2}} \quad$ **($\boldsymbol{N = 4}$ has $\boldsymbol{X = x - 2}$)**

Subtracting $\boldsymbol{m}$ is "centering" or "detrending". The mean of $\boldsymbol{X}$ is zero.

Dividing by $\boldsymbol{\sigma}$ is "normalizing" or "standardizing". The variance of $\boldsymbol{X}$ is 1.

It is fun to see the Central Limit Theorem giving the right answer at the center point $X = 0$. At that point, the factor $e^{-X^2/2}$ equals 1. We know that the variance for N coin flips is $\sigma^2 = N/4$. The center of the bell-shaped curve has height $1/\sqrt{2\pi\sigma^2} = \sqrt{\mathbf{2/N\pi}}$.

What is the height at the center of the coin-flip distribution p_0 to p_N (the binomial distribution)? For $N = 4$, the probabilities for $0, 1, 2, 3, 4$ heads come from $\left(\frac{1}{2} + \frac{1}{2}\right)^4$.

$$\textbf{Center probability} = \frac{\mathbf{6}}{\mathbf{16}} \qquad \left(\frac{1}{2} + \frac{1}{2}\right)^4 = \frac{1}{16} + \frac{4}{16} + \frac{\mathbf{6}}{\mathbf{16}} + \frac{4}{16} + \frac{1}{16} = 1. \tag{9}$$

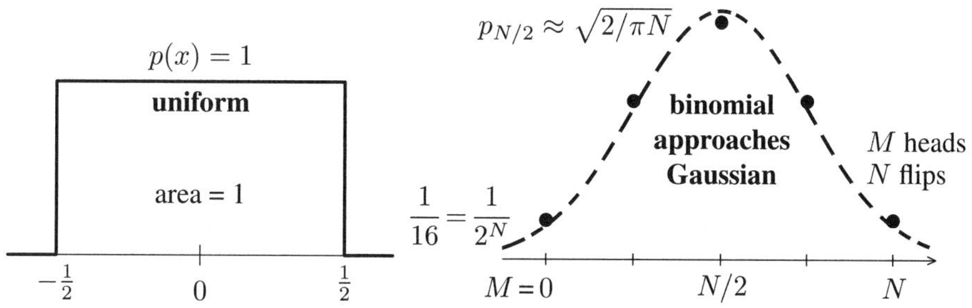

Figure 8.13: The probabilities $p = (1, 4, 6, 4, 1)/16$ for the number of heads in 4 flips. These p_i approach a Gaussian distribution with variance $\sigma^2 = N/4$ centered at $m = N/2$. For X, the Central Limit Theorem gives convergence to the normal distribution $\mathbf{N}(0, 1)$.

The binomial $\left(\frac{1}{2} + \frac{1}{2}\right)^N$ tells us the probabilities for $0, 1, \ldots, N$ heads:

The center term is the probability of $\frac{N}{2}$ heads, $\frac{N}{2}$ tails $\quad \boldsymbol{\dfrac{1}{2^N} \dfrac{N!}{(N/2)!\,(N/2)!}}$

For $N = 4$, those factorials produce $4!/2!\,2! = 24/4 = 6$. For large N, *Stirling's formula* $\sqrt{2\pi N}(N/e)^N$ *is a close approximation to* $N!$. Use this formula for N and twice for $N/2$:

$$\begin{array}{l}\textbf{Limit of coin flip}\\ \textbf{Center probability}\end{array} \qquad p_{N/2} \approx \frac{1}{2^N} \frac{\sqrt{2\pi N}(N/e)^N}{\pi N(N/2e)^N} = \frac{\sqrt{2}}{\sqrt{\pi N}} = \frac{1}{\sqrt{2\pi}\sigma}. \tag{10}$$

The last step used the variance $\sigma^2 = N/4$ for coin-tossing. The result $1/\sqrt{2\pi}\sigma$ matches the center value (above) for the Gaussian. The Central Limit Theorem is true:

The centered binomial distribution approaches the normal distribution $\boldsymbol{p(x)}$ as $\boldsymbol{N \to \infty}$.

Covariance Matrices and Joint Probabilities

Linear algebra enters when we run M different experiments at once. We might measure age and height ($M = 2$ measurements of N children). Each experiment has its own mean value. So we have a vector $\boldsymbol{m} = (m_1, m_2)$ containing two mean values. Those could be *sample means* of age and height. Or m_1 and m_2 could be *expected values* of age and height based on known probabilities.

A matrix becomes involved when we look at variances. Each experiment will have a sample variance S_i^2 or an expected $\sigma_i^2 = \mathrm{E}\left[(x_i - m_i)^2\right]$ based on the squared distance from its mean. Those M numbers $\sigma_1^2, \ldots, \sigma_M^2$ will go on the main diagonal of the "variance-covariance matrix". So far we have made no connection between the M parallel experiments. They measure different random variables, but the experiments are not necessarily independent!

If we measure age and height for children, the results will be strongly correlated. Older children are generally taller. Suppose the means m_a and m_h are known. Then σ_a^2 and σ_h^2 are the separate variances in age and height. **The new number is the covariance $\boldsymbol{\sigma_{ah}}$, which measures the connection of each possible age to each possible height.**

$$\textbf{Covariance} \quad \sigma_{ah} = \mathbf{E}\left[(\textbf{age} - \textbf{mean age})\ (\textbf{height} - \textbf{mean height})\right]. \tag{11}$$

This definition needs a close look. To compute σ_{ah}, it is not enough to know the probability of each age and the probability of each height. We have to know the **joint probability $\boldsymbol{p_{ah}}$ of each pair (age and height)**. This is because age is related to height.

p_{ah} = probability that a random child has age $= \boldsymbol{a}$ **and** height $= \boldsymbol{h}$: both at once

p_{ij} = **probability that experiment 1 produces $\boldsymbol{x_i}$ and experiment 2 produces $\boldsymbol{y_j}$**

Suppose experiment 1 (age) has mean m_1. Experiment 2 (height) has its own mean m_2. The covariance between experiments 1 and 2 looks at **all pairs** of ages x_i and heights y_j. We multiply by the joint probability p_{ij} of that age-height pair.

Expected value of $(\boldsymbol{x} - \boldsymbol{m_1})(\boldsymbol{y} - \boldsymbol{m_2})$

$$\textbf{Covariance} \quad \sigma_{12} = \sum_{\text{all } i,j} \sum p_{ij}(x_i - m_1)(y_j - m_2) \tag{12}$$

To capture this idea of "joint probability p_{ij}" we begin with two small examples.

Example 4 **Flip two coins separately**. With 1 for heads and 0 for tails, the results can be $(1,1)$ or $(1,0)$ or $(0,1)$ or $(0,0)$. Those four outcomes all have probability $\left(\frac{1}{2}\right)^2 = \frac{1}{4}$. For independent experiments we multiply probabilities: *The covariance is zero.*

$$p_{ij} = \textbf{Probability of } (i,j) = (\textbf{Probability of } i) \textbf{ times } (\textbf{Probability of } j).$$

Example 5 *Glue the coins together*, facing the same way. The only possibilities are $(1,1)$ and $(0,0)$. Those have probabilities $\frac{1}{2}$ and $\frac{1}{2}$. The probabilities p_{10} and p_{01} are zero. $(1,0)$ and $(0,1)$ won't happen because the coins stick together: both heads or both tails.

Joint probability matrices for Examples 1 and 2 $\quad \boldsymbol{P_1} = \begin{bmatrix} \frac{1}{4} & \frac{1}{4} \\ \frac{1}{4} & \frac{1}{4} \end{bmatrix}$ and $\boldsymbol{P_2} = \begin{bmatrix} \frac{1}{2} & 0 \\ 0 & \frac{1}{2} \end{bmatrix}$.

Let me stay longer with P, to show it in good matrix notation. The matrix shows the probability p_{ij} of each pair (x_i, y_j)—starting with $(x_1, y_1) =$ (heads, heads) and $(x_1, y_2) =$ (heads, tails). Notice the row sums p_1, p_2 and column sums P_1, P_2 and the total sum $= 1$.

Probability matrix $\quad P = \begin{bmatrix} p_{11} & p_{12} \\ p_{21} & p_{22} \end{bmatrix} \qquad \begin{matrix} p_{11} + p_{12} = \boldsymbol{p_1} \\ p_{21} + p_{22} = \boldsymbol{p_2} \end{matrix} \begin{pmatrix} \text{first} \\ \text{coin} \end{pmatrix}$

(second coin) column sums $\quad \boldsymbol{P_1} \quad \boldsymbol{P_2} \qquad$ 4 entries add to 1

Those sums p_1, p_2 and P_1, P_2 are the **marginals** of the joint probability matrix P:

$p_1 = p_{11} + p_{12} =$ chance of heads from **coin 1** (coin 2 can be heads or tails)
$P_1 = p_{11} + p_{21} =$ chance of heads from **coin 2** (coin 1 can be heads or tails)

Example 1 showed *independent* random variables. Every probability p_{ij} equals p_i times p_j $\left(\frac{1}{2}\right.$ times $\frac{1}{2}$ gave $p_{ij} = \frac{1}{4}$ in that example$\left.\right)$. In this case **the covariance $\boldsymbol{\sigma_{12}}$ will be zero.** Heads or tails from the first coin gave no information about the second coin.

Zero covariance $\boldsymbol{\sigma_{12}}$ for independent trials	$V = \begin{bmatrix} \sigma_1^2 & 0 \\ 0 & \sigma_2^2 \end{bmatrix} =$ **diagonal covariance matrix** V.

Independent experiments have $\sigma_{12} = 0$ because every p_{ij} equals $(p_i)(p_j)$ in equation (12).

$$\sigma_{12} = \sum_i \sum_j (p_i)(p_j)(x_i - m_1)(y_j - m_2) = \left[\sum_i (p_i)(x_i - m_1)\right]\left[\sum_j (p_j)(y_j - m_2)\right] = [0][0].$$

Example 6 The glued coins show perfect correlation. Heads on one means heads on the other. The covariance σ_{12} moves from 0 to $\boldsymbol{\sigma_1}$ **times** $\boldsymbol{\sigma_2}$. This is the largest possible value of σ_{12}. Here it is $\left(\frac{1}{2}\right)\left(\frac{1}{2}\right) = \sigma_{12} = \left(\frac{1}{4}\right)$, as a separate computation confirms:

$$\textbf{Means} = \frac{\mathbf{1}}{\mathbf{2}} \qquad \boldsymbol{\sigma_{12}} = \frac{\mathbf{1}}{\mathbf{2}}\left(1 - \frac{1}{2}\right)\left(1 - \frac{1}{2}\right) + \mathbf{0} + \mathbf{0} + \frac{\mathbf{1}}{\mathbf{2}}\left(0 - \frac{1}{2}\right)\left(0 - \frac{1}{2}\right) = \frac{\mathbf{1}}{\mathbf{4}}$$

Heads or tails from coin 1 gives complete information about heads or tails from coin 2:

Glued coins give largest possible covariances **Singular covariance matrix: determinant = 0**	$V_{\text{glue}} = \begin{bmatrix} \sigma_1^2 & \sigma_1\sigma_2 \\ \sigma_1\sigma_2 & \sigma_2^2 \end{bmatrix}$

Always $\sigma_1^2\sigma_2^2 \geq (\sigma_{12})^2$. Thus σ_{12} is *between* $-\sigma_1\sigma_2$ *and* $\sigma_1\sigma_2$. The matrix V is **positive definite** (or in this singular case of glued coins, V is **positive semidefinite**). Those are important facts about all M by M covariance matrices V for M experiments.

Note that the **sample covariance matrix S** from N trials is certainly semidefinite. Every sample X = (age, height) contributes to the **sample mean $\overline{X}$** $= (m_a, m_h)$. Each rank-one term $(X_i - \overline{X})(X_i - \overline{X})^{\mathrm{T}}$ is positive semidefinite and we just add to reach the matrix S. No probabilities in S, use the actual outcomes:

$$\boldsymbol{\overline{X}} = \frac{X_1 + \cdots + X_N}{N} \qquad \boldsymbol{S} = \frac{(X_1 - \overline{X})(X_1 - \overline{X})^{\mathrm{T}} + \cdots + (X_N - \overline{X})(X_N - \overline{X})^{\mathrm{T}}}{N-1} \tag{13}$$

The Covariance Matrix V is Positive Semidefinite

Come back to the *expected* covariance σ_{12} between two experiments 1 and 2 (two coins):

$$\begin{aligned} \sigma_{12} &= \text{expected value of } [(\textit{output}\,1 - \textit{mean}\,1) \text{ times } (\textit{output}\,2 - \textit{mean}\,2)] \\ \boldsymbol{\sigma_{12}} &= \textstyle\sum\sum \boldsymbol{p_{ij}\,(x_i - m_1)\,(y_j - m_2)}. \text{ The sum includes all } i, j. \end{aligned} \tag{14}$$

$p_{ij} \geq 0$ is the probability of seeing outputs x_i in experiment 1 **and** y_j in experiment 2. Some pair of outputs must appear. Therefore the N^2 joint probabilities p_{ij} add to 1.

$$\textbf{Total probability (all pairs) is 1} \qquad \sum_{\text{all } i,j}\sum p_{ij} = 1. \tag{15}$$

Here is another fact we need. *Fix on one particular output x_i in experiment 1. Allow all outputs y_j* in experiment 2. Add the probabilities of $(x_i, y_1), (x_i, y_2), \ldots, (x_i, y_n)$:

$$\textbf{Row sum } \boldsymbol{p_i} \textbf{ of } \boldsymbol{P} \qquad \sum_{j=1}^{n} \boldsymbol{p_{ij}} = \textbf{probability } \boldsymbol{p_i} \textbf{ of } \boldsymbol{x_i} \textbf{ in experiment 1.} \tag{16}$$

Some y_j must happen in experiment 2 ! Whether the two coins are completely separate or glued, we get the same answer $\frac{1}{2}$ for the probability $p_H = p_{HH} + p_{HT}$ that coin 1 is heads:

$$\text{(separate) } P_{HH} + P_{HT} = \frac{1}{4} + \frac{1}{4} = \frac{\mathbf{1}}{\mathbf{2}} \qquad \text{(glued) } P_{HH} + P_{HT} = \frac{1}{2} + 0 = \frac{\mathbf{1}}{\mathbf{2}}.$$

That basic reasoning allows us to write one matrix formula that includes the covariance σ_{12} along with the separate variances σ_1^2 and σ_2^2 for experiment 1 and experiment 2. We get the whole covariance matrix V by adding the matrices V_{ij} for each pair (i, j):

$$\begin{array}{l}\textbf{Covariance matrix} \\ \boldsymbol{V} = \textbf{sum of all } \boldsymbol{V_{ij}}\end{array} \qquad \boldsymbol{V} = \sum_{\text{all } i,j}\sum p_{ij} \begin{bmatrix} (x_i - m_1)^2 & (x_i - m_1)(y_j - m_2) \\ (x_i - m_1)(y_j - m_2) & (y_j - m_2)^2 \end{bmatrix} \tag{17}$$

Off the diagonal, this is equation (12) for the covariance σ_{12}. On the diagonal, we are getting the ordinary variances σ_1^2 and σ_2^2. I will show in detail how we get $V_{11} = \sigma_1^2$ by using equation (16). Allowing all j just leaves the probability p_i of x_i in experiment 1 :

$$\boldsymbol{V_{11}} = \sum_{\text{all } i,j}\sum p_{ij}(x_i - m_1)^2 = \sum_{\text{all } i} (\text{probability of } x_i)\,(x_i - m_1)^2 = \boldsymbol{\sigma_1^2}. \qquad (18)$$

Please look at that twice. It is the key to producing the whole covariance matrix by one formula (17). The beauty of that formula is that it combines 2 by 2 matrices V_{ij}. And the matrix V_{ij} in (17) for each pair of outcomes i, j is **positive semidefinite** :

V_{ij} has diagonal entries $p_{ij}(x_i - m_1)^2 \geq 0$ and $p_{ij}(y_j - m_2)^2 \geq 0$ and $\det(V_{ij}) = 0$.

That matrix V_{ij} has rank 1. Equation (17) multiplies p_{ij} *times column* $\boldsymbol{U}$ *times row* $\boldsymbol{U}^{\mathrm{T}}$:

$$\begin{bmatrix} (x_i - m_1)^2 & (x_i - m_1)(y_j - m_2) \\ (x_i - m_1)(y_j - m_2) & (y_j - m_2)^2 \end{bmatrix} = \begin{bmatrix} x_i - m_1 \\ y_j - m_2 \end{bmatrix} \begin{bmatrix} x_i - m_1 & y_j - m_2 \end{bmatrix} \qquad (19)$$

Every matrix $\boldsymbol{p_{ij} U U^{\mathrm{T}}}$ *is positive semidefinite.* So the whole matrix V (the sum of those rank 1 matrices) is **at least semidefinite**—and probably V is definite.

The covariance matrix V is positive definite unless the experiments are dependent.

Now we move from two variables x and y to M variables like age-height-weight. The output from each trial is a vector X with M components. (Each child has an age-height-weight vector $\boldsymbol{X}$ with 3 components.) The covariance matrix V is now M by M. The matrix $\boldsymbol{V}$ is created from the output vectors $\boldsymbol{X}$ and their average $\overline{\boldsymbol{X}} = \mathbf{E}[\boldsymbol{X}]$:

Covariance matrix	$\boldsymbol{V} = \mathbf{E}\left[(\boldsymbol{X} - \overline{\boldsymbol{X}})(\boldsymbol{X} - \overline{\boldsymbol{X}})^{\mathrm{T}}\right]$	$\boldsymbol{V_{ij}} = \sum p_{ij}\,(\boldsymbol{X_i} - \boldsymbol{m_i})\,(\boldsymbol{X_j} - \boldsymbol{m_j})$	(20)

Remember that $\boldsymbol{X}\boldsymbol{X}^{\mathrm{T}}$ and $\overline{\boldsymbol{X}}\,\overline{\boldsymbol{X}}^{\mathrm{T}} =$ (column) (row) are M by M matrices.

For $M = 1$ (one variable) you see that $\overline{X}$ is the mean m and V is the variance σ^2. For $M = 2$ (two coins) you see that $\overline{X}$ is (m_1, m_2) and V matches equation (7). The expectation always adds up outputs times their probabilities. For age-height-weight the output could be $X =$ (5 years, 31 inches, 48 pounds) and its probability is $\boldsymbol{p_{5,31,48}}$.

Now comes a new idea. *Take any linear combination* $\mathbf{c}^{\mathrm{T}}\boldsymbol{X} = c_1X_1 + \cdots + c_MX_M$. With $\mathbf{c} = (6, 2, 5)$ this would be $\mathbf{c}^{\mathrm{T}}\boldsymbol{X} = 6 \times \text{age} + 2 \times \text{height} + 5 \times \text{weight}$. By linearity we know that its expected value $\mathrm{E}[\boldsymbol{c}^{\mathrm{T}}\boldsymbol{X}]$ is $\boldsymbol{c}^{\mathrm{T}}\mathrm{E}[\boldsymbol{X}] = \boldsymbol{c}^{\mathrm{T}}\overline{\boldsymbol{X}}$:

$$\mathbf{E}[\boldsymbol{c}^{\mathrm{T}}\boldsymbol{X}] = \boldsymbol{c}^{\mathrm{T}}\mathbf{E}[\boldsymbol{X}] = 6\,(\text{expected age}) + 2\,(\text{expected height}) + 5\,(\text{expected weight}).$$

More than the mean of $c^{\mathrm{T}}X$, we also know its *variance* $\boldsymbol{\sigma^2 = c^{\mathrm{T}}Vc}$:

$$\text{Variance of } \boldsymbol{c}^{\mathrm{T}}\boldsymbol{X} = \boldsymbol{c}^{\mathrm{T}}\mathbf{E}\left[\left(\boldsymbol{X}-\overline{\boldsymbol{X}}\right)\left(\boldsymbol{X}-\overline{\boldsymbol{X}}\right)^{\mathrm{T}}\right]\boldsymbol{c} = \boldsymbol{c}^{\mathrm{T}}\,\boldsymbol{V}\boldsymbol{c} \qquad (21)$$

Now the key point: *The variance of* $\boldsymbol{c}^{\mathrm{T}}\boldsymbol{X}$ *can never be negative.* So $\boldsymbol{c^{\mathrm{T}}Vc \geq 0}$. New proof: *The covariance matrix V is positive semidefinite by the energy test* $\boldsymbol{c^{\mathrm{T}}Vc \geq 0}$.

Covariance matrices V open up the link between probability and linear algebra: $\boldsymbol{V}$ **equals** $\boldsymbol{Q\Lambda Q^{\mathrm{T}}}$ with eigenvalues $\lambda_i \geq 0$ and orthonormal eigenvectors $\boldsymbol{q}_1$ to $\boldsymbol{q}_M$.

Diagonalizing the covariance matrix V means finding M *independent* experiments as combinations of the original M experiments.

The Covariance Matrix for $Z = AX$

Here is a good way to see σ_z^2 when $z = x + y$. Think of (x, y) as a column vector $\boldsymbol{X}$. Think of the 1 by 2 matrix $\boldsymbol{A} = \begin{bmatrix} 1 & 1 \end{bmatrix}$ multiplying that vector $\boldsymbol{X} = (x, y)$. Then $\boldsymbol{AX}$ is the sum $z = x + y$. The variance $\boldsymbol{\sigma_z^2}$ goes into matrix notation as

$$\boldsymbol{\sigma_z^2} = \begin{bmatrix} 1 & 1 \end{bmatrix}\begin{bmatrix} \sigma_x^2 & \sigma_{xy} \\ \sigma_{xy} & \sigma_y^2 \end{bmatrix}\begin{bmatrix} 1 \\ 1 \end{bmatrix} \quad \text{which is} \quad \boldsymbol{\sigma_z^2 = AVA^{\mathrm{T}}}. \qquad (22)$$

Now for the main point. The vector $\boldsymbol{X}$ could have M components coming from M experiments (instead of only 2). Those experiments will have an M by M covariance matrix $\boldsymbol{V_X}$. The matrix A could be K by M. Then $A\boldsymbol{X}$ is a vector with K combinations of the M outputs (instead of one combination $x + y$ of 2 two outputs).

That vector $\boldsymbol{Z} = A\boldsymbol{X}$ of length K has a K by K covariance matrix $V_{\boldsymbol{Z}}$. Then the great rule for covariance matrices—of which equation (22) was only a 1 by 2 example—is this beautiful formula: The covariance matrix of $\boldsymbol{AX}$ is $\boldsymbol{A}$ **(covariance matrix of** $\boldsymbol{X}$**)** $\boldsymbol{A}^{\mathrm{T}}$:

$$\textbf{The covariance matrix of } Z = AX \qquad \boxed{V_Z = AV_XA^{\mathrm{T}}} \qquad (23)$$

To me, this neat formula shows the beauty of matrix multiplication. I won't prove this formula, just admire it. It is constantly used in applications.

The Correlation ρ

Correlation ρ_{xy} is closely related to covariance σ_{xy}. They both measure dependence or independence. Start by rescaling or "standardizing" the random variables x and y **The new** $\boldsymbol{X = x/\sigma_x}$ **and** $\boldsymbol{Y = y/\sigma_y}$ **have variance** $\boldsymbol{\sigma_X^2 = \sigma_Y^2 = 1}$. This is just like dividing a vector $\boldsymbol{v}$ by its length to produce a unit vector $\boldsymbol{v}/||\boldsymbol{v}||$ of length 1.

The correlation of x and y is the covariance of X and Y. If the original covariance of x and y was σ_{xy}, then rescaling to X and Y gives correlation $\boldsymbol{\rho_{xy} = \sigma_{xy}/\sigma_x\sigma_y}$.

$$\boxed{\textbf{Correlation } \rho_{xy} = \frac{\sigma_{xy}}{\sigma_x\sigma_y} = \textbf{covariance of } \frac{x}{\sigma_x} \textbf{ and } \frac{y}{\sigma_y}} \qquad \textbf{Always } -1 \leq \rho_{xy} \leq 1$$

Problem Set 8.4

1 If all 24 samples from a population produce the same age $x = 20$, what are the sample mean μ and the sample variance S^2 ? What if $x = 20$ or 21, 12 times each ?

2 Add 7 to every output x. What happens to the mean and the variance ? What are the new sample mean, the new expected mean, and the new variance ?

3 We know: $\frac{1}{3}$ of all integers are divisible by 3 and $\frac{1}{7}$ of integers are divisible by 7. What fraction of integers will be divisible by 3 or 7 or both ?

4 Suppose you sample from the numbers 1 to 1000 with equal probabilities $1/1000$. What are the probabilities p_0 to p_9 that the last digit of your sample is $0, \ldots, 9$? What is the expected mean m of that last digit? What is its variance σ^2 ?

5 Sample again from 1 to 1000 but look at the last digit of the sample *squared.* That square could end with $x = 0, 1, 4, 5, 6$, or 9. What are the probabilities $p_0, p_1, p_4, p_5, p_6, p_9$? What are the (expected) mean m and variance σ^2 of that number x?

6 (a little tricky) Sample again from 1 to 1000 with equal probabilities and let x be the *first* digit ($x = 1$ if the number is 15). What are the probabilities p_1 to p_9 (adding to 1) of $x = 1, \ldots, 9$? What are the mean and variance of x?

7 Suppose you have $N = 4$ samples $157, 312, 696, 602$ in Problem 5. What are the first digits x_1 to x_4 of the squares? What is the sample mean μ ? What is the sample variance S^2 ? Remember to divide by $N - 1 = 3$ and not $N = 4$.

8 Equation (4) gave a second equivalent form for S^2 (the variance using samples):

$$\boldsymbol{S^2} = \frac{1}{N-1} \text{ sum of } (x_i - m)^2 = \frac{1}{N-1} \left[\left(\text{sum of } x_i^2 \right) - Nm^2 \right].$$

Verify the matching identity for the expected variance σ^2 (using $m = \Sigma\, p_i\, x_i$):

$$\boldsymbol{\sigma^2 = \textbf{sum of } p_i\, (x_i - m)^2 = \left(\textbf{sum of } p_i\, x_i^2 \right) - m^2.}$$

9 Computer experiment: Find the average $A_{1000000}$ of a million random 0-1 samples ! What is your value of the standardized variable $X = \left(A_N - \frac{1}{2}\right) / 2\sqrt{N}$?

10 For any function $f(x)$ the expected value is $\mathrm{E}[f] = \sum p_i\, f(x_i)$ or $\int p(x)\, f(x)\, dx$ (discrete or continuous probability). The function can be x or $(x - m)^2$ or x^2. If the mean is $\mathrm{E}[x] = m$ and the variance is $\mathrm{E}[(x - m)^2] = \sigma^2$, **what is** $\mathbf{E}[\boldsymbol{x^2}]$?

11 Show that the standard normal distribution $p(x)$ has total probability $\int p(x)\, dx = 1$ as required. A famous trick multiplies $\int p(x)\, dx$ by $\int p(y)\, dy$ and computes the integral over all x and all y ($-\infty$ to ∞). The trick is to replace $dx\, dy$ in that double integral by $r\, dr\, d\theta$ (polar coordinates with $x^2 + y^2 = r^2$). Explain each step :

$$2\pi \int_{-\infty}^{\infty} p(x)\, dx \int_{-\infty}^{\infty} p(y)\, dy = \iint_{-\infty}^{\infty} e^{-(x^2+y^2)/2} dx\, dy = \int_{\theta=0}^{2\pi} \int_{r=0}^{\infty} e^{-r^2/2}\, r\, dr\, d\theta = \boldsymbol{2\pi}.$$

A1 The Ranks of AB and $A + B$

This page establishes key facts about ranks: **When we multiply matrices, the rank cannot increase**. You will see this by looking at column spaces and row spaces. And there is one special situation when the rank cannot decrease. Then you know the rank of AB. Statement 4 will be important when data science factors a matrix into UV or CR.

Here are five key facts in one place: inequalities and equalities for the rank.

1 **Rank of $AB \leq$ rank of A**   **Rank of $AB \leq$ rank of B**
2 **Rank of $A + B \leq$ (rank of A) + (rank of B)**
3 **Rank of $A^{\mathrm{T}}A$ = rank of AA^{T} = rank of A = rank of A^{T}**
4 **If A is m by r and B is r by n**—both with rank r—**then AB also has rank r**

Statement 1 involves the column space and row space of AB:

$$\mathbf{C}(AB) \text{ is contained in } \mathbf{C}(A) \qquad \mathbf{C}((AB)^{\mathrm{T}}) \text{ is contained in } \mathbf{C}(B^{\mathrm{T}})$$

Every column of AB is a combination of the columns of A (*matrix multiplication*)
Every row of AB is a combination of the rows of B (*matrix multiplication*)

Remember from Section 1.4 that **row rank = column rank.** We can use rows or columns. *The rank cannot grow when we multiply AB.* Statement **1** in the box is frequently used.

Statement 2 Each column of $A + B$ is the sum of (column of A) + (column of B).

rank $(A + B) \leq$ rank (A) + rank (B) is always true. It combines bases for $\mathbf{C}(A)$ and $\mathbf{C}(B)$.

rank $(A + B) =$ rank (A) + rank (B) is not always true. It is certainly false if $A = B = I$.

Statement 3 A and $A^{\mathrm{T}}A$ both have n columns. **They also have the same nullspace.** (See Problem 4.1.9.) So $n - r$ is the same for both, and *the rank r is the same for both.* Then rank$(A^{\mathrm{T}}) \geq$ rank$(A^{\mathrm{T}}A) =$ rank(A). Exchange A and A^{T} to show their equal ranks.

Statement 4 We are told that A and B have rank r. By Statement **3**, $A^{\mathrm{T}}A$ and BB^{T} have rank r. Those are r by r matrices so they are invertible. So is their product $A^{\mathrm{T}}ABB^{\mathrm{T}}$. Then

$$r = \textbf{rank of } (A^{\mathrm{T}}ABB^{\mathrm{T}}) \leq \textbf{ rank of } (AB) \text{ by Statement 1 : } A^{\mathrm{T}}, B^{\mathrm{T}} \text{can't increase rank}$$

We also know rank $(AB) \leq$ rank $A = r$. So we have proved that **AB has rank exactly r.**

Note This does not mean that every product of rank r matrices will have rank r. Statement **4** assumes that A has exactly r columns and B has r rows. BA can easily fail.

$$A = \begin{bmatrix} 1 \\ 1 \\ 1 \end{bmatrix} \qquad B = \begin{bmatrix} 1 & 2 & -3 \end{bmatrix} \qquad \textbf{AB has rank 1} \qquad \textbf{But } BA \textbf{ is zero!}$$

A2 Eigenvalues and Singular Values : Rank One

A rank one matrix has the simple form $A = \boldsymbol{x}\boldsymbol{y}^{\mathrm{T}}$. Its singular vectors $\boldsymbol{u}_1, \boldsymbol{v}_1$ and its only nonzero singular value σ_1 are incredibly easy to find :

$$\boldsymbol{u}_1 = \frac{\boldsymbol{x}}{||\boldsymbol{x}||} \qquad \boldsymbol{v}_1 = \frac{\boldsymbol{y}}{||\boldsymbol{y}||} \qquad \sigma_1 = ||\boldsymbol{x}||\,||\boldsymbol{y}||.$$

You see immediately that $A = \boldsymbol{x}\boldsymbol{y}^{\mathrm{T}} = \boldsymbol{u}_1\sigma_1\boldsymbol{v}_1^{\mathrm{T}}$. One nonzero in the $m \times n$ matrix Σ. All other columns of the orthogonal matrices U and V are perpendicular to $\boldsymbol{u}_1$ and $\boldsymbol{v}_1$. The decomposition $A = U\Sigma V^{\mathrm{T}}$ reduces to the first term $A = \boldsymbol{u}_1\sigma_1\boldsymbol{v}_1^{\mathrm{T}}$ because rank $= 1$.

Eigenvalues and eigenvectors are not quite that easy. Of course the matrix A must be square. To make life simple we continue with a 2 by 2 matrix $A = \boldsymbol{x}\boldsymbol{y}^{\mathrm{T}}$. Certainly $\boldsymbol{x}$ is an eigenvector !

$$A\boldsymbol{x} = \boldsymbol{x}\boldsymbol{y}^{\mathrm{T}}\boldsymbol{x} = \lambda_1\boldsymbol{x} \;\text{ so }\; \lambda_1 \;\text{ is the number }\; \boldsymbol{y}^{\mathrm{T}}\boldsymbol{x}. \tag{1}$$

The other eigenvalue is $\lambda_2 = 0$ since A is singular (rank $= 1$). The eigenvector $\boldsymbol{x}_2 = \boldsymbol{y}^{\perp}$ must be perpendicular to $\boldsymbol{y}$, so that $A\boldsymbol{x}_2 = \boldsymbol{x}\boldsymbol{y}^{\mathrm{T}}\boldsymbol{y}^{\perp} = \boldsymbol{0}$. If $\boldsymbol{y} = (a, b)$ then $\boldsymbol{y}^{\perp}$ is its 90° rotation $(b, -a)$.

The transpose matrix $A^{\mathrm{T}} = \boldsymbol{y}\boldsymbol{x}^{\mathrm{T}}$ has the same eigenvalues $\boldsymbol{y}^{\mathrm{T}}\boldsymbol{x}$ and 0. Its eigenvectors are the "left eigenvectors" of A. They will be $\boldsymbol{y}$ and $\boldsymbol{x}^{\perp}$ (because $\boldsymbol{x}\boldsymbol{y}^{\mathrm{T}}$ has eigenvectors $\boldsymbol{x}$ and $\boldsymbol{y}^{\perp}$). The only question is the scaling that decides the eigenvector lengths.

The requirement is (left eigenvector)^{T}(right eigenvector) $= 1$. Then the left eigenvectors are the rows of X^{-1} when the right eigenvectors are the columns of X : *perfection* ! In our case those dot products of eigenvectors now stand at $\boldsymbol{y}^{\mathrm{T}}\boldsymbol{x}$ and $(\boldsymbol{x}^{\perp})^{\mathrm{T}}\boldsymbol{y}^{\perp}$. Divide both left eigenvectors $\boldsymbol{y}$ and $\boldsymbol{x}^{\perp}$ by the number $\boldsymbol{y}^{\mathrm{T}}\boldsymbol{x}$, to produce $X^{-1}X = XX^{-1} = I$:

$$\boxed{\boldsymbol{x}\boldsymbol{y}^{\mathrm{T}} = \boldsymbol{X\Lambda X^{-1}} = \begin{bmatrix} \boldsymbol{x} & \boldsymbol{y}^{\perp} \end{bmatrix} \begin{bmatrix} \boldsymbol{y}^{\mathrm{T}}\boldsymbol{x} & 0 \\ 0 & 0 \end{bmatrix} \begin{bmatrix} \boldsymbol{y} \\ \boldsymbol{x}^{\perp} \end{bmatrix} \Big/ \boldsymbol{y}^{\mathrm{T}}\boldsymbol{x}} \tag{2}$$

Finally there is one more crucial possibility, that $\boldsymbol{y}^{\mathrm{T}}\boldsymbol{x} = \boldsymbol{0}$. Now the eigenvalues of $A = \boldsymbol{x}\boldsymbol{y}^{\mathrm{T}}$ are **zero and zero**. A has only one line of eigenvectors, because $\boldsymbol{y}^{\perp}$ is in the same direction as $\boldsymbol{x}$. The diagonalization (2) breaks down because the eigenvector matrix X becomes *singular*. We cannot divide by its determinant $\boldsymbol{y}^{\mathrm{T}}\boldsymbol{x} = \boldsymbol{0}$.

This shows how eigenvectors can go into a death spiral (or a fatal embrace $\boldsymbol{x} = \boldsymbol{y}^{\perp}$). Of course the pairs of singular vectors $\boldsymbol{x}, \boldsymbol{x}^{\perp}$ and $\boldsymbol{y}, \boldsymbol{y}^{\perp}$ remain orthogonal.

Question In equation (2), verify that $\begin{bmatrix} \boldsymbol{y}^{\mathrm{T}} \\ \boldsymbol{x}^{\perp\mathrm{T}} \end{bmatrix} \begin{bmatrix} \boldsymbol{x} & \boldsymbol{y}^{\perp} \end{bmatrix} = (\boldsymbol{y}^{\mathrm{T}}\boldsymbol{x}) \begin{bmatrix} 1 & 0 \\ 0 & 1 \end{bmatrix}$.

Question When does $A = \boldsymbol{x}\boldsymbol{y}^{\mathrm{T}}$ have orthogonal eigenvectors ?

A3 Counting Parameters in the Basic Factorizations

$$A = LU \quad A = QR \quad S = Q\Lambda Q^{\mathrm{T}} \quad A = X\Lambda X^{-1} \quad A = QS \quad A = U\Sigma V^{\mathrm{T}}$$

This is a review of key ideas in linear algebra. The ideas are expressed by those factorizations and our plan is simple: *Count the parameters in each matrix.* We hope to see that in each equation like $A = LU$, the two sides have the same number of parameters.

For $A = LU$, both sides have n^2 parameters.

$\boldsymbol{L}$: **Triangular** $n \times n$ matrix with 1's on the diagonal	$\frac{1}{2}\boldsymbol{n(n-1)}$
$\boldsymbol{U}$: **Triangular** $n \times n$ matrix with free diagonal	$\frac{1}{2}\boldsymbol{n(n+1)}$
$\boldsymbol{Q}$: **Orthogonal** $n \times n$ matrix	$\frac{1}{2}\boldsymbol{n(n-1)}$
$\boldsymbol{S}$: **Symmetric** $n \times n$ matrix	$\frac{1}{2}\boldsymbol{n(n+1)}$
$\boldsymbol{\Lambda}$: **Diagonal** $n \times n$ matrix	$\boldsymbol{n}$
$\boldsymbol{X}$: $n \times n$ matrix of independent **eigenvectors**	$\boldsymbol{n^2 - n}$

Comments are needed for Q. Its first column $\boldsymbol{q}_1$ is a point on the unit sphere in $\mathbf{R}^n$. That sphere is an $\boldsymbol{n-1}$-dimensional surface, just as the unit circle $x^2 + y^2 = 1$ in $\mathbf{R}^2$ has only one parameter (the angle θ). The requirement $||\boldsymbol{q}_1|| = 1$ has used up one of the n parameters in $\boldsymbol{q}_1$. Then $\boldsymbol{q}_2$ has $n-2$ parameters—it is a unit vector and it is orthogonal to $\boldsymbol{q}_1$. The sum $(n-1) + (n-2) + \cdots + 1$ equals $\frac{1}{2}\boldsymbol{n(n-1)}$ free parameters in Q.

The eigenvector matrix X has only $n^2 - n$ parameters, not n^2. If $\boldsymbol{x}$ is an eigenvector then so is $c\boldsymbol{x}$ for any $c \neq 0$. We could require the largest component of every $\boldsymbol{x}$ to be 1. This leaves $n-1$ parameters for each eigenvector (and no free parameters for X^{-1}).

The count for the two sides now agrees in all of the first five factorizations.

For the SVD, use the reduced form $\boldsymbol{A}_{m\times n} = \boldsymbol{U}_{m\times r}\boldsymbol{\Sigma}_{r\times r}\boldsymbol{V}^{\mathrm{T}}_{r\times n}$ (known zeros are not free parameters!) Suppose that $m \leq n$ and A is a full rank matrix with $\boldsymbol{r = m}$. The parameter count for A is $\boldsymbol{mn}$. *So is the total count for U, Σ, and V.* The reasoning for orthonormal columns in U and V is the same as for orthonormal columns in Q.

$$U \text{ has } \frac{1}{2}\boldsymbol{m(m-1)} \quad \Sigma \text{ has } \boldsymbol{m} \quad V \text{ has } (n-1)+\cdots+(n-m) = \boldsymbol{mn} - \frac{1}{2}\boldsymbol{m(m+1)}$$

Finally, suppose that A is an *m by n matrix of rank r.* **How many free parameters in a rank r matrix?** We can count again for $U_{m\times r}\Sigma_{r\times r}V^{\mathrm{T}}_{r\times n}$:

$$U \text{ has } (m-1)+\cdots+(m-r) = \boldsymbol{mr} - \frac{1}{2}\boldsymbol{r(r+1)} \quad V \text{ has } \boldsymbol{nr} - \frac{1}{2}\boldsymbol{r(r+1)} \quad \Sigma \text{ has } \boldsymbol{r}$$

The total parameter count for rank r is $\boldsymbol{(m+n-r)\,r}$.

We reach the same total for $A = CR$ in Section 1.4. The r columns of C were taken directly from A. The row matrix R includes an r by r identity matrix (not free!). Then the count for CR agrees with the previous count for $U\Sigma V^{\mathrm{T}}$, when the rank is r:

C has $\boldsymbol{mr}$ parameters $\qquad R$ has $\boldsymbol{nr - r^2}$ parameters $\qquad$ Total $\boldsymbol{(m+n-r)\,r}$.

A4 Codes and Algorithms for Numerical Linear Algebra

LAPACK	is the first choice for dense linear algebra codes.
ScaLAPACK	achieves high performance for very large problems.
COIN/OR	provides high quality codes for the optimization problems of operations research.

Here are sources for specific algorithms.

Direct solution of linear systems	
Basic matrix-vector operations	BLAS
Elimination with row exchanges	LAPACK
Sparse direct solvers (UMFPACK)	SuiteSparse, SuperLU
QR by Gram-Schmidt and Householder	LAPACK
Eigenvalues and singular values	
Shifted QR method for eigenvalues	LAPACK
Golub-Kahan method for the SVD	LAPACK
Iterative solutions	
Preconditioned conjugate gradients for $S\boldsymbol{x} = \boldsymbol{b}$	Trilinos
Preconditioned GMRES for $A\boldsymbol{x} = \boldsymbol{b}$	Trilinos
Krylov-Arnoldi for $A\boldsymbol{x} = \lambda x$	ARPACK, Trilinos, SLEPc
Extreme eigenvalues of S	see also BLOPEX
Optimization	
Linear programming	CLP in COIN/OR
Semidefinite programming	CSDP in COIN/OR
Interior point methods	IPOPT in COIN/OR
Convex Optimization	CVX, CVXR
Randomized linear algebra	
Randomized factorizations via pivoted QR	users.ices.utexas.edu/
$A = CMR$ columns/mixing/rows	~pgm/main_codes.html
Fast Fourier Transform	FFTW.org
Repositories of high quality codes	GAMS and Netlib.org
ACM Transactions on Mathematical Software	TOMS
Deep learning software	
Deep learning in Julia	Fluxml.ai/Flux.jl/stable
Deep learning in MATLAB	Mathworks.com/learn/tutorials/deep–learning–onramp.html
Deep learning in Python and JavaScript	Tensorflow.org, Tensorflow.js
Deep learning in R	Keras, KerasR

A5 Matrix Factorizations

1. $\boldsymbol{A} = \boldsymbol{CR} =$ (basis for column space of A) (basis for row space of A)

   Requirements: C is m by r and R is r by n. Columns of A go into C if they are not combinations of earlier columns of A. R contains the nonzero rows of the reduced row echelon form $\boldsymbol{R_0} = \textbf{rref}(\boldsymbol{A})$. Those rows begin with an r by r identity matrix, so R equals $\begin{bmatrix} I & F \end{bmatrix}$ times a column permutation P.

2. $\boldsymbol{A} = \boldsymbol{CMR^*} \begin{pmatrix} C = \text{first } r \\ \text{independent columns} \end{pmatrix} \begin{pmatrix} W = \text{first } r \text{ by } r \\ \text{invertible submatrix} \end{pmatrix}^{-1} \begin{pmatrix} R^* = \text{first } r \\ \text{independent rows} \end{pmatrix}$

   Requirements: C and R^* come directly from A. Those columns and rows meet in the r by r matrix $W = M^{-1}$ (Section 3.2) : $M =$ mixing matrix. The first r by r invertible submatrix W is the intersection of the r columns of C with the r rows of R^*.

3. $\boldsymbol{A} = \boldsymbol{LU} = \begin{pmatrix} \text{lower triangular } L \\ \text{1's on the diagonal} \end{pmatrix} \begin{pmatrix} \text{upper triangular } U \\ \text{pivots on the diagonal} \end{pmatrix}$

   Requirements: No row exchanges as Gaussian elimination reduces square A to U.

4. $\boldsymbol{A} = \boldsymbol{LDU} = \begin{pmatrix} \text{lower triangular } L \\ \text{1's on the diagonal} \end{pmatrix} \begin{pmatrix} \text{pivot matrix} \\ D \text{ is diagonal} \end{pmatrix} \begin{pmatrix} \text{upper triangular } U \\ \text{1's on the diagonal} \end{pmatrix}$

   Requirements: No row exchanges. The pivots in D are divided out from rows of U to leave 1's on the diagonal of U. If A is symmetric then U is L^{T} and $\boldsymbol{A} = \boldsymbol{LDL^{\mathrm{T}}}$.

5. $\boldsymbol{PA} = \boldsymbol{LU}$ (permutation matrix P to avoid zeros in the pivot positions).

   Requirements: A is invertible. Then P, L, U are invertible. P does all of the row exchanges on A in advance, to allow normal LU. Alternative: $A = L_1 P_1 U_1$.

6. $\boldsymbol{S} = \boldsymbol{C^{\mathrm{T}}C} =$ (lower triangular) (upper triangular) with $\sqrt{D}$ on both diagonals

   Requirements: S is symmetric and positive definite (all n pivots in D are positive). This *Cholesky factorization* $C = \mathsf{chol}(S)$ has $C^{\mathrm{T}} = L\sqrt{D}$, so $S = C^{\mathrm{T}}C = LDL^{\mathrm{T}}$.

7. $\boldsymbol{A} = \boldsymbol{QR} =$ (orthonormal columns in Q) (upper triangular matrix R).

   Requirements: A has independent columns. Those are *orthogonalized* in Q by the Gram-Schmidt or Householder process. If A is square then $Q^{-1} = Q^{\mathrm{T}}$.

8. $\boldsymbol{A} = \boldsymbol{X\Lambda X^{-1}} =$ (eigenvectors in X) (eigenvalues in Λ) (left eigenvectors in X^{-1}).

   Requirements: A must have n linearly independent eigenvectors.

9. $\boldsymbol{S} = \boldsymbol{Q\Lambda Q^{\mathrm{T}}} =$ (orthogonal matrix Q) (real eigenvalue matrix Λ) (Q^{T} is Q^{-1}).

   Requirements: S is *real and symmetric*: $S^{\mathrm{T}} = S$. This is the Spectral Theorem.

10. $\boldsymbol{A = BJB^{-1}}$ = (generalized eigenvectors in B) (Jordan blocks in J) (B^{-1}).

Requirements: A is any square matrix. This *Jordan form* J has a block for each linearly independent eigenvector of A. Every block has only one eigenvalue.

11. $\boldsymbol{A = U\Sigma V^{\mathrm{T}}} = \begin{pmatrix}\text{orthogonal}\\ U \text{ is } m\times m\end{pmatrix}\begin{pmatrix} m\times n \text{ singular value matrix}\\ \sigma_1,\ldots,\sigma_r \text{ on its diagonal}\end{pmatrix}\begin{pmatrix}\text{orthogonal}\\ V \text{ is } n\times n\end{pmatrix}.$

Requirements: None. This ***Singular Value Decomposition*** (SVD) has the eigenvectors of AA^{T} in U and eigenvectors of $A^{\mathrm{T}}A$ in V; $\sigma_i = \sqrt{\lambda_i(A^{\mathrm{T}}A)} = \sqrt{\lambda_i(AA^{\mathrm{T}})}$.

Those singular values are $\sigma_1 \geq \sigma_2 \geq \cdots \geq \sigma_r > 0$. By column-row multiplication

$$\boldsymbol{A = U\Sigma V^{\mathrm{T}}} = \sigma_1 \boldsymbol{u}_1\boldsymbol{v}_1^{\mathrm{T}} + \cdots + \sigma_r \boldsymbol{u}_r\boldsymbol{v}_r^{\mathrm{T}}.$$

If S is symmetric positive definite then $U = V = Q$ and $\Sigma = \Lambda$ and $S = Q\Lambda Q^{\mathrm{T}}$.

12. $\boldsymbol{A^+ = V\Sigma^+U^{\mathrm{T}}} = \begin{pmatrix}\text{orthogonal}\\ n\times n\end{pmatrix}\begin{pmatrix} n\times m \text{ pseudoinverse of } \Sigma\\ 1/\sigma_1,\ldots,1/\sigma_r \text{ on diagonal}\end{pmatrix}\begin{pmatrix}\text{orthogonal}\\ m\times m\end{pmatrix}.$

Requirements: None. The *pseudoinverse* A^+ has A^+A = projection onto row space of A and AA^+ = projection onto column space. $A^+ = A^{-1}$ if A is invertible. The shortest least-squares solution to $A\boldsymbol{x} = \boldsymbol{b}$ is $\boldsymbol{x}^+ = A^+\boldsymbol{b}$. This solves $A^{\mathrm{T}}A\boldsymbol{x}^+ = A^{\mathrm{T}}\boldsymbol{b}$.

13. $\boldsymbol{A = QS}$ = (orthogonal matrix Q) (symmetric positive definite matrix S).

Requirements: A is invertible. This *polar decomposition* has $S^2 = A^{\mathrm{T}}A$. The factor S is semidefinite if A is singular. The reverse polar decomposition $A = KQ$ has $K^2 = AA^{\mathrm{T}}$. Both have $Q = UV^{\mathrm{T}}$ from the SVD.

14. $\boldsymbol{A = U\Lambda U^{-1}}$ = (unitary U) (eigenvalue matrix Λ) (U^{-1} which is $U^{\mathrm{H}} = \overline{U}^{\mathrm{T}}$).

Requirements: A is *normal*: $A^{\mathrm{H}}A = AA^{\mathrm{H}}$. Its orthonormal (and possibly complex) eigenvectors are the columns of U. Complex λ's unless $S = S^{\mathrm{H}}$: Hermitian case.

15. $\boldsymbol{A = QTQ^{-1}}$ = (unitary Q) (triangular T with λ's on diagonal) ($Q^{-1} = Q^{\mathrm{H}}$).

Requirements: *Schur triangularization* of any square A. There is a matrix Q with orthonormal columns that makes $Q^{-1}AQ$ triangular: Section 6.3.

16. $\boldsymbol{F_n} = \begin{bmatrix} I & D\\ I & -D\end{bmatrix}\begin{bmatrix}\boldsymbol{F_{n/2}} & \\ & \boldsymbol{F_{n/2}}\end{bmatrix}\begin{bmatrix}\text{even-odd}\\ \text{permutation}\end{bmatrix}$ = one step of the recursive **FFT**.

Requirements: F_n = Fourier matrix with entries w^{jk} where $w^n = 1$: $F_n\overline{F}_n = nI$. D has $1, w, \ldots, w^{n/2\,-\,1}$ on its diagonal. For $n = 2^\ell$ the *Fast Fourier Transform* will compute $F_n\boldsymbol{x}$ with only $\frac{1}{2}n\ell = \frac{1}{2}n\log_2 n$ multiplications from ℓ stages of D's.

A6 The Column-Row Factorization of a Matrix

Abstract

The active ideas in linear algebra are often expressed by matrix factorizations: $S = Q\Lambda Q^{\mathrm{T}}$ for symmetric matrices (the spectral theorem) and $A = U\Sigma V^{\mathrm{T}}$ for all matrices (singular value decomposition). Far back near the beginning comes $A = LU$ for successful elimination: Lower triangular times upper triangular. This paper is one step earlier, with bases in $A = CR$ for the column space and row space of any matrix—and a proof that column rank = row rank. The echelon form of A and the pseudoinverse A^+ appear naturally. The "proofs" are mostly "observations".

Introduction

An introduction is hardly necessary for so short a paper. But I can explain the background.

In teaching linear algebra, the course often begins slowly. The idea of a vector space waits until Chapter 3. The highly important topic of singular values is squeezed into the final week or completely omitted. A new plan is needed.

I now start the course in a different way. The multiplication $A\boldsymbol{x}$ produces a combination of the columns of A. All combinations fill the *column space of* A—a key idea to visualize. Simple examples of $A\boldsymbol{x} = \mathbf{0}$ show the idea of linear dependence. Starting with column 1, we create a matrix C with a full set of independent columns—a basis for the column space.

I believe that this "fast start" is also a better start. Every column of A is a combination of the columns of C. Introducing matrix multiplication, *that fact becomes* $A = CR$. We have a natural factorization of A, to be followed by $A = LU$ (elimination) and $A = QR$ (Gram-Schmidt) and $S = Q\Lambda Q^{\mathrm{T}}$ (eigenvalues in the spectral theorem) and $A = U\Sigma V^{\mathrm{T}}$ (singular values in the SVD). The course has a structure that students can follow. A new textbook called "Linear Algebra for Everyone" is in preparation.

The key point for this paper is that the matrix R in $A = CR$ is already famous. R is the *reduced row echelon form of* A, with any zero rows removed. It has a simple "formula" $R = \begin{bmatrix} I & F \end{bmatrix} P$ which the mechanics of elimination will execute. And it has a "meaning" that is hidden in those row operations on A. R tells us the combinations of independent columns in C, which produce all the columns of A.

The Factorization $A = CR$

A is a real matrix with m rows and n columns. Some of those columns might be linear combinations of previous columns. Here is a natural way, working from left to right, to find a complete set of independent columns.

If column 1 is not zero, put it into C.
If column 2 of A is not a multiple of column 1, put it into C.
If column 3 is not a linear combination of columns 1 and 2, put it into C. *Continue.*

At the end, C will have r independent columns. Those columns will be a basis for the column space of A. Every column of A is a combination of columns of C, and the coefficients in those combinations go into columns of R:

$$A = \begin{bmatrix} 1 & 4 & 7 \\ 2 & 5 & 8 \\ 3 & 6 & 9 \end{bmatrix} = \begin{bmatrix} 1 & 4 \\ 2 & 5 \\ 3 & 6 \end{bmatrix} \begin{bmatrix} 1 & 0 & -1 \\ 0 & 1 & 2 \end{bmatrix} = CR \text{ with } r = 2$$

The matrix R contains an r by r identity matrix in the columns that correspond to independent columns of A. Column 3 of R tells us that the columns of A have $\boldsymbol{a}_3 = -\boldsymbol{a}_1 + 2\boldsymbol{a}_2$. A first observation: *$R$ is the "reduced row echelon form" of A, without its $m - r$ zero rows.*

A second observation: *Every row of A is a combination of the rows of R.* That comes directly from the matrix multiplication $A = CR$. In this example row 1 of A equals row 1 of R plus 4 times row 2 of R. The coefficients like 1 and 4 in those linear combinations are in the rows of C. And the rows of R are independent, because r by r identity matrix is a submatrix of R. Then the rows of R are a basis for the row space of A.

This matrix R is usually computed by row operations on A, to reach the "echelon form". Here R appears after the column basis in C.

A third observation comes from $A = CR = (m \times r)\,(r \times n)$: *The column rank of A equals the row rank of A.* The same number r counts independent columns in C and independent rows in R.

A fourth observation is a "formula" for the reduced row echelon form $R_0 = \textbf{rref}(A)$. Normally this matrix with $m - r$ zero rows is constructed directly by row operations on A, and C does not appear. A direct description of R_0 could be as follows.

Suppose the basic columns in C are columns $n_1 < n_2 < \cdots < n_r$ of A. The other $n - r$ columns of A are combinations $N = CF$ of those r basic columns (in order). Then the reduced row echelon form of A with $m - r$ zero rows is R_0:

$$R_0 = \begin{bmatrix} I & F \\ 0 & 0 \end{bmatrix} P$$

The permutation P puts the n columns of C and N into their correct order in A, and $I = r \times r$ identity matrix.

Note that $A = C \begin{bmatrix} I & F \end{bmatrix} P = \begin{bmatrix} C & N \end{bmatrix} P$ has the columns of A in their original order, thanks to P. In the formula above, R_0 is constructed directly from A and its uniqueness is clear. Eric Grinberg has invented the name "*gauche basis*" for the columns of C—a brilliant suggestion that reinforces the left-to-right construction of this basis for the column space of A.

Uniqueness of the reduced row echelon form seems to be a moot question when there is an explicit formula for that matrix R_0. The formula cannot be new, but I don't know a reference. Observations 1-3 are definitely not new.

A Mixing Matrix

Here is a variation on the matrix factorization $A = CR$. The matrix C contains actual columns of A but the matrix R does not contain rows of A. For symmetric perfection, we might prefer the matrix R^* containing the r uppermost linearly independent rows taken directly from A. Generally $A = CR^*$ will not be true. To recover a correct factorization of A, we need to include a **mixing matrix** M between C and R^*. Then $\boldsymbol{A = CMR^*}$. (The idea of a mixing matrix has become widespread in numerical linear algebra. The symbol U is often chosen instead of M, but U is needed in the important factorizations LU and $U\Sigma V^{\mathrm{T}}$. We would like to nominate the letter M and the adjective *mixing*.)

Does M have a simple formula? *Yes.* M^{-1} is the r by r submatrix at the intersection of the r columns of C with the r rows of R^*. If those happen to be the first r columns and the first r rows, it is easy to see that MR^* will produce the familiar r by r identity matrix that begins the reduced row echelon form R. Then $A = CMR^*$ is identified with $A = CR$.

The Pseudoinverse

Factorizations like $A = CR$ are familiar to algebraists. It is not surprising that they connect to other constructions. An example in linear algebra is the **pseudoinverse of $\boldsymbol{A}$**.

We write the pseudoinverse as A^+. It inverts the mapping (multiplication by A) from row space to column space. The pseudoinverse is *zero* on the nullspace of A^{T}. Thus it inverts A where inversion is possible: $\boldsymbol{AA^+A = A}$.

$$\begin{bmatrix} 2 & 0 & 0 \\ 0 & 0 & 0 \end{bmatrix}^+ = \begin{bmatrix} \frac{1}{2} & 0 \\ 0 & 0 \\ 0 & 0 \end{bmatrix} \quad \text{and} \quad A^+ = (U\Sigma V^{\mathrm{T}})^+ = V\Sigma^+ U^{\mathrm{T}}$$

The pseudoinverse connects perfectly to the rank r factors in $A = CR$. The pseudoinverse of C is its left inverse $C^+ = (C^{\mathrm{T}}C)^{-1}C^{\mathrm{T}}$. The pseudoinverse of R is its right inverse $R^+ = R^{\mathrm{T}}(RR^{\mathrm{T}})^{-1}$. Then the pseudoinverse of $A = CR$ is $\boldsymbol{A^+ = R^+C^+}$, because all ranks are equal to r.

This short paper was submitted to the *Journal of Convex Analysis.* A further note about these factorizations is in progress with Daniel Drucker and Alexander Lin. Our plan is to post both papers on the arXiv website in 2020.

A7 The Jordan Form of a Square Matrix

We know that some square matrices A with repeated eigenvalues do not have n independent eigenvectors. Therefore they can't be diagonalized by $X\Lambda X^{-1}$: *X is not invertible.* Jordan established a *nearly diagonal form* $J = BAB^{-1}$ that has k Jordan blocks $J_1, \dots J_k$ when A has k independent eigenvectors:

$$J = BAB^{-1} = \begin{bmatrix} J_1 & & & \\ & \cdot & & \\ & & \cdot & \\ & & & J_k \end{bmatrix} \text{ has Jordan blocks } J_i = \begin{bmatrix} \lambda_i & 1 & 0 & 0 \\ & \cdot & 1 & 0 \\ & & \cdot & 1 \\ & & & \lambda_i \end{bmatrix}$$

If A can be diagonalized, then $k = n$ and $B = X$ and J is the eigenvalue matrix Λ (*all n blocks are* 1 *by* 1). If A can't be diagonalized, then one or more blocks will be larger. Each block J_i of size n_i has only one true eigenvector $\boldsymbol{x}_i = (1, 0, \dots, 0)$ and one eigenvalue λ_i. The matrix B contains eigenvectors of A along with "generalized eigenvectors". Here is an example rather than a proof.

Example $J = \left[\begin{array}{cc|c} 3 & 1 & 0 \\ 0 & 3 & 0 \\ \hline 0 & 0 & 3 \end{array}\right]$ has $\lambda = 3, 3, 3$ with only two genuine eigenvectors $\begin{bmatrix} 1 \\ 0 \\ 0 \end{bmatrix}$ and $\begin{bmatrix} 0 \\ 0 \\ 1 \end{bmatrix}$ and $A = B^{-1}JB$.

This Jordan form makes $A^N = B^{-1}J^NB$ and $e^{tA} = B^{-1}e^{tJ}B$ as simple as possible to compute. For powers of J, we just compute block by block:

$$\begin{bmatrix} 3 & 1 \\ 0 & 3 \end{bmatrix}^N = \begin{bmatrix} 3^N & N3^{N-1} \\ 0 & 3^N \end{bmatrix} \qquad \exp\left(\begin{bmatrix} 3 & 1 \\ 0 & 3 \end{bmatrix} t\right) = \begin{bmatrix} e^{3t} & te^{3t} \\ 0 & e^{3t} \end{bmatrix}$$

That exponential formula is telling us the missing solution to the differential equation $d\boldsymbol{U}/dt = J\boldsymbol{U}$ (and also $d\boldsymbol{u}/dt = A\boldsymbol{u}$). The usual solution has e^{3t}. We can't just use that twice, when $\lambda = 3$ is repeated. The missing solution is $\boldsymbol{te^{3t}}$. And a triple eigenvalue $\lambda = 3$ with only one eigenvector (and one Jordan block) would involve t^2e^{3t}.

The Cayley-Hamilton Theorem

"Every matrix A satisfies its own characteristic equation $p(A) = 0$." The determinant of $A - \lambda I$ is a polynomial $p(\lambda)$, and the n solutions to $p(\lambda) = 0$ are the eigenvalues of A. Our example above has $p(\lambda) = (\lambda - 3)^3$ with a triple eigenvalue $\lambda = 3, 3, 3$. Then Cayley-Hamilton says that $p(A) = (A - 3I)^3$ has to be the zero matrix. Jordan makes this easy, because $p(A) = Bp(J)B^{-1}$ and $p(J)$ is certainly zero:

Example $p(J) = (J - 3I)^3 = \begin{bmatrix} 0 & 1 & 0 \\ 0 & 0 & 0 \\ 0 & 0 & 0 \end{bmatrix}^3 = \begin{bmatrix} 0 & 0 & 0 \\ 0 & 0 & 0 \\ 0 & 0 & 0 \end{bmatrix}$ and then $(A - 3I)^3 = 0$.

A8 Tensors

In linear algebra, a tensor is a multidimensional array. (To Einstein, a tensor was a function that followed certain transformation rules.) A matrix is a 2-way tensor. *A 3-way tensor T is a stack of matrices.* Its elements have three indices: row number i and column number j and "tube number" k.

An example is a color image. It has 3 slices corresponding to red-green-blue. A slice of T shows the density of one of those primary colors RGB ($k = 1$ to 3), at each pixel (i, j) in the image.

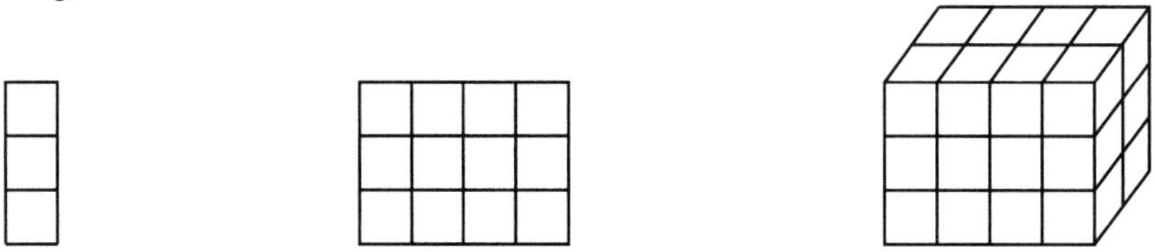

vector **matrix (2-way array)** **tensor (3-way array)**

Another example is a joint probability tensor. Now p_{ijk} is the probability that a random individual has (for example) age i and height j and weight k. The sum of all those numbers p_{ijk} will be 1. For $i = 9$, the sum of all p_{9jk} would be the fraction of individuals that have age 9—the sum over one slice of the tensor.

A fundamental problem—with tensors as with matrices—is to decompose the tensor T into simpler pieces. For a matrix A that was accomplished by the SVD. The pieces that add to A are matrices (we should now say 2-tensors), with the special property that each piece is a rank-one matrix $\boldsymbol{u}\boldsymbol{v}^{\mathrm{T}}$. Linear algebra allowed us to require that the $\boldsymbol{u}$'s from different pieces were orthogonal, and the $\boldsymbol{v}$'s were also orthogonal and that there were only r pieces ($r \leq m$ and $r \leq n$).

Sad to say, this SVD format is not possible for a 3-way tensor. We can still ask for R rank-one pieces that approximately add to T:

$$\textbf{\textit{CP} Decomposition} \qquad T \approx \boldsymbol{a}_1 \circ \boldsymbol{b}_1 \circ \boldsymbol{c}_1 + \cdots + \boldsymbol{a}_R \circ \boldsymbol{b}_R \circ \boldsymbol{c}_R. \tag{1}$$

Orthogonality of the $\boldsymbol{a}$'s and of the $\boldsymbol{b}$'s and of the $\boldsymbol{c}$'s is generally impossible. The number of pieces is not set by T (its "rank" is not well defined). But an approximate decomposition of this kind is still useful in computations with tensors. One option is to solve alternately for the $\boldsymbol{a}_i$ (with fixed $\boldsymbol{b}_i$ and $\boldsymbol{c}_i$) and then for the $\boldsymbol{b}_i$ (fixed $\boldsymbol{a}_i$ and $\boldsymbol{c}_i$) and then for the $\boldsymbol{c}_i$ (fixed $\boldsymbol{a}_i$ and $\boldsymbol{b}_i$). Those subproblems can be reduced to least squares. Other approximate decompositions of T are possible.

The theory of tensor decompositions (multilinear algebra) is driven by applications. We must be able to compute with T. So the algorithms are steadily improving, even without the orthogonality properties of an SVD.

A9 The Condition Number

The condition number measures the ratio of (*change in solution*) to (*change in data*). The most common problem is to solve n linear equations $A\boldsymbol{x} = \boldsymbol{b}$ in n unknowns $\boldsymbol{x}$. In this case the data is $\boldsymbol{b}$ and the solution is $\boldsymbol{x} = A^{-1}\boldsymbol{b}$. The matrix A is fixed. The change in the data is $\Delta\boldsymbol{b}$ and the change in the solution is $\boldsymbol{\Delta x} = A^{-1}\boldsymbol{\Delta b}$. We have to decide the meaning of the word "change". Do we compute *the absolute change* $||\Delta\boldsymbol{b}||$ *or the relative change* $||\Delta\boldsymbol{b}|| \,/\, ||\boldsymbol{b}||$? That decision for the data $\boldsymbol{b}$ will bring a similar decision for the solution $\boldsymbol{x}$.

$$\begin{matrix}\textbf{Absolute}\\ \textbf{condition}\end{matrix} = \max_{\boldsymbol{b},\,\Delta\boldsymbol{b}} \frac{||\Delta\boldsymbol{x}||}{||\Delta\boldsymbol{b}||} = \max \frac{||A^{-1}\Delta\boldsymbol{b}||}{||\Delta\boldsymbol{b}||} \qquad \begin{matrix}\textbf{Relative}\\ \textbf{condition}\end{matrix} = \max_{\boldsymbol{b},\,\Delta\boldsymbol{b}} \frac{||\Delta\boldsymbol{x}|| \,/\, ||\boldsymbol{x}||}{||\Delta\boldsymbol{b}|| \,/\, ||\boldsymbol{b}||} \tag{1}$$

The absolute choice looks good but it has a problem. If we divide the matrix A by 10, we are multiplying A^{-1} by 10. The absolute condition goes up by 10. But solving $A\boldsymbol{x} = \boldsymbol{b}$ is not 10 times harder. The relative condition number is the right choice.

$$\boxed{\mathsf{cond}(A) = \max_{\boldsymbol{b},\,\Delta\boldsymbol{b}} \frac{||A^{-1}\Delta\boldsymbol{b}||}{||\Delta\boldsymbol{b}||}\,\frac{||A\boldsymbol{x}||}{||\boldsymbol{x}||} = ||\boldsymbol{A^{-1}}||\;||\boldsymbol{A}|| = \frac{\boldsymbol{\sigma_{\max}}}{\boldsymbol{\sigma_{\min}}}} \tag{2}$$

If A is the simple diagonal matrix Σ with entries $\sigma_1 \geq \cdots \geq \sigma_r = \sigma_{\text{min}}$, then its norm is $\sigma_{\text{max}} = \sigma_1$. The norm of A^{-1} is $1/\sigma_{\text{min}}$. The orthogonal matrices U and V in the SVD leave the norms unchanged. So the ratio $\sigma_{\text{max}} \,/\, \sigma_{\text{min}}$ is $\text{cond}(A)$. We are using the usual measure of length $||\boldsymbol{x}||^2 = x_1^2 + \cdots + x_n^2$.

Notice that σ_{min} (*not* λ_{min}) measures the distance from A to the nearest singular matrix. At first we might expect to see $A - \lambda_{\text{min}} I$, bringing the smallest eigenvalue to zero. *Wrong*. The nearest singular matrix to $A = U\Sigma V^{\mathrm{T}}$ is $U(\Sigma - \sigma_{\text{min}} I)V^{\mathrm{T}}$ because the orthogonal matrices U and V^{T} don't affect the norm. Bring the smallest singular value to zero.

The eigenvalues of A have different condition numbers. Suppose λ is a simple root (not a repeated root) of the equation $\det(A - \lambda I) = 0$. Then $A\boldsymbol{x} = \lambda\boldsymbol{x}$ and $A^{\mathrm{T}}\boldsymbol{y} = \lambda\boldsymbol{y}$ for unit eigenvectors $||\boldsymbol{x}|| = ||\boldsymbol{y}|| = 1$. **The condition number of $\boldsymbol{\lambda}$ is $\boldsymbol{1/|y^{\mathrm{T}}x|}$.** In other words it is $1/|\cos\theta|$, where θ is the angle between the right eigenvector $\boldsymbol{x}$ and the left eigenvector $\boldsymbol{y}$. (The name comes from the equation $\boldsymbol{y}^{\mathrm{T}}A = \lambda\boldsymbol{y}^{\mathrm{T}}$, with $\boldsymbol{y}^{\mathrm{T}}$ on the left side of A.)

Notice that a symmetric matrix A will have $\boldsymbol{y} = \boldsymbol{x}$ with $\cos\theta = 1$. The eigenvalue problem is perfectly conditioned for symmetric matrices, just as $A\boldsymbol{x} = \boldsymbol{b}$ was perfectly conditioned for orthogonal matrices with $||Q||\;||Q^{-1}|| = 1$.

The formula $1/|\boldsymbol{y}^{\mathrm{T}}\boldsymbol{x}|$ comes from the change $\Delta\lambda \approx \boldsymbol{y}^{\mathrm{T}}\,\Delta A\,\boldsymbol{x}/\boldsymbol{y}^{\mathrm{T}}\boldsymbol{x}$ in the eigenvalue created by a small change ΔA in the matrix.

A10 Markov Matrices and Perron-Frobenius

This appendix is about positive matrices (all $a_{ij} > 0$) and nonnegative matrices (all $a_{ij} \geq 0$). Markov matrices M are important examples, when *every column of M adds to* 1. Positive numbers adding to 1 makes you think immediately of probabilities.

A useful fact about any Markov matrix M: *The largest eigenvalue is always* $\lambda = 1$. We know that every column of $M - I$ adds to zero. So the rows add to the zero row, and $M - I$ is not invertible: $\lambda = 1$ is an eigenvalue. Here are two examples:

$$A = \begin{bmatrix} 0.8 & 0.3 \\ 0.2 & 0.7 \end{bmatrix} \text{ has eigenvalues 1 and } \frac{1}{2} \qquad B = \begin{bmatrix} 0 & 1 \\ 1 & 0 \end{bmatrix} \text{ has eigenvalues 1 and } -1$$

That matrix A is typical of Markov. The eigenvectors are $\boldsymbol{x}_1 = (0.6, 0.4)$ and $\boldsymbol{x}_2 = (1, -1)$:

$$\begin{bmatrix} 0.8 & 0.3 \\ 0.2 & 0.7 \end{bmatrix}\begin{bmatrix} 0.6 \\ 0.4 \end{bmatrix} = \begin{bmatrix} 0.6 \\ 0.4 \end{bmatrix} \text{ is a steady state}$$

$$\begin{bmatrix} 0.8 & 0.3 \\ 0.2 & 0.7 \end{bmatrix}\begin{bmatrix} 1 \\ -1 \end{bmatrix} = \frac{1}{2}\begin{bmatrix} 1 \\ -1 \end{bmatrix} \text{ is a "transient" that disappears}$$

Our favorite example is based on rental cars in Chicago and Denver. We start with 100 cars in Chicago and no cars in Denver: $\boldsymbol{y}_0 = (100, 0)$. Every month we multiply the current vector $\boldsymbol{y}_n$ by A to find $\boldsymbol{y}_{n+1}$: the number in Chicago and Denver after $n + 1$ months:

$$\boldsymbol{y}_0 = \begin{bmatrix} 100 \\ 0 \end{bmatrix} \quad \boldsymbol{y}_1 = \begin{bmatrix} 80 \\ 20 \end{bmatrix} \quad \boldsymbol{y}_2 = \begin{bmatrix} 70 \\ 30 \end{bmatrix} \quad \boldsymbol{y}_3 = \begin{bmatrix} 65 \\ 35 \end{bmatrix} \cdots \boldsymbol{y}_\infty = \begin{bmatrix} \mathbf{60} \\ \mathbf{40} \end{bmatrix}.$$

That steady state $(60, 40)$ is an eigenvector of A for $\lambda = 1$. If we had started with $\boldsymbol{y}_0 = (60, 40)$ then we would have stayed there forever. Starting at $(100, 0)$ we needed to get 40 cars to Denver. You see that number 40 at time zero reduced to 20 at time 1, 10 at time 2, and 5 at time 3. That is the effect of the other eigenvalue $\lambda = \frac{1}{2}$, dividing its eigenvector by 2 at every step:

$$\boldsymbol{y}_0 = \begin{bmatrix} 60 \\ 40 \end{bmatrix} + \begin{bmatrix} 40 \\ -40 \end{bmatrix} \quad \boldsymbol{y}_1 = \begin{bmatrix} 60 \\ 40 \end{bmatrix} + \begin{bmatrix} 20 \\ -20 \end{bmatrix} \cdots \boldsymbol{y}_n = \begin{bmatrix} 60 \\ 40 \end{bmatrix} + \left(\frac{1}{2}\right)^n \begin{bmatrix} 40 \\ -40 \end{bmatrix}.$$

This is $\boldsymbol{y}_n = A^n \boldsymbol{y}_0$ coming from the single step equation $\boldsymbol{y}_{n+1} = A\boldsymbol{y}_n$. In matrix notation, A^n approaches rank one!

$$A^n = (X\Lambda X^{-1})^n = X\Lambda^n X^{-1} = \begin{bmatrix} 0.6 & 1 \\ 0.4 & -1 \end{bmatrix}\begin{bmatrix} 1 & \\ & (\frac{1}{2})^n \end{bmatrix}\begin{bmatrix} 1 & 1 \\ 0.4 & -0.6 \end{bmatrix} \to A^\infty = \begin{bmatrix} 0.6 & 0.4 \\ 0.4 & 0.4 \end{bmatrix}$$

You have now seen a typical Markov matrix with $\lambda_{\text{max}} = 1$. Its eigenvector $(0.6, 0.4)$ is the survivor as time goes forward. All small eigenvalues have $\lambda^n \to 0$. But our second Markov example has a second eigenvalue $\lambda = -1$. *Now we don't approach a steady state*:

$$B = \begin{bmatrix} 0 & 1 \\ 1 & 0 \end{bmatrix} \text{ has eigenvalue } \lambda_1 = \mathbf{1} \text{ with } \boldsymbol{x}_1 = \begin{bmatrix} 1 \\ 1 \end{bmatrix} \text{ and } \lambda_2 = \mathbf{-1} \text{ with } \boldsymbol{x}_2 = \begin{bmatrix} 1 \\ -1 \end{bmatrix}.$$

The zeros in B allow that second eigenvalue $\lambda_2 = -1$ to have the same size as $\lambda_1 = 1$. All cars switch cities every month (Chicago to Denver, Denver to Chicago). If we start at $\boldsymbol{y}_0 = (60, 40)$ then the next month has $\boldsymbol{y}_1 = B\boldsymbol{y}_0 = (40, 60)$:

$$\boldsymbol{y}_0 = \begin{bmatrix} 60 \\ 40 \end{bmatrix} = \begin{bmatrix} 50 \\ 50 \end{bmatrix} + \begin{bmatrix} 10 \\ -10 \end{bmatrix} \quad \boldsymbol{y}_1 = \begin{bmatrix} 40 \\ 60 \end{bmatrix} = \begin{bmatrix} 50 \\ 50 \end{bmatrix} - \begin{bmatrix} 10 \\ -10 \end{bmatrix} \quad \boldsymbol{y}_2 = \boldsymbol{y}_0 \quad \boldsymbol{y}_3 = \boldsymbol{y}_1 \cdots$$

No steady state because $\lambda_2 = -1$ also has size $|\lambda_2| = 1$. This will not happen when the Markov matrix A has all $a_{ij} > 0$. It *might happen* when B has some $B_{ij} = 0$.

Perron found the proof in the first case $A_{ij} > 0$. Then Frobenius allowed $B_{ij} = 0$. In this short appendix we stay with Perron. Every positive matrix A is allowed.

Theorem (Perron) All numbers in $A\boldsymbol{x} = \lambda_{\max}\,\boldsymbol{x}$ are strictly positive.

Proof Start with $A > 0$. The key idea is to look at all numbers t such that $A\boldsymbol{x} \geq t\boldsymbol{x}$ for some nonnegative vector $\boldsymbol{x}$ (other than $\boldsymbol{x} = \mathbf{0}$). We are allowing inequality in $A\boldsymbol{x} \geq t\boldsymbol{x}$ in order to have many small positive candidates t. For the **largest value $t_{\max}$** (which is attained), we will show that ***equality holds***: $A\boldsymbol{x} = t_{\max}\,\boldsymbol{x}$. Then $t_{\max}$ is our eigenvalue $\lambda_{\max}$ and $\boldsymbol{x}$ is the positive eigenvector—which we now prove.

If $A\boldsymbol{x} \geq t_{\max}\boldsymbol{x}$ is not an equality, multiply both sides by A. Because $A > 0$, that produces a strict inequality $A^2\boldsymbol{x} > t_{\max}A\boldsymbol{x}$. Therefore the positive vector $\boldsymbol{y} = A\boldsymbol{x}$ satisfies $A\boldsymbol{y} > t_{\max}\,\boldsymbol{y}$. This means that $t_{\max}$ could be increased. This contradiction forces the equality $A\boldsymbol{x} = t_{\max}\boldsymbol{x}$, and *we have an eigenvalue*. Its eigenvector $\boldsymbol{x}$ is positive because on the left side of that equality, $A\boldsymbol{x}$ is sure to be positive.

To see that no eigenvalue can be larger than $t_{\max}$, suppose $A\boldsymbol{z} = \lambda\boldsymbol{z}$. Since λ and $\boldsymbol{z}$ may involve negative or complex numbers, we take absolute values: $|\lambda||\boldsymbol{z}| = |A\boldsymbol{z}| \leq A|\boldsymbol{z}|$ by the "triangle inequality." This $|\boldsymbol{z}|$ is a nonnegative vector, so this $|\lambda|$ is one of the possible candidates t. Therefore $|\lambda|$ cannot exceed $t_{\max}$—which must be $\lambda_{\max}$.

Many Markov examples start with a zero in the matrix (Frobenius) but then A^2 or some higher power A^m is strictly positive (Perron). So these "primitive matrices" also have one steady state eigenvector from $\lambda = 1$.

The big example is the **Google matrix** G that is the basis for the PageRank algorithm. (A major company survives entirely on linear algebra.) The matrix starts with $A_{ij} = 1$ when page j has a link to page i. Then divide each column j of A by the number of outgoing links: then column sums $= 1$. Finally $G_{ij} = \alpha A_{ij} + (1-\alpha)/N$ so that every $G_{ij} > 0$ (Perron). See Wikipedia and the book by Amy Langville and Carl Meyer (which is quickly found using Google).

Reference

Amy Langville and Carl Meyer, *Google's Page Rank and Beyond: The Science of Search Engine Rankings*, Princeton University Press (2011).

Index

Index of Symbols

$(AA^{\mathrm{T}})A = A(A^{\mathrm{T}}A)$, 261
$(AB)\,C = A\,(BC)$, 30, 38
$(AB)\boldsymbol{x} = A(B\boldsymbol{x})$, 33
$(AB)^{\mathrm{T}} = B^{\mathrm{T}}A^{\mathrm{T}}$, 67
$(AB)^{-1} = B^{-1}A^{-1}$, 51
A^{-1}, 50, 57
$(A\boldsymbol{x})^{\mathrm{T}}\boldsymbol{y} = \boldsymbol{x}^{\mathrm{T}}(A^{\mathrm{T}}\boldsymbol{y})$, 68
$(A^{-1})^{\mathrm{T}} = (A^{\mathrm{T}})^{-1}$, 67
$(A^{-1})_{ij} = C_{ji}/\det A$, 180, 186
$A = BJB^{-1}$, 339
$A = CMR^*$, 90, 95, 338, 342
$A = CR$, 32, 85, 338, 340
$A = CW^{-1}R^*$, 89, 90
$A = LDU$, 63
$A = LU$, 58–60, 338
$A = QR$, 171, 281, 338
$A = QS$, 280
$A = U\Sigma V^{\mathrm{T}}$, 259, 339
$A = U_r\Sigma_r V_r^{\mathrm{T}}$, 260
$A = X\Lambda X^{-1}$, 215, 338
AB, 29, 34, 35
AB and BA, 225, 262
ABC, 318
$AV_r = U_r\Sigma_r$, 260
$A\boldsymbol{x} = \boldsymbol{b}$, 27, 40
$A\boldsymbol{x} = \lambda\boldsymbol{x}$, 202
$A\boldsymbol{v}$, 3
$A\boldsymbol{v} = \sigma\boldsymbol{u}$, 259
$A^{\mathrm{T}}A$ and AA^{T}, 262
$A^{\mathrm{T}}A\widehat{\boldsymbol{x}} = A^{\mathrm{T}}\boldsymbol{b}$, 147, 148
A^+, 133, 280
$A^k = X\Lambda^k X^{-1}$, 216
$AB = \boldsymbol{a}_1\boldsymbol{b}_1^* + \cdots + \boldsymbol{a}_n\boldsymbol{b}_n^*$, 34
CP Decomposition, 344
$E = ||A\boldsymbol{x} - \boldsymbol{b}||^2$, 154, 155
$E^{-1} = L$, 52, 53
E_{ij}, 49
$P = A(A^{\mathrm{T}}A)^{-1}A^{\mathrm{T}}$, 147
$P = QQ^{\mathrm{T}}$, 176
$PA = LU$, 65, 338
$P^2 = P$, 146
QR, 171, 280, 338
$Q^{\mathrm{T}}Q = I$, 165
$Q^{\mathrm{T}} = Q^{-1}$, 166
$[Q, R] = \mathbf{qr}(A)$, 176
R, 33
$R = \begin{bmatrix} I & F \end{bmatrix} P$, 88, 89, 340
R_0, 85
$R_0 = \begin{bmatrix} I & F \\ 0 & 0 \end{bmatrix} P$, 89
$R_0\boldsymbol{x} = \boldsymbol{d}$, 96
$S = A^{\mathrm{T}}A$, 231, 265
$S = C^{\mathrm{T}}C$, 338
$S = LDL^{\mathrm{T}}$, 231, 237
$S = P^{\mathrm{T}} \begin{bmatrix} -F \\ I \end{bmatrix}$, 89
$S = Q\Lambda Q^{\mathrm{T}}$, vii, 227, 237, 338
$S^{\mathrm{T}} = S$, 69
$T(c\boldsymbol{v} + d\boldsymbol{w}) = c\,T(\boldsymbol{v}) + d\,T(\boldsymbol{w})$, 192
$T = d/dx$, 194
T_{ijk}, 344
$V\Sigma^+U^{\mathrm{T}}$, 280
$X^{-1}AX = \Lambda$, 215
$\boldsymbol{u}(t) = e^{\lambda t}\boldsymbol{x}$, 244
$\boldsymbol{F}(\boldsymbol{x}, \boldsymbol{v})$, viii, 289
$\mathbf{S}^{\perp}$, 141

Six Great Theorems of Linear Algebra

Dimension Theorem All bases for a vector space have the same number of vectors.

Counting Theorem Dimension of column space + dimension of nullspace = number of columns.

Rank Theorem Dimension of column space = dimension of row space. This is the rank.

Fundamental Theorem The row space and nullspace of A are orthogonal complements in $\mathbf{R}^n$.

SVD There are orthonormal bases ($\boldsymbol{v}$'s and $\boldsymbol{u}$'s for the row and column spaces) so that $A\boldsymbol{v}_i = \sigma_i \boldsymbol{u}_i$.

Spectral Theorem If $A^{\mathrm{T}} = A$ there are orthonormal $\boldsymbol{q}$'s so that $A\boldsymbol{q}_i = \lambda_i \boldsymbol{q}_i$ and $A = Q\Lambda Q^{\mathrm{T}}$.

Linear Algebra in a Nutshell

((*The matrix* A *is* n *by* n))

Nonsingular	**Singular**
A is invertible	A is not invertible
The columns are independent	The columns are dependent
The rows are independent	The rows are dependent
The determinant is not zero	The determinant is zero
$A\boldsymbol{x} = \mathbf{0}$ has one solution $\boldsymbol{x} = \mathbf{0}$	$A\boldsymbol{x} = \mathbf{0}$ has infinitely many solutions
$A\boldsymbol{x} = \boldsymbol{b}$ has one solution $\boldsymbol{x} = A^{-1}\boldsymbol{b}$	$A\boldsymbol{x} = \boldsymbol{b}$ has no solution or infinitely many
A has n (nonzero) pivots	A has $r < n$ pivots
A has full rank $r = n$	A has rank $r < n$
The reduced row echelon form is $R = I$	R has at least one zero row
The column space is all of $\mathbf{R}^n$	The column space has dimension $r < n$
The row space is all of $\mathbf{R}^n$	The row space has dimension $r < n$
All eigenvalues are nonzero	Zero is an eigenvalue of A
$A^{\mathrm{T}}A$ is symmetric positive definite	$A^{\mathrm{T}}A$ is only semidefinite
A has n (positive) singular values	A has $r < n$ singular values

Linear Algebra and Learning from Data

See math.mit.edu/learningfromdata This is the new textbook for the applied linear algebra course 18.065 at MIT. It starts with the basic factorizations of a matrix :

$$A = CR \quad A = LU \quad A = QR \quad A = X\Lambda X^{-1} \quad S = Q\Lambda Q^{\mathrm{T}} \quad A = U\Sigma V^{\mathrm{T}}$$

The goal of deep learning is to find patterns in the training data. Matrix multiplication is interwoven with the nonlinear ramp function ReLU $(x) = \mathbf{max}\,(0, x)$. The result is a learning function that can interpret *new data*. The textbook explains how and why this succeeds—even in the classroom. Linear algebra and student projects are the keys.

$\boldsymbol{i} = \begin{bmatrix} 1 \\ 0 \end{bmatrix}, \boldsymbol{j} = \begin{bmatrix} 0 \\ 1 \end{bmatrix}$, 110
$\boldsymbol{p} = A\widehat{\boldsymbol{x}}$, 147
$\boldsymbol{p} = QQ^{\mathrm{T}}\boldsymbol{b}$, 167
$\boldsymbol{u}_k = A^k \boldsymbol{u}_0$, 221
$\boldsymbol{x}^+ = \boldsymbol{A}^+\boldsymbol{b}$, 159
$\boldsymbol{x} = \boldsymbol{x}_p + \boldsymbol{x}_n$, 97
$\boldsymbol{x} = A\backslash\boldsymbol{b}$, 45, 97
$\boldsymbol{x} * \boldsymbol{v}$, 300
$\boldsymbol{x}^{\mathrm{T}}S\boldsymbol{x}/\boldsymbol{x}^{\mathrm{T}}\boldsymbol{x}$, 265
$\widehat{\boldsymbol{x}} = R^{-1}Q^{\mathrm{T}}\boldsymbol{b}$, 171
$\overline{\boldsymbol{x}}^{\mathrm{T}}\boldsymbol{y}$, 236
$\boldsymbol{y}^{\perp}$, 335
$\boldsymbol{u} = \boldsymbol{v}/||\boldsymbol{v}||$, 12, 281
$\boldsymbol{v} + \boldsymbol{w}$, 2, 5
$\boldsymbol{v} - \boldsymbol{w}$, 4
$\boldsymbol{v} \cdot \boldsymbol{w}/||\boldsymbol{v}||\,||\boldsymbol{w}|| = \cos\theta$, 15
$\boldsymbol{v} \cdot \boldsymbol{w} = \boldsymbol{v}^{\mathrm{T}}\boldsymbol{w}$, 2, 12
$|\boldsymbol{v} \cdot \boldsymbol{w}| \leq ||\boldsymbol{v}||\,||\boldsymbol{w}||$, 15
$||\boldsymbol{x}||^2 = \overline{\boldsymbol{x}}^{\mathrm{T}}\boldsymbol{x}$, 236
$||\boldsymbol{x}||_1$, 283
$||\boldsymbol{v} + \boldsymbol{w}|| \leq ||\boldsymbol{v}|| + ||\boldsymbol{w}||$, 15
$||\boldsymbol{v}||^2 = \boldsymbol{v} \cdot \boldsymbol{v}$, 11
$c\boldsymbol{v} + d\boldsymbol{w}$, 3, 7
$\det AB = (\det A)(\det B)$, 184, 197
$\det P = 1$ or -1, 179
$\det(A - \lambda I) = 0$, 203
$\det(A^{\mathrm{T}}) = \det(A)$, 184
eig(A), 206
ℓ^1 norm, 283, 284
$\ell^2 + \ell^1$, 285
$\frac{1}{2}N\log_2 N$, 66
ReLU, viii, 286, 289, 290, 297, 356
null(A), 142
rref(A), 35, 341
$\mathbf{C}(A)$, 23
$\mathbf{C}(A^{\mathrm{T}})$, 78
$\mathbf{N}(A)$, 83
$\mathbf{N}(A^{\mathrm{T}})$, 121
$\mathbf{N}(A^{\mathrm{T}}) \perp \mathbf{C}(A)$, 136
$\mathbf{N}(A) \perp \mathbf{C}(A^{\mathrm{T}})$, 136
$\mathbf{N}(A^{\mathrm{T}}A) = \mathbf{N}(A)$, 141, 149, 152
$\mathbf{V}^{\perp}$, 137
cond $= \sigma_1/\sigma_n$, 345
$\mathbf{V} + \mathbf{W}$, 120
$\mathbf{V} \cap \mathbf{W}$, 82, 120
$\mathbf{V} \cup \mathbf{W}$, 120
$\mathbf{Z}$, 113
$\begin{bmatrix} A & \boldsymbol{b} \end{bmatrix}$, 81, 96
$\sigma_k = \sqrt{\lambda_k}$, 261
$\begin{bmatrix} A\boldsymbol{x} & A\boldsymbol{y} & A\boldsymbol{z} \end{bmatrix}$, 29
Rank$(AB) \leq$ rank(A), 129, 334
Rank$(AB) \leq$ rank(B), 129, 334
$\frac{1}{3}n^3$ multiplications, 58
$|\lambda| \leq \sigma_1$, 264
$||A - B||$, 260, 274
$||A\boldsymbol{x}||/||\boldsymbol{x}||$, 265
$||A||$, 274, 345
$||A^{-1}||$, 345
$||Q\boldsymbol{x}|| = ||\boldsymbol{x}||$, 167, 280
e^{At}, 249
$e^A e^B \neq e^B e^A$, 255
$e^{At} = X\, e^{\Lambda t}\, X^{-1}$, 250
mnp steps, 34
LAPACK, 172
randn, 19, 267